Policy Shock

Policy Shock examines how policy-makers in industrialized democracies respond to major crises. After the immediate challenges of disaster management, crises often reveal new evidence or frame new normative perspectives that drive reforms designed to prevent future events of a similar magnitude. Such responses vary widely – from cosmetically masking inaction, to creating stronger incentive systems, requiring greater transparency, reorganizing government institutions, and tightening regulatory standards. This book situates post-crisis regulatory policy-making through a set of conceptual essays written by leading scholars from the fields of economics, psychology, and political science, which probe the latest thinking about risk analysis, risk perceptions, focusing events, and narrative politics. It then presents 10 historically rich case studies that engage with crisis events in three policy domains: offshore oil, nuclear power, and finance. It considers how governments can prepare to learn from crisis events – by creating standing expert investigative agencies to identify crisis causes and frame policy recommendations.

Edward J. Balleisen is Vice Provost for Interdisciplinary Studies and Professor of History and Public Policy at Duke University. His latest book, *Fraud: An American History from Barnum to Madoff* (2017), chronicles the history of regulatory governance in the United States. He received his Ph.D. from Yale University.

Lori S. Bennear is the Juli Plant Grainger Associate Professor of Energy Economics and Policy at Duke University, where she is also Co-Director of the Rethinking Regulation Program at the Kenan Institute for Ethics and Associate Director for Educational Programs at the Energy Initiative. Her works span environmental domains including toxics, drinking water, and energy. She received her Ph.D. from Harvard University.

Kimberly D. Krawiec is the Kathrine Robinson Everett Professor of Law at Duke University, where she teaches courses on corporate law and financial regulation. Her research centers on the regulation of financial markets, "taboo" or contested markets, and business organizations. She received her J.D. from Georgetown University.

Jonathan B. Wiener is the Perkins Professor of Law and Professor of Environmental Policy and Public Policy at Duke University. He is Co-Director of the Rethinking Regulation Program at Duke. He is a University Fellow of Resources for the Future (RFF). He served as President of the Society for Risk Analysis in 2008 and as Co-Chair of the World Congress on Risk in 2012. He received his J.D. from Harvard University.

Policy Shock

Recalibrating Risk and Regulation after Oil Spills, Nuclear Accidents, and Financial Crises

Edited by

EDWARD J. BALLEISEN, LORI S. BENNEAR,
KIMBERLY D. KRAWIEC, AND JONATHAN B. WIENER

Duke University

CAMBRIDGE
UNIVERSITY PRESS

University Printing House, Cambridge CB2 8BS, United Kingdom

One Liberty Plaza, 20th Floor, New York, NY 10006, USA

477 Williamstown Road, Port Melbourne, VIC 3207, Australia

4843/24, 2nd Floor, Ansari Road, Daryaganj, Delhi – 110002, India

79 Anson Road, #06–04/06, Singapore 079906

Cambridge University Press is part of the University of Cambridge.

It furthers the University's mission by disseminating knowledge in the pursuit of education, learning, and research at the highest international levels of excellence.

www.cambridge.org
Information on this title: www.cambridge.org/9781107140219
DOI: 10.1017/9781316492635

© Cambridge University Press 2017

This publication is in copyright. Subject to statutory exception and to the provisions of relevant collective licensing agreements, no reproduction of any part may take place without the written permission of Cambridge University Press.

First published 2017

Printed in the United States of America by Sheridan Books, Inc.

A catalogue record for this publication is available from the British Library.

Library of Congress Cataloging-in-Publication Data
NAMES: Balleisen, Edward J., editor. | Bennear, Lori S., editor. | Krawiec, Kimberly D., editor. | Wiener, Jonathan Baert, 1962– editor.
TITLE: Policy shock : recalibrating risk and regulation after oil spills, nuclear accidents and financial crises / edited by Edward Balleisen, Lori S. Bennear, Kimberly D. Krawiec, and Jonathan Wiener.
DESCRIPTION: Cambridge, United Kingdom ; New York, NY : Cambridge University Press, 2017. | Includes bibliographical references and index.
IDENTIFIERS: LCCN 2017028166 | ISBN 9781107140219 (alk. paper)
SUBJECTS: | MESH: Disaster Planning | Risk Management | Radioactive Hazard Release – prevention & control | Petroleum Pollution – prevention & control | Policy Making
CLASSIFICATION: LCC HV553 | NLM WA 295 | DDC 363.34/8–dc23
LC record available at https://lccn.loc.gov/2017028166

ISBN 978-1-107-14021-9 Hardback

Cambridge University Press has no responsibility for the persistence or accuracy of URLs for external or third-party Internet websites referred to in this publication and does not guarantee that any content on such websites is, or will remain, accurate or appropriate.

Contents

List of Figures	*page* viii
List of Tables	xi
List of Boxes	xiii
List of Contributors	xiv
Acknowledgments	xvi

1 Introduction 1
   Edward J. Balleisen, Lori S. Bennear, Kimberly D. Krawiec, and Jonathan B. Wiener

PART I THE CONCEPTUAL TERRAIN OF CRISES AND RISK PERCEPTIONS

2 Economic Analysis, Risk Regulation, and the Dynamics of Policy Regret 43
   Lori S. Bennear

3 Revised Risk Assessments and the Insurance Industry 58
   Carolyn Kousky

4 Understanding Public Risk Perception and Responses to Changes in Perceived Risk 82
   Elke U. Weber

5 Focusing Events, Risk, and Regulation 107
   Thomas A. Birkland and Megan K. Warnement

6 The Story of Risk: How Narratives Shape Risk Communication, Perception, and Policy 129
   Frederick W. Mayer

PART II CASE STUDIES ON OFFSHORE OIL SPILLS

7   From Santa Barbara to the Exxon Valdez: Policy Learning
    and the Emergence of a New Regime for Managing
    Oil Spill Risk   151
    Marc Allen Eisner

8   The Nordic Model of Offshore Oil Regulation: Managing
    Crises through a Proactive Regulator   181
    Ole Andreas Engen and Preben H. Lindøe

9   Reform in Real Time: Evaluating Reorganization as
    a Response to the Gulf Oil Spill   204
    Christopher Carrigan

PART III CASE STUDIES ON NUCLEAR ACCIDENTS

10 Recalibrating Risks of Nuclear Power: Reactions
    to Three Mile Island, Chernobyl, and Fukushima   245
    Elisabeth Paté-Cornell

11 Nuclear Accidents and Policy Responses in Europe:
    Comparing the Cases of France and Germany   269
    *Kristian Krieger, Ortwin Renn, M. Brooke Rogers,
    and Ragnar Löfstedt*

12 Public Attitudes and Institutional Changes in Japan
    following Nuclear Accidents   305
    Atsuo Kishimoto

PART IV CASE STUDIES OF FINANCIAL CRISES

13 Regulatory Responses to the Financial Crises of the Great
    Depression: Britain, France, and the United States   349
    Youssef Cassis

14 Financial Decommodification: Risk and the Politics
    of Valuation in US Banks   371
    Bruce G. Carruthers

15 Euro Area Risk (Mis)management   395
    Barry Eichengreen

16 The Regulatory Responses to the Global Financial Crisis:
    Some Uncomfortable Questions   435
    Stijn Claessens and Laura Kodres

PART V CONCLUSIONS

17 Institutional Mechanisms for Investigating the Regulatory Implications of a Major Crisis: The Commission of Inquiry and the Safety Board 485
Edward J. Balleisen, Lori S. Bennear, David Cheang, Jonathon Free, Megan Hayes, Emily Pechar, and A. Catherine Preston

18 Recalibrating Risk: Crises, Learning, and Regulatory Change 540
Edward J. Balleisen, Lori S. Bennear, Kimberly D. Krawiec, and Jonathan B. Wiener

Index 562

Figures

1.1	The impact of Lehman Brothers' bankruptcy ripples through world markets, chronicled by the American press	page 2
1.2	A shorebird covered in crude oil from the BP Deepwater Horizon oil spill	7
1.3	*Business Week* shows a Japanese girl being tested for radiation after the Fukushima reactor meltdowns	7
1.4	Our three types of crisis-response case studies	18
1.5	Map of potential pathways from crisis to policy change	19
1.6	Schematic of policy options in response to crisis	20
4.1	Perceived risk versus actual risk	85
5.1	Risk perception matrix and the competing influences of terror	119
5.2	Two models of the influence of focus events on policy actors and levels	123
7.1	Petroleum imports, 1960–2000	157
7.2	Offshore petroleum production, 1960–2000	158
7.3	Total volume of spills by spill size, 1973–2009	174
7.4	Number of significant spills by spill size, 1973–2009	174
8.1	Two modes of risk regulations	187
9.1	Gallup Poll results measuring public preference for economic growth or environmental protection (1984–2012)	210
9.2	*New York Times* and *Washington Post* article mentions of offshore oil and gas (January 2002–December 2012)	211
9.3	Timeline of select congressional bills proposing reorganization of MMS and related functions (2008–2012)	229
10.1	History of the global nuclear power industry	254

List of Figures

11.1	France, fuel mix, total production, 2010	273
11.2	Germany, fuel mix, total production, 2010	274
11.3	Nuclear reactors in operation, France and Germany, 1955–2025	275
11.4	French fuel mix, total energy consumption, 1970–2006	276
11.5	German fuel mix, total energy consumption, 1970–2006	276
12.1	Simplified event model	307
12.2	Transitions in the administrative systems that govern nuclear power	307
12.3	Time series variations of public opinion surveys and the five major nuclear accidents	310
12.4	Transition in the power supply share in Japan prior to 2011 (billion kWh)	313
12.5	Public opinion before and after the *Mutsu* accident	316
12.6	Public opinion on nuclear power before and after the TMI accident	319
12.7	Public perception of safety measures before and after the radiation leakage from the Tsuruga Nuclear Power Station	321
12.8	Change in public opinion shortly after the Fukushima accident	334
12.9	Scope of the new requirements compared with former requirements	336
12.10	Shift in the share of votes in proportional representation	340
15.1	Bank leverage ratios in the Euro Area and the United States	397
15.2	Sovereign spread compression in the Euro Area (1990–2011)	399
15.3	Divergence and adjustment of unit labor costs	401
15.4	Change in private sector credit in the Euro Area and the United States	402
15.5	Convergence of Euro Area countries (1960–92)	411
15.6	Internet Incidence of Euro crisis-related terms (Google Trends)	420
16.1	Dimensions of regulatory reform	473
17.1	Spectrum of independence possessed by policy investigatory bodies in the United States	486
17.2	Forms of national commissions of inquiry in the five case countries	490
17.3	Summary of the commission of inquiry process	492

17.4	Organizational charts, National Transportation Safety Board (2014) and Chemical Safety and Hazard Investigation Board (2013)	506
17.5	Results of NTSB recommendations, 1967–2000	507
17.6	Results of NTSB recommendations averaged by year, 1967–2000	507
17.7	History of independent investigatory bodies in American transportation and chemical safety	512
17.8	Relationship maps between interested parties, NTSB and CSB	518
17.9	Mapping economic sectors by frequency of crisis events and degree of perceived community of fate	530

Tables

1.1	Offshore oil crises and responses	page 22
1.2	Nuclear reactor crises and responses	27
5.1	Key issues addressed in aviation security legislation	117
5.2	Levels of policy codification	122
7.1	State oil pollution liability regimes	168
9.1	Subject matter of Congressional Hearings in which MMS or successor agency personnel testified by topic (1982–2012)	208
9.2	Reforms enacted in response to Gulf oil spill (May 2010–May 2013)	216
10.1	Partial history of subduction-plate earthquakes along the Sanriku and Sendai coasts	263
11.1	Key nuclear policy actors established between 1945 and 1979	279
11.2	Selected nuclear policies and regulations between 1945 and 1979	280
11.3	Key nuclear policy actors established between 1986 and 2011	283
11.4	Selected nuclear policies and regulations between 1986 and 2011	285
11.5	Nuclear actors, policies and regulations after 2011	290
11.6	Internationally comparable opinion polls on nuclear energy, 1982–2011	292
12.1	Impacts of the five accidents on public opinion and regulatory policies in Japan	311

12.2	Accidents and investigatory commissions and their reports	342
15.1	Determinants of 10-year government bond spreads relative to German bunds 2005Q1–2011Q4	415
15.2	Determinants of 10-year government bond spreads relative to German bunds, 2005Q1–2011Q4 with post-2008Q2 interaction effects	415
15.3	First mention of crisis phrases	417
17.1	Major commissions appointed in response to crisis events (1960–2010)	493

Boxes

4.1	Hormone Replacement Therapy	*page* 83
4.2	Nuclear Power	84
4.3	2008 Subprime Mortgage Crisis	85
16.1	Overall approaches to determine specific reforms	449

Contributors

Edward J. Balleisen, Duke University
Lori S. Bennear, Duke University
Thomas A. Birkland, North Carolina State University
Christopher Carrigan, The George Washington University
Bruce G. Carruthers, Northwestern University
Youssef Cassis, European University Institute
David Cheang, Duke University
Stijn Claessens, Bank for International Settlements
Barry Eichengreen, University of California, Berkeley
Marc Allen Eisner, Wesleyan University
Ole Andreas Engen, University of Stavanger
Jonathon Free, Duke University
Megan Hayes, Duke University
Atsuo Kishimoto, Osaka University
Laura Kodres, International Monetary Fund
Carolyn Kousky, University of Pennsylvania
Kimberly D. Krawiec, Duke University

List of Contributors

Kristian Krieger, King's College London

Preben H. Lindøe, University of Stavanger

Ragnar Löfstedt, King's College London

Frederick W. Mayer, Duke University

Elisabeth Paté-Cornell, Stanford University

Emily Pechar, Duke University

A. Catherine Preston, Duke University

Ortwin Renn, Institute for Advanced Sustainability Studies (IASS), Potsdam

M. Brooke Rogers, King's College London

Megan K. Warnement, North Carolina State University

Elke U. Weber, Princeton University

Jonathan B. Wiener, Duke University

Acknowledgments

We are grateful for the generous support of numerous people and organizations in bringing the ideas behind this book to fruition. Duke University provided both financial and logistical support, through the Kenan Institute for Ethics, the Bass Connections program, and the Office of the Provost. KIE's co-directors, Noah Pickus and Suzanne Shanahan, provided an institutional home for Duke's Rethinking Regulation Program, through which we generated the questions and ideas that blossomed into this volume. In addition to their work as co-authors on Chapter 17, A. Catherine Preston and Jonathan Free also provided help as research assistants, as did Christopher Geary. The Smith Richardson Foundation provided financial support for this project, and we particularly thank Mark Steinmeyer at the Smith Richardson Foundation for his advice and support.

We thank all of the contributors for attending multiple authors' meetings and working diligently to ensure that the chapters built upon one another. We are particularly grateful to Granger Morgan, Susan Webb-Yackee, and Christie Ford, who served as external reviewers of the manuscript. Additional crucial feedback on earlier drafts came from the many individuals who participated in sessions on this project at Duke's Rethinking Regulation Program, including many of our Duke faculty colleagues, former Congressman Brad Miller, former administrators of OIRA Sally Katzen and John Graham, and former North Carolina environmental regulator Robin Smith; and from participants at the conference organized by the International Risk Governance Council (led by Marie-Valentine Florin, with assistance from Marcel Burkler and Claire Mays) and the Organization for

Economic Cooperation and Development (OECD) (including the Regulatory Policy Division, led by Nick Malyshev, and the High Level Risk Forum, led by Stéphane Jacobzone) in Paris in October 2014. We are also very appreciative of Lew Bateman and all the editors at Cambridge University Press who helped guide this book to publication, and to the two anonymous reviewers who gave instructive comments on all aspects of the book.

Finally, we are grateful to our families and friends for their support and sacrifice during the long course of this project.

Edward J. Balleisen

Lori S. Bennear

Kimberly D. Krawiec

Jonathan B. Wiener

1

Introduction

Edward J. Balleisen, Lori S. Bennear, Kimberly D. Krawiec, and Jonathan B. Wiener

Crises punctuate our world. Their causes and consequences are woven through complex, interconnected social and technological systems. Consider these three recent events, each of which dramatically upended expectations about risk:

- In the fall of 2008, the global financial system experienced a full-blown panic. Credit flows seized up, ushering in the worst global recession since the 1930s and leading newspapers to convey the resulting "shocks" to financial markets.
- In April 2010, a blowout at the British Petroleum Deepwater Horizon drilling platform killed eleven workers and triggered a three-month-long oil spill, sending nearly five million barrels of crude into the Northern Gulf of Mexico, which fouled beaches, estuaries, and fishing grounds.
- In March 2011, an earthquake and a resulting tsunami killed 20,000 people in Japan. The natural disaster also caused reactor meltdowns at the Fukushima nuclear power plant, forcing the evacuation of tens of thousands of people, unleashing a long-term leak of radioactive water into the Pacific Ocean and creating a daunting set of challenges as officials sought to stabilize pools of spent fuel rods and protect local populations from radioactive fallout.

Each of these three recent events attracted extraordinary attention from the media and the global public, raising concerns about dangers that may lurk within the complex technological and social systems on which we depend to sustain our economy and way of life. They also generated criticisms of the regulatory systems that were supposed to

FIGURE 1.1 The impact of Lehman Brothers' bankruptcy ripples through world markets, chronicled by the American press
Source: http://businessjournalism.org/2013/09/5-year-recession-after-wall-streets-crash-and-a-look-at-401k-trends/

prevent such failures, as well as demands for new regulatory actions to reduce the risks that the crises had brought into sharp relief. In the aftermath, policy elites and the broader public ponder the meaning of such events and look for appropriate responses. Once a consensus emerges that they indeed constitute crises (and sometimes even before), government agencies, legislative committees, think tanks, citizens' groups, scholars, and often official commissions begin to investigate their causes, consider whether better policy might have prevented them, and debate what regulatory adjustments governments should adopt, if any.

In this multidisciplinary volume, we examine how people and policymakers respond to crises. This exercise requires care in defining what we mean by "regulatory crises," which, for us, are events that create substantial social damage and capture public attention. They may thereby call into question existing mechanisms for controlling and managing risk, and generate widespread proposals for adjustments in regulatory policy. Regulatory crises are distinct from many non-crisis events that may have substantial social damage but fail to capture public attention and lead to review of existing regulations. An example of such a non-crisis event would be particulate matter pollution from coal-fired power plants, which kills many people but does so slowly and without much public outcry. We also distinguish regulatory crises from global catastrophes, such as a major asteroid collision, that threaten the existence of all life, or all human life, or civilization, and hence would not be amenable to subsequent regulatory responses after they occurred (Wiener 2016).

After the immediate challenges of disaster management, crises may – sometimes – reveal new evidence or frame new normative perspectives that drive new policies designed to prevent future crises. The resulting policy responses may vary widely – for example, tightening regulatory standards, creating stronger incentive systems, requiring greater transparency, reorganizing government institutions, or cosmetically masking inaction. We delve into a series of enduring puzzles about the relationships between crises and regulatory decision-making, exploring the following questions:

- How do crises change the risk perceptions of the general public, policy elites, or both?
- How do changes in the risk perceptions of policy elites, or those of the wider public, result in different policy responses?
- How do the narratives that emerge about crises shape the policy response (or inaction) that ensues?
- When crises do generate regulatory responses, how and why do those responses vary? How do differing features of crises, and of the social and political systems in which they occur, influence the adoption of different policy instruments and strategies?
- To the extent that it is possible to tell, when do crisis-driven regulatory changes lead to desirable reforms, as opposed to hasty overreactions or policy mismatches?
- How might governments (both elected officials and regulatory policy-makers), businesses, citizens' groups, and scholars do a better job of both learning to prepare for crises and preparing to learn from crises, so that regulatory responses are more successful?

We have shaped our exploration of these matters with two broad, intersecting audiences in mind. The first encompasses the many scholarly communities that study regulatory governance. Our goals for that readership are to synthesize current research findings on crisis-driven regulatory policy from many fields of knowledge, to provide extensive new evidence about regulatory policy-making in some especially salient contexts of crisis, and to lay out the most important issues deserving additional scholarly attention. The second intended audience comprises policy elites, especially within regulatory agencies and the offices of elected officials, regulated businesses, and non-governmental organizations. For these readers, we offer a conceptual framework for how to make sense of crises as they unfold and especially how to assess options for reforming risk regulation in their aftermath. On this last point, we pay especially close attention to best practices for crisis investigatory bodies, suggesting ways that governments can prepare to learn from "policy shocks" – events that few policy-makers anticipated, or that policy-makers presumed to be so rare as not to justify significant efforts to prevent them or mitigate their impacts.

We remain too close in time to the Global Financial Crisis of 2008, or BP Deepwater Horizon, or Fukushima to have a clear sense of their long-term implications for regulatory policy-making. But they collectively raise the sorts of questions that we identify above about the relationship between "crisis" and policy formulation. One common tendency in thinking about these questions, articulated by several observers in the wake of these recent disasters, is that crisis episodes can dramatically reconfigure perceptions of reality, which then at least sometimes, and perhaps often, drive major policy changes. One can certainly point to many examples that fit this pattern. Indeed, significant turning points in the history of regulatory governance frequently have been triggered by crisis.

HISTORICAL CONTEXT

Over the last two centuries, and across the globe, far-reaching changes in regulatory policy have often (though of course not always) represented responses to sudden, largely unexpected and damaging events (Percival 1998; Birkland 2006; Repetto 2006; D. Carpenter and Sin 2007; Wuthnow 2010). To be sure, not all crises lead to regulation, and not all regulations derive from crises (Kahn 2007). Consider the following list of episodes – hardly exhaustive, but lengthy enough to suggest the wide range of contexts in which crises have generated major shifts in regulatory policy.

We begin with some examples from the regulation of health and safety, whether in specific industries or across the wider environment. As early as 1838, exploding boilers on American river steamboats brought forth a congressionally mandated safety inspection regime (Burke 1966). On both sides of the Atlantic, the introduction of modern public health regulation during the nineteenth century ensued in the wake of infectious disease epidemics (Rosenberg 1987; Bourdelais 2006). In late nineteenth-century Europe and the United States, the imposition of new safety protocols for coal mines followed mining disasters that dramatized the dangers of deep-level mineral extraction (Reid 1986; Aldrich 1997). Harrowing industrial workplace tragedies, like the 1911 fire at New York City's Triangle Shirtwaist Factory, frequently gave rise to tougher safety rules and inspection regimes (Pool 2012). Significant changes in the twentieth-century regulation of pharmaceuticals often occurred only after some vivid demonstration of an unsafe drug's terrible impact (the American deaths caused by ethyl glycol-infused antibiotics in 1937; the European birth defects caused by thalidomide in the early 1960s), or because of the widely covered death toll from a new disease (HIV/AIDS in the 1980s). In several countries, significant movement to limit industrial air pollution arose after dramatic episodes like the killer fogs that beset London in the 1950s. Clean water laws were enacted after incidents such as the Cuyahoga River catching fire. The Seveso dioxin accident of 1976 gave rise to new European directives on chemical facility safety, just as the discovery of hazardous waste at Love Canal, New York, spurred the 1980 US CERCLA Superfund cleanup law, and the 1984 Bhopal, India, chemical plant disaster encouraged the refashioning of safety regimes throughout the global chemical industry (King and Lenox 2000; Lenox and Nash 2003) as well as enactment of the US Emergency Planning and Community Right-to-Know Act in 1986. This dynamic, moreover, could operate at the global as well as national level. Thus the identification of the stratospheric ozone hole in 1985–86 (along with other factors such as a shift in industry lobbying and a favorable cost-benefit analysis within the government) helped trigger the US government's adoption of the 1987 Montreal Protocol to phase out chlorofluorocarbons (Litfin 1994).

Another policy terrain strongly marked by crisis-driven regulation involves oversight of corporate governance and the financial markets. Governments tended to adopt tougher rules on corporate governance and accounting after well-publicized corporate scandals – the South Sea Bubble in the 1720s, the over-issue of stock at the New York and New Haven Railroad in the 1850s; the collapse of several American insurance companies in the early 1870s (Harris 1994; Shaw 1978; *The International Review (1874–1883)* 1877). New schemes of financial regulation also tended to emerge in the aftermath of economy-wide financial panics (as with much tighter capital requirements for American trust companies and the creation of the Federal Reserve after the Panic of 1907, and the dramatic refashioning of securities regulation in the wake of the 1929 stock market crash) (Tallman and Moen 1990; Seligman 2004).

The apparent connection among crisis events, reshaped risk perceptions, and regulatory policy change has continued over the last quarter-century. One might point to American contexts such as the savings & loan crisis of the late 1980s, which led to a reversal of some banking deregulation; the fraud-related bankruptcies at Enron, Worldcom, and Tyco in the late 1990s, which engendered a new regime for corporate accounting and the 2002 US Sarbanes-Oxley Act (Rockness and Rockness 2005); or the 9/11 terror attacks in the United States, which prompted a massive expansion of the national security apparatus, its reorganization into a new cabinet-level Department of Homeland Security, and two wars (Cohen, Cuéllar, and Weingast 2006). Similarly, one might look to European events such as a series of food safety crises in the late 1980s and 1990s, notably mad cow disease and foot and mouth disease, which undermined public confidence and added momentum for various food safety policies; or the volcanic ash crisis of 2011, which encouraged a centralization of air traffic management (Alemanno 2011). Or one might stress events in emerging economies, including recent episodes in China concerning unsafe milk, toys, and other products that generated pressures for tougher regulatory oversight of manufacturing standards (Bamberger and Guzman 2008); or even more recent accidents in South Asian clothing factories that elicited new avenues of workplace safety regulation both nationally and through global supply chains (Venkatesan 2013).

FRAMING "CRISIS" AND THE DESIRABILITY OF CRISIS-DRIVEN REGULATORY RESPONSE

Scholars of regulatory governance have long noted the salience of crisis episodes in reshaping policy agendas and forging political environments conducive to significant regulatory change, especially once modern media outlets existed to spread public awareness of these events and to shape public perception of them. Graphic newspaper descriptions and mass-produced prints brought the human impact of nineteenth-century disasters to a wide audience. The advent of photography, radio, cinema, television, round-the-clock cable news networks, and then the Internet and social media platforms only further expanded the avenues for conveying the social costs of crisis events in captivating, personal terms, such as the pictures in Figures 1.2 and 1.3. This news coverage often generated compassion for innocent victims and outrage directed toward culpable villains. In democratic societies, such coverage can generate strong political pressures for governmental action – both to redress the wrongs

FIGURE 1.2 A shorebird covered in crude oil from the BP Deepwater Horizon oil spill
Source: http://archive.boston.com/bigpicture/2010/06/caught_in_the_oil.html

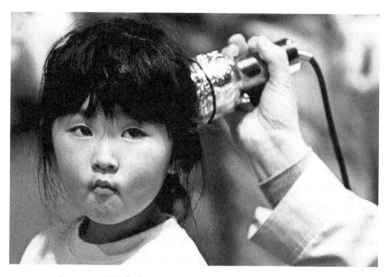

FIGURE 1.3 *Business Week* shows a Japanese girl being tested for radiation after the Fukushima reactor meltdowns
Source: www.wsj.com/articles/SB10001424053111903639404576515890556446756

already inflicted on victims and to reduce the risk of recurrences. By galvanizing general public concern, crises can also curb the capacity of business groups to stymie policy changes that they viewed as inimical to their interests, or generate a rethinking of how best to conceptualize what those interests are.

Accidents and disasters, however, do not automatically produce politically salient crises. Instead, proponents of new directions in regulatory policy always have had to shape the public meaning of such events, convincing decision-makers that their immediate consequences and long-term portents were so great as to demand priority consideration despite all the other issues and interests clamoring for attention. Sometimes the key actors in shaping such evaluations have been policy entrepreneurs within the state; in other circumstances, corporations, interest groups, social movements, or even lone experts have taken the lead. Usually, the manufacturing of a full-blown crisis depends on the emergence of a coalition that shares policy goals, if not necessarily motivations or rationales.

In modern industrialized, democratic societies, such coalitions typically have faced a common set of challenges. First, they must effectively personalize the social and economic losses resulting from the event, while developing a persuasive explanation of its causes. Second, they have to imagine a cluster of proposed solutions that allow policy-makers and the wider public to see the problems exposed by the event as amenable to action. Finally, they have to chart a plausible path for the formulation and implementation of the proposed solutions, taking account of the social, political, and institutional contexts that mediate policy-making. Absent such clearly defined and realistic reform programs, politicians and ordinary citizens are far more likely to indulge in some form of psychological denial, refusing to see a need for prompt action (Campbell and Kay 2014). These are potentially daunting obstacles. And yet the basic recipe of crisis-driven policy initiatives has become sufficiently well known that savvy political operators remain alert for situations that lend themselves to labeling as a crisis, thereby opening up avenues for policy reforms. As President Obama's chief of staff Rahm Emmanuel framed this sensibility after the 2008 financial crash, "you never want a serious crisis to go to waste."

That reality is a key reason that the desirability of crisis-driven regulatory change remains a hotly contested question. All too often, some academics worry, policy-makers may take advantage of public clamor to enact their own pet programs, potentially unrelated to the risks exposed by the crisis (Romano 2005; Coglianese and Carrigan 2012). More generally, critics argue that the emotional punch associated with crisis-driven regulatory reform encourages overreaction or policy mismatches (Kuran and Sunstein 1999; Wuthnow 2010). On this view, when legislators and regulatory officials face insistent popular demands to take action, whipped up by news coverage of a rare event that brought harm to innocent individuals, they tend to adopt aggressive policies that impose heavy costs, induce new perils, and sometimes do little to prevent the risks at issue. For example, although the 2011 tsunami killed more than 20,000 people in Japan, the subsequent evacuation of the area around the damaged Fukushima Daiichi nuclear power plant apparently saved few or no lives from radiation exposure while costing an estimated 1,600 lives through dislocation of frail residents (Johnson 2015).

Skeptics of crisis-driven policy change often advocate reliance on strong regulatory oversight mechanisms to ensure deliberative analysis of the wide range of risks facing society, and the pros and cons of policy proposals (Breyer 1993). Such oversight mechanisms include the United States Office of Information and Regulatory Affairs (OIRA), the European Union's Impact Assessment Board (IAB) (renamed the Regulatory Scrutiny Board in 2015), and a growing number of similar institutions around the world (Wiener 2006; Wiener 2013; Wiener and Ribeiro 2016). OIRA and the IAB can serve as institutional brakes on hasty regulatory decision-making, using regulatory impact assessments (RIAs) to facilitate sober evaluation of both the risks highlighted by a crisis event and the advisability of proposed reforms. Conceptually similar "think before you act" laws have also been enacted to protect the environment against hasty construction projects via environmental impact assessments. Analogous proposals have called for regulations to undergo multiple stages of legislative scrutiny so that final passage of new regulatory constraints only occurs well after crisis-related passions have cooled; or even for automatic sunset provisions so that crisis-generated regulations later require legislative reaffirmation, presumably on the basis of a considered evaluation of policy impacts (Romano 2005). Some scholarly critics of crisis-driven regulation also argue that policy-makers too often neglect needed regulation of important risks that do not come to the fore, because they do not give rise to dramatic episodes that would attract the media's cameras and emotion-laden narratives of avoidable suffering (Slovic, Flynn, and Kunreuther 2013; Weber 2006). Everyday risks that take many lives, such as tobacco smoking, traffic accidents, gun violence, influenza and malaria, or slow-developing harms such as climate change, receive far less notice than immediate crisis events – from the news media, the public, legislators, and regulators – and yet may deserve greater regulatory attention.

On the other hand, some observers of regulatory governance argue that crises can present rare opportunities for needed reform. They stress the capacity of concentrated interest groups to resist regulatory proposals during ordinary political times, and of policy-makers to deflect the popular pressures created by a crisis event. As a result, they contend that bold new policies need to be adopted in the wake of crisis, to seize the political opportunity created by public outcry, and to overcome the foreseeable moderation of these policies as powerful interest groups later influence the less well-covered details of bureaucratic implementation (Coffee 2012). Meanwhile, other scholars point out that good policy analysis and centralized oversight mechanisms such as OIRA, the IAB, and executive orders can be used not only to brake but also to prompt new regulatory policies (Kagan 2001; Graham 2007). Such prompts might be warranted when accumulating evidence or a sudden crisis strengthens the case for new policies that regulatory agencies had not yet pursued.

To be sure, the historical record also makes clear that crises do not necessarily motivate significant regulatory responses or shift political agendas and the parameters of policy debates. In retrospect, the oil crises of the 1970s did not generate dramatic reorientation of regulatory policies to foster American investment in alternative energy. Nor did Hurricane Katrina (2005) prompt serious regulations of greenhouse gases as a means of mitigating global climate change. In some instances, policy-makers defuse popular demands through study commissions or minor concessions, perhaps rearranging institutional deck chairs or adopting new rules without much attention to enforcement. In still other contexts, governments implement new policies without crises, such as the pioneering sulfur dioxide (SO_2) allowance trading system enacted in the US in 1990 to reduce acid rain, and the major 1996 reforms to the American Safe Drinking Water Act.

The inescapable conclusion is that the desirability of crisis-driven policy change varies enormously. Sometimes such policy change is hasty and misguided. But other times it reflects justified and well-conceived boldness. Often it incorporates both problematic and effective features. In at least some instances, crises reveal new dimensions of a complex problem, create channels for overcoming seemingly intractable political impasses, and lead to sensibly crafted regulatory policies that mitigate risks at reasonable cost. Such laudable reforms might occur either by creating new political support for long-germinating policies with much to recommend them, or by spurring policy learning – genuine reconsideration of strategic approaches and the appropriate mix of policy tools – which leads to well-designed and effective regulatory innovations.

In short, we are skeptical of attempts to identify a simple crisis-regulatory response dynamic. Instead, we remain impressed by the variety of regulatory responses to moments of crisis – both in terms of the degree of policy change, and the diverse array of policy mechanisms and strategies that policy-makers have adopted when they choose to respond to crises. We are struck by the dearth of research about precisely which events receive the label of regulatory "crises," exactly how those crises shape new regulatory agendas, how the specific characteristics of crises influence the selection of differing regulatory reforms, and whether those reforms achieve their intended goals. We are equally convinced that governments could prepare more effectively to learn from crises when they occur, either by developing norms to govern special Commissions of Inquiry or by setting up independent, standing investigatory bodies to study crisis events and draw out policy recommendations for regulatory reform.

THE PLAN OF THE BOOK

This volume reconsiders the links among crisis, public perceptions, and regulatory policy. We wish to draw a better analytical map of crisis-driven risk regulation, achieving greater clarity about when and especially how "crisis" generates regulatory change. This endeavor begins in Part I with several conceptual essays that convey the most recent multidisciplinary knowledge about risk perceptions, risk analysis, and institutional responses to the sorts of disasters that shake up political debate and policy agendas. The book then proceeds to case study clusters (Parts II–IV) about policy responses to the sorts of crises highlighted at the outset: offshore oil spills, nuclear accidents, and financial crashes. Our case studies, written by leading experts, examine several incidents of each of these types of crisis, comparing policy responses in the United States, Europe, Japan, and elsewhere. In two concluding chapters (Part V), we first examine the role of "disaster investigation bodies" as mechanisms for learning from crises and recommending needed reforms. Then we distill an agenda for further research, especially with regard to how different types of crises occurring in different contexts influence choices among different types of regulatory responses. We recommend ways to reduce "policy regret" and enhance "policy resilience," building learning into regulatory systems to reduce hasty policies that distort regulatory responses and yield dysfunction, and increase reflective understanding and policy improvement.

Thus, we hope to unpack the black box of crisis-driven regulation: we examine not just whether crises spur policy change, but which kinds of regulatory responses policy-makers adopt, why they make those choices, and what consequences those choices have. Hence the cases that we study all involve crises that have generated some policy responses; we are not testing whether all crises spur policy responses, or whether all policy changes require a crisis. We focus on what goes on at the regulatory institutions – the complex interactions among expert analysis, interest group pressures, political pressures, legal rules, turf, prior history, preparedness (including policy ideas already "on the shelf"), agility to learn, institutional culture, and other factors. We examine which regulatory responses officials choose when crises suddenly disturb the pre-existing institutional ecosystem. We then assess, where sufficient time has passed, how these choices among different policy options yield different outcomes. Our three clusters of cases illustrate significant variation across topical subject matter (oil, nuclear power, and financial markets), time period, country, agency, and prevailing context (political, economic, and social).

PART I: THE CONCEPTUAL TERRAIN OF RISK AND RISK PERCEPTIONS

In recent decades, scholars from across the social sciences have learned a great deal about risk assessment and risk perceptions. Researchers also have made progress in understanding some aspects of how policy-makers respond to crisis. But scholars have tended to examine these vital questions in isolation from one another. In the first part of our volume, five conceptual essays convey the most recent scholarly findings about these issues, furnishing an analytic foundation for the detailed case studies that follow.

The extensive literature in this arena addresses several key lines of inquiry. One concerns how experts should undertake risk assessments and policy analyses, and how government officials should use them as a basis for decision-making. A second investigates how private firms, markets, and related actors make sense of and respond to risk. A third examines how individuals in the general public form risk perceptions and how those perceptions change over time. And a fourth considers the role of crises in the news media and the policy process, both in terms of influencing what gets on the policy agenda and in shaping the narratives that frame public appraisal of events and subsequent policy responses.

The first conceptual essay, written by environmental economist and volume co-editor Lori S. Bennear, examines scholarship on the use of risk assessment and economic analysis in regulatory policy-making, and on the impact that shifting risk perceptions have on such analyses. Risk assessments are critical components of Regulatory Impact Analysis (RIA), which the US federal government has required in some form since the 1970s and which governments across the globe have increasingly required as well (Wiener et al. 2011; Wiener and Ribeiro 2016). Traditional risk assessment requires an informed estimate of the probabilities of various outcomes when some hazard threatens a population (Renn 2008). Scholars have done extensive work to determine how best to elicit these probabilities when the distribution of outcomes is difficult to characterize (Helmer-Hirschberg 1967; Dalkey, Brown, and Cochran 1969; Morgan and Henrion 1990) and how to perform statistical analysis to incorporate uncertainty in such probability distributions, developing statistical techniques that are now fairly standard in RIAs (OMB 2003). But rare extreme events pose risks whose probabilities are particularly hard to assess. More recent scholarship has challenged some of the underlying components of the neoclassical model of economic analysis under uncertainty, by examining hyperbolic discounting (Laibson 1997), fat-tailed risks (Weitzman 2011; Weitzman 2009), and discounting under long time horizons (Gollier and Weitzman 2010; Newell and Pizer 2003). Economic sociologists have further emphasized the importance of holistic approaches to risk assessment that take account of how multiple risks interact within complicated systems, such as financial networks or tightly coupled, complex technologies such as offshore oil drilling or nuclear power plants (Perrow 2007; Renn 2008; Schneiberg and Bartley 2010). These features further complicate efforts to anticipate crises. The second essay, written by the economist Carolyn Kousky of the University of Pennsylvania, considers the role of private market actors, particularly insurance and reinsurance companies, as alternatives or supplements to risk regulation. Insurance plays a critical role in catastrophe management. But as our understanding of risks improves, particularly with regard to fat-tailed or tail-correlated risks, insurance becomes more complicated. Many natural disasters have been shown to exhibit fat-tailed risks (Schoenberg, Peng, and Woods 2003; Newman 2005). As a result, managing "value-at-risk" requirements to prevent insurance company insolvency requires charging large premiums (Jaffee and Russell 1997) and for many catastrophic risks, premiums would need to be so high that nobody would insure (Kousky and Cooke 2012).

Together, these two chapters offer analytical tools for making sense of how policy-makers develop baseline risk assessments, analyze rare extreme risks, and allocate risk management between public institutions and the private sector. They also raise a crucial issue for regulatory officials who confront a crisis: how does one tell if the crisis event represents an unfortunate outcome within the range of prevailing risk estimates (a "bad draw"), or if it represents an unanticipated disaster that reveals new information and demands that we recalibrate the relevant risk profile? This question is a pivotal one, because, as Bennear explains, crises that result in significant costs and harms often prompt a psychological response that she terms "policy regret" – a wish that decision-makers had been better able to evaluate the risks that generated the crisis, and thus adopt policies to prevent it.

We next turn to public perceptions of risk. Cognitive psychologists have closely scrutinized the nature of human risk perception. One important finding has been that public perception of risks often differs from expert appraisal of risks (Slovic 1987; Slovic 2010). Notably, the public may be less concerned (than experts) about familiar frequent risks (Slovic 1987) and about potentially catastrophic risks that are too rare or novel to be memorable in recent experience (Weber 2006; Gilbert and Wilson 2007; Wiener 2016), while the public may be more concerned than experts about unfamiliar but occasionally occurring events that remain salient – or in the terminology of cognitive psychologists, "available" – in recent memory (Kuran and Sunstein 1999). Non-experts also tend to pay greater attention to risks associated with events that involve clearly identifiable individual victims or villains, rather than risks affecting large, but anonymous, groups (Small and Loewenstein 2003; Small, Loewenstein, and Slovic 2007).

The important question of what drives changes in risk perceptions, however (either increasing or decreasing concern, and whether among policy elites, the general public, or both), remains less well understood. The third conceptual essay in our volume, written by Elke Weber, assesses what we know about cognitive perceptions of societal risks, focusing especially on how people process new evidence that some activity entails much greater risk than they previously assumed. Weber, a professor of psychology and codirector of the Center for Decision Sciences at the Columbia Business School, has been a leading scholar in this area for more than two decades. Her chapter draws extensively on the literature in psychology and neuroscience. It considers important differences between the ways that individuals process new information about personal risks and how organizations process new information about societal risks. A key point raised by Weber's chapter is the role of highly visible or salient events in influencing public perception of risks. She discusses cognitive decision research showing that experienced events have a larger influence on public attitudes than descriptions of potential future events. This finding helps to explain the public outcry that follows many crisis events and the relative public passivity in the face of many expert warnings. It also helps to account for the greater political impact of crises that hit, in some or way or another, closer to home.

Weber's chapter also explains the brain's different modes of processing alarms (quickly) versus making decisions (more slowly), and the corresponding implications for public acceptance of familiar risks coupled with public demand for prompt policy responses to shocking events. She further comments on research that explores the magnitude of alarms needed to be perceived as shocking, relative to the baseline of other risks. This essay shows how the dynamics of human risk perception inform public and political reactions to the case studies addressed later in the book, as well as other events such as terrorist attacks and natural disasters.

Dramatic shifts in risk perceptions neither guarantee significant shifts in regulatory policy nor predetermine specific choices when policy-makers choose to respond substantively to crisis. Over the past quarter-century, a number of political scientists have closely examined the dynamics of "focusing events," occurrences that profoundly command the attention of policy elites and the broader public. Our volume's fourth conceptual essay, authored by political scientists Thomas A. Birkland and Meg K. Warnement, offers a trenchant overview of this scholarship. Birkland and Warnement invoke John Kingdon's "streams metaphor" of the policy process to highlight the factors that contribute to policy change (Kingdon 2003). These include: "the Problem Stream," which contains ideas about policy problems; the "Politics Stream," which contains the ebb and flow of electoral politics, public opinion, and the like; and the "Solutions Stream," which contains the set of ideas about how governments might concretely address problems (Kingdon 2003; Sabatier 1988; Birkland 1998; D. Carpenter and Sin 2007; Baumgartner and Jones 2010).

Birkland and Warnement observe that opportunities for policy change tend to emerge when events bring two or more of these streams together in a way that makes the matching of solutions with problems more likely.

They also detail a number of more specific factors that typically influence the path from "event" to "regulatory change": the magnitude of the event (defined as the extent to which it aggregates harms in one place and time (this point overlaps with Weber's stress on the relative significance of baselines in determining what is shocking); its rarity; the emergence of a series of related crisis events that burnish the credibility of reformers and allow for refinement of policy proposals; and the presence of policy entrepreneurs, who can build constituencies for change, not least through the marshaling of cultural symbols that shape media coverage and commentary.

The "focusing event" literature helps to explain how crises can overcome policy inertia to spur regulatory change. But it does not fully predict *which* regulatory strategies officials will select or develop in response to crises, nor the particular bundle of regulatory instruments and policies that they will adopt. Frederick W. Mayer's essay, on framing techniques and moral narratives, presents some potential avenues of explanation for such policy choices. Mayer, a professor of public policy, examines how framing techniques and moral narratives mediate focusing events. He stresses that moral framing, often in the form of stories, powerfully shapes the public definition of events as crises (or non-crises), influences the resulting adjustment (or non-adjustment) of risk assessments, and guides the decision-making of policy elites charged with formulating institutional responses to such events (Satterfield, Slovic, and Gregory 2000; Jones and McBeth 2010). His chief example, the public discourse about how to evaluate and respond to climate change, emphasizes the role that narratives can play in deflecting regulatory action. But this methodological approach also promises to elucidate how political entrepreneurs build support for particular regulatory solutions – especially how they translate policy learning into salient and persuasive arguments for new approaches (Sabatier and Jenkins-Smith 1993; D. P. Carpenter 2001).

Each of the literatures discussed in these conceptual chapters offers crucial insights about the nature of risk regulation before the advent of particular crises, and the patterns of policy responses to crises. Through the clusters of case studies on nuclear accidents, oil spills, and financial crises, our project builds on these scholarly frameworks and links them together to develop a richer understanding of how past crises have influenced redirections of regulatory policy. We study how crises affected the choice among alternative policy responses, when policy innovations effectively achieved their objectives (or did not), and what other outcomes ensued. The case study authors apply these various conceptual frameworks through their examinations of how regulatory officials in a variety of societies and policy arenas responded to several major crises.

PARTS II THROUGH IV: CASE STUDIES ON OFFSHORE OIL SPILLS, NUCLEAR ACCIDENTS, AND FINANCIAL CRASHES

The three types of crises that we have chosen for our case studies – offshore oil spills, nuclear accidents, and financial crashes – all involve industries characterized by tightly coupled complex systems, in the sense discussed by the sociologist Charles Perrow (Perrow 2007; Perrow 1999). In the case of deep sea oil drilling and the production of electricity through nuclear fission, complexity and tight coupling result from a mix of technological and social parameters. These businesses depend on sophisticated engineering, with interconnected technological subsystems that relate to one another in complex, non-linear ways, and that rely on complicated monitoring systems that allow human operators to see developments within subsystems, but that sometimes obscure how events are affecting the technological system as a whole. Technology, primarily in the form of complex computer securities platforms and trading algorithms, has also contributed greatly to the risks in modern financial systems. Here a great deal of risk also resides in the dense networks of credit that link so many far-flung financial institutions, which create the possibility of a chain reaction of debt defaults, bank runs, and collapsing asset values.

Our three clusters of case studies, however, also include a number of instructive variations. These include differences in industry structures (e.g. degree of vertical integration, outsourcing, globalization), dominant regulatory tools (e.g. prescriptive, performance-based, self-regulatory), significant political institutions (e.g. regulatory agencies, legislative committees), and relevant non-governmental actors (e.g. environmentalists, consumer protection advocates, corporations, insurance companies). These areas also raise complex dilemmas about how to balance conflicting policy goals – the trade-offs among economic growth, energy independence, worker safety, national security, and environmental protection; or among economic growth, public confidence in financial institutions, the protection of taxpayers, and stability of the financial system.

We might have included a wider range of substantive policy arenas and truly transformative "focusing events," such as chemical safety and the Bhopal disaster, food safety and disease outbreaks, or homeland security and the terrorist attacks of 9/11. Similarly, we might have incorporated contexts in which scientific uncertainty, ferocious political opposition,

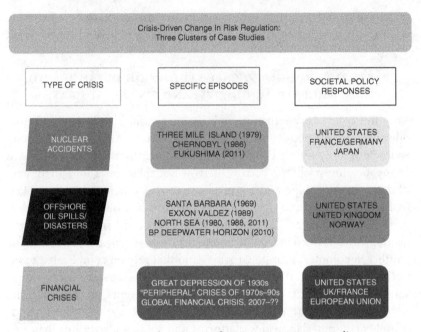

FIGURE 1.4 Our three types of crisis-response case studies

a lack of a sufficiently damaging event, or some combination thereof, has stymied or delayed consensus for regulatory action (e.g. global climate change). Alternatively, we might have zeroed in on just one policy arena, digging in even more deeply to historical context and comparative/transnational experience. We have sought to strike a balance among these various alternatives, thereby gaining more than just one view of regulatory response in a given policy arena, comparing across diverse policy contexts, and yet limiting the number of variables at play to enable us to draw more robust inferences. Figure 1.4 offers an overview of the case studies in the volume.

Through the narrative analysis of these case studies, we investigate the regulatory changes spurred by a given crisis event. Figure 1.5 offers a stylized depiction of these potential consequences – with regard to risk perceptions, policy debates, adoption and implementation of new regulatory policies, and the ramifications of those new policies. At each stage in this causal chain, of course, developments are mediated by the configuration of relevant interest groups, politics, and policy institutions. Along with media coverage and public opinion, those contexts also profoundly shape the emergence of a societal context that a given event indeed represents a "crisis" demanding the attention of policy elites.

Introduction

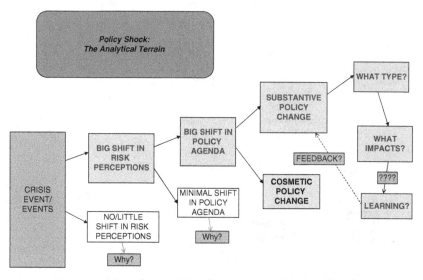

FIGURE 1.5 Map of potential pathways from crisis to policy change

The overarching goal of each set of case studies is to illustrate how regulatory institutions in different countries made sense of and responded to different crises. Figure 1.6 depicts an analytical framework for the policy choices evident in the various case studies. These three schematics lay out a range of options for crisis-driven policy change. They distinguish between strategic posture, tactical posture, and the menu of more particular policy instruments that policy-makers may adopt, which tend to cluster around particular strategic and tactical orientations. With regard to strategic considerations, governments may respond to a crisis by leaving the basic structure of the relevant industry alone, or seeking to reconstruct it (encouraging consolidation or rather breaking up oligopoly). They may take a notably broad or narrow view of the relevant issues raised by the crisis. They may focus on reducing risks, or making systems more resilient.

Tactically, governments may respond to a crisis through institutional reorganization or substantive reform of regulatory rules and standards. They may delegate more or less to administrative agencies, courts, and/or non-governmental actors such as corporations and NGOs. They may choose more rigid, prescriptive approaches to policy or to punitive sanctions. Alternatively, they may adopt more flexible and cooperative policy frameworks, or attempt to harness market-based incentives as means toward new regulatory ends. Specific strategic and tactical orientations

STRATEGIC POSTURE

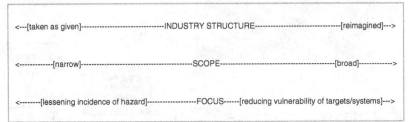

TACTICAL POSTURE

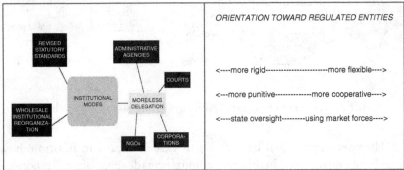

POLICY TOOLS AND INSTRUMENTS

RESHAPING INDUSTRY STRUCTURE	LEVERAGING MARKET FORCES	PUBLIC REGULATORY OVERSIGHT
Deregulation/Privatization	Disclosure Regime	Prescriptive Design Standards
Market Partitioning	Liability Regime	Orchestration:
Technological Phase out	Ratings Regime	Guide to Best Practices
Nationalization/Public Option	Tradable Permits	Certification Scheme
	Increased or Reduced Subsidies/Taxes	Inspection Regime:
		Regular Monitoring
		Accident and Near-Miss Inquiries

FIGURE 1.6 Schematic of policy options in response to crisis

lend themselves to particular bundles of regulatory instruments. A determination that a crisis requires reconfiguration of sectoral structure will focus policy decisions, as will a decision to rely primarily on a redirection of market incentives, a preference to maximize cooperation from regulated parties, or a focus on modes of public regulatory oversight.

The case studies additionally take note of the outcomes of crisis-driven regulatory policies, especially when sufficient time has passed to make evaluations of those outcomes plausible. Pivotal issues about outcomes include whether such policies prove to be durable or ephemeral; whether they appear to produce their intended results, and/or unintended consequences of significance; and whether policy institutions effectively learn from these impacts.

OFFSHORE OIL SPILLS

Offshore oil production began off the coast of California in the late nineteenth century, when oil speculators invested in oil drilling rigs that extended along a long wooden pier extending out into the Pacific Ocean. Since that time, offshore oil production has grown steadily and now constitutes a little over a third of all global crude production. In addition, over two-thirds of the world's hydrocarbons are transported from the point of extraction to the refining or consumption location by ship. As a result, large quantities of extracted crude oil, natural gas, and natural gas liquids are now always present in the ocean environment (World Ocean Review 3 2014).

The policies that govern offshore oil and gas production and oceanic transportation of hydrocarbons must balance tensions among multiple competing policy objectives. On one side, lie the goals of economic growth (spurred by lower fuel prices), energy independence, political security, and free enterprise, which tend to align with increased oil and gas production. On the other side lie the goals of worker safety, environmental protection, and the conservation of scarce resources, which generally suggest the need for caution and restraint.

Despite the best efforts of policy-makers to balance these often conflicting aims, the history of offshore oil production and oil transport has been characterized by several major accidents, which have resulted in injuries, loss of life, and/or damages from the spilled crude oil. These accidents frequently, though not always, attract public attention and expose the inherent tensions in energy policy, often leading to significant regulatory change. In order to better understand how these accidents affect risk perceptions and how these changes in perceptions affect regulatory policy, we commissioned three case studies that collectively examine accidents in American waters and the North Sea, two especially significant contexts for oil and gas extraction and transport. Some key features of these incidents, such as the type of incident, major consequences, and the resulting policy changes are provided in Table 1.1.

TABLE 1.1 *Offshore oil crises and responses*

	Santa Barbara	Ekofisk Bravo	Alexander Kielland	Piper Alpha	Exxon Valdez	Snorre A	Gullfaks C	Deepwater Horizon
Time Place	1969 California, USA	1977 North Sea, Norway	1980 North Sea, Norway	1988 North Sea, UK	1989 Alaska, USA	2004 North Sea, Norway	2009–2010 North Sea, Norway	2010 Gulf of Mexico, USA
Type of accident	Oil spill from drilling platform	Oil spill from drilling platform	Collapse of rig used as hotel for workers	Explosion and fire on drilling platform	Oil spill from tanker	Gas leak; near miss of more serious blowout	Gas leak	Oil spill from drilling platform
Impacts	Ecosystem Tourism	Ecosystem	Lives lost	Lives lost	Ecosystem Tourism	Lives lost	Lives lost	Lives lost Ecosystem Tourism
Policy response and type: USA	New liability rules; strict liability established	None	None	None	Oil Pollution Act of 1990 (Legislation)	None	None	Organizational change Modified regulatory approach; Safety and Environmental Management system
Policy response and type: Europe	None	Environmental safety of drilling enters policy agenda (Norway)	New regulations regarding internal control (Norway)	New regulatory approach; Safety Case (UK)	None	Limited; Investigation (Norway)	Limited; Independent investigation (Norway)	Reviews of regulatory systems

The political scientist Marc Allen Eisner's essay focuses on American policy responses to a pair of accidents in US waters – the 1969 drilling blowout in Santa Barbara Bay and the 1989 Exxon Valdez oil spill that resulted from a tanker grounding off the coast of Alaska. Eisner traces the immediate impact of these events on public perceptions and policy, while also analyzing the evolution of policy between these two extreme events, which developed in response to the dynamic growth and expansion of the US offshore oil and gas industry. Thus his essay combines analysis of accident-driven policy with analysis of policy evolution that is not directly tied to a specific crisis, but remains informed by policy choices made in prior crises.

The chapter by Ole Andreas Engen, a safety engineer, and Preben H. Lindøe, a sociologist, investigates policy responses to accidents in the North Sea by the major European oil producers – the UK and Norway. Engen and Lindøe show that early policy in Europe was powerfully influenced by three accidents in the North Sea – the 1977 blowout at the Phillips Petroleum-owned Ekofisk Bravo platform; the 1980 collapse of the Alexander Kielland rig also owned by Phillips Petroleum; and the 1988 explosion that occurred on the Occidental Petroleum-owned Alpha platform operating in the Piper oil field (and so referred to as Piper Alpha) off the UK Continental Shelf. The Ekofisk Bravo accident made the potential environmental consequences of offshore drilling salient, despite the fact that high levels of evaporation and turbulence lead to almost no directly noticeable environmental damage. Nonetheless, the accident prompted adoption of regulatory reforms in Norway. The Kielland disaster resulted from gale force winds that caused one of the rig's legs to give way. The platform tilted to 35 degrees and capsized within one minute, killing 123 of the 212 people onboard. This accident prompted significant reforms in Norway and the development of the so-called Nordic Model, a unique hybrid regime that balanced limited command-and-control regulations with more prominent performance-based regulations managed in tandem by government, industry, and labor officials. The Piper Alpha explosion killed 167 of 229 people onboard, and remains the deadliest accident in offshore oil/gas history. Following this accident, the UK government commissioned an investigation that produced the Cullen Report, which recommended significant changes to the UK regulatory regime for offshore oil/gas facilities and moved this regulatory regime closer to the approach adopted in Norway after the Kielland accident. Engen and Lindøe examine the policy responses to each of these accidents (and other smaller ones) and highlights how these accidents shaped policy in

Europe in ways significantly different from how policy developed in the United States. The essay also examines the robustness of the European approach to regulation of offshore oil and gas in the face of more recent accidents and argues for the importance of a proactive regulator, one who actively seeks to anticipate and manage problems, as a key factor in the robustness of the regime.

The third chapter in the section, by the political scientist Christopher Carrigan, explores responses to the most recent large-scale offshore oil disaster, the 2010 explosion at the BP Deepwater Horizon drilling rig, which killed 11 workers and dumped nearly 5 million barrels of oil into the Gulf of Mexico. Like Exxon Valdez, this accident prompted enormous public concern, because of intensive news coverage and the visible nature of the spill's environmental impacts. But with Deepwater Horizon, the drama of attempts over several months to plug the underwater gusher accentuated public interest. Carrigan focuses particularly on the subsequent institutional reorganization of American regulatory agencies after Deepwater Horizon, as policy-makers sought to heighten safety without curbing oil and gas output.

Several themes emerge from these three cases. First, regulatory policy in this domain has been significantly shaped by crises that changed risk perceptions both among the policy elite and the broader public. This is evidenced by significant reforms following the Santa Barbara accident, the Exxon Valdez, and the accidents in the North Sea. These events precipitated paradigm shifts in how the public and the policy elites perceived the risks associated with offshore oil drilling in both the US and Europe, as well as how governments tried to manage them. However, the response to the Deepwater Horizon blowout has, at least thus far, proved more limited in scope, perhaps because the this accident, while large, did not fundamentally change risk perceptions among the policy elite. Second, regulatory responses to oil spills seem to be highly conditional on the geographic location of the accident, even though risks from offshore oil drilling are similar in different locations. European responses followed accidents in the North Sea, while US responses followed accidents in the US. In this policy domain, the depth of political concern and policy response seem to vary directly with the distance from the location of the disaster. A third theme that emerges from these cases is that there is a great deal of time-dependence in policy responses. Once a policy regime is set in place, employing a particular set of approaches, subsequent crises may generate refinements of the already adopted approach rather than large-scale and more radical shifts in regulatory regimes (which may require an even more dramatic crisis).

MAJOR NUCLEAR POWER ACCIDENTS

After physicists in the 1930s recognized the possibility of chain-reaction nuclear fission and then deployed it in the 1940s to create atomic bombs, interest soon turned to designing slower controlled fission reactions that would generate electricity for civilian energy. The first commercial nuclear power plants began operation in the 1950s, when reactors were commissioned in the US, the Soviet Union, the UK, and France, followed soon by Germany. Japan started its first commercial reactor in the 1960s. Over the subsequent two decades, the technology spread to several other countries, aided by technical and financial assistance from international agencies like the International Atomic Energy Agency and the World Bank.

Today there are just under 100 operating commercial nuclear power reactors in the US, providing about 19 percent of the country's electricity generation. In 2011, Japan had just over 50 commercial reactors, which generated about 30 percent of its electricity, but all of Japan's nuclear reactors were closed following the Fukushima Daiichi disaster in 2011. Thus in 2014, Japan's nuclear power plants generated no electricity, and by the end of 2015, only two of its 45 operable reactors had been brought back online (Nuclear Energy Institute, n.d.(a)). In France, almost 60 reactors generated over 75 percent of the country's power needs in 2014, while nine German reactors generated about 16 percent of its electricity in 2014 (Nuclear Energy Institute n.d.(b); IEA 2017).

Nuclear fission promised a cleaner energy source, in the sense that, unlike fossil fuels, it would not emit air pollutants (such as sulfur oxides, nitrogen oxides, particulate matter, mercury, and greenhouse gases). These air pollutants (especially fine particulate matter) are associated with tens of thousands of premature deaths each year in the US, hundreds of thousands of premature deaths each year in Europe, and millions worldwide (Lelieveld et al. 2015). Nuclear power emits no greenhouse gases and thus helps avoid climate change. But no fuel eliminates all risks; every energy choice confronts a risk-risk trade-off. The nuclear power fuel cycle has posed concerns regarding, for example, uranium mill tailings, low-level radiation from power plant operation, disposal and leakage of spent fuel wastes, diversion of fissile material to make weapons, and the dispersion of radiation resulting from any accidents leading to reactor core meltdowns.

Advocates also hoped that nuclear power would generate electricity at very low cost and offer energy independence rather than reliance on imported fossil fuels. But high capital costs for plant construction, delays due to siting disputes and regulatory hurdles, and the question of waste disposal, as well as other factors, have kept prices higher than advocates anticipated. In the US, cheap electricity from coal, and more recently from shale gas fracturing, have garnered large shares of the electricity market, while the cost of renewable sources such as solar and wind continue to drop. US federal laws have attempted to reduce the costs of US nuclear reactors, such as by limiting the industry's liability for a major accident (the Price-Anderson Act of 1957), and by designating a federal repository for commercial reactor waste (ostensibly at Yucca Mountain, Nevada, though still not approved to operate and still in continuing litigation).

Studying accidents at nuclear reactors can contribute in several ways to our understanding of crises, perceptions, and policy changes. A nuclear power plant accident raises the specter of radiation exposure to the surrounding, downwind, and downstream population. Awareness of these risks may in turn arouse public concern and opposition in advance of plant construction, and spark public panic and outcry if an accident occurs. Nuclear power reactors are complex technological systems with myriad interconnections and potential causes of failure. They are managed and monitored by fallible human beings. Moreover, the regulatory institutions governing nuclear power may face limits on their personnel, expertise, funding, information, and legal authority; and they may harbor conflicting objectives such as promoting the industry while also regulating safety. If an accident galvanizes public opinion and yields policy change, it may drive several different types of response, including tightening regulatory standards, refocusing regulatory attention (such as from reactor design to reactor operation), reorganizing government agencies (such as by splitting promotion from oversight), and delegating monitoring and/or standard-setting tasks to industry.

Three case study chapters in this volume address nuclear power accidents. Each considers the influence on perceptions and policy responses of three major crisis events: the accidents at Three Mile Island (1979), Chernobyl (1986), and Fukushima (2011). Some key features of these three incidents, such as the magnitude of radioactive fallout and the resulting health effects (which remain highly contested questions) are sketched in Table 1.2. Each chapter's case study examines the response to these three accidents (and others) in a different regulatory system(s). The chapter on the responses in the US, authored by Elisabeth

TABLE 1.2 *Nuclear reactor crises and responses*

	Three Mile Island	Chernobyl	Fukushima
Time	1979	1986	2011
Place	Pennsylvania, USA	USSR/Ukraine	Japan
Type of accident	Partial core meltdown, [release of radioactive steam]	Power surge, vessel rupture, explosion and fire, release of airborne radioactive plume [which spread across Europe]	Three reactor meltdowns, [airborne release?], leakage of spent fuel, and release of radioactive coolant water into ocean and groundwater
Impacts • Workers • Residents • Ecosystems	Low to none?	Reactor accident: from ~30 to 4,000 deaths (WHO 2011) (some NGOs estimate higher) 150,000 residents evacuated, others resettled. Wildlife benefits: closed site became nature preserve.	Tsunami: about 20,000 deaths. Reactor accident: Few to none? (WHO 2013; UN Special Committee 2013; IAEA August 2015). 140,000 residents evacuated for many months to years. Ocean pollution.
Policy response & type: USA	De facto moratorium on new licensing, 1980–2012	Minimal. Distinguished Chernobyl as different design of reactor and containment.	Concern re older reactors; impetus for newer, small modular reactors.
Policy response & type: Germany	Little	Minority party (SDP) called for phase out by 2030; enacted under SDP government in 1998, then relaxed under CDP government in 2008.	CDP government accelerated phase out by 2022; increased use of coal and wind.
Policy response & type: France	Little	Little. Distinguished Chernobyl as different design of reactor and containment.	Little. Concern that even the expert Japanese engineers could have an accident.
Policy response & type: Japan	Minimal. Instead, response to *Mutsu* accident (1974): reorganized agencies, created NSC.	Minimal. Distinguished Chernobyl as different design of reactor and containment. Instead, response to JCO Tokaimura accident (1999): reorganized agencies, created NISA.	Shut down all ~50 reactors in Japan; increased imports of fossil fuels; ongoing debate over restart; reorganized agencies, created NRA to replace NISA.

Paté-Cornell, a professor of management science at Stanford University, illustrates the strong US response to TMI (tightening standards for licensing, resulting in a de facto moratorium on new licenses for three decades, while also delegating some safety monitoring to an industry consortium, INPO), the less significant US policy responses to Chernobyl or Fukushima, and the limited reorganization of US agencies. The chapter on responses in Japan, authored by Atsuo Kishimoto, a risk expert at the national industrial safety institute and now at Osaka University, shows that significant policy changes were made in Japan (notably large-scale reorganizations of regulatory oversight), mainly in response to accidents that occurred in Japan – smaller events in the 1970s and 1990s, and then Fukushima – but with little response to TMI and Chernobyl. A third chapter on nuclear policy in Germany and France authored by Kristian Krieger, Ortwin Renn, M. Brooke Rogers, and Ragnar Löfstedt, risk management scholars in Potsdam and London, recounts the major responses of German policy to Chernobyl and Fukushima (by adopting a phase out of nuclear power, initially by 2030, then relaxed, and then after Fukushima accelerated to 2022), but meanwhile the relatively smaller response of French policy to these crisis events.

These case studies on nuclear accidents illustrate the varying influence that crisis events can have on public perceptions and policy responses. Not all the nuclear reactor crises affected all policymakers: some countries responded strongly to local accidents, others responded significantly to distant accidents, and still others did not respond much to any of the accidents. Some policy changes arose without crises. The degree of influence of a nuclear reactor crisis on policy appeared to depend, among other factors, on the magnitude of event, geographic proximity, public attention and framing, similarity of technology and accident cause, availability and cost of alternatives, confidence in elite leadership, and structure and role of governance institutions. These cases also highlight the trade-offs involved in responding to a crisis in one technology (such as nuclear power) when key alternatives also pose risks (such as air pollution and carbon dioxide emissions from coal, groundwater pollution and methane releases from shale gas fracturing, dependence on imports of foreign fuels, and cost and reliability questions about renewables). And these cases indicate the variation in type of regulatory response, which ranged from agency reorganization, the adoption of a new scheme of industry self-regulation, and tighter standards, to a moratorium on new reactors and a phase out of existing reactors.

Introduction

FINANCIAL CRASHES

Dozens of major financial panics have occurred since the advent of modern banking, public finance, and corporate securities markets, touching nearly every society that has embraced the basic institutions of capitalism. These events proved especially frequent in the nineteenth century, before the introduction of publicly mandated deposit insurance, capital requirements, and schemes of prudential regulation, occurring somewhere in the North Atlantic world essentially every decade. Crises originated in many financial markets, sometimes engulfing the market for sovereign debt, sometimes initially battering stock markets, sometimes centering on housing finance. The most affected financial institutions varied as well, ranging from commercial banks and insurance companies to trust companies and brokerage houses. In some cases, the reach of crises never extended beyond regional or national boundaries, but in others they cut a much wider geographic swath.

Nonetheless, as the economic historian Charles Kindleberger has argued, significant financial crises share some important characteristics, regardless of the type of market at the epicenter of crisis or the extent of its geographic reach. Financial panics typically occur after a period of significant prosperity, which fosters extensive optimism among investors and business leaders and an associated ready supply of credit. Often a specific stretch of good times is bound up with extensive investment in emerging industries and widespread reliance on new financial instruments and strategies, whose complexities, especially concerning their implications for vulnerabilities in the overall financial system, were not well understood by the great majority of market participants. The onset of crisis almost always begins with some triggering event (or cluster of events) that leads investors and the managers of firms to reassess the risks that they face – to engage, in the modern language of risk analysis, in "Bayesian updating" – and therefore to sell assets.

Before the 1930s, this widespread re-evaluation of economic prospects and values usually was triggered by the failure of a significant bank or other financial institution. After the American New Deal, and the subsequent global diffusion of public deposit schemes and government backstops of their nationally significant financial institutions, financial crises became far less common for several decades. But under the twin pressures of post-1975 globalization and financial deregulation, they have re-emerged, with a much wider array of triggers. The key event in 2008, for example, was not so much the failure of a key financial institution – Lehman Brothers – as it was the realization that the American government would not bail it out (while the anticipated prospect of government bailouts may have encouraged even more of the risk-taking that helped cause the crisis).

Regardless of how financial crises begin, they typically unleash a rapid transition from free-flowing credit to general distrust of counterparties. Fears of collateral bankruptcies produce a rush to liquidity, as firms, investors, and speculators seek to protect themselves from anticipated losses by calling in debts and selling financial instruments. This general chasing after cash then causes dramatic declines in asset values, exacerbating the pressures on debtors and generating a sharp spike in business failures as well as dramatic declines in economic output.

As is the case with each of the case study clusters contained in this book, one challenge faced by regulators in responding to financial crises stems from the competing policy goals at stake. Governments typically want to maintain public confidence in financial markets, protect investors (especially unsophisticated investors) from deception or imposition, and, if public insurance or implicit public guarantees exist as a means of forestalling bank runs, protect taxpayers from the costs of bailouts. But policy-makers also wish to facilitate innovation, encourage the flow of capital to businesses and consumers, and protect the competitive position of financial centers. These goals are often in conflict. Legislators and financial regulators may seek to balance the desire to facilitate economic growth against the imperative to maintain systemic stability. Alternatively, policy-makers may weigh the need to facilitate innovation in financial products against the desire to protect investors from fraud, overreaching, or investments perceived as too risky or complex for many investors. And innovations in financial institutions, such as have occurred with the investment strategies and financial products used by hedge funds and other "shadow banking" institutions, may conflict with the imperatives to control systemic risk and protect taxpayers from bailouts.

These tensions invariably shape policy-making both before and after a crisis period. After a lengthy period of stability, financial regulations designed to promote the stability of the financial system may strike decision-makers as a burdensome and unnecessary impediment to growth. After a crisis, when economic conditions remain precarious, regulators may feel that they have to balance some of their preferred regulatory responses against fears that aggressive containment of important economic actors and markets will stifle recovery. Thus the financial regulator's job with respect to crisis avoidance is not simply to choose the policy option most likely to avoid a future crisis (a hard enough job, in most cases). Rather, the regulator must also weigh crisis avoidance against other competing policy goals, rendering the task both more substantively and politically difficult.

Further complicating policy-making with respect to crises is the need to address various affected constituencies. In the case of financial regulation, the financial industry represents a large, concentrated, and highly organized constituency that spends vast resources on influencing policy at the legislative, agency, and judicial levels, in both national and international forums. Although not a monolithic whole – the policy preferences of smaller, community banks, for example, may diverge from those of larger mega-banks – there are nonetheless many issues on which industry preferences largely converge. More importantly, perhaps, is the lack of a traditional constituency for whom financial stability is a high-salience issue. NGOs expressly dedicated to issues of financial stability are a relatively new phenomenon. Moreover, even experts often struggle to comprehend the causes of and cures for financial crises, with such matters typically occasioning vigorous differences of opinion. Educating the public and sustaining policy focus on reforms that seek to foster long-term stability can be a challenge.

Our case studies of regulatory responses to financial crashes consider three significant episodes of widespread financial instability. The chapter by historian Youssef Cassis examines the regulatory response to the great depression by the world's leading financial powers of the time – Britain, France, and the United States – finding great variation in regulatory response, which he attributes to a combination of factors, including: the nature and violence of the shock, its perception, the weight of public opinion, the demand for change, the role of experts, the power struggle between interest groups, and the process of regulatory change. Cassis explores these factors and their impact of resulting regulation, with particular attention to differences across the three countries in the severity, length, and depth of the crisis, as well as the way in which the crisis was perceived and framed in each country.

Three chapters grapple with the regulatory consequences of the most recent Global Financial Crisis of 2007–08, which began in the American mortgage market before spreading to other American debt markets, housing markets in several other countries, and the market for European sovereign debt. The chapter by the sociologist Bruce G. Carruthers analyzes the push by the American Bankers' Association (ABA) for the relaxation of mark to market accounting in favor of mark to model valuation in the wake of the 2008 crisis. Carruthers considers this event as a form of "decommodification" that is analogous in some ways to decommodification elements of the labor market, such as minimum wage laws, unemployment insurance, workplace safety rules, pensions, and the like. He also places the fair value accounting debate in historical context, looking at the move away from market-based valuation methods under New Deal reforms toward measures of "intrinsic value," including bond ratings.

The chapter by the economists Stijn Claessens and Laura Kodres survey the voluminous regulatory responses in Europe and North America. They argue that risks and uncertainties will always remain within financial systems despite attempts to gain better information about them. This is, in part, because of a (conscious) risk-return trade-off and in part due to (unknown) unknowns. Because risks will thus remain, they conclude that crisis management needs to be an integral part of financial system design, not improvisations after the fact. Claessens and Kodres conclude that the most important conceptual and practical challenge identified by their analysis is that policy-makers (and market participants) need to think in more systemic ways when engaging in risk monitoring efforts and financial system reforms.

The essay by the economist Barry Eichengreen explores the regulatory implications of the European sovereign debt crisis. Eichengreen contends that the views of European policy-makers were heavily shaped by history, by faith in the operation of financial markets, and by expert opinion from rating agencies. These three factors caused them to focus on a narrow and specific set of risks, namely that excessive budget deficits and their monetization would give rise to high inflation. This was at the expense of attention to other sources of risk, including private debt and high bank leverage. Although the crisis has spawned much discussion of institutional reform, Eichengreen argues that policy-makers continue to focus, to a remarkable extent, on excessive deficits at the expense of other relevant risks.

Both the Great Depression and the 2008 Global Financial Crisis occurred after the creation of pivotal modern regulatory institutions such as central banks, and after the widespread recognition of the need for minimum capital reserves for financial institutions. Each episode engulfed several major economies, exposing global linkages in the financial system and prompting searching policy debates about the nature of systemic financial risks and the most appropriate means of addressing them. These particular crises further reveal the contingent nature of policy responses. In each case, the experience gave rise to a policy trajectory rather than just an immediate, single set of legislative or administration actions. Debates over reform continued for some years, as did the jockeying of interest groups to define the meaning of the crisis, as well the process of adapting regulatory policies to perceived new realities. And in each case, affected societies often diverged significantly in their interpretations of the crisis, and thus the sorts of regulatory reforms that their leaders deemed appropriate.

PART V: MECHANISMS FOR LEARNING FROM CRISES

In the first of two concluding chapters, two of the co-editors, along with five Duke student collaborators, focus on the institutional design of crisis investigation bodies. This chapter examines the strengths and limitations of special Commissions of Inquiry, which governments often convene to establish an official explanation for crisis events and propose policy recommendations to prevent reoccurrences. It also takes a close look at standing independent accident investigation agencies, such as the United States National Transportation Safety Board, and considers the potential for extending this model of inquiry to the sorts of crises that we analyze in the remainder of the volume.

In the final chapter, the co-editors develop an overarching descriptive analysis of how different types of crises (and associated changes in perceptions) influence the selection of different regulatory responses, and offer a normative analysis of crisis-driven regulatory policy and how to improve it through better learning. We discuss how to evaluate when and how crisis-driven policy innovations are desirable, including whether they effectively address the risks highlighted by crisis events, impose acceptable costs, and induce other risks. To reduce "policy regret" and enhance "policy resilience," we recommend both "learning to prepare" for crises and "preparing to learn" from crises. Along these lines, we suggest ways to improve the actual responses of policy-makers in the future – their responses to crises and the accompanying demands for regulatory reform that the crises tend to generate.

We have no pretension of being able to furnish policy-makers who might confront as yet unseen future crises with a formulaic handbook for crisis-related regulatory decision-making. But we do think that we can provide those officials with better historical understanding of the promise and pitfalls of past crisis-driven policy-making, and a more sophisticated and useful set of questions to ask themselves and the experts on whom they rely for analysis.

The provision of such guidance strikes us as of increasingly critical importance. As our technologies and economic systems become more complex and interconnected, even as ordinary risks are reduced and human longevity increases, extreme crises that impose large costs may become more common or more consequential – or at least more shocking against the baseline of otherwise declining ordinary risks. Ironically, in a safer society, smaller events may be perceived as more upsetting (than in a past riskier society), and so even more policy may become crisis-driven. We have good reason to think carefully about how governments have responded, are responding, and should respond to such events, so that policy-makers can learn to prepare for crises and prepare to learn from crises – to reduce policy regret and increase policy resilience.

References

Aldrich, Mark. 1997. "The Perils of Mining Anthracite: Regulation, Technology and Safety, 1870–1945." *Pennsylvania History: A Journal of Mid-Atlantic Studies* 64 (3): 361–83.

Alemanno, Alberto. 2011. *The Challenges of Emergency Risk Regulation – Beyond the Volcanic Ash Crisis*. Northampton, MA: Edward Elgar. https://hal-hec.archives-ouvertes.fr/hal-00638652.

Bamberger, Kenneth A., and Andrew T. Guzman. 2008. "Keeping Imports Safe: A Proposal for Discriminatory Regulation of International Trade." *California Law Review* 96 (6): 1405–45.

Baumgartner, Frank R., and Bryan D. Jones. 2010. *Agendas and Instability in American Politics*. Chicago, IL: University of Chicago Press.

Birkland, Thomas A. 1998. "Focusing Events, Mobilization, and Agenda Setting." *Journal of Public Policy* 18 (01): 53–74.

Birkland, Thomas A. 2006. *Lessons of Disaster: Policy Change After Catastrophic Events*. Washington, DC: Georgetown University Press.

Bourdelais, Patrice. 2006. *Epidemics Laid Low: A History of What Happened in Rich Countries*. Baltimore, MD: Johns Hopkins University Press.

Breyer, Stephen G. 1993. *Breaking the Vicious Circle: Toward Effective Risk Regulation*. Cambridge, MA: Harvard University Press.

Burke, John G. 1966. "Bursting Boilers and the Federal Power." *Technology and Culture* 7 (1): 1–23. doi:10.2307/3101598.

Campbell, Troy H., and Aaron C. Kay. 2014. "Solution Aversion: On the Relation between Ideology and Motivated Disbelief." *Journal of Personality and Social Psychology* 107 (5): 809–24.

Carpenter, Daniel P. 2001. *The Forging of Bureaucratic Autonomy: Reputations, Networks, and Policy Innovation in Executive Agencies, 1862–1928*. Princeton, NJ: Princeton University Press.

Carpenter, Daniel, and Gisela Sin. 2007. "Policy Tragedy and the Emergence of Regulation: The Food, Drug, and Cosmetic Act of 1938." *Studies in American Political Development* 21 (02): 149–80.

Coffee, John. 2012. "The Political Economy Of Dodd-Frank: Why Financial Reform Tends To Be Frustrated And Systemic Risk Perpetuated." *Cornell Law Review* 97: 1019–82.

Coglianese, Cary, and Christoper Carrigan. 2012. "Oversight in Hindsight: Assessing the U.S. Regulatory System in the Wake of Calamity." In *Regulatory Breakdown: The Crisis of Confidence in U.S. Regulation*, edited by Cary Coglianese. Philadelphia, PA: University of Pennsylvania Press, 1–20.

Cohen, Dara Kay, Mariano-Florentino Cuéllar, and Barry R. Weingast. 2006. "Crisis Bureaucracy: Homeland Security and the Political Design of Legal Mandates." *Stanford Law Review* 59 (3): 673–759.

Dalkey, Norman Crolee, Bernice B Brown, and Samuel Cochran. 1969. *The Delphi Method: An Experimental Study of Group Opinion*. Vol. 3. Santa Monica, CA: Rand Corporation.

Gilbert, Daniel T., and Timothy D. Wilson. 2007. "Prospection: Experiencing the Future." *Science* 317 (5843): 1351–54.

Gollier, Christian, and Martin L. Weitzman. 2010. "How Should the Distant Future Be Discounted When Discount Rates Are Uncertain?" *Economics Letters* 107 (3): 350–53.

Graham, John D. 2007. "The Evolving Regulatory Role of the U.S. Office of Management and Budget." *Review of Environmental Economics and Policy* 1 (2): 171–91. doi:10.1093/reep/rem013.

Harris, Ron. 1994. "The Bubble Act: Its Passage and Its Effects on Business Organization." *The Journal of Economic History* 54 (3): 610–27.

Helmer-Hirschberg, Olaf. 1967. "Analysis of the Future." Product Page. www.rand.org/pubs/papers/P3558.html.

International Atomic Energy Agency. 2015. *The Fukushima Daiichi Accident, Report of the Director*. Vienna, Austria: International Atomic Energy Agency.

International Energy Agency. 2017. "OECD – Electricity and Heat Generation," *IEA Electricity Information Statistics* (database). doi: http://dx.doi.org/10.1787/data-00457-en

Jaffee, Dwight M., and Thomas Russell. 1997. "Catastrophe Insurance, Capital Markets, and Uninsurable Risks." *Journal of Risk and Insurance*, 64 (2): 205–30.

Johnson, George. 2015. "When Radiation Isn't the Real Risk." *New York Times*, September 22: D3.
Jones, Michael D., and Mark K. McBeth. 2010. "A Narrative Policy Framework: Clear Enough to Be Wrong?" *Policy Studies Journal* 38 (2): 329–53.
Kagan, Elena. 2001. "Presidential Administration." *Harvard Law Review* 114 (8): 2245–2385. doi: 10.2307/1342513.
Kahn, Matthew E. 2007. "Environmental Disasters as Risk Regulation Catalysts? The Role of Bhopal, Chernobyl, Exxon Valdez, Love Canal, and Three Mile Island in Shaping U.S. Environmental Law." *Journal of Risk and Uncertainty* 35: 17–43.
King, Andrew A., and Michael J. Lenox. 2000. "Industry Self-Regulation without Sanctions: The Chemical Industry's Responsible Care Program." *The Academy of Management Journal* 43 (4): 698–716.
Kingdon, John W. 2003. *Agendas, Alternatives, and Public Policies*. New York: Longman Classics in Political Science.
Kousky, Carolyn, and Roger Cooke. 2012. "Explaining the Failure to Insure Catastrophic Risks." *The Geneva Papers on Risk and Insurance – Issues and Practice* 37 (2): 206–27.
Kuran, Timur, and Cass R. Sunstein. 1999. "Availability Cascades and Risk Regulation." *Stanford Law Review*, 51 (4): 683–768.
Laibson, David. 1997. "Golden Eggs and Hyperbolic Discounting." *The Quarterly Journal of Economics*, 112 (2): 443–77.
Lelieveld, J., J. S. Evans, M. Fnais, D. Giannadaki, and A. Pozzer. 2015. "The Contribution of Outdoor Air Pollution Sources to Premature Mortality on a Global Scale." *Nature* 525 (7569): 367–71.
Lenox, Michael J., and Jennifer Nash. 2003. "Industry Self-regulation and Adverse Selection: A Comparison across Four Trade Association Programs." *Business Strategy and the Environment* 12 (6): 343–56.
Litfin, Karen T. 1994. *Ozone Discourses*. New York: Columbia University Press.
Morgan, M. Granger, and M. Henrion. 1990. *Uncertainty: A Guide to Dealing with Uncertainty in Quantitative Risk and Policy Analysis*. Cambridge, UK: Cambridge University Press.
Newell, Richard G., and William A. Pizer. 2003. "Discounting the Distant Future: How Much Do Uncertain Rates Increase Valuations?" *Journal of Environmental Economics and Management* 46 (1): 52–71.
Newman, Mark E. J. 2005. "Power Laws, Pareto Distributions and Zipf's Law." *Contemporary Physics* 46 (5): 323–51.
Nuclear Energy Institute. n.d.(a) "Japan Nuclear Updates." www.nei.org/News-Media/News/Japan-Nuclear-Update.

Nuclear Energy Institute. n.d.(b) "World Nuclear Power Plants In Operation." https://www.nei.org/Knowledge-Center/Nuclear-Statistics/World-Statistics/World-Nuclear-Power-Plants-in-Operation.
Office of Management and Budget (OMB) 2003. Circular A-4. September 17. https://obamawhitehouse.archives.gov/omb/circulars_a004_a-4/.
Percival, Robert V. 1998. "Environmental Legislation and the Problem of Collective Action." *Duke Environmental Law and Policy Forum* 9: 9–28.
Perrow, Charles. 1999. *Normal Accidents*. Princeton, NJ: Princeton University Press.
Perrow, Charles. 2007. "Disasters Ever More? Reducing US Vulnerabilities." In *Handbook of Disaster Research*, edited by Havidán Rodríguez, Enrico L. Quarantelli, and Russell R. Dynes. New York, NY: Springer, 521–33.
Pool, Heather. 2012. "The Politics of Mourning: The Triangle Fire and Political Belonging." *Polity* 44 (2): 182–211. doi:10.1057/pol.2011.23.
Reid, Donald. 1986. "Putting Social Reform into Practice: Labor Inspectors in France, 1892–1914." *Journal of Social History* 20 (1): 67–87.
Renn, Ortwin. 2008. *Risk Governance: Coping with Uncertainty in a Complex World*. London, UK: Earthscan.
Repetto, Robert, ed. 2006. *Punctuated Equilibrium and the Dynamics of US Environmental Policy*. New Haven, CT: Yale University Press.
Rockness, Howard, and Joanne Rockness. 2005. "Legislated Ethics: From Enron to Sarbanes-Oxley, the Impact on Corporate America." *Journal of Business Ethics* 57 (1): 31–54. doi:10.1007/s10551-004-3819-0.
Romano, Roberta. 2005. "The Sarbanes-Oxley Act and the Making of Quack Corporate Governance." *Yale Law Journal* 114: 1521–1611.
Rosenberg, Charles E. 1987. *The Cholera Years: The United States in 1832, 1849, and 1866*. Chicago, IL: University of Chicago Press.
Sabatier, Paul A. 1988. "An Advocacy Coalition Framework of Policy Change and the Role of Policy-Oriented Learning Therein." *Policy Sciences* 21 (2-3): 129–68.
Sabatier, Paul A., and Hank C. Jenkins-Smith. 1993. *Policy Change and Learning: An Advocacy Coalition Approach*. Boulder, CO: Westview Press.
Satterfield, Terre, Paul Slovic, and Robin Gregory. 2000. "Narrative Valuation in a Policy Judgment Context." *Ecological Economics* 34 (3): 315–31.
Schneiberg, Marc, and Tim Bartley. 2010. "Regulating or Redesigning Finance? Market Architectures, Normal Accidents, and Dilemmas of

Regulatory Reform." In *Markets on Trial: The Economic Sociology of the Finance Crisis: Part A*, edited by Michael Lounsbury and Paul M. Hirsch. Bingley, UK: Emerald Group, 281–308.

Schoenberg, Frederic Paik, Roger Peng, and James Woods. 2003. "On the Distribution of Wildfire Sizes." *Environmetrics* 14 (6): 583–92.

Seligman, Joel. 2004. *Fundamentals of Securities Regulation*. Frederick, MD: Aspen Publishers Online.

Shaw, Robert B. 1978. *A History of Railroad Accidents, Safety Precautions, and Operating Practices*, 2nd ed. Binghamton, NY: Vail-Ballou Press.

Slovic, Paul. 1987. "Perception of Risk." *Science* 236 (4799): 280–85.

Slovic, Paul. 2010. *The Feeling of Risk: New Perspectives on Risk Perception*. New York, NY: Earthscan.

Slovic, Paul, James Flynn, and Howard Kunreuther. 2013. *Risk, Media and Stigma: Understanding Public Challenges to Modern Science and Technology*. New York, NY: Earthscan.

Small, Deborah A., and George Loewenstein. 2003. "Helping a Victim or Helping the Victim: Altruism and Identifiability." *Journal of Risk and Uncertainty* 26 (1): 5–16.

Small, Deborah A., George Loewenstein, and Paul Slovic. 2007. "Sympathy and Callousness: The Impact of Deliberative Thought on Donations to Identifiable and Statistical Victims." *Organizational Behavior and Human Decision Processes* 102: 143–53.

Tallman, Ellis W., and Jon R. Moen. 1990. "Lessons from the Panic of 1907." *Economic Review – Federal Reserve Bank of Atlanta* 75 (3): 2–13.

The International Review (1874–1883). 1877. "The Life Insurance Question: What Is Wrong in Life Insurance?," May.

United Nations Scientific Committee. 2013. *Report of the United Nations Scientific Committee on the Effects of Atomic Radiation*. New York, NY: United Nations.

Venkatesan, Rashmi. 2013. "Clothing Garment Workers in Safety: The Case of Bangladesh." *Economic and Political Weekly*. www.epw.in/web-exclusives/clothing-garment-workers-safety-case-bangladesh.html.

Weber, Elke U. 2006. "Experience-Based and Description-Based Perceptions of Long-Term Risk: Why Global Warming Does Not Scare Us (Yet)." *Climatic Change* 77 (1–2): 103–20. doi:10.1007/s10584-006-9060-3.

Weitzman, Martin L. 2009. "On Modeling and Interpreting the Economics of Catastrophic Climate Change." *The Review of Economics and Statistics* 91 (1): 1–19.

Weitzman, Martin L. 2011. "Fat-Tailed Uncertainty in the Economics of Catastrophic Climate Change." *Review of Environmental Economics and Policy* 5 (2): 275–92.

Wiener, Jonathan B. 2006. "Better Regulation in Europe." *Current Legal Problems* 59 (January): 447–518.

Wiener, Jonathan B. 2013. "The Diffusion of Regulatory Oversight." In *The Globalization of Cost-Benefit Analysis in Environmental Policy*, edited by Michael A. Livermore and Richard L. Revesz. Cambridge, MA: Cambridge University Press, 123–41.

Wiener, Jonathan B. 2016. "The Tragedy of the Uncommons: On the Politics of Apocalypse." *Global Policy* 7: 67–80.

Wiener, Jonathan B., and Daniel Ribeiro. 2016. "Impact Assessment: Diffusion and Integration." In *Comparative Law and Regulation*, edited by Francesca Bignami and David Zaring. Northampton, MA: Edward Elgar, 159–89.

Wiener, Jonathan B., Michael D. Rogers, James K. Hammitt, and Peter H. Sand, eds. 2011. *The Reality of Precaution: Comparing Risk Regulation in the United States and Europe*. Washington, DC: RFF Press.

World Health Organization. 2013. *Health Risk Assessment from the Nuclear Accident after the 2011 Great East Japan Earthquake and Tsunami, Based on a Preliminary Dose Estimation*. Geneva, Switzerland: World Health Organization.

World Ocean Review (WOR) 3. 2014. "Marine Resources—Opportunities and Risks." Hamburg, Germany: Maribus. http://worldoceanreview.com/wp-content/downloads/wor3/WOR3_english.pdf.

Wuthnow, Robert. 2010. *Be Very Afraid: The Cultural Response to Terror, Pandemics, Environmental Devastation, Nuclear Annihilation, and Other Threats*. Oxford, UK: Oxford University Press.

PART I

THE CONCEPTUAL TERRAIN OF CRISES AND RISK PERCEPTIONS

2

Economic Analysis, Risk Regulation, and the Dynamics of Policy Regret

Lori S. Bennear

In the wake of a crisis, a common refrain is "Why didn't we do more to prevent this terrible disaster?" All too frequently, ex-post analysis suggests that policy did not sufficiently account for extreme events, or that the level of regulation imposed or enforced was insufficient. For example, after Hurricane Katrina a bipartisan commission found that the levee system had been designed for "standard" hurricanes and not for the most severe storms (US House of Representatives 2006). The bipartisan commission investigating the Deepwater Horizon oil spill found that inspections and regulations were insufficient to prevent an oil spill of that significant magnitude (National Commission 2011). Similarly, the Financial Crisis Inquiry Commission found that "widespread failures in financial regulation and supervision proved devastating to the stability of the nation's financial markets" (Financial Crisis Inquiry Commission 2011). For the sake of simplicity, I will refer to this phenomenon as "policy regret." Policy regret is characterized by a post-crisis sentiment that decision-makers had not sufficiently identified the probability or seriousness of risks before the onset of crisis and so had insufficiently acted to prevent it. This sentiment tends to be strongly correlated with an impulse to engage in significant post-crisis policy reforms.

One of the key features of policy regret seems to be the belief, ex-post, that the problems with the current regulations should have been known and corrected in advance. This characteristic implies systematic failures in the regulatory process that hinder implementation of programs and policies that would actually have been welfare enhancing. The focus of this chapter is on the role of economic analysis in the regulatory process and what role, if any, these analyses play in reducing or fostering policy regret.

Economic analyses of risk regulations of some form have been required since the Nixon administration (Copeland 2011). However, most people associate requirements for economic analysis, and particularly benefit-cost analysis, with Executive Order 12291 issued in 1981 by President Ronald Reagan. This executive order required benefit-cost analysis for all regulations with an annual cost of $100 million or more (Copeland 2011). More specifically, it required agencies to monetize all benefits and costs and to choose both regulatory objectives and regulatory alternatives (levels) that maximized net benefits to society. This executive order was maintained by President George H.W. Bush and was replaced by a similar executive order (EO 12866) by President Bill Clinton. Executive Order 12866 maintained the requirement of benefit-cost analysis on economically significant rules (those with costs greater than $100 million annually), but it also extended this requirements to other "significant" rules that created conflict among agencies, changed entitlements, or raised novel legal issues (Copeland 2011). These same principles have been expanded to cover independent regulatory agencies by President Barack Obama's Executive Orders 13563 and 13579 (Copeland 2011). In order to help regulatory agencies comply with these executive orders, the Office of Management and Budget's Office of Information and Regulatory Affairs published a series of guidance documents with specific instructions for how to conduct benefit-cost analysis (Office of Management and Budget 2003).

This chapter examines the economic theory underlying benefit-cost analysis for risk regulations and its application for US policy, and then examines the phenomenon of policy regret in light of these analyses. The next section describes the neoclassical economic theory of decision-making under uncertainty. This theory was developed for individual decisions. The following section describes how American regulators (now emulated by policy-makers around the world) applied these principles to collective decision-making through benefit-cost analysis. The chapter then discusses the role of economic analysis in preventing or fostering policy regret, noting how this phenomenon has shaped the regulatory responses to crises in this volume's case studies.

NEOCLASSICAL ECONOMIC VIEW OF INDIVIDUAL DECISION-MAKING UNDER RISK

In understanding how economic analyses model decisions under risk and uncertainty, it is important to delineate the distinction between risk and uncertainty. Decisions under risk are decisions made when the probabilities of different outcomes are known. In contrast, uncertainty characterizes situations in which probabilities of different outcomes are not known and/or the exact set of outcomes is not known.

Neoclassical economic theory models decisions under risk and uncertainty in an expected utility framework, focusing on the dynamics of individual choice (Mas-Colell, Whinston, and Green 1995; von Neumann and Morgenstern 1944). Imagine that there is a set of possible outcomes. One can also think of outcomes as states of the world, so a set of three outcomes might be lose $50, lose $10, win $100. One can then contrast various lotteries that specify a set of probabilities over the set of outcomes. Two lotteries might be $L_1=(0.5, 0, 0.5)$ and $L_2=(0, 0.90, 0.10)$. Expected utility theory predicts that individuals will choose the lottery with the highest expected payoff. For simplicity, assume individuals expected payoff is equivalent to the monetary outcome. In this case, expected utility theory predicts that an individual will choose L_1 because the expected payoff of that lottery is $25 while the expected payoff of L_2 is $0.

However, expected utility theory does not always predict that individuals will choose the lottery with the highest weighted average monetary payoff because some individuals exhibit risk aversion. An individual is said to be risk averse if they prefer a sure payoff of $X to a lottery that gives $X in expectation. If the individual is indifferent between the sure bet and the lottery with the same expected payoff, the individual is risk neutral. If the individual prefers the lottery to the sure bet, s/he is said to be risk loving. We can measure risk aversion using the concept of a certainty equivalent (Mas-Colell, Whinston, and Green 1995). For any lottery, the certainty equivalent is the largest payoff, received with certainty, that the individual would trade for the lottery. Going back to our examples, L_1 has an expected payoff of $25. If an individual was willing to accept $20 with certainty in lieu of the lottery, but would choose the lottery over a certain payoff of $21, then that person's certainty equivalent is $20. The lower the certainty equivalent, the more risk averse the individual. In general, individuals tend to exhibit non-increasing relative risk aversion. That is, they are willing to take on equal or greater risk (as a percentage of total income) as income rises (Mas-Colell, Whinston, and Green 1995). A poor person might be willing to except only $10 with certainty against Lottery 1 above (which pays $25 in expectation), while Bill Gates would likely choose the lottery over any payment less than $25.

So far we have been focused on monetary payoffs. Most individual decisions about risk do not involve straightforward monetary payoffs. Consider an individual's decision about how often to wear a seat belt while driving. People have different utility from wearing the seat belt always, most of the time, some of the time, or never. Importantly, these utilities are combinations of the convenience costs of fastening the seat belt, the reduction in risk of serious injury from wearing the seat belt, the likelihood of a fine if caught not wearing the seatbelt, and so on. Because individuals have different preferences, they are likely to weigh these costs and benefits differently. Thus, even if we have well-defined probabilities of the likelihood of injuries and death as a function of frequency of seat belt use, we might still expect some heterogeneity in individual decisions about how often to wear a seat belt.

Our discussion thus far has focused on situations of risk, where the probabilities of different outcomes (or states of the world) are well known. In many situations, these probabilities are not well known, so these are situations of uncertainty. Even when probabilities are not well known, individuals have to form some expectations about those probabilities in order to make decisions. We call these subjective probabilities (Anscombe and Aumann 1963; Mas-Colell, Whinston, and Green 1995). If we return to the seat belt example, it is likely that individuals do not know precisely the probability of death or injury if they wear a seat belt all the time, most of the time, rarely or never. Instead they use some beliefs about those probabilities in making decisions. With some additional technical refinements, expected utility theory functions the same way with subjective probabilities as with technical probabilities – individuals are predicted to make the decision that yields the highest expected utility where the expectations are formed using subjective probabilities as weights (Mas-Colell, Whinston, and Green 1995).

An interesting body of research has developed around how individuals form and update these subjective probabilities. The traditional view is that individuals are rational and update their beliefs in the face of new information using Bayes' rule, which provides a specific mathematical formula for how one should update beliefs based on new information. For example, you have never seen a swan before, and the first swan you see is white. You wonder if the next swan will be white or some other color and place equal probability on both outcomes. This is akin to placing one white marble and one black marble in a bag. You expect that the likelihood of drawing a white marble from the bag is 50 percent. You then encounter another white swan and put another white marble in the bag. Now you expect the likelihood of drawing a white marble to be 67.78 percent rather than 50 percent. If you continue adding a white marble every time you see a white swan, you eventually come to the belief that, with near certainty, swans are white. However, many researchers in psychology and economics have demonstrated that individuals do not typically update beliefs in a Bayesian fashion, but rather use heuristics. These heuristics are discussed in more detail in Chapter 4, which focuses on how humans perceive risk and how those perceptions change over time.

One additional complication of expected utility theory is that many decisions involve payoffs that occur over different time horizons. The neoclassical approach to addressing this issue is to reduce payoffs over different time horizons to the *present value* of the stream of payoffs. In order to understand the concept of present value, we first need to introduce the concept of a discount rate. An individual's pure rate of time preference, or discount rate, can be thought of as the rate of return required to make the individual indifferent between receiving some amount of money today and some amount in one year. For example, if an individual is indifferent between receiving $100 today or $110 one year from now, his/her discount rate is 10 percent. Higher discount rates reflect a stronger preference for present consumption over future consumption. Note that the discount rate can be positive absent inflation or risk purely because humans prefer consumption today to consumption tomorrow.

The present value then represents the dollar amount an economically rational individual would accept today in exchange for a stream of future payments. For example, if your grandmother sends you a check for $50 every year on your birthday for 18 years and your discount rate is 5 percent, then the present value of this stream of $50 payments is roughly $584. If someone offered to trade you more than $584 dollars today in exchange for the stream of birthday checks from Grandma, you would make the trade. If they offered you less than $584, you would prefer to keep Grandma's annual checks.

To see how the concept of present value and discounting can be connected back to expected utility, imagine that a risk-neutral individual can purchase a birthday lottery ticket for $100. This birthday lottery will yield a stream of $10 payments on the individual's birthday every year for 10 years with probability 0.90 or a stream of $50 payments every year for 10 years with a probability of 0.10. Should the individual buy the lottery ticket? To answer that question using expected utility theory, we need to first calculate the present value of each of the two streams of payments.

If the individual has a 5 percent discount rate, then the stream of $10 payments over 10 years has a present value of roughly $77 and the stream of $50 payments over 10 years has a present value of $386. The expected present value of the lottery is then roughly $108 (0.9*77 + 0.1*386 = 107.9). The expected present value of the payments from the lottery exceeds the lottery purchase price and so the individual should buy the lottery ticket. If instead the individual had a 7 percent discount rate, the expected present value of the lottery is reduced to $98 and the individual should not buy the lottery ticket.

Expected utility theory is the foundation for most economic analyses of risk, and it may be a useful theory for individuals thinking about how they should make decisions under risk. However, there are several well-known critiques of expected utility theory as a *predictor* of human behavior (Allais 1953; Kahneman and Tversky 1979). In fact, there are so many examples of how expected utility theory fails to accurately predict human behavior that an entire subfield called behavioral economics has been developed to illuminate these contradictions and offer alternative theoretical approaches (Wilkinson and Klaes 2012; DellaVigna 2009; Camerer, Loewenstein, and Rabin 2003). Sociologists and cultural anthropologists similarly argue that much behavior reflects social routines and cultural norms rather than individual calculation. Some who are loyal to the neoclassical approach are dismissive of behavioral economics and appeals to social patterns, arguing that robust definitions of utility can account for all apparent inconsistencies. The behavioralists counter that expanding the definition of utility to accommodate all apparent paradoxes misses the point. A sufficiently expansive definition of utility may explain everything, but be able to predict nothing. Regardless of the ongoing controversy, the neoclassical approach to individual decision-making remains the foundation for analysis for social decision-making described in the next section.

ECONOMIC ANALYSIS FOR COLLECTIVE DECISION-MAKING UNDER RISK

Thus far the discussion has focused exclusively on individual decision-making. The focus of this book, however, is on collective decision-making. We are exploring how governments, regulators, industry, and so forth, make decisions under uncertainty, how they make sense of and respond to crisis events that challenge pre-existing assumptions about risks, and how they *should* learn from and respond to such events. In order to extend expected utility theory to collective decision-making, we need to develop aggregates for four key parameters in the individual problem: (1) the utility function, (2) discount rate, (3) risk aversion, and (4) subjective probabilities.

The neoclassical view of the collective utility function is measured in purely monetary terms. Basically all costs and benefits are monetized, and the present value of net benefits represents the collective value or utility of a given project/regulation/etc. Ideally, the dollar values associated with benefits and costs are derived from individual decisions, thereby revealing underlying individual preferences. However, some benefits are difficult or impossible to measure through market interactions, and these benefits values are frequently derived from surveys that seek to elicit individual preferences. Not all benefits can be monetized, and the guidelines for economic analysis in the United States and elsewhere frequently allow for documentation of non-monetizable benefits. However, the quantitative economic and risk analyses focus exclusively on the monetized value of benefits.

Representing the aggregate utility function as the present value of net benefits from a policy has its normative foundations in the Pareto principle. The Pareto principle argues that a policy should be adopted if at least one person can be made better off and nobody is made worse off. However, in practice, it is never the case that a policy creates at least one winner and no losers. Rather there are benefits and costs, and the distribution of those benefits and costs may not be uniform across the population, generating some winners and some losers. The modified Pareto principle (also known as the Kaldor–Hicks principle) argues that a policy should be adopted if the present value of net benefits is positive, because in that circumstance, it is possible, with transfers, to distribute those net benefits in ways that at least one person is better off and nobody is worse off. This is a far weaker normative foundation, which states, in essence, that if the pie is larger, it is *possible* to make everyone better off. Of course, transfers are never enforced to ensure that the pie is distributed in a way that makes at least one person's slice bigger and nobody's slice smaller. But that is the normative foundation for the positive net-benefit criterion, for better or worse.

In order to calculate the present value of the stream of future benefits and costs that accrue to different members of society, one needs something akin to an average social discount rate. Ideally, one wants a discount rate that reflects individual citizens' actual rates of time preference, and markets can provide data that allow us to infer average rates of time preference.

There are two competing theories on the appropriate social discount rate. The first argues that net benefits of public projects should be discounted at a rate that reflects the population's pure time rate of preference without risk (Arrow and Lind 1970). Proponents of this view argue that the government holds a fully diversified portfolio of investments, and thus, the marginal return on this investment is risk free. Ideally the government discount rate should reflect how an "average" person trades after-tax money today for after-tax money tomorrow with no risk. This parameter is approximated using short-term US treasury bonds rates and subtracting out the capital gains tax rate. Using historical rates of return on three-month treasury bonds and capital gains tax rates yields an estimate of the pure rate of time preference on the order of 2 percent.

A second theory of the appropriate social discount rate argues that social projects should compete for capital on the same terms that private projects compete for capital (Hirshleifer 1964) and that the government is no more diversified against risk than other major market players. Tax dollars spent on public projects are dollars not spent on other private investments. The cost of using capital for those public projects is the rate of return we could have gotten from private investment. General capital market indices such as the S&P 500 or Wilshire 5000 provide a good indication of what the average rate of return on private investment is. Using historical rates of return from these indices suggests a social discount rate of approximately 7 percent.

Extending the individual decision problem to social decisions when the probabilities of different states are unknown requires something akin to subjective probabilities. The most well-established method for eliciting social subjective probabilities is the Delphi method. The Delphi method was established by RAND Corporation and consists of a series of structured survey questions designed to derive probabilities of certain event from experts in a field (Helmer-Hirschberg 1967; Dalkey 1969). A group of experts is asked to estimate the probability of particular events (or the value of an unknown parameter) individually without knowing who else is in the expert group. Each expert then receives information on the distribution of estimates from all the group members and is allowed to change his/her response and provide a justification if the estimate is outside the 25th–75th percentile of the group's responses. In the next round the experts receive the new probability distribution and the justifications (if any) for extreme positions and are allowed to refine their estimates again. The process is sometimes repeated one more time. The end result is a distribution of probabilities that can reasonably be viewed as expert consensus.

In situations where multiple parameters are uncertain, Monte Carlo analysis can then be used to derive distributions of expected net benefits (Office of Management and Budget 2003). In Monte Carlo analysis, researchers run thousands of computer simulations, with each simulation randomly choosing a different set of parameters from their respective distributions. At the end of each run the net benefits are recorded, and the results over thousands of runs provides a distribution of net benefits, given the underlying distributions of the unknown parameters. It is critical to also model correlations in uncertainty among certain key parameters. For example, if the occurrence of a hurricane is highly correlated with flooding, then one would not want the Monte Carlo analysis to independently draw probabilities of hurricanes and flood events. The Delphi method (or something similar) can also be used to derive correlations among parameters when insufficient data exist to directly estimate them.

Monte Carlo analysis provides both an estimate of the full distribution of net benefits. So the analyst can examine not only the expected value of net benefits (the mean), but also the probability that net benefits falls within a certain range. For instance, two rules may both have expected net benefits of $100 million, but one may have a 50 percent probability of having net benefits greater than zero while the other has a 75 percent probability of having net benefits greater than zero. In this case society is better off choosing the second rule, which yields the same expected net benefits but with higher odds of positive net benefits.

At this stage it is worth noting the mixture of "democratic" and "technocratic" features of expected utility theory applied to social decision-making. While the underlying dollar values of net benefits and the discount rate are derived from aggregations of individuals' behavior in markets, and might therefore be thought to be democratic representations of preferences, the measures of uncertainty are generally derived from expert elicitation and might be thought of as technocratic. We will discuss implications of this dichotomy in situations where risk perceptions shift suddenly in detail in the next section, but it is fairly clear already that large deviations from public perception of probabilities associated with particular events and those perceptions held by experts may give rise to significant differences between public and expert views on the desirable policy or level of regulatory stringency.

IMPLICATIONS FOR SHIFTING RISK PERCEPTIONS

Let's return now to situations where a disaster occurs that shifts public perceptions of the risks associated with a particular activity. As illustrated in the examples in the Introduction, these situations are frequently accompanied by widely expressed accusations that officials did not take sufficiently aggressive or well-conceived preventative actions. Policymakers accordingly face the question "How could this terrible event have happened?"

The practice of systematic economic analysis described in the prior section is designed to provide the type of information necessary to make policy decisions that do not lead to this type of regret – a careful weighing of benefits and costs of action using the best information available. Sometimes, however, officials eschew serious benefit-cost analysis when fashioning regulatory policy. In some cases, political considerations and conflicts of interest prevent data-driven, technocratic risk assessment and risk management. The role of politics in regulation and regulatory failures will be highlighted in this volume's coverage of financial crashes (Chapters 13–16). Those chapters suggest that regulatory processes in finance have been heavily influenced by the goals and objectives of the regulated industry itself and often do not reflect a more objective weighing of costs and benefits for society as a whole.

Related explanations for the existence of ex-post regret, despite requirements for systematic risk and economic analysis, are that officials did not correctly or completely undertake it, or did not sufficiently update analysis in response to new information. There are many possible errors in the application of these analytic methods that could lead to policy regret, including using the wrong distribution to model key parameters; modeling parameters as independent when they are correlated or otherwise not appropriately accounting for correlations; not fully measuring and monetizing all costs and benefits over long time horizons; rushing analysis because of statutory requirements to issue rules quickly; not revising regulations in light of new information due to statutory requirements that inhibit revision of rules. These issues arise in all three sets of case studies in this book. For offshore oil, the belief that requiring many redundant safety systems meant that if one system failed, there would be many others that would not fail, assumed that the probabilities of failure were independent. Recent events have suggested these failure probabilities are often highly correlated. Similar issues arise in the nuclear case studies. In the case studies on financial crashes the issue is often that firms take on too much risk because their models incorrectly assume that probabilities of bad events are either very unlikely or not correlated across investments. Both of these assumptions have proven incorrect in recent crashes.

Both of these explanations for policy regret – that pre-crisis decision-making did not reflect careful benefit-cost analysis, or resulted from poorly executed benefit-cost analysis – suggest a straightforward (if not easily implemented) solution. The answer to such shortcomings is surely more and better economic analysis. But these two explanations do not account for most policy regret; indeed, even with an ideal economic analysis, policy regret may be inevitable.

As alluded to in the prior section, economic analysis of risk regulation relies on expert estimates of uncertain parameters, including the probabilities of extreme events, the magnitudes of extreme events, and correlations among extreme events. There is a large body of literature illustrating that public perceptions of risk often differ, sometimes dramatically, from expert perceptions of risk (Slovic et al. 1995; Slovic 1987; Cole and Withey 1981; Wright and Bolger 1992; Sandman, Weinstein, and Klotz 1987), although some recent scholarship debates those differences (Rowe and Wright 2001). To the extent that the public views certain activities as more risky or less risky than the experts whose opinions were used in the economic analysis, the results of the analysis, namely the level of regulation, may not align with citizens' preferences. Whether this is a good or bad thing depends on one's view of the democratic process. These differences in risk perception play a role in different countries' responses to nuclear power accidents, as will be evident in the cases in Chapters 10–12.

A related issue is that in the case of rare occurrences of extreme outcomes, witnessing one may not actually affect expectations of the probability distributions much if individuals or experts are using Bayesian updating. Returning to our white swan example, if after years of seeing only white swans, one encounters a single black swan, this may not change one's expectation of the likelihood that the next swan one sees will be white. However, if individuals use the availability heuristic or other cognitive heuristics (see Chapter 4) to update beliefs, then extreme events may be more salient and may have a larger impact on subjective probabilities than Bayes' rule would suggest. While there is no reason to think that any individual expert or lay citizen is more or less likely to form expectations using Bayes' rule or cognitive heuristics, the Delphi method used to elicit expert judgments is designed to minimize these cognitive biases. Thus one might expect that witnessing a large oil spill or nuclear accident doesn't change the experts' views of the underlying probabilities of these extreme events as much as it changes the lay public's expectations. This sort of pattern seems to be operating for many key actors discussed in Chapters 7–12 of this volume. This disconnect necessarily leads to policy regret in the face of extreme events and puts the regulators in the awkward position of defending expert risk perceptions as more valid or rational than the public's.

Similarly, for extreme (or even moderately extreme) events, it may take time to know whether the observation of an extreme event is an indication that the distributions used in the economic analysis are actually incorrect and require updating or whether the extreme event was just a bad draw from the existing distribution. If one takes climate change as an example, there is natural variation in weather, so observing a hot summer or an unusually active hurricane season, in and of itself, might not indicate a fundamental shift in the climate. By the time the data are sufficient to reveal a shift in climate, there may be significant regret that more was not done sooner. The case studies in this volume also provide some support for this longitudinal view. Sometimes the first incident does not result in significant policy change, but rather it prompts some experts, interest groups, and elected officials to rethink prevailing risk assessments. After the second, third, or nth incident, there is more momentum for change as the argument that the first incident was just a bad draw loses salience.

A final source of policy regret even with well-done economic analysis results from the tendency of economic analyses to have a static character. That is, they typically capture expectations of the present value of net benefits of different types of actions at one point in time. However, all of the parameters, and particularly the uncertain parameters associated with likelihood of different types of outcomes, are inherently dynamic processes. Policies that are promulgated with the best available economic analysis may still be revealed to be suboptimal in future years because some of the values that went into the analysis turned out to be incorrect due to unexpected technological change, economic growth, or numerable other factors. Most government agencies, even in the most highly industrialized societies, lack the capacity to constantly optimize regulatory policy in the face of new information. As a result, even with the best economic analysis, some policy regret is inevitable.

CONCLUSIONS

Even in a policy environment that relies heavily on systematic economic and risk analysis, there are several possible explanations for the existence of policy regret. These explanations fall broadly into two categories – problems with the analysis itself and problems with public's and policy-makers' understanding about what the analysis can and cannot tell us.

The first of these two categories concerns problems with the analysis itself that might lead to inaccurate projections and expectations about risk. This problem can arise from extending an economic model of choice under uncertainty that was designed for an individual to choices made under uncertainty by entire societies. Certainly some the key parameters in that social decision-making framework under risk have been heavily debated in the literature. The extension to a social framework also requires the acceptance of a utilitarian philosophy on social welfare, namely that society is better off if the pie is made bigger. There can also be problems that result from improper modeling of key risk parameters. In particular, models that assume that the probability of bad events is very small (thin tail risk, see Chapter 3) or that the probability of failures across different systems are uncorrelated (uncorrelated tail risk, see Chapter 3) have been associated with dramatically underestimated social risks in finance, hazard insurance, offshore oil safety, and other policy domains.

The second category contains problems with understanding what sophisticated benefit-cost analysis can and cannot tell us. Significant gaps can arise from differences in risk perceptions, particularly between the policy-makers and the public. Problems can arise from failure to update beliefs or differences in the way the public and policy-makers update beliefs. These tensions can be exacerbated when the events are themselves infrequent. When bad events happen infrequently, it is hard for policy-makers to know if the observation of a bad event is just a bad draw from a known risk distribution or represents "news" that the previous estimations of the distribution were incorrect. It often takes time and more than one bad event to make this distinction. But as we will see in several of the case studies, the political pressure for action following a crisis often does not allow for extended study and analysis to figure out whether action is required. By contrast, entrenched interests and official skepticism can also delay the adoption of reforms developed by experts who have become convinced that pre-crisis risk assessments were not sufficiently pessimistic.

In the end, economic and risk analyses are useful tools that can inform public policy and help to reduce policy regret. However, some amount of policy regret is inevitable and not necessarily the result of poor analysis (although it could be). The key to producing better responses to crisis events is to learn how better to manage policy regret; to know when it is justified because of badly designed pre-crisis policy and when it is not. The case studies that follow provide extensive guidance on how best to approach this vexing but crucial policy question.

References

Allais, M. 1953. "Le Comportement de L'homme Rationnel Devant Le Risquye, Critique Des Postulats et Axioms de L'ecole Americaine." *Econometrica* 21: 503–46.

Anscombe, F., and R. Aumann. 1963. "A Definition of Subjective Probability." *Annals of Mathematical Statistics* 34: 199–205.

Arrow, K. J., and R. C. Lind. 1970. "Uncertainty and the Evaluation of Public Investment Decisions." *American Economic Review* 60: 364–78.

Camerer, C. F., G. Loewenstein, and M. Rabin. 2003. *Advances in Behavioral Economics*. Princeton, NJ: Princeton University Press.

Cole, G. A., and S. B. Withey. 1981. "Perspectives on Risk Perceptions." *Risk Analysis* 1: 143–63.

Copeland, C. W. 2011. *Cost-Benefit and Other Analysis Requirements in the Rulemaking Process*. #R41974. www.fas.org/sgp/crs/misc/R41974.pdf.

Dalkey, N. C. 1969. *The Delphi Method: An Experimental Study of Group Opinion*. #RM-5888-PR. www.rand.org/pubs/research_memoranda/RM5888.html.

DellaVigna, S. 2009. "Psychology and Economics: Evidence from the Field." *Journal of Economic Literature* 47 (2): 315–72.

Financial Crisis Inquiry Commission. 2011. *The Finanical Crisis Inquiry Report: Final Report of the National Commission on the Causes of the Finanical And Economic Crisis in the United States*. Washington, DC. http://fcic-static.law.stanford.edu/cdn_media/fcic-reports/fcic_final_report_full.pdf.

Helmer-Hirschberg, O. 1967. *Analysis of the Future: The Delphi Method*. #P-3558. www.rand.org/pubs/papers/P3558.html.

Hirshleifer, J. 1964. "Efficient Allocation of Capital in an Uncertain World." *American Economic Review* 54: 77–85.

Kahneman, D., and A. Tversky. 1979. "Prospect Theory: An Analysis of Decision under Risk." *Econometrica: Journal of the Econometric Society* 47 (2): 263–91.

Mas-Colell, A., M. D. Whinston, and J. R. Green. 1995. *Microeconomic Theory*. New York: Oxford University Press.

National Commission, on the BP Deepwater Horizon Oil Spill and Offshore Drilling. 2011. "Deep Water: The Gulf Oil Disaster and the Future of Offshore Drilling." www.oilspillcommission.gov/.

Office of Management and Budget. 2003. Circular A-4, Regulatory Analysis. Washington, DC. https://georgewbush-whitehouse.archives.gov/omb/circulars/a004/a-4.html.

Rowe, G., and G. Wright. 2001. "Differences in Expert and Lay Judgments of Risk: Myth or Reality?" *Risk Analysis* 21 (2): 341–56.

Sandman, P. M., N. D. Weinstein, and M. L. Klotz. 1987. "Public Response to Risk from Geological Radon." *Journal of Communication* 37 (3): 93–108.

Slovic, P. 1987. "Perception of Risk." *Science* 236: 280–85.

Slovic, P., T. Malmfors, D. Krewski, C. K. Mertz, N. Neil, and S. Bartlett. 1995. "Intuitive Toxicology II. Expert and Lay Judgments of Chemical Risks in Canada." *Risk Analysis15* 15: 661–75.

US House of Representatives. 2006. A Failure of Initiative: Final Report of the Select Bipartisan Committee to Investigate the Preparation for and Response to Hurricane Katrina. Washington, D.C. www.c-span.org/pdf/katrinareport.pdf.

von Neumann, J., and O. Morgenstern. 1944. *Theory of Games and Economic Behavior*. Princeton, NJ: Princeton University Press.

Wilkinson, N., and M. Klaes. 2012. *An Introduction to Behavioral Economics*. London, UK: Palgrave Macmillan.

Wright, G., and F. Bolger. 1992. "Introduction." In *Expertise and Decision Support*, edited by G. Wright and F. Bolger. New York: Plenum, 1–8.

3

Revised Risk Assessments and the Insurance Industry

Carolyn Kousky

3.1 INTRODUCTION

Insurance plays an important role in the management of disaster risk.[1] First and foremost, it can protect individuals and businesses from financial losses that would be too severe for them to handle on their own. Insurance can also make funds available for rebuilding quickly after an event, helping to increase the resiliency of a community. By limiting the risk exposure of individuals or firms, insurance allows for certain business to occur, which may be too risky otherwise, whether it is the adoption by small farm-holders of a higher-yielding crop that is less resistant to drought, or the willingness of doctors to perform surgeries in which human error is possible. Insurance does not lower actual losses, but it is a way to manage remaining risk after all cost-effective risk-mitigation measures have been taken.[2]

Insuring disaster or catastrophe risks, as opposed to more well-behaved risks, such as automobile accidents, can be challenging, particularly if the insurance company does not fully understand the risk or if the risk changes significantly over time. This chapter examines catastrophe insurance with a particular focus on events that alter risk assessments in the insurance industry. This chapter addresses two overarching questions:

---

[1] I would like to thank Will Rafey for his research assistance. I would also like to thank Andy Castaldi and Robert Muir-Wood for helpful feedback on this chapter.

[2] It is possible that insurance premiums could be altered to encourage investment in risk reduction. As noted by Woo (1999), however, the market for insurance is inefficient and cyclical. In soft market conditions, when prices are low, it may discourage risk reduction, and when prices are high, many may simply forgo disaster insurance entirely. Insurance, then, is not likely to be the best tool for incentivizing hazard mitigation.

1. What types of events lead the (re)insurance industry to update risk assessments?
2. How do companies, consumers, and the government respond to updated risk assessments?

This chapter focuses, in general, on the United States, although many of the findings, particularly related to the private (re)insurance industry, will likely be globally applicable.

The next section of the chapter discusses the challenges of insuring disaster risks. Section 3.3 discusses the role and impact of disaster events on the insurance industry, even in the absence of changing risk assessments. Section 3.4 turns to the question of what types of events lead to a revision in risk estimates. Sections 3.5, 3.6, and 3.7 then outline the response to such changes by insurance companies, consumers, and the government. Section 3.8 concludes.

3.2 INSURING CATASTROPHES

The theory of insurance rests on the notion that the insured is risk averse and would pay more than the expected value of the loss to transfer the risk. Many authors have discussed the idealized conditions for insurability of a risk, or for risk transfer to an insurance company to occur (e.g. Swiss Re 2005; Charpentier 2008). For present purposes, I identify five conditions:

1. a degree of randomness to loss occurrences and their magnitude;
2. independent, thin-tailed, and quantifiable risks;
3. determinable losses;
4. no adverse selection or moral hazard; and
5. demand meets supply (the market clears).

First, losses must be unintentional and random – that is, not known with certainty. This requirement is perhaps obvious; insurance cannot be provided for events occurring with certainty or over which the insured has complete control. The second criterion allows for a benefit from pooling risks. Independent risks are those that are not correlated with each other. For example, when one person gets into a car accident, it does not make it more likely that their neighbors will, as well. Thin-tailed risks are those for which the probability of extremely large losses becomes negligible. With independent and thin-tailed risks, the law of large numbers and the central limit theorem ensure benefits to the insurance company from combining the risks of many insureds: the average claim converges to the expected value, and the tails of the aggregation become thin (that is, normally distributed). As more policies are written, insurers can charge a rate closer to the expected loss (Cummins 2006). The third criterion is that losses must be measurable and verifiable for settlement of claims to occur. The fourth condition has to do with the information and incentives of the insured. Adverse selection occurs when the insured knows more about his or her risk than the insurance company, such that higher-risk individuals or firms are more likely to insure, but the company cannot price for this. Moral hazard occurs when insurance induces the insured to fail to take loss-reduction efforts or to intentionally cause a loss. Finally, insurance will not be sold if it cannot be offered at a price that is profitable for the insurance company and that insureds are willing and able to pay.

Violation of the second and fifth of these criteria[3] can make insuring disaster risks challenging. With catastrophes, claims are often spatially correlated, such that all of the structures in a neighborhood may be damaged at once, as when an earthquake or hurricane occurs. Catastrophes are also characterized by fat-tailed losses. With fat-tailed loss distributions, the probability of an event declines slowly, relative to its severity. Put simply, very large losses are possible. Indeed, many natural catastrophes, from earthquakes to wildfires, are fat tailed (e.g. Schoenberg et al. 2003; Newman 2005; Holmes et al. 2008).

With correlated and fat-tailed risks, claims will be more volatile from year to year than for other lines of insurance, with companies on occasion experiencing very severe losses. With non-catastrophic risks, premiums received in a given year can largely cover losses experienced in that year. For catastrophic risks, on the other hand, insurance firms must solve an intertemporal smoothing problem of trying to match regular premium payments, insufficient in any given year to cover a large loss, with the need for enormous sums of capital in the catastrophe years (Jaffee and Russell 1997). Given this, firms are managed to keep the probability of insolvency below a certain level (so-called value-at-risk requirements)[4] will be required to charge very large premiums to cover catastrophic risks because much more capital is needed to be able to pay claims when a disaster strikes. This can lead to rates that seem overly high in "good"

---

[3] They could also violate the third criterion, as demonstrated by the wind–water controversy following Hurricane Katrina.

[4] This type of management means that the firm ensures access to enough capital to cover up to a certain percentile of the aggregate loss distribution (thus guaranteeing solvency up to that point). Regulations in the EU (Solvency II), for example, have set the annual probability of insolvency at 0.5 percent.

years, but which are needed to account for the "bad" years. Take the case of Florida. From 1993 to 2003, the rate of return on net worth for homeowner insurers was 2.8 percent nationwide but a much larger 25 percent for Florida. When 2004 and 2005, two years with powerful storms, are included, the situation reverses dramatically, with the national return being −0.7 percent, and the return in Florida being a devastating −38.1 percent (Insurance Information Institute 2009).

There are multiple ways that insurance companies can meet capital-holding requirements, which are set either internally, by rating agencies, or by regulation. One is holding capital themselves. Another is the purchase of reinsurance. Reinsurance is essentially insurance for insurers and is a global market. Reinsurance can be expensive (much higher than expected losses), but it also stabilizes the potential losses of insurance companies and allows for increased capacity (Cummins et al. 2008). Reinsurance markets have been observed to cycle through hard and soft conditions (discussed further in the next section), which impacts prices. Reinsurance prices can also be driven by external international factors. Prices in reinsurance may then passed through down to the purchaser of an insurance policy. In this way, the premiums for disaster polices can be influenced by reinsurance market conditions.

The higher rates that need to be charged to cover the need for high levels of capital to cover catastrophic lines could be so high as to lead to a breakdown in the insurance market. That is, there may be catastrophic risks that a company cannot insure at a price that insureds would either be willing or able to pay (Kousky and Cooke 2012). For instance, Munich Re executives have noted that they did not provide more reinsurance to companies with exposure to the 2011 Japanese earthquake because there was no demand at the price they deemed necessary (Munich Re 2012). (In Japan, the government established a reinsurance program for the residential earthquake insurance market due to the low market capacity and high cost (see Mahul and White 2013).) In such cases, homeowners and firms will either forgo insurance or the government may need to step in to provide coverage (see Section 3.7).

In addition, insuring risks is not a yes–no proposition. Risks exist along a continuum, from the easy-to-insure to the difficult. Automobiles are largely easy to insure.[5] A private market for coverage from nuclear

---

[5] Insurance companies also limit their risk (and moral hazard) in creative ways. For instance, in Germany, insurance companies have refused to pay for accidents caused by excessive speeding, and this has been upheld in court (Stahel 2003).

terrorist attacks is almost nonexistent because it could have unfathomable losses, and losses would be highly correlated across lines of insurance and with the broader economy. Most risks are somewhere in between, and insurance companies have multiple strategies for increasing their ability to cover a risk. For example, reinsurance companies can often diversify over greater geographic areas and over a wider range of perils, such that ceding some portion of risk to a reinsurer can allow a primary insurance company to offer more coverage. Or, if alternative risk-transfer instruments allow insurance companies to place some risk in the capital markets, it could help them access more capital in large loss years. Companies also limit their exposure in various ways to assume a portion of catastrophe risk that they can comfortably manage, such as increasing deductibles and copayments, enacting policy limits, carefully limiting underwriting, segmenting markets, or working cooperatively with government on hazard mitigation (Swiss Re 2005).

These strategies also suggest that the concept of insurability of risks is dynamic (Swiss Re 2005). New strategies can alter a firm's appetite for covering disaster risks, and broader economic and political conditions can also change insurability over time. In addition, new information on a risk could lead (re)insurers to reconsider the insurability of risks or the conditions under which risks could be covered. An extreme event could be one such source of new information (the main focus of this chapter), or it could come from improved science, updated risk modeling, or from changes in regulation and policy that influence the size of claims (notable for lines such as liability insurance). The rest of this chapter focuses on what types of events lead (re)insurers to alter their assessments of a risk, how they respond when this happens, and how consumers and governments respond.

3.3 THE ROLE OF DISASTER EVENTS IN THE INSURANCE INDUSTRY

The insurance industry is in the risk business. Insurance firms undertake detailed analysis and modeling of the perils they choose to cover. One event will thus not usually lead to any dramatic reassessment of risks, as long as it is seen as a draw from a reasonably well-characterized distribution. For instance, after the large 2011 tornado losses, State Farm stated: "catastrophe claims experience tends to be aberrational. We evaluate claims experience over longer periods of time than we do other types of claims experience" (Berkowitz 2012: n.p.). The question is when an event makes an insurance company believe that it is indicative of a new, changing, or previously mischaracterized risk.

Following major disaster events, changes are often observed in the insurance market. Prices for insurance policies may increase, or insurance companies may alter coverage conditions. For example, hard markets – meaning that prices increase and supply is scarce – have been observed following many disasters, such as Hurricane Andrew in 1992, the Northridge Earthquake in 1994, and the September 11, 2001 terrorist attacks (Cummins 2006). More recent events have also seemed to trigger price increases. For example, following the terrible flooding in Thailand in 2011, which cost the industry over $10 billion, companies raised rates around 20 percent and restricted flood coverage (Ng 2012).[6]

It is difficult to untangle the causes of these observed changes in the market, however, and insurance companies could increase premiums or decrease supply following a large event for multiple reasons that are *not* due to a revision in risk estimates. First, the disaster could coincide with other market changes. For example, a 2012 *Wall Street Journal* article highlighted increasing prices for homeowners insurance, but part of the reason was that insurers have seen a low return on their investments (Andriotis 2012), not because they revised their estimates of disaster risk following the large losses in 2011.

Also, although firms ensure access to more capital to cover catastrophic lines, a large event will deplete their surplus.[7] Several studies suggest that companies may not be able to adequately smooth the payments of catastrophe losses over time because of institutional and political constraints, even when they are aware of the risk (Jaffee and Russell 1997). Due to the difficulty of preventing negative capital shocks in catastrophic lines, firms may raise rates after an event to recoup lost capital. In a sense, they spread the disaster costs over later years. Capital could be raised from several different sources. A visible disaster, however, may make it more difficult for insurance companies to raise external capital, either directly because of the shock to surplus (investors may not want their funds going to pay off debt) or because investors change their risk perceptions (Jaffee and Russell

---

[6] Indeed, 2011 was a huge loss year for the industry globally. For example, the year reduced profits for Catlin, an insurance and reinsurance company, by 82 percent, leading it to increase prices on catastrophe lines by 9 percent for primary insurance and 17 percent for reinsurance in the United States and 12 percent for reinsurance elsewhere (Gray 2012).

[7] The extent of the decline in surplus will depend, of course, in part on how much reinsurance they had purchased.

2003). Although theory suggests that investors' inability to evaluate risks can make it expensive to raise capital from them, thus providing one rationale for reinsurance purchases, reinsurers may also charge high prices and restrict supply post-disaster, for similar reasons discussed here (Froot and O'Connell 1997). This could cause a primary insurance company to raise premiums as a way to increase the capital it cannot obtain (or obtain as cheaply) from other sources. Depleted capital could also cause a firm to be downgraded by rating agencies, and the firm could then raise premiums or limit exposure in the highest-risk areas to reestablish a higher rating and maintain its financial position.

Yet another reason for the observed market changes after large disaster events may have to do with the politics of rate setting. Some have suggested that state regulators (discussed further below) tend to suppress disaster insurance premiums and, as such, insurance companies may use a major event as an excuse to raise prices, even if the event did not alter their risk perceptions (Kunreuther and Pauly 2005). Though difficult to observe, a scandal following Hurricane Andrew suggests that this may occur. Major newspapers reported that, while the storm was raging over Florida, a memo was circulated to American International Group (AIG) senior managers that the company could use the storm as an excuse to raise rates (Angbazo and Narayanan 1996). Text from the memo quoted by the *Los Angeles Times* suggests that AIG felt the industry could not absorb the loss without rate increases, and interview comments from AIG executives indicate that they felt the insurance was not priced high enough prior to Andrew (Mulligan 1992).

Finally, a disaster may also lead to changes in demand, as opposed to supply. If demand increases after an event as consumers reevaluate the risk or their need for insurance, higher prices could be observed in the market post-disaster (discussed further in Section 3.6). Unlike other industries, insurance often experiences an upward shock to demand concurrent with a downward shock to supply because large events can both increase demand and deplete insurance capital (Swiss Re 2005). This joint effect can lead to large price increases after events and can make price and quantity for disaster insurance very "spiky."

Disentangling all of the impacts of a disaster that are not due to risk-perception changes from the adjustments that occur when a catastrophe does alter the risk perception of insurers is difficult empirically, and very little work addresses this question. The next sections summarize what is known about this topic and offer some initial findings and hypotheses for future investigation.

3.4 WHAT EVENTS LEAD TO A CHANGE IN RISK PERCEPTIONS?

An extreme event could cause an insurance company to revise its estimation of the risk in any of three situations. The first is when the distribution of the risk is unknown, often due to very limited data on the risk. In this case, one extreme event carries more informational content for the insurance company, or leads to more updating, than when a large amount of data on a risk is available. The second situation is when the loss distribution is changing over time, such that an extreme event can be indicative of a change in the risk. The third is when an event identifies a new risk that had not been previously identified or incorporated into decision-making.

In practice, however, these need not be mutually exclusive categories. A risk may be unknown and changing over time, and new aspects of the risk may emerge post-event. Take the case of terrorism. The probability distribution of terrorism damages is clearly a distribution that is unknown. Getting an improved understanding of terrorism risk is also limited by lack of access to classified governmental information. The risk is continually changing over time as counterterrorism policies evolve, as terrorists shift their targets and intentions, and as both government security agencies and terrorists react to and anticipate each other. The changing nature of the risk can also lead to new aspects of the risk emerging, such as new types of attacks. It is thus no surprise that the attacks on September 11, 2001, provided new information that led to dramatic updates of terrorism risk by (re)insurance companies.

Insurance companies were quite aware of the risk of terrorism to the insurance lines they offered before 2001, having dealt with previous events both within and outside of the United States. Still, prior to September 11th, many firms simply had not considered an attack of the magnitude experienced. At the time of September 11th, it was the costliest insured event in history (coming in just behind Hurricane Andrew in 1992).[8] As Warren Buffet wrote of General Re in his letter to shareholders in 2001: in "setting prices and also in evaluating aggregation of risk, we had either overlooked or dismissed the possibility of large-scale terrorism losses" (Buffet 2002, p. 8). Swiss Re revealed similar sentiments, noting that the September 11th attacks demonstrated a "staggering, previously inconceivable scale of threat scenarios and loss potentials" (Schaad 2002, p. 2).

---

[8] It has since been surpassed by Hurricane Katrina, the 2011 Japanese earthquake, and Hurricane Sandy in terms of insured losses.

Academics and some government officials had discussed potentially catastrophic terrorism scenarios well before September 11, 2001, and yet insurance companies did not exclude such coverage from their contracts and appeared surprised by the magnitude of the losses, as noted by Buffet, Swiss Re, and others. In interviews of insurance officials, it was found that a total loss of the World Trade Center was simply not considered by most companies; their probable maximum loss[9] may have included a plane taking out a few floors and a resulting fire, but not destruction of both towers (Ericson and Doyle 2004, pp. 216 and 221). The 1993 World Trade Center bombing was manageable for companies in terms of losses, and it seemed inconceivable that a much greater magnitude loss was possible. In addition, prior to September 11th, terrorist attacks were not uncommon on US interests overseas, but only 0.5 percent of all terrorist actions between 1991 and 1998 were in North America (Cummins and Lewis 2003). The September 11th attacks thus changed insurers' view of terrorism risk: it became something that could be more common within the United States, and attacks could be much larger in magnitude than had been previously believed.

September 11th also led to the identification of a new aspect of terrorism risk or a new type of risk. The event drove home for insurance companies that losses across lines of business could be highly correlated in a manner not previously appreciated (Ericson and Doyle 2004). The same single event caused losses in property, business interruption, liability, workers' compensation, aviation, life, and health insurance lines. Companies also realized after September 11th that a large claim-generating event could also lead to investment losses, further stressing the company, such that disaster events were not necessarily independent of market conditions. In his shareholder letter the year after the attacks, Buffet (2002, p. 7) stresses that insurance companies should "accept only those risks that they are able to properly evaluate" and "ceaselessly search for possible correlation among seemingly-unrelated risks." These correlations were made plain by September 11th.

A similar updating of risk perceptions occurred after Hurricane Katrina, when, again, the types of losses that can be sustained following a large event were surprising. Hurricane Katrina generated the highest level of insured losses in history. Insurance companies often rely on catastrophe models – probabilistic models of different perils that take

---

[9] Probable maximum loss refers to the largest modeled loss for which the company is pricing.

into account a company's portfolio to estimate losses – when evaluating their pricing and underwriting. Three firms do the vast majority of such modeling: Risk Management Solutions (RMS), AIR, and EQECAT. These models generally underpredicted losses from Katrina. The storm led RMS modelers to realize that when disaster events are extreme, they can result in what RMS (RMS 2005, Foreword) called "super cats," or events with such large amounts of damage that they "give rise to nonlinear loss amplification, correlation, and feedback." Super cats result in impacts for many more lines of business and generate damage of types not previously modeled, such as infrastructure collapse, looting and crime, water contamination, prolonged business interruption, and delayed economic recovery (NAPCO LLC 2006). Adding these possibilities – along with a range of other lessons learned from Katrina – into the RMS model increased damage estimates by up to 125 percent (NAPCO LLC 2006). All firms recalibrated their models after the 2004 and 2005 seasons, increasing probable maximum losses by between 10 percent to over 100 percent (Fleckenstein 2006).

Hurricane Katrina also provided new information on how hurricane risks may be evolving over time from climate change. Before Katrina, the RMS hurricane model had relied on hurricane events over the past 100 years, using a representative selection of such storms to model losses. This did not account for any cyclical variations, such as El Niño and La Niña events, nor did it account for increasing sea-surface temperature. In a new 2006 model, RMS drew on a panel of scientific experts to instead take a forward-looking view of hurricane activity, determining that higher sea temperatures would lead to increased activity and incorporating the role of the Atlantic multi-decadal oscillation in affecting hurricane activity (NAPCO LLC 2006). This led to increases in modeled losses of about 40 percent for Florida and between 25 percent and 30 percent for the Mid-Atlantic.

This makes clear that insurers at times will be incorporating updated risk assessments by the risk modeling companies on which they rely. Beyond risk modelers, insurers also have to respond to updated requirements of rating agencies. After the 2004 and 2005 hurricane seasons, for example, rating agencies increased their capital requirements and adopted more rigorous stress testing for catastrophe lines, forcing insurers to raise more capital or reduce exposure in order to maintain their ratings (Fleckenstein 2006). This was in addition to insurers realizing after the series of storms that many homeowners were underinsured, leading to high claim payments even for partially damaged structures. When companies increased their insurance to value checks, this led to an increase in exposure, which required a change in capital requirements. Combining all these factors created a shock to the market.

Large events can provide new information to insurers about a risk, particularly for risks that were poorly understood to begin with or those that are changing over time, or it can indicate risks not previously identified. The September 11th attacks and Hurricane Katrina are both examples of this type of risk updating by insurance companies. This suggests a model of learning about risks that is quite jumpy. In practice, catastrophic risks are continuously changing. Human populations and development patterns change over time; our technologies to mitigate hazards evolve; and other stresses, such as climate change or political conditions, shift. Although to some extent, all of these potential drivers are constantly modeled by the insurance industry, often it takes an extreme event to give deeper insight into the manner and magnitude of changes to risks.

On the other hand, however, insurance companies, though constantly evaluating risks and presumably more skilled at doing so than some other industries, may still be biased in their evaluation of risks, particularly after a very extreme event. They could overweight recent events, for example, or charge higher premiums when risks are unknown, a practice termed ambiguity aversion (Kunreuther et al. 2013). In addition, investor pressure after an extreme loss may induce insurance companies to overreact, or managers may otherwise be incentivized to react in this fashion.

3.5 INSURER REACTION TO CHANGING RISK PERCEPTIONS

When an extreme event causes an insurer to estimate a risk as higher than it previously did, the company may alter its pricing, coverage conditions, underwriting strategies, and/or capital management in response. Often, companies adjust all of these simultaneously, although this section discusses each in turn.

Large premium increases have been observed after many of the largest events the insurance industry has experienced. After the September 11th attacks, many companies revised their probable maximum loss to include the complete loss of the buildings they were insuring (Ericson and Doyle 2004, p. 221), which led to a large jump in prices. Reinsurance companies made similar revisions, and the supply of reinsurance for terrorism was dramatically curtailed after 9/11; what was available was priced at extremely high rates (CBO 2007), which had impacts on the pricing of primary insurance. Some primary companies raised rates dramatically, and others refused to offer the coverage at all. As an example, one Midwestern airport saw its liability premium increase 280 percent in 2002, along with the exclusion of terrorism coverage (Swiss Re 2005).

Following a reassessment of the risk, companies also often change the conditions of coverage to make the risk more insurable. A notable example of this is hurricane deductibles, introduced after Hurricane Andrew. Following the storm, in many coastal states, companies introduced deductibles specific to either hurricanes or wind damage, often set as a percentage of the home, ranging from 1 percent to 5 percent (McChristian 2012). These have been a lasting legacy of the storm and expanded beyond Florida.

Other examples come from the 2011 tornadoes. Those events led some companies to seek broader geographic diversification in order to reduce the amount of exposure they had in tornado-prone areas, to increase deductibles, and to change pricing based on new factors, such as the age and quality of a home's roof (Berkowitz 2012). The *Wall Street Journal* noted that Allstate, for example, was introducing a new policy in Kansas and Oklahoma that would not pay the full cost of roof replacement (Andriotis 2012).

If companies revise their estimates of the risk upward sufficiently, they may believe that they need to lessen the amount of exposure to the risk in their overall portfolios. In the extreme, companies may abandon a location or line of business altogether. As already mentioned, after September 11th, many (re)insurers dramatically restricted the supply of terrorism coverage. It was widely reported in the months after the attack that commercial terrorism coverage was unavailable, even at very high prices (Swiss Re 2005). A 2002 Swiss Re publication noted that, in response to the attacks, "the insurance industry is compelled to fundamentally review its risk acceptance position, and to reduce and limit coverages granted to avoid unmanageable exposures in the future. In fact, the question whether terrorism risk is insurable at all must be fundamentally reviewed" (Schaad 2002, p. 5).

Insurance companies also pulled back along the Gulf Coast after Hurricane Katrina, although not as severely as was seen in the terrorism market after September 11th. The trimming of hurricane exposure, however, has continued over many years since Katrina, suggesting that companies fundamentally do not consider the risk to be improving. In February 2010, for instance, Farmer's announced that it would drop more than 10,000 policyholders in the coastal counties of Alabama, joining Allstate, Alfa Mutual, and State Farm in shedding policies in Baldwin and Mobile counties (Amy 2010).

Reducing coverage is not independent of the prices firms can charge. In 2007, State Farm stopped writing new homeowners policies in Florida (Stroud 2012). Part of the difficulty was a struggle over rates. Florida was preventing State Farm from raising rates to the level the company thought was needed. Negotiations with the state followed, and, in December 2009, State Farm agreed to drop 125,000 high-risk policies instead of exiting the state entirely, and Florida allowed the company to raise its rates by 14.8 percent (Fineout 2010).

A pullback of insurance companies after the Northridge earthquake led to a crisis in California. Following the earthquake, insurance companies paid out $15 billion in claims but had collected only $3.4 billion in earthquake premiums in the preceding 25 years (GAO 2007). The Northridge Earthquake generated far higher claims than insurers had deemed plausible. Claims analysis revealed a large amount of underinsurance, appurtenant structures that were more vulnerable and represented a greater share of property value than had been assumed, and zip-code level data that masked important heterogeneity (RMS 2004). Concerned about their ability to charge risk-based rates and the challenge in diversifying their earthquake risk, insurance companies began shedding their earthquake exposure throughout the state. California state law, however, required that insurance companies offer earthquake coverage if they wrote residential coverage, so companies began to quit writing all residential coverage. The California Department of Insurance estimated that in the summer of 1996, 95 percent of the homeowners insurance market in the state had stopped or dramatically limited the sale of new policies (CEA 2010). This triggered a housing crisis that led the state to intervene (discussed below).

Finally, insurance companies may alter the management of their capital. They may increase reserves, for instance, or purchase more reinsurance. Following Hurricane Andrew and the Northridge Earthquake, strategies were developed to transfer risk to the capital markets, such as through catastrophe bonds (Kunreuther and Michel-Kerjan 2005). This opened up a new source of capital to firms; for many reasons, the market remains small, although it is growing.

Some amount of "risk contagion" is also apparent. In other words, when a bad disaster of one type occurs, prices go up for other disaster lines as well. For example, the prices for earthquake coverage increased after Hurricane Katrina, and the price of catastrophe bonds for natural disasters increased after the September 11th terrorist attacks (Froot 2008). This could be due to the general decreases in surplus. It could also be, however, that as insurance companies revise their risk estimates of one catastrophe line, it prompts them to reconsider their evaluations of other catastrophe risks as well. Hurricane Katrina prompted a reexamination of the catastrophe models, as discussed earlier, and whereas some changes were unique to windstorms, some were broadly applicable to a range of catastrophes, leading to price increases on other catastrophic lines, like earthquake coverage (Advisen 2006).

What happens after the immediate post-disaster changes? In some cases, changes may be temporary. Insurance companies may find new tools to help insure the risk and access new pools of capital to expand their underwriting. As prices increase and quantity is restricted, it can attract new capital to the market, helping to reestablish lower prices and more availability. Terrorism coverage has improved since September 11th, although this is partly due to federal legislation, as discussed below. It may also be due to improved modeling of terrorism losses, such as the model developed by RMS (RMS 2008).

In other cases, however, the new assessments of the risk may be such that the market changes after the event are permanent. For example, since Katrina, companies have continued to push for higher prices along the coast and have continued to tightly manage their exposure to hurricane losses. Katrina suggested not only that companies had fundamentally underestimated potential hurricane losses, but also that the risk may be worsening.

It is also worth remembering that insurance companies cannot perfectly predict changing risks and cannot possibly guard against every catastrophe scenario. Rare events that stress the industry will always occur. For example, one executive noted that, although EU regulations (Solvency II) require insurance companies to reserve for the one-in-200-year event, that means that, with 100 large insurance companies, roughly one will go bankrupt every two years (Munich Re 2012).

3.6 CONSUMER REACTION TO CHANGING RISK PERCEPTIONS

An extreme event may not only lead insurance companies to update their risk assessments, but also cause consumers to update their risk perceptions. If individuals assess a risk as higher after an extreme event then they did before, it may lead them to change their insurance purchases. Findings from behavioral economics suggest that individuals can often be poor evaluators of risks, using mental shortcuts and rules of thumb (Kahneman et al. 1982). For example, after an event, individuals often assess the risk as higher because it is now salient for them. This has been termed the availability heuristic (Tversky and Kahneman 1973). When this happens, it could increase insurance demand. This increase in demand could come at the same time the market is hardening, exacerbating the changes post-disaster in both the price and quantity of available insurance coverage.

Supporting these theoretical arguments, several studies have found empirical evidence that individuals are often more likely to purchase insurance in the wake of a disaster. Browne and Hoyt (2000) find that flood insurance purchases at a state-level increase when flood damages the previous year are higher. Looking only at flood insurance purchases in Florida, Michel-Kerjan and Kousky (2010) find that, after the 2004 hurricanes, the number of policies-in-force statewide jumped 6 percent, compared to increases of only 1 percent to 2 percent in other years. They also find that more homeowners in areas hit by the hurricanes lowered their deductibles and chose the maximum coverage level.

September 11th also had impacts on insurance demand. A month after the attacks, CNN reported a 30 percent spike in the purchase of travel insurance, even though the total number of travelers was still below pre-attack levels (Max 2001). As another example, following the Deepwater Horizon oil spill, firms engaged in drilling began to demand greater levels of insurance coverage (Booz Allen Hamilton 2010). Finally, some empirical evidence suggests that catastrophes lead to greater purchases of life insurance (Fier and Carson 2010). Demand can also fall as salience declines. There is some indication that, as the time since the last earthquake increases in California, more people drop earthquake insurance (Wilkinson 2009).

3.7 GOVERNMENT REACTION TO CHANGING RISK PERCEPTIONS

The government has intervened in insurance markets in the United States at both a state and federal level following extreme events. Government intervention is often motivated by a perception that the insurance market has broken down, with private companies choosing not to offer coverage for a risk or only at prices that are beyond what many consumers can afford. Governments can intervene in insurance markets in a number of ways. The insurance industry is regulated at the state level by insurance commissioners. Commissioners oversee licensing of insurance companies (and thus a firm's ability to sell insurance within a state), pricing, solvency, underwriting, marketing, and claims handling, among other things. Beyond regulation, many states have created their own insurance programs for consumers who cannot find a policy in the private market. Finally, the federal government also runs several (re)insurance programs. This section discusses these various interventions.

After an extreme event, state insurance commissioners often work to prevent steep price increases or the exit of firms in an attempt to protect consumers. Following Hurricane Andrew, insurance regulators in Florida prohibited dramatic rate increases and only let companies gradually increase prices over a decade, and the state legislature passed a moratorium on policy cancellations (Klein 2009). Part of this, however, was due to the AIG controversy mentioned earlier. In response to it, the Florida insurance commissioner froze AIG premiums and warned all companies that unjustified rate increases would not be permitted (Angbazo and Narayanan 1996). The state also put in place a three-year moratorium that limited how quickly firms could reduce their market share in the state (McChristian 2012). Following Hurricane Katrina, the state allowed an initial wave of price increases that were generally highest in coastal areas but then began disapproving them in 2006 (Klein 2009). Other states have taken similar measures. After Katrina, the Louisiana Department of Insurance prohibited cancellation or nonrenewal of residential dwelling and commercial property insurance for structures damaged by Hurricane Katrina or Rita until 60 days after all repair and reconstruction had been completed (Klein 2009).

More dramatically, states have set up their own insurance programs after observing a severe hardening of the insurance market (Kousky 2011). These state programs, often called residual market mechanisms, are created for residents who cannot find policies in the voluntary market. They take a variety of forms, including state FAIR (Fair Access to Insurance Requirements) plans,[10] state wind pools or "beach plans" that provide wind-only coverage in certain high-risk areas, hybrid programs that write both hazard-specific policies and complete dwelling coverage, an earthquake program in California, and a reinsurance fund in Florida. Many programs have eligibility requirements intended to ensure that policies are

---

[10] Following riots and civil disorder in many urban areas, federal legislation in 1968 made federal riot insurance available to states that enacted FAIR plans that offer homeowner coverage to residents who cannot find policies in the voluntary market. Initially, these plans offered coverage only for fire, but many have expanded, and some even offer wind coverage.

purchased only as a last resort. Almost all of these programs were adopted in reaction to an extreme event and insurer reactions to that event.[11]

The CEA is a good example. It was established following the housing crisis triggered by the Northridge earthquake. Insurance companies can comply with the state mandate to offer earthquake coverage with residential dwelling coverage by participating in the CEA. Upon joining the CEA, insurers make a capital contribution and are able to be assessed after an event if available capital and reinsurance do not cover all claims. The CEA began operating in late 1996; within a year, almost all insurance companies were again operating in the state (CEA 2010). The CEA is the largest earthquake insurer in California. Historically, the coverage offered was quite limited, with low caps and high deductibles. Recently, however, the CEA has introduced new policies with broader coverage and more choice. Still, take-up rates for earthquake coverage are only around 10%, leaving many homeowners vulnerable to earthquake losses.

The CEA is well prepared to cover claims from a severe earthquake, but other state programs are not. Florida's insurance program, initially created after Hurricane Andrew, has been a source of controversy, for example, with many observers noting that its pricing is too low, such that it will be unable to pay claims from a severe hurricane,[12] and artificially low prices could cause perverse incentives that may not occur in a private market. The state has taken steps recently to raise rates and improve the standing of the program.

State intervention could be used as a temporary measure to relieve market hardening after a disaster. Hawaii's program exemplifies this possibility. The program was established in 1992 after Hurricane Iniki. By 2000, private insurance companies were writing their own policies in Hawaii, and the state program has not written any new policies over the last decade. Such short-term intervention is useful when the private market

---

[11] For instance, Louisiana Citizens and the Alabama Beach Pool were created in response to Hurricane Camille, Hawaii's fund was created after Hurricane Iniki, the Texas Windstorm Insurance Association was created following Hurricane Celia, the precursor to Florida Citizens was created following Hurricane Andrew, and the California Earthquake Authority was created after the Northridge earthquake.

[12] When reserves are not enough to cover claims, the program would be funded by post-event assessments on policyholders in the program. Following policyholder assessments, all property and casualty insurance companies in the state would be assessed, and finally an emergency assessment on all property and casualty policies. This distributes the risk beyond only the at-risk homeowners in the program. Recouping these assessments would take time, however, making the program reliant on issuing post-event bonds.

simply needs time to revise strategies and rebuild capital. It will not be possible if the private market fundamentally reassesses a risk as uninsurable. Florida Citizens, for example, is the largest insurer in the state; it will probably remain a fixture of the Florida insurance market. Indeed, many state programs have grown in recent years and play a vital role that is unlikely to be replaced completely by the private market in the near future.

The federal government has also created its own insurance programs. Building in part on a widespread perception that flood insurance was unavailable in the private market, Congress created the National Flood Insurance Program (NFIP) in 1968. Now housed in the Federal Emergency Management Agency (FEMA), the program makes flood coverage, up to certain limits, available to residents of participating communities that adopt baseline floodplain management regulations. Purchase of a flood insurance policy is mandatory for homeowners in a FEMA-mapped 100-year floodplain with a mortgage from a federally backed or regulated lender.

To encourage the purchase of insurance, the NFIP historically discounted the premiums of some policies. This prevented it from building a catastrophe reserve or purchasing reinsurance, and the losses of Hurricane Katrina sent the program deeply into debt to the US Treasury. Characteristic of a catastrophic risk, the NFIP paid out more in claims after Katrina than it had over the life of the program to that point. This debt generated many discussions on reform of the program, which culminated in the passage of the Biggert-Waters Flood Insurance Reform Act of 2012 and the Homeowners Flood Insurance Affordability Act of 2014. Under these laws, rates for some previously discounted policies are now increasing. This, coupled with new hazard mapping in some locations, has raised concerns over the affordability of flood insurance (Kousky and Kunreuther 2013), a tension that is found in many government insurance programs. Political pressure for lower rates often exists, but subsidizing costs could theoretically lead to an overinvestment in risky locations and underinvestments in hazard mitigation.

Most recently, the federal government has intervened in the terrorism insurance market.[13] As mentioned above, reinsurance companies began excluding terrorism coverage in 2002 renewals. This prompted 45 states

---

[13] Other countries have intervened in the terrorism insurance market. France, for instance, developed a terrorism insurance pool with a state guarantee after September 11th. French law from the mid-1980s made coverage for terrorism risks mandatory and, as such, the government had to intervene when the reinsurance market collapsed after the attacks. Britain already had a pooling scheme with a government backstop developed after terrorism incidents in London in the early 1990s.

to allow primary insurers to exclude terrorism coverage in policies as they came up for renewal (Brown et al. 2004). Terrorism coverage, with a few exceptions, was not available, and commentators warned of negative impacts on the economy. Some also argued that the federal government could be more effective at the intertemporal diversification required for terrorism risks and that the federal government had a hand in managing terrorism risk and thus should intervene in the market.

The federal government responded by passing the Terrorism Risk Insurance Act (TRIA) in fall 2002. It created a federal backstop for terrorism insurance for US property–casualty insurers. Under the act, the federal government covers certain declared terrorism losses for insurance companies above a deductible and up to a limit (the insurance companies also have a coinsurance amount). The firms pay nothing up front, but the government has the option to recoup costs ex post with a surcharge on commercial insurance policies that cannot exceed 3 percent. TRIA was intended to be temporary, with the thought that the private market would recover, but it has been extended multiple times.

The legislation mandated that firms make terrorism coverage available in exchange for the federal government assuming some of the risk. This did lead to more supply after its passage. And some evidence indicates that many insurers would not offer terrorism coverage if TRIA were not in place. Cummins (2006) notes that in 2004, 90 percent of insurers wrote coverage that was covered by TRIA, but only 40 percent wrote coverage for terrorism acts that TRIA did not reinsure. This law did not cap prices, however, and many companies appear to have gone without terrorism coverage rather than pay high prices. That said, premiums for terrorism coverage have fallen and demand for primary terrorism coverage has increased substantially since TRIA's passage (CBO 2007). Take-up for terrorism insurance was 27 percent in 2003 but grew to 61 percent in 2009 (Marsh 2010). This is despite the fact that the 2005 extension increased the deductibles and coinsurance borne by primary insurers. Although the purchase of reinsurance by primary insurance companies has increased, it is only a quarter of primary insurers' deductible under TRIA and thus still quite limited (CBO 2007).

The question of when and how governments should intervene in insurance markets is a difficult one. If the price increases and declines in availability of insurance post-disaster are influenced largely by a negative shock to capital, the market should re-equilibrate after a short period of time as capital is restored. Government intervention, however, is often more lasting than initially intended, and in these cases may be counterproductive. Recent events, however, have led insurance companies to rethink the insurability of some catastrophic lines, such as terrorism and hurricane coverage, sending prices upward and available quantities downward. Governments often see their role as maintaining both availability and affordability of insurance coverage. As we have seen, however, catastrophe risks are expensive to insure. Often government intervention in the name of affordability simply transfers the costs to others. The overall question of who should cover the costs of catastrophic events is a difficult one worthy of more public discussion and quite beyond the scope of this chapter.

3.8 CONCLUSION

The insurance industry and the modeling companies that support it often have some of the best evaluations of risk available. Even in the absence of full information about a risk, such as when risks are unknown, uncertain, or are evolving over time, (re)insurance companies often still offer coverage. Although done at a price insurance companies feel can justify assumption of the risk, this provides needed risk management for many homeowners and firms. For unknown and changing loss distributions, however, extreme events can provide important new information on the nature of the peril. When this occurs, insurance firms will reevaluate their pricing, exposure, underwriting policies, and capital management. Severe events that lead to this type of risk updating, however, also often lead to updating on the part of consumers and governments. This chapter has traced how all three sectors may respond when they assess a risk as higher than they did previously. This may simply lead to temporary adjustments, or it can cause new equilibrium conditions in the market or permanent government interventions.

References

Advisen (2006). Earthquake: The Other Insurance Crisis. Advisen LtD. www.cybersure.com/Documents/The_Other_Crisis.pdf.

Amy, J. (2010). "Farmers insurance to drop wind coverage from 10,000 coastal policyholders." *Al.com* (Birmingham, AL). http://blog.al.com/live/2010/02/farmers_insurance_group_to_dro.html.

Andriotis, A. M. (2012). "Home Insurance Goes Through the Roof." *The Wall Street Journal Market Watch*, March 9. http://blogs.marketwatch.com/realtimeadvice/2012/03/09/home-insurance-goes-through-the-roof/.

Angbazo, L. A. and R. Narayanan (1996). "Catastrophic Shocks in the Property-Liability Insurance Industry: Evidence on Regulatory and Contagion Effects." *The Journal of Risk and Insurance* 63(4): 619–37.

Berkowitz, B. (2012). "Insurers Forced to Rethink Tornado Coverage." *Chicago Tribune*, March 7. www.reuters.com/article/us-insurance-tornadoes-idUSTRE8261X320120307.

Booz Allen Hamilton (2010). The Offshore Oil and Gas Industry Report in Insurance – Part One, Report funded by the Department of Energy's National Energy Technology Laboratory.

Brown, J. R., J. D. Cummins, C. M. Lewis and R. Wei (2004). "An Empirical Analysis of the Economic Impact of Federal Terrorism Reinsurance." *Journal of Monetary Economics* 51: 861–98.

Browne, M. J. and R. E. Hoyt (2000). "The Demand for Flood Insurance: Empirical Evidence." *Journal of Risk and Uncertainty* 20(3): 291–306.

Buffet, W. E. (2002). Letter to Shareholders. Berkshire Hathaway Inc. 2001 Annual Report.

CBO (2007). *Federal Reinsurance for Terrorism Risks: Issues in Reauthorization*. Washington, DC, Congressional Budget Office.

CEA (2010). Annual Report to the Legislature and Insurance Commissioner on Program and Operations. Sacramento, CA, California Earthquake Authority.

Charpentier, A. (2008). "Insurability of Climate Risks." *The Geneva Papers* 33: 91–109.

Cummins, J. D. (2006). "Should the Government Provide Insurance for Catastrophes?" *Federal Reserve Bank of St. Louis Review* July/August: 337–56.

Cummins, J. D. and C. M. Lewis (2003). "Catastrophic Events, Parameter Uncertainty and the Breakdown of Implicit Long-Term Contracting: The Case of Terrorism Insurance." *The Journal of Risk and Uncertainty* 26(2/3): 153–78.

Cummins, J. D., G. Dionne, R. Gagne and A. Nouira (2008). *The Costs and Benefits of Reinsurance*. CIRRELT-2008-26. Montreal, Université de Montréal, Interuniversity Research Centre on Enterprise Networks, Logistics, and Transportation.

Ericson, R. V. and A. Doyle (2004). *Uncertain Business: Risk, Insurance, and the Limits of Knowledge*. Toronto, University of Toronto Press.

Fier, S. G. and J. M. Carson (2010). Catastrophes and the Demand for Life Insurance. Working Paper, available at SSRN: http://ssrn.com/abstract=1333755.

Fineout, G. (2010). "Truce Reached with State Farm Florida; Citizens Property Insurance Corp. Could Grow Due to Deal." *Florida Underwriter*. 27(1): 24. https://insurancenewsnet.com/oarticle/Truce-Reached-With-State-Farm-Florida-Citizens-Property-Insurance-Corp-Could-a-152331.

Fleckenstein, M. (2006). Rating Agency Recalibrations. In *The Review: Cedant's Guide to Renewals 2006*. London: Informa UK Limited: 40–43.

Froot, K. A. (2008). "The Intermediation of Financial Risks: Evolution in the Catastrophe Reinsurance Market." *Risk Management and Insurance Review* 11(2): 281–94.

Froot, K. A. and P. G. J. O'Connell (1997). On the Pricing of Intermediated Risks: Theory and Application to Catastrophe Reinusurance. NBER Working Paper 6011. Cambridge, MA, National Bureau of Economic Research.

GAO (2007). *Public Policy Options for Changing the Federal Role in Natural Catastrophe Insurance*. Washington, DC, United States Government Accountability Office

Gray, A. (2012). "Catlin Warns of Disaster Insurance Rates." *Financial Times*, February 9.

Holmes, T. P., R. J. Huggett, Jr. and A. L. Westerling (2008). Statistical Analysis of Large Wildfires. In *The Economics of Forest Disturbances: Wildfires, Storms, and Invasive Species*, edited by T. P. Holmes, J. P. Prestemon and K. L. Abt. Springer Science: 59–77.

Insurance Information Institute (2009). *Catastrophes: Insurance Issues. Issues Updates*. New York, NY, Insurance Information Institute.

Jaffee, D. M. and T. Russell (1997). "Catastrophe Insurance, Capital Markets, and Uninsurable Risks." *The Journal of Risk and Insurance* 64(2): 205–30.

Jaffee, D. M. and T. Russell (2003). Markets Under Stress: The Case of Extreme Event Insurance. In *Economics for an Imperfect World: Essays in Honor of Joseph E. Stiglitz*, edited by R. Arnott, B. Greenwald, R. Kanbur and B. Nalebuff. Cambridge, MA, MIT Press: 35–52.

Kahneman, D., P. Slovic and A. Tversky, eds. (1982). *Judgment under Uncertainty: Heuristics and Biases*. Cambridge, UK, Cambridge University Press.

Klein, R. W. (2005). *A Regulator's Introduction to the Insurance Industry*. Kansas City, MO, National Association of Insurance Commissioners.

Klein, R. W. (2009). *Hurricane Risk and the Regulation of Property Insurance Markets*. Atlanta, GA, Center for RMI Research, Georgia State University.

Kousky, C. (2011). "Managing the Risk of Natural Catastrophes: The Role and Functioning of State Insurance Programs." *Review of Environmental Economics and Policy* 5(1): 153–71.

Kousky, C. and R. Cooke (2012). "Explaining the Failure to Insure Catastrophic Risks." *The Geneva Papers* 37: 206–27.

Kousky, C. and H. Kunreuther (2013). Addressing Affordability in the National Flood Insurance Program. RFF Issue Brief 13-02. Washington, DC, Resources for the Future.

Kunreuther, H. and E. Michel-Kerjan (2005). *Insurability of (Mega-) Terrorism Risk: Challenges and Perspectives. Terrorism Risk: Insurance in OECD Countries*. Organisation for Economic Co-operation and Development. Paris, France, OECD Publishing.

Kunreuther, H. and M. Pauly (2005). *Insurance Decision-Making and Market Behavior. Foundation and Trends in Microeconomics*, vol 1, no 2. Hanover, MA, Now Publishers, Inc.: 63–127.

Kunreuther, H. C., M. V. Pauly and S. McMorrow (2013). *Insurance & Behavioral Economics: Improving Decisions in the Most Misunderstood Industry*. New York, NY, Cambridge University Press.

Mahul, O. and E. White (2013). Earthquake Risk Insurance. Knowledge Note 6-2: The economics of disaster risk, risk management, and risk financing, Washington, DC, World Bank. https://openknowledge.worldbank.org/bitstream/handle/10986/16149/793930BRI0dr m000Box377374B00Public0.pdf?sequence=1.

Marsh (2010). *The Marsh Report: Terrorism Risk Insurance 2010*. New York, NY, Marsh.

Max, S. (2001). "Is Travel Insurance Worth It?" *CNN*, October 9. http://money.cnn.com/2001/10/09/insurance/travel_insurance/.

McChristian, L. (2012). *Hurricane Andrew and Insurance: The Enduring Impact of an Historic Storm*. Tampa, FL, Insurance Information Institute.

Michel-Kerjan, E. and C. Kousky (2010). "Come Rain or Shine: Evidence on Flood Insurance Purchases in Florida." *Journal of Risk and Insurance* 77(2): 369–97.

Mulligan, T. S. (1992). "Florida Agency Orders Freeze on Insurer's Rates." *Los Angeles Times*, September 10. http://articles.latimes.com/1992-09-10/business/fi-453_1_hurricane-andrew.

Munich Re (2012). 'This Is Totally Different from Anything We Have Ever Experienced'. *Topics Geo* 2011: 12–19.

NAPCO LLC (2006). *The Impact of Changes to the RMS U.S. Hurricane Catastrophe Model*. Iselin, NJ, NAPCO LLC.

Newman, M. E. J. (2005). "Power Laws, Pareto Distributions and Zipf's Law." *Contemporary Physics* 46(5): 323–51.

Ng, J. (2012). "Insurance to Cost More in Areas Hit by Natural Disasters." *Asia One: The Business Times*, February 5. http://news.asiaone.com/print/News/AsiaOne%2BNews/Business/Story/A1Story20120203-325702.html.

RMS (2004). *Northridge Earthquake 10-Year Retrospective*. Newark, CA, Risk Management Solutions.

RMS (2005). *Hurricane Katrina: Profile of a Super Cat: Lessons and Implications for Catastrophe Risk Management*. Newark, CA, Risk Management Solutions.

RMS (2008). *Terrorism Risk: 7-Year Retrospective, 7-Year Future Perspective*. Newark, CS, Risk Management Solutions.

Schaad, W. (2002). *Terrorism – Dealing With the New Spectre*. Zurich, Swiss Reinsurance Company.

Schoenberg, F. P., R. Peng and J. Woods (2003). "On the Distribution of Wildfire Sizes." *Environmetrics* 14(6): 583–92.

Stahel, W. R. (2003). "The Role of Insurability and Insurance." *The Geneva Papers on Risk and Insurance* 28(3): 374–81.

Stroud, M. (2012). "As Weather Gets Biblical, Insurers Go Missing." *Reuters*, April 11. www.reuters.com/article/us-insurance-disasters-idUSBRE83911S20120411.

Swiss Re (2005). *Innovating to Insure the Uninsurable. Sigma*. Zurich, Swiss Reinsurance Company.

Tversky, A. and D. Kahneman (1973). "Availability: A Heuristic for Judging Frequency and Probability." *Cognitive Psychology* 5: 207–32.

Wilkinson, C. (2009). *The California Earthquake Authority*. New York, NY, Insurance Information Institute.

Woo, G. (1999). *The Mathematics of Natural Catastrophes*. London, Imperial College Press.

4

Understanding Public Risk Perception and Responses to Changes in Perceived Risk

Elke U. Weber

INTRODUCTION

This chapter will introduce scholars and policy-makers interested in the public's perception of risk and its effect on individual and collective responses to some psychological literature on these topics. Depending on the reader, the sections below will contain either too much or too little information. The latter (too little information) can be remedied by consulting the provided references to specific facts, theories, or arguments for more information. The former (too much information) can be solved by reading just the remainder of this introduction and the conclusions and by sampling from the intervening sections as needed.

The section "Function of Risk Perception" argues that the general public only distinguishes between a small number of risk levels (present/absent or negligibly low/need to monitor/unacceptably high) and that triggering events shift their perceptions from one level to another. This results in either an under- or overreaction to existing risks in most situations, as illustrated in Figure 4.1. Real-world examples of such rapid shifts in reaction are provided in Boxes 4.1 to 4.3.

"Individual Risk Perception" provides more background on several psychological processes that give rise to the discrepancy between actual and perceived levels of risk. Risk often is a feeling rather than a statistic based on objective outcomes, and this feeling is influenced by reactions (dread, feeling out of control) that may not closely connect to objective loss or risk statistics. Importantly, it is this feeling of being at risk that motivates action. However, this flag goes down when some action has been taken, giving rise to a single action bias that discourages sustained attention to complex risks.

> BOX 4.1 *Hormone Replacement Therapy*
>
> For several decades until recently, hormone replacement therapy (HRT) was a widely used medical intervention for menopausal women in the US that was designed to decrease the risks of coronary heart disease and osteoporosis, but it could also result in possible increases in the risk of breast cancer. HRT was a $3.3 billion market in 2001. Even though the possible risks of HRT were always known and evidence about their magnitude accumulated gradually, press coverage and public narrative about its risk-benefit ratio showed a far less gradual trajectory, with (in hindsight) initial underestimates of the actual risks and more recently probably occasional overestimates (Katz 2011). Perceived risk increased dramatically following the results of a randomized control clinical trial by the Women's Health Initiative of the National Institute of Health in 2002, which showed reduced incidence of colorectal cancer and bone fractures, but also larger incidence of breast cancer, heart attacks, and strokes, concluding that the benefits did not outweigh the risks. This assessment was confirmed in a large national study done in the UK in 2004. The number of women taking hormone replacement therapy has dropped steeply as a result. NIH's Women's Health Initiative and the United States Preventive Task Force have drastically reduced recommendations for such therapy.

"Social Amplification of Risk" makes the point that social, institutional, and cultural processes also play a role in the perception of risk, typically by amplifying individual responses to triggering events that increase the perceptions of risk. Social amplification provides one or multiple narratives for why risk levels have changed and what needs to be done in response. Agreement on a narrative contributes to action. Multiple competing narratives may increase perceptions of risk, but they discourage corrective action.

"Individual Action and Choice under Risk and Uncertainty" argues that normative models of action from economics or finance fail to capture much observed behavior. Risk-return models for the pricing of risky investment options in finance can be adapted to describe the observation that responses to risk and risky situations involve a trade-off between the expected returns and the perceived riskiness of the situation. However, in contrast to the finance models that price risky investment options as

> BOX 4.2 *Nuclear Power*
>
> Nuclear power accidents provide a good example of the type of event that leads to a rapid step-function increase in perceived risk, as the result of a perceived loss of control over possible adverse catastrophic consequences. The American public's opposition to nuclear power in the late 1950s triggered the investigation of the psychological risk dimensions discussed above. That is, the nuclear power industry commissioned psychologists to explain why public risk perceptions of nuclear power generation (compared to the use of other fuels like coal) were so different from engineering estimates. Presumably it was the better understanding of the sources of these public fears that led to greater public acceptance of nuclear power, and during the 1970s the number of reactors under construction increased continuously. The Three Mile Island accident in 1979, a partial nuclear meltdown, put a halt to that, despite the fact that only small amounts of radioactive gases and radioactive iodine were released into the environment (International Atomic Energy Association 2008). Public fear about insufficient understanding and control over a dangerous and complex technology was expressed and amplified by the media, who used sorcerer's apprentice storylines in movies like "The China Syndrome." The 1986 Chernobyl accident reinforced concern about gaps in our understanding of the risks of the technology, and the recent Fukushima Daiichi accident showed that existing backup plans to provide coolant to reactor cores had dangerous gaps under conditions of natural disasters in the form of tsunamis. Long-term regulatory and political reactions to the most recent nuclear power accident are still unclear in most countries, but at least in one major Western democracy, namely Germany, it has lead to a public decision to phase out nuclear power by 2022, and countries like Japan, Switzerland, and Italy have announced reductions in their reliance on nuclear power generation (Clayton 2011).

a trade-off between option outcomes' expected value and variance, psychophysical models of risk-return trade-offs model expected returns and perceived risk as a psychological variable (a feeling) that can vary between individuals or groups as a function of cultural beliefs or expectations or past experience.

> BOX 4.3 *2008 Subprime Mortgage Crisis*
>
> The 2008 subprime mortgage crisis is an example of an unexpected regime change. Prior to the crisis, US investors widely believed that real estate prices would or could never fall, a belief that had been supported by over 30 years of steady and sometimes quite dramatic price increases (www.census.gov/const/uspriceann.pdf). Once this belief was challenged by empirical events to the contrary, panic resulted, as existing models and narratives no longer provided guidance for action. When real estate prices started to fall in 2007/2008, the worst financial crisis since the great depression emerged.

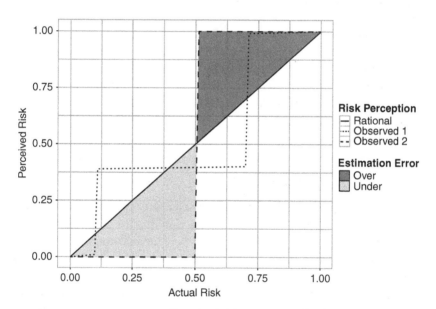

FIGURE 4.1 Perceived risk versus actual risk

"Regulatory Responses" reviews similarities and differences in the responses by members of the general public vs. domain experts and policymakers to risks and changes in perceived risk. One large difference is the way in which the two groups receive and seek out information about the possible outcomes of different risks or risky choice options, either by trial-and-error exposure and personal experience of consequences over time (decisions from experience, more common among the general public) or by statistical description of possible consequences based on existing data or theories and models (decisions from description, more common among technical experts). Decisions from experience put a lot of weight on recent experiences, which explains some of the apparent fickleness on the part of the general public in their perceptions of risk. Some of the biases that describe individual responses to risk and uncertainty, in particular the single action bias (i.e. the tendency to take one protective action against a perceived risk, but to then take down the flag that indicates that some action is necessary) may also explain some failings on the part of policy-makers, for example, the absence of spontaneous ex-post regulatory review. Theories like the garbage can model of organizational decision-making question the application of rational choice model assumptions at corporate or governmental levels and substitute them with more anarchic processes.

"Causes of Changes in Perceived Risk" discusses two different attributions of apparent changes in perceived risk, namely previous failure to accurately assess risk levels or a change in external circumstances (i.e. a regime change). Different attributions call for different solutions. "Reactions to Changes in Perceived Risk" suggests that the presence or absence of trust in regulatory agencies or other groups that can help control risks will play an important role.

FUNCTION OF RISK PERCEPTION

Human perception, judgment, and choice are complex. They have evolved to allow decision-makers to function in a broad range of environments that change with seasons or political regimes and over time, and to satisfy a great variety of often-contradictory goals. Thus human actors need to procure sustenance on a regular basis, which may require exploration and risk-taking, but at the same time need to ensure safety and survival, which require protection and caution. Within the rational actor framework, dominant in policy circles, actions are modeled as the product of deliberations that involve analyses of the probabilistic costs and benefits of different response options (Becker 1976). In contrast, psychological models of action selection allow for a far broader range of triggers and processes (Weber and Johnson 2009), including emotional responses to situations. Proverbial emotions of greed vs. fear or subjective feelings of confidence vs. caution give rise to exploration or novelty seeking vs. retreat to the known and familiar, respectively. In many situations it is the perception of an imminent threat that triggers a switch from exploration and opportunity seeking to self-protective behavior.

Change is inevitable, and adaptation to change is necessary and happens eventually, either on individual or evolutionary time scales. Nevertheless, change is effortful and poses risks, and people dislike it ex-ante, giving rise to a strong status quo bias (Samuelson and Zeckhauser 1988). Motivation to overcome this status quo bias typically comes in the form of a strong affective signal that business as usual is no longer an option, which at an analytical level may be encoded as an increase in perceived risk of the status quo. Bracha and Weber (2012) argue that the human need for predictability and control is central to a psychological account of confidence and to an understanding of perceptions of risk, fear, and panics. Confidence in a system, say in a technology or a financial market, results when citizens or investors believe they understand how things work, which leads to a sense of predictability and being in control (Einhorn 1986) and the perception of low risk (E. U. Weber, Siebenmorgen, and Weber 2005). A feeling of control legitimizes opportunity seeking, that is, reaping benefits without fear of catastrophic losses.

The presence of risk or potential for adverse consequences is a continuous variable, as shown in Figure 4.1. Without professional training in probability theory and probabilistic reasoning, people have a tendency to simplify probabilistic situations and may distinguish only between two (absent or present) or three (absent, moderate, severe) different levels of risk, also shown in Figure 4.1. What this means is that for low levels of risk, existing risks are given less attention and weight than their probability warrants (Hertwig et al. 2004). Events that destroy the sense of predictability and control tend to trigger rapid and drastic shifts in the perception of risk, taking it up to the next or other level of concern, often resulting in overreactions or panics. Behavior at the individual level includes retreat to safe and familiar choice options (be they investment vehicles or technologies) to minimize exposure to perceived danger, until a new account/model of how things work has been established (Bracha and Weber 2012). Real-world examples of such responses can be found in Boxes 4.1 to 4.3. Reassessments of risk at the individual level may, in turn, trigger expert reassessment at the regulatory level to examine whether prior evaluations of risk levels were correct in light based on the new information (the triggering accident or crisis) and/or to diagnose a regime change and identify its causes.

INDIVIDUAL RISK PERCEPTION

Uncertainty refers to a state in which decision-makers are unable to predict what exactly will happen if they engage in a given action. The degree of this uncertainty can vary, with endpoints on a continuum that ranges from no to partial to full information about outcomes and their probabilities (Knight 1921). In economics this is represented by the probability distribution over future states of the world, where "decision under risk" refers to decision made when the probability distribution over future states of the world is known and "decision under uncertainty" or "ambiguity" refers to decisions made when this probability distribution is unknown. The less that is known about future probability distributions of outcomes based on past experience, that is, the greater the degree of uncertainty, the more room there is for individual and situational differences in the assessment of existing risk, as the result of differential attention to either the upside potential of an uncertain outcome distribution (wishful thinking or optimism) or the downside potential (precaution or pessimism) (E. U. Weber 2010).

RISK AS A FEELING

In contrast to the economic or engineering mathematical assessments of risk as the likelihood and severity of adverse events, psychology depicts *risk perception* as an intuitive assessment of such events and their consequences. Popular uses of the term risk often also refer to either the probability of an aversive event (the risk of rain) or its severity (value at risk), rather than some combination of the two. Evidence from cognitive, social, and clinical psychology indicates that risk perceptions are influenced by past experiences that result in associations – connections between objects or events contiguous in space or time, resembling each other, or having some causal connection (Hume 2000) – and affective responses – feelings or emotions – and that these influence risk perception as much or even more than analytic processes (E. U. Weber 2010). Kahneman (initially in his 2003 Nobel address (Kahneman 2003) and more extensively in his 2011 book) has captured decades of behavioral research by characterizing two modes of thinking, called System 1 and System 2. The associative and affective processes that give rise to intuitive perceptions of risk are typical of System 1 thinking, which operates automatically and quickly with no effort or voluntary control, and is available to everyone from an early age. Analytic assessments of risk, on the other hand, are typical of System 2 processes, which work by algorithms and rules such as probability calculus, Bayesian updating, and formal logic. System 2 processes must be taught explicitly and require conscious effort and control, and thus operates more slowly. Even though these two processing systems do not map cleanly onto distinct regions of the brain and often operate cooperatively and in parallel (E. U. Weber and Johnson 2009), Kahneman (2011) argues that the distinction between System 1 and 2 helps to make clear the tension between automatic and largely involuntary processes and effortful and more deliberate processes in the human mind. Psychological research over the past decade has documented the prevalence of System 1 processes in the intuitive assessment of risk, depicting them as essentially effort-free inputs that orient and motivate adaptive behavior, especially under conditions of uncertainty (Finucane et al. 2000; Loewenstein et al. 2001; Peters et al. 2006).

PSYCHOLOGICAL RISK DIMENSIONS

Puzzled by the American public's perception of the riskiness of nuclear power that did not coincide with engineering or public safety estimates of morbidity or mortality risks associated with nuclear-generated vs. other (carbon-based) sources of power, the nuclear power industry in the 1970s commissioned several psychologists to investigate this discrepancy. Slovic identified two psychological risk dimensions that influence people's intuitive perceptions of health and safety risks in ways common across numerous studies in multiple countries and that explain differences between the risk perceptions of members of the general public vs. those of technical experts (Slovic 1987). The first dimension, *dread risk*, captures emotional reactions to hazards like nuclear reactor accidents, or nerve gas accidents. That is, things that make people anxious because of a perceived lack of control over exposure to these events and because their consequences may be catastrophic. The second dimension, *unknown risk*, refers to the degree to which a risk (e.g. DNA technology) is seen as new, with a perceived lack of control due to unforeseeable consequences. Responsiveness to these factors shows that the human processing system maps both the severity and the uncertainty component of the risk of future events into affective responses and represents risk as a feeling rather than as a statistic (Loewenstein et al. 2001), consistent with System 1 processing.

The fact that *dread* and the *unknowability* of a risk increase risk perception provides an explanation for the moderating effect of familiarity on the perceptions of the risk of a hazard or risky choice option, holding objective information about the probability distributions of possible outcomes constant (e.g. E. U. Weber, Siebenmorgen, and Weber 2005). Knowing a certain product, investment, person, or environment gives rise to the feeling of familiarity. Empirical research shows that familiarity not only breeds liking, but also breeds greater comfort, that is, reduces dread and feelings of risk, and increase the feeling of control (E. U. Weber et al. 2005). The association between familiarity and lower levels of risk can be legitimate, when greater personal experience with a risky option (e.g. 20 years of working in a nuclear power plant without any accidents) provides a more reliable data base to assess existing danger), but it may also be spurious, as when a stock simply has a familiar name or is of a local firm (see Huberman 2001).

DETECTING CHANGES IN RISK

Detecting changes in risk can be challenging for multiple reasons. The fact that humans habituate to changes in magnitude or intensity makes gradual change very hard to detect. Ernst Heinrich Weber's Law (1834) specifies the magnitude of a just noticeable difference (JND) for sensory perception and finds that the increase in magnitude necessary to perceive a JND is proportional to the starting value, meaning that greater increments are necessary to detect increases at higher levels.

People's default mental model, at least during periods of stability, is one of perseverance of conditions, meaning that people require a very strong signal to believe that there has been a regime change, that is, that conditions have changed to a regime with either greater or lesser risks.

Few events are deterministic, and the fact that outcomes are often probabilistic makes the detection of regime changes more difficult, as a more negative or positive outcome than expected can also be simply an extreme draw from the distribution of outcomes under the old regime. People's expectations of change (or stability) are important in their ability to detect trends in probabilistic environments, as illustrated by a historic climate example (Kupperman 1982, reported in E. U. Weber 1997). English settlers who arrived in North America in the early colonial period assumed that climate was a function of latitude. Newfoundland, which is south of London, was thus expected to have a moderate climate. Despite repeated experiences of far colder temperatures and resulting deaths and crop failures, colonists clung to their expectations based on latitude and generated ever-more complex explanations for these deviations from expectations. In another example, farmers in Illinois were asked to recall salient growing season temperature or precipitation statistics for seven preceding years (E. U. Weber 1997). Farmers who believed that their region was undergoing climate change recalled temperature and precipitation trends consistent with this expectation, whereas farmers who believed in a constant climate, recalled temperatures and precipitations consistent with that belief. Similarly, Leiserowitz and colleagues (2008) found that differences in political ideology between segments of the US population, associated with beliefs about climate stability shape climate change perceptions.

SOCIAL AMPLIFICATION OF RISK

Social, institutional, and cultural processes have been shown to amplify individual responses to a risk (Kasperson et al. 1988). Such amplification by scientists or engineers who communicate the risk assessment, news media, interpersonal networks, and other groups and institutions occur in the transfer of information about the risk and in the protective response mechanisms of society (Weinstein et al. 2000; Taylor 1983). Evidence from the health literature, the social psychological literature, and the risk communication literature suggests that these social and cultural processes serve to modify perceptions of risk in ways that can both augment or decrease response in ways that are presumably socially adaptive and that constitute a battle for individual and public attention. Media and public attention, just like individual attention, tends to be focused on adverse events ("if it bleeds, it leads;" E. U. Weber 1994), and thus most social amplification has the effect of increasing perceptions of risk. Only occasionally does the absence of widely anticipated adverse events (such as the millennium or Y2K computer program bug that was erroneously predicted to bring down computer systems around the world) garner media attention, leading to an amplification of decreases in perceived risk.

Fundamental worldviews also shape how people select some risks for attention and ignore others. Douglas and Wildavsky (1982) identified five distinct "cultures" (labeled hierarchical, individualist, egalitarian, fatalist, and hermitic) that are said to differ in their patterns of interpersonal relationships in ways that affect perceptions of risk. Hierarchists tend to perceive industrial and technological risks as opportunities and thus less risky, whereas egalitarians see them as threats to their social structure (Dake 1991). Leiserowitz (2006) provides evidence for the value of this approach to understanding group differences in the US in their perceptions of climate change risks. Other researchers trace differences in risk perceptions to differences in fundamental value priorities, following the work of Schwartz (1992) or in worldviews such as the New Ecological Paradigm (Dunlap and Van Liere 1978).

Social amplification can be seen as a process that provides one or multiple narratives for why risk levels have changed and what needs to be done in response. As Frederick W. Mayer's conceptual essay (Chapter 6) for this volume argues, agreement on a narrative contributes to action and shapes the specific dimensions of any response. Multiple competing narratives may increase perceptions of risk but discourage corrective action.

INDIVIDUAL ACTION AND CHOICE UNDER RISK AND UNCERTAINTY

Perception of risk or detection of changes in risk are not an end in themselves, but are signals to motivate protective action, which often suggest changes from business as usual to counteract the status quo bias (Samuelson and Zeckhauser 1988). Different models of risky choice put a different emphasis on the role of perceptions of risk. Neither objective nor subjective risk perception plays any role in expected utility theory (von Neumann and Morgenstern 1944/47), the ruling normative model of risky choice in economics, or in its widely known, behaviorally more descriptive psychological version, prospect theory (Kahneman and Tversky, 1979). However, the risk – return framework of finance provides such a role for the perception of risk. Markowitz (1952) proposed to model people's willingness to pay (WTP) for risky option X as a trade-off between the option's return $V(X)$ and its risk $R(X)$, with the assumption that people will try to minimize level of risk for a given level of return:

$$WTP(X) = V(X) + bR(X)$$

Traditional risk-return models in finance equate $v(X)$ with the expected value of outcomes that can occur under option X and $R(X)$ with the variance of these possible outcomes. Model parameter b describes the precise nature of the trade-off between the maximization of return and minimization of risk and serves as an individual difference index of risk aversion. This model is widely used in finance, for example, in the Capital Asset Pricing Model (Sharpe 1964).

Behavioral extensions of this normative risk – return framework (Sarin and Weber 1993) question the equating of risk with outcome variance. Psychological studies have examined the perception of risk, both directly – by assessing people's judgments or rankings of the riskiness of risky options and modeling these, often using axiomatic measurement models – and indirectly – by inferring the best fitting metric of riskiness from observed choices under the assumption of risk-return trade-offs (see E. U. Weber 2001). These studies are unanimous in their verdict that the variance or standard deviation of outcomes fails to account for perceived risk, that is, for the intuitive feeling of being at risk that people can quantify by judging riskiness of different choice options or action alternatives, for example, on a scale from 0 to 100. Risk judgments deviate from the variance or standard deviation of possible choice outcomes for a variety of reasons. First, deviations above and below the mean contribute symmetrically to the mathematically defined variance, whereas perceptions of riskiness tend to be affected far more by downside variation (e.g. Luce and Weber 1986). Second, variability in outcomes is perceived relative to average returns. A standard deviation of +/- $100 is huge for a risky option with a mean return of $50 and amounts to rounding error for a risky option with a mean return of $1 million. The coefficient of variation (CV), defined as the standard deviation that has been standardized by dividing by the EV, that is,

$$CV(X) = SD(X)/EV(X)$$

provides a relative measure of risk, that is, risk per unit of return. The most important implication of using the CV as a measure of perceived risk for the current discussion is the fact that increases in risk will be harder to detect, the larger the average level of existing risk is, following E. H. Weber's (1834) psychophysical law. The CV is used in many applied domains and provides a vastly superior fit to the risk-taking data of foraging animals and people who make decisions from experience, as discussed in the next section. E. U. Weber, Shafir, and Blais 2004). E. U. Weber et al. (2004) show that simple reinforcement learning models that describe choices in such learning environments predict behavior that is proportional to the CV and not the variance.

Psychophysical risk – return models thus allow for both return (V) and risk (R) to be psychological variables that may need to be directly assessed from members of the general public and that may show individual, group, cultural, or situational differences, rather than just being objective and immutable attributes of risky options (E. U. Weber, Blais, and Betz 2002). Whereas technically trained scientists and policy experts may use the formalizations of risk used in the normative models of their discipline, members of the general public can be expected to assess risk more intuitively and thus be influenced more by the psychological risk dimensions described earlier. Psychophysical risk – return models agree with the risk-return models used in finance on the assumption that preference (including willingness-to-pay) for risky option X is a trade-off between its expected value and its perceived risk:

$$WTP(X) = V(X) - bR(X)$$

This assumption has been verified in numerous studies and forms the basis of a psychometric scale (the Domain SPEcific Risk Taking, or DOSPERT scale, E. U. Weber, Blais, and Betz 2002) that outperforms expected utility-based measures of risk attitude (e.g. the Holt–Laury measure of risk attitude) and other psychometric scales in a broad range of lab and real-world settings (Enkavi et al. 2014).

DECISIONS FROM DESCRIPTION VS. FROM EXPERIENCE

There are important differences in the way people make decisions when information about uncertain choice options comes from repeated personal experience rather than a statistical (numeric or graphic) description of possible outcomes and their likelihood (E. U. Weber, Shafir, and Blais 2004). This distinction between learning about risk and risky outcome distributions from experience versus from description matters because ostensibly the same information about events and their likelihoods can lead to very different perceptions and actions (Hertwig et al. 2004), as the result of the engagement of different psychological processes. Learning from repeated personal experience involves the System 1 associative and affective processes described above, that are fast and automatic, while learning from statistical descriptions requires analytic processing and cognitive effort. Possibly for this reason, when given the choice between attending to information provided in the form of statistical summaries or to information provided by personal experience, personal experience is more likely to capture most people's attention, and its impact dominates statistical information, even though the latter is often far more reliable (Erev and Barron 2005). These findings help to explain a key pattern identified by Thomas A. Birkland and Megan K. Warnement in their conceptual essay in this volume (Chapter 5) – that a repetition of similar crisis events in a given society has a much bigger impact than a single event.

There is evidence that individuals draw different lessons from experience than from description, especially when small probability events are involved. Decisions from description are described well by prospect theory (Kahneman and Tversky 1979; Tversky and Kahneman 1992), which is based on hundreds of studies of choices between described risky options, typically in the form of monetary lotteries. In such choices, decision-makers tend to overweight the impact of small probability events, especially when such events have large positive or negative valence (e.g. a .001 chance of making $5,000, or a .005 chance of brain damage as the side effects of vaccinating against measles). In contrast, decisions from experience follow classical reinforcement learning that gives recent events more weight than distant events (E. U. Weber, Shafir, and Blais 2004). Such updating is adaptive to dynamic environments where circumstances might change. Because rare events (e.g. large financial losses) have a smaller probability of having occurred recently, they tend (on average) to have a *smaller* impact on the decision than their objective likelihood of occurrence would warrant. When they do occur, however, they have a much *larger* impact on related decisions than warranted by their probability. This makes learning and decisions from experience more volatile across respondents and past outcome histories than learning and decisions from description (Yechiam, Barron, and Erev 2005). These reinforcement learning models and their predicted more volatile responses to small probability risks seem to describe the general public's dynamic and fluctuating reactions to small probability risks far better than rational choice models or their psychological extensions like prospect theory.

REGULATORY RESPONSES

The social environment, including its social structures and formal or informal institutions, can be seen as a way of extending individual capabilities and/or correcting for existing individual-level problems or biases (Ostrom 1990; Boyd and Richerson 1985). Given human finite attention and processing capacity and the resulting tendency to allocate scarce capacity to decisions and events close in time and space (E. U. Weber and Johnson 2009), human society has developed a division of labor whereby different individual or social problems that require longer time horizons or greater attention are assigned to designated professional groups, who are charged with developing the required professional knowledge and expertise and with using their available attention to monitor opportunities as well as risks within their designated sphere. Thus epidemiologists and medical researchers are in charge of health risks, climatologists in charge of climate risks, and so on. Similarly, government agencies like the Department of Health or the EPA are designated to take action on behalf of and in the interest of their citizens in their designated domain, in situations where individual knowledge, interest, or attention is deemed inadequate or where individual action would be insufficient to address social problems, as in common-pool resource dilemmas.

Domain-specific regulatory oversight of different sources of risks to individual citizens and the social collective (i.e. some divide and conquer strategy) is as good an idea at the social level as it is at the individual level (Clemens 1997). At the individual level, it is very hard to keep some (let alone an optimal level of) attention on all possible sources of risk. Instead, as worry increases about one type of risk, concern about other risks has been shown to go down, as if people had only so much capacity for worry or a finite pool of worry (E. U. Weber 2006). Illustrations of the finite pool of worry effect are provided by the observation that increases in the concern of the American public about terrorism post-9/11 resulted in decreased concern about other issues such as restrictions of civil liberties as well as climate change (Leone and Anrig 2003), or that the 2008 financial crisis reduced concern about climate change and environmental degradation (Pew Research Center 2009).

Regulatory guidance and oversight in areas of important societal risks is needed from a behavioral decision theoretical perspective not only to supplement individual perceptions of risk, but also to supplement individual action. E. U. Weber (1997) coined the phrase *single action bias* for the following phenomenon observed in contexts ranging from medical diagnosis to farmers' reactions to climate change. Decision-makers are very likely to take a single action to reduce a risk that they are concerned about, but are much less likely to take additional steps that would provide incremental protection or risk reduction. The single action taken is not necessarily the most effective one, nor is it the same for different decision-makers. Regardless of which single action was taken first, decision-makers tend not to take further action, presumably because the first action reduces the feeling of worry or vulnerability. E. U. Weber (1997) found that farmers who showed concern about global warming in the early 1990s were likely to change either something in their production practice (e.g. irrigate), their pricing practice (e.g. ensure crop prices through the futures market), or lobbied for government interventions (e.g. ethanol taxes), but hardly ever engaged in more than one of those actions, even though a portfolio of protective actions might have been advisable. The fear of climate change seemed to set a "flag" that some action was required, but remained in place only until one such action was taken, that is, any single protective action had the effect of taking down the "impending danger flag." While such behavior might have served us well in our evolutionary history, when single actions generally sufficed to contain important risks, in more complex environments where a portfolio of risk management actions is advised, purely affect-driven, single action biased responses may not be sufficient. Hansen, Marx, and Weber (2004) found evidence for the single action bias in farm practices that can be interpreted as protective actions against climate change and/or climate variability. Thus, farmers who indicated that they had the capacity to store grain on their farms were significantly less likely to indicate that they used irrigation or that they had signed up for crop insurance, even though all three actions in combination would provide greater protection against climate risks.

The relationship between public and technocratic perceptions of risk as well as responses to risks or to perceived changes in risk is complex. E. U. Weber and Stern (2011) describe some of the differences between the risk assessment by scientists and non-scientists. Scientists use multiple methods to guard against error in their assessment of causal relationships and uncertainty, including observations and experiments, systematic observation and measurement, mathematical models that incorporate theories and observational data and are tested against new data, systems of checking measurements and peer-reviewing research studies to catch errors, and scientific debate and deliberation about the meaning of the evidence, with special attention given to new evidence that calls previous ideas into question. Scientific communities sometimes organize consensus processes such as those used by the IPCC and in NRC studies to clarify which conclusions are robust and which remain in dispute. Although these methods do not prevent all error, the scientific methods identify unresolved issues and allow for continuing correction of error. Non-scientists' ways of perceiving risks and responding to risk and uncertainty, briefly reviewed above, leave them more vulnerable to systematic misunderstanding. Personal experience can easily mislead (E. U. Weber 1997), mental models of causal relationships can be too simple or wrongly applied (Bostrom et al. 1994), judgment can be driven more by affect, values, and worldviews than by evidence (Slovic 1987), and attention and response can be very selective and incomplete (E. U. Weber and Johnson 2009).

In situations where expert and public perceptions of risk disagree, the way in which the two perceptions affect each other is also complex and typically does not follow a rational model of influence. In other words, while one would expect that people would let their personal perception of risk be informed and influenced by the more comprehensive and systematic expert risk assessment, which they have at least indirectly commissioned, public and media attention and response to risk is typically swayed less by diagnostic personal exposure and memorable events than by statistical summaries or theoretical arguments or models (E. U. Weber 2006; Zaval et al. 2014). At the same time, regulatory bodies often need to respond to public perceptions or changes in public perceptions of risk, even when domain experts disagree with these assessment, because public fear, even when unfounded, has negative consequences for public health and creates barriers to responses or non-responses that might be advocated by technical experts.

Some of the less than functional behavioral patterns in the face of risk described above for members of the general public may also help to explain the behavior of regulatory agencies. The single action bias, for example, may be responsible for the preference or regulators to solve problems in single steps, rather than by a sequencing of gradual interventions, where the results of Step 1 interventions are observed and used to inform Step 2 or 3 interventions. Even if interventions are intended to be an initial step in a sequence of regulatory actions, regulatory attention wanders away from the risk, and the issue becomes "cold." Similar dynamics may contribute to the absence of spontaneous ex-post regulatory review, necessitating the existence of meta-regulatory agencies like the Office of Information and Regulatory Affairs (OIRA) and Presidential Executive Orders (e.g. #13,563 or #13,610).

Some descriptive frameworks for organizational decision-making embrace the assumption that the actions of companies, regulatory agencies, or governing bodies may not be best described as the result of rational deliberation and belief updating as new information about the risks and benefits of different courses of action becomes available. The garbage can model (Cohen et al. 1972), for example, breaks the causal chain between problems and their solutions and assumes instead that organizations tend to produce "solutions" which may be discarded because no appropriate problem currently exists or because the solution is not acceptable in the current political climate. However problems may eventually arise for which a search of the garbage might yield fitting solutions. As Birkland and Warnement (Chapter 5, this volume) emphasize, focusing events that change public and regulatory perceptions of risk, such as Hurricane Katrina in 2005, may have long-term effects, such as the National Flood Insurance Reform Act of 2012.

REASONS FOR CHANGE IN PERCEIVED RISK

There appears to be some asymmetry in the mechanisms and thus the speed with which perceptions of perceived risk change in the direction of decreased vs. increased risks. Perceived risk tends to decrease slowly and steadily, in a continuous fashion, as people fail to experience adverse consequences when engaging in potentially risky activities or when being exposed to potentially risky environments. The mechanism for such decreases in perceived riskiness is the absence of negative feedback in decisions from experience and the increasing familiarity with the sources of potential risk. Familiarity not only breeds liking (and increased choice), but also decreased perceptions of risk (E. U. Weber, Siebenmorgen, and Weber 2005).

Increases in perceived risk, on the other hand, tend to be far more rapid and typically not gradual. Major accidents or financial, public health, or other crises can send a strong signal that prior assessments of risk were too low, either as the result of insufficient information about existing dangers or because the "regime" has changed. Such rapid increases in perceived risk tend to be mediated by emotional rather than analytic assessment, supporting the notion that risk typically is a feeling, rather than a statistic (Loewenstein et al. 2001). Perceived risk increases when the ability to predict and control outcomes in probabilistic environments is put into question.

The need for control is a basic human need (Maslow 1954). Persistent failures to do so can lead to depression and learned helplessness (Seligman 1975), while having a sense of control is associated with better health (Plous 1993). The illusion of control refers to the human tendency to believe we can control or at least influence outcomes, even when these outcomes are the results of chance events. For example, individuals often believe they can control the outcome of rolling dice in a game of craps – throwing the dice hard for large numbers and softly for low numbers (Langer 1975). Outside of the casino, most outcomes require a combination of skill and chance, but the illusion of control also gets people to overestimate their degree of control over adverse consequences in such situations, believing for example that driving is a safer means of transportation than air travel, contrary to accident statistics (Slovic 1987).

The illusion of control is more commonly found in familiar situations and in situations associated with the exercise of skill, for example, situations that provide involvement in the choice and competition (Langer and Roth 1975), and in stressful and competitive situations, including financial trading (Fenton-O'Creevy et al. 2003). Social psychologists argue that the illusion of control is adaptive, since it motivates people to persist at tasks when they might otherwise give up and because there is evidence that it is more common in mentally healthy than in depressed individuals (Taylor and Brown 1988).

New and complex environments or technologies are potential threats, and we manage the perceived risks by forming a mental model of how the new technology and/or environment work. This model gets tested by repeated trial-and-error exposure, that is, by engaging with risky and also rewarding options and by observing resulting outcomes and consequences. The absence of negative consequences and the occurrence of essentially predicted outcomes make us confident in our understanding of how things work and our ability to control adverse consequences. However, both the complexity and riskiness of these new technologies or environments may be underestimated in the face of positive feedback.

Events suggesting that existing mental models might be incomplete or faulty and that beliefs of control are therefore illusory – when individuals or groups realize that they can no longer predict and hence control important (financial, social, or technological) events and outcomes in their lives – trigger rapid and drastic shifts in the perception of risk, often resulting in panics. Such emotional reactions can be seen as an adaptive early warning system, evolution's way to jolt us out of our habitual way of doing things, counteracting our strong status quo bias (Samuelson and Zeckhauser 1988).

Black swan events, that is, the occurrence of something previously considered outside of the plausible range of events, are a signal that our current mental model of the risky or uncertain processes is inadequate or faulty. Hence a reassessment of risks and benefits of different choice options is necessary and short-term protective action may be required. Such a fear or panic reaction in response to a signal indicating that we do not have a correct model of how things work and hence are not able to control consequences essentially reactivates the second psychological risk dimension, discussed above, fear of the unknown, which previously may have been assuaged by repeated personal successful experience with the risky choice options.

Increases in our perceptions of risk are aversive events that motivate us to turn away from newly dangerous technologies or environments and to turn to the old and familiar, whether this means embracing a known technology like coal-generated power with its known risks of climate-changing emissions, or moving from mortgage-backed securities to holding gold. Just as social processes amplify individual responses and reactions during periods of perceived control and (over)confidence, so do social processes also amplify the perceived loss of control and feelings of panic (Kasperson et al. 1988).

REACTIONS TO CHANGES IN PERCEIVED RISK

People's reactions to trigger events indicating that some activity entails much greater risk than previously assumed will depend at least in part on the attribution of this change in perceived risk, that is, whether it is seen as an indication of a regime shift (i.e. that something important in the environment has changed) or as an indication that existing knowledge and control over the potential risk is smaller than previously assumed. The distinction between regime shifts (something changing in the external environment that may or may not have been predictable) and the revelation of incomplete or faulty mental models of the situation is of course not clear cut, but more on a continuum, as a complete and omniscient mental model of the situation would anticipate multiple regimes as well as the reasons and timing of regime shifts.

In addition to withdrawal to known and safe choice options, as discussed above, in either situation, a response to insufficient knowledge about the situation and inadequate appreciation of its risk or its complexity (including the existence of regime changes) will be a public request for safe guards on the one hand and additional research into the existing risk, until better predictive models are in place. If trust has not been irreplaceably lost as the result of the triggering accident or crisis, the general public will turn to the regulatory bodies to which it has outsourced vigilance and action in this particular content domain to provide the necessary remedial research and regulation. If trust has been lost, other more general institutions may come into play, like investigating commissions staffed by trusted organizations like national academies.

CONCLUSIONS

This chapter proposes a psychological account of perceived risk and changes in the perception of risk that applies dual process theory, where

System 1 associative, motivational, and emotional processes (e.g. wishful thinking or fear) influence and often compete with System 2 analytic or deliberative processes (Kahneman 2011). Psychological accounts of the public's and policy experts' responses to risk and changes in perceived risk allow for a much broader range of individual and situational differences in response than the rational actor framework of economics. At the individual level, decision-makers appear to be too strongly influenced by recent events and their perceptions and responses thus too volatile compared to responses based on a rational analysis of events that should trigger reassessments of risk. Social amplification of risk tends to contribute to the discrepancy between public and expert assessments of risk, rather than to a solution. Narratives of sources of risk and of control play a larger role than rational choice models would predict, and thus need to be considered more explicitly, both as obstacles and as tool, in regulatory responses to changes in perceived risk.

References

Becker, G. S. (1976). *The Economic Approach to Human Behavior*. Chicago: University of Chicago Press.

Bostrom, A., Morgan, M. G., Fischhoff, B., and Read, D. (1994). What Do People Know About Global Climate Change? 1. Mental Models. *Risk Analysis* 14, 959–70.

Boyd, R., and Richerson, P. J. (1985). *Culture and the Evolutionary Process*. Chicago: University of Chicago Press.

Bracha, A., and Weber, E. U. (2012). A Psychological Perspective on Control, Uncertainty, Risk, and Panic. Working Paper, Center for Decision Sciences.

Clayton, M. (2011). Germany to Phase Out Nuclear Power. Could the US Do the Same? *Christian Science Monitor*. June 7. www.csmonitor.com/USA/2011/0607/Germany-to-phase-out-nuclear-power.-Could-the-US-do-the-same/.

Clemens, R. T. (1997). *Making Hard Decisions: An Introduction to Decision Analysis*. Boston, MA: Duxbury Press.

Cohen, M. D., March, J. G., and Olsen, J. P. (1972). A Garbage Can Model of Organizational Choice. *Administrative Science Quarterly* 17 (1): 1–25.

Dake, K. (1991). Orienting Dispositions in the Perception of Risk: An Analysis of Contemporary Worldviews and Cultural Biases. *Journal of Cross-Cultural Psychology* 22: 61–82.

Douglas, M., and Wildavsky, A. (1982). *Risk and Culture: An Essay on the Selection of Technological and Environmental Dangers.* Berkeley: University of California Press.

Dunlap, R. E., and Van Liere, K. D (1978). The New Environmental Paradigm: A Proposed Measuring Instrument and Preliminary Results, *Journal of Environmental Education* 9: 10–19.

Einhorn, H. J. (1986). Accepting Error to Make Less Error. *Journal of Personality Assessment* 50(3): 387–95.

Erev, I., and Barron, G. (2005). On Adaptation, Maximization, and Reinforcement Learning among Cognitive Strategies. *Psychological Review* 112(4): 912–31.

Fenton-O'Creevy, M., Nicholson, N., Soane, E., and Willman, P. (2003). Trading on Illusions: Unrealistic Perceptions of Control and Trading Performance. *Journal of Occupational and Organizational Psychology* 76(1):53–68.

Finucane, M. L., Alhakami, A., Slovic, P., and Johnson, S. M. (2000). The Affect Heuristic in Judgements of Risks and Benefits. *Journal of Behavioral Decision Making* 13: 1–17.

Hansen, J., Marx, S., and Weber, E. U. (2004). *The Role of Climate Perceptions, Expectations, and Forecasts in Farmer Decision Making: The Argentine Pampas and South Florida.* Technical Report 04–01. Palisades, NY: International Research Institute for Climate Prediction (IRI).

Hertwig, R., Barron, G., Weber, E. U., and Erev, I. (2004). Decisions from Experience and the Effect of Rare Events in Risky Choice. *Psychological Science* 15(8): 534–39.

Huberman, G. (2001). Familiarity Breeds Investment. *Review of Financial Studies* 14 (3): 659–80.

Hume, D. (2000). *An Enquiry Concerning Human Understanding: A Critical Edition.* Oxford University Press.

International Atomic Energy Association. (2008). 50 Years of Nuclear Energy. www.iaea.org/About/Policy/GC/GC48/Documents/gc48inf-4_ftn3.pdf.

Kahneman, D. (2003). "Maps of Bounded Rationality: Psychology for Behavioral Economics." *American Economic Review* 93(5): 1449–75.

Kahneman, D. (2011). *Thinking, Fast and Slow.* New York: Farrar, Straus and Giroux.

Kahneman, D., and Tversky, A. (1979). Prospect Theory: An Analysis of Decision under Risk. *Econometrica* 47(2): 263–92.

Kasperson R. E., Renn, O., Slovic, P., Brown, H. S., Emel, J., Goble, R., Kasperson, J. X., and Ratick, S. (1988). The Social Amplification of Risk: A Conceptual Framework. *Risk Analysis* 8: 177–87.

Katz, D. (2011). Hormone-Replacement Follies: A Brief History. *Huffington Post*, April 13. www.huffingtonpost.com/david-katz-md/hormone-replacement-_b_847884.html.

Knight, F. H. (1921). *Risk, Uncertainty, and Profit*. Washington, DC: Beard Books (reprinted 2002).

Kupperman, K. O. (1982). The Puzzle of the American Climate in the Early Colonial Period. *The American Historical Review* 87: 1262–89.

Langer, E. J. (1975). The Illusion of Control. *Journal of Personality and Social Psychology* 32(2): 311–28.

Langer, E. J., and Roth, J. (1975). Heads I Win, Tails it's Chance: The Illusion of Control as a Function of the Sequence of Outcomes in a Purely Chance Task. *Journal of Personality and Social Psychology* 32(6): 951–55.

Leiserowitz, A. (2006). Climate Change Risk Perception and Policy Preferences: The Role of Affect, Imagery, and Values. *Climatic Change* 77: 45–72.

Leiserowitz, A., E. Maibach, C. Roser-Renouf, Y.U.S. of Forestry, E. Studies, Y.P. on C. Change, and G.M.U.C. for C.C. Communication (2008). *Global Warming's" Six America": An Audience Segmentation*. New Haven: Yale School of Forestry and Environmental Studies.

Loewenstein, G., Weber, E. U., Hsee, C. K., and Welch, N. (2001). Risk as Feelings. *Psychological Bulletin* 127(2): 267–86.

Leone, R. C. and Anrig, Jr., G. (2003). *The War on Our Freedoms: Civil Liberties in an Age of Terrorism*. New York: Century Foundation.

Luce, R. D., and Weber, E. U. (1986). An Axiomatic Theory of Conjoint, Expected Risk. *Journal of Mathematical Psychology* 30: 188–205

Markowitz, H. M. (1952). Portfolio Selection. *Journal of Finance* 7: 77–91.

Maslow, A. H. (1954). *Motivation and Personality*. New York: Harper.

Ostrom, E. (1990). *Governing the Commons: The Evolution of Institutions for Collective Action*. Cambridge, UK: Cambridge University Press.

Peters E., Västfjäll, D., Gärling, T., and Slovic, P. (2006). Affect and Decision Making: A "Hot" Topic. *Journal of Behavioral Decision Making* 19: 79–85. doi: 10.1002/bdm.528.

Pew Research Center (2009). *Public Praises Science; Scientists Fault Public, Media*. Washington, DC: Pew Research Center.

Plous, S. (1993). *The Psychology of Judgment and Decision Making*. New York: McGraw-Hill.

Samuelson, W., and Zeckhauser, R. (1988). Status Quo Bias in Decision Making. *Journal of Risk and Uncertainty* 1(1): 7–59.

Sarin, R. K., and Weber, M. (1993). Risk–Value Models. *European Journal of Operations Research* 70: 135–49.

Schwartz, S. H. (1992). Universals in the Content and Structure of Values: Theory and Empirical Tests in 20 Countries. In M. Zanna (ed.), *Advances in Experimental Social Psychology* (Vol. 25). New York: Academic Press, pp. 1–65.

Seligman, M. E. P. (1975). *Helplessness: On Depression, Development and Death*. San Francisco: Freeman.

Sharpe, W. F. (1964). Capital Asset Prices: A Theory of Market Equilibrium under Conditions of Risk, *Journal of Finance*, 19(3): 425–42.

Slovic, P. (1987). Perception of Risk. *Science* 236: 280–85.

Taylor, S. E. (1983), Adjustment to Threatening Events: A Theory of Cognitive Adaptation, *American Psychologist*, 38 (11): 1161–73.

Taylor, S. E., and Brown, J. D. (1988). Illusion and Well-Being: A Social Psychological Perspective On Mental-Health. *Psychological Bulletin* 103(2): 193–210.

Tversky, A., and Kahneman, D. (1992). Advances in Prospect Theory: Cumulative Representation of Uncertainty. *Journal of Risk and Uncertainty* 5(4): 297–323.

von Neumann, J., and Morgenstern, O. (1944/1947). *Theory of Games and Economic Behavior*. Princeton: Princeton University Press.

Weber, E. H. (1834). *De pulsu, resorptione, auditu et tactu. Anatationes anatomicae et physiologicae*. Leipzig: Koehler.

Weber, E. U. (1994). From Subjective Probabilities to Decision Weights: The Effect of Asymmetric Loss Functions on the Evaluation of Uncertain Outcomes and Events. *Psychological Bulletin*, 115: 228–42.

Weber, E. U. (1997). Perception and Expectation of Climate Change: Precondition for Economic and Technological Adaptation. In M. Bazerman, D. Messick, A. Tenbrunsel, and K. Wade-Benzoni (eds.), *Psychological Perspectives to Environmental and Ethical Issues in Management*. San Francisco, CA: Jossey-Bass, pp. 314–41.

Weber, E. U. (2001). Decision and Choice: Risk, Empirical Studies. In N. J. Smelser and P. B. Baltes (eds.), *International Encyclopedia of the Social and Behavioral Sciences*. Oxford, UK: Elsevier Science Limited, pp. 13347–51.

Weber, E. U. (2006). Experience-Based and Description-Based Perceptions of Long-Term Risk: Why Global Warming Does Not Scare Us (Yet). *Climatic Change* 77: 103–20.

Weber, E. U. (2010). Risk Attitude and Preference. *Wiley Interdisciplinary Reviews: Cognitive Science* 1(1): 79–88.

Weber, E. U., Blais, A. -R., and Betz, N. (2002). A Domain-Specific Risk-Attitude Scale: Measuring Risk Perceptions and Risk Behaviors. *Journal of Behavioral Decision Making* 15: 263–90

Weber, E. U., and Johnson, E. J. (2009). Mindful Judgment and Decision Making. *Annual Review of Psychology* 60: 53–86.

Weber, E. U., Shafir, S., and Blais, A.-R. (2004). Predicting Risk-Sensitivity in Humans and Lower Animals: Risk as Variance or Coefficient of Variation. *Psychological Review* 111(2): 430–45.

Weber, E. U., Siebenmorgen, N., and Weber, M. (2005). Communicating Asset Risk: How Name Recognition and the Format of Historic Volatility Information Affect Risk Perception and Investment Decisions. *Risk Analysis* 25(3): 597–609.

Weber E. U., and Stern, P. C. (2011). Public Understanding of Climate Change in the United States. *American Psychologist* 66: 315–28.

Weinstein, N. D., Lyon, J. E., Rothman, A. J. and Cuite, C. L. (2000). Preoccupation and Affect as Predictors of Protective Action Following Natural Disaster. *British Journal of Health Psychology*, 5: 351–63.

Yechiam E., G. Barron, and I. Erev (2005). The Role of Personal Experience in Contributing to Different Patterns of Response to Rare Terrorist Attacks. *Journal of Conflict Resolution* 49: 430–39.

Zaval, L., Keenan, E. A., Johnson, E. J., and Weber, E. U. (2014). How Warm Days Lead to Increased Belief in Global Warming. *Nature Climate Change*. 4(2): 143–47.

5

Focusing Events, Risk, and Regulation

Thomas A. Birkland and Megan K. Warnement

FOCUSING EVENTS IN THE AGENDA-SETTING PROCESS

In recent years, several major disasters, crises, and catastrophes have occurred in the United States and around the world. The September 11 attacks are the most consequential event of the twenty-first century thus far, changing the course of national and international history and policy in a way unseen since the outset of World War II. These attacks spurred a series of far-reaching statutory, regulatory, and political initiatives the effects of which continue to be discussed and debated to this day. But we need not look only to September 11 to see crises that generated heightened attention to particular regulatory problems and greater pressure for policy change. Events like Hurricane Katrina and the 2008 Global Financial Crisis prompted intense public interest in debates over how to mitigate, respond to, and recover from so-called natural disasters, and how to prevent macro-financial instability. The case studies in this volume all engage with similar highly consequential events that reshaped policy discourse and often generated far-reaching transformations in policy.

In the public policy literature, these sudden crises are often called *focusing events*. The purpose of this chapter is to analyze the agenda-setting properties of focusing events and to consider their potential effect on legislative and regulatory policy change. We illustrate this concept through a discussion of civil aviation regulation, which nicely conveys that the way that a series of focusing events can structure the evolution of regulatory policy over a generational time span. The chapter begins by teasing out the key characteristics of focusing events. We then present empirical applications of the focusing event idea to aviation incidents, and include a risk-based explanation of focusing events. Finally we conclude with the effects such events have the selection of policy alternatives.

FOCUSING EVENTS AND AGENDA SETTING

Focusing events help to structure the agenda-setting process in any complex policy-making organization, as some issues gain and others lose attention among policy-makers and the public. As several political scientists have observed, agenda-setting matters greatly because "the definition of the alternatives is the supreme instrument of power" (Schattschneider 1975, 66), whether "alternatives" refers to issues/problems that require attention or the range of acceptable solutions to those issues/problems. In democratic societies, groups – or *advocacy coalitions* (Sabatier, Jenkins-Smith, and Lawlor 1996; Sabatier and Jenkins-Smith 1999) engage in rhetorical battles, in different venues, to define dominant policy agendas, while simultaneously attempting to deny or limit such influence by other actors (Cobb and Ross 1997; Hilgartner and Bosk 1988). Group competition tends to be fierce because the agenda space is limited by individual and organizational constraints on information processing, so that no system can accommodate all issues and ideas (Walker 1977; Baumgartner and Jones 1993; Cobb and Elder 1983). The jockeying for influence occurs with regard to the identification of which problems are most important, and over what causes and solutions surround any one problem (Hilgartner and Bosk 1988; Lawrence and Birkland 2004; Birkland and Lawrence 2009). The agenda-setting process therefore acts as an institutional filter, sifting issues, problems, and ideas, and implicitly assigning priorities to these issues. John Kingdon (2003) argues that agenda change is driven by two broad phenomena: changes in indicators of underlying problems, which lead to debates over whether and to what extent a problem exists and is worthy of action; and *focusing events*, or sudden shocks to policy systems that lead to attention and potential policy change.

For Kingdon the policy process in industrialized democracies is comprised of three conceptual streams: a *problem* stream, which contains ideas about various problems in society to which public policy might be applied; the *politics* stream, containing the ebb and flow of electoral politics, public opinion, and the like; and the *policy* stream, which contains a set of ideas about how problems could be addressed, either in general or with regard to some specific issue area. In all three streams, problems and solutions, and the means by which solutions can be implemented are not self-evident; they are subject to considerable debate and conflict. Policy change opportunities arise when two or more streams come together at a moment in time where problems are matched with solutions, and where politics aligns in a way that makes this matching more likely.

Focusing events provide an urgent, symbol-rich example of what many would argue is an obvious policy failure, requiring what pro-change forces would argue is the necessity for rapid policy change to prevent the recurrence of the recent disaster. Of course, other interests may argue that an event is atypical, that existing systems can and will bring the acute problems under control, and that sweeping change is not necessary. Regardless of the ultimate outcome, we can argue that pressure for policy change does increase in most focusing events, and that debate over ideas increases, both among policy-makers and in the attentive public, to the extent that an attentive public exists. Focusing events may trigger greater attention to problems and solutions because they increase the likelihood of more influential and powerful actors entering the conflict on the side of policy change (Schattschneider 1975; Baumgartner and Jones 1993), by way of greater claims of policy failure and a more active search for solutions, leading to a greater likelihood of policy change (Birkland 2006). Often this increasing pressure for action is coordinated by one or more policy entrepreneurs, who seize on the event to push long-standing policy preferences, or who become newly sensitized to issues highlighted by the event.

For Kingdon focusing events represent any "little push" to the policy process, which could be "... a crisis or disaster that comes along to call attention to the problem, a powerful symbol that catches on, or the personal experience of a policy-maker." At times, Kingdon has argued that focusing events gain their agenda-setting power by aggregating their harms in one place and time: one plane crash that kills 200 people will get more attention than 200 single fatal accidents; one big terrorist attack in one place will get more attention than several hundred smaller events, particularly if they are far away. In this strand of his thinking, sudden events "simply bowl over everything standing in the way of prominence on the agenda" (Kingdon 2003, 96),

At other times, however, Kingdon portrays focusing events as being more varied and subtle, and includes in his definition personal experiences of policy-makers with matters of personal interest, such as diseases that affect them or their families. Kingdon also argues that "the emergence and diffusion of a powerful symbol" is a focusing event that "acts ... as reinforcement for something already taking place" (2003, 97). When symbols of events propagate – an elderly woman in New Orleans sheltering herself from the rain with an American flag, the raising of the flag at Ground Zero in a manner reminiscent of the raising of the flag at Iwo Jima, or the images of oil-soaked wildlife after oil spills – these symbols would not have their power if it were not for *the event itself*. In other words, the symbols alone may have some power, but their shared interpretation is deeply shaped by the event itself and the discourses that surround it. The propagation of the symbol amplifies the focusing power of the event, particularly if the symbol is particularly evocative (Birkland and Lawrence 2002).

Perhaps because of Kingdon's variable uses of "focusing event," the term appears with many guises in scholarship on agenda setting. Social scientists often conflate sudden crises and shocks with individual experience and symbol propagation that are often reflective of what Peter May (1992) calls political learning about more effective policy arguments. For example, Kingdon's inclusion of political events as focusing events, such as the 1963 March on Washington for Jobs and Freedom, conflates the role of political mobilization in the politics stream with the revelation and depiction of problems in the problem stream. It makes more sense to see "shocking" and sudden focusing events as not purposefully caused, and as exogenous to a policy community or domain, even if the policies and practices adopted by key members of that community make such events more likely.

Birkland (1997; 1998) has developed a more defensible definition of a *potential focusing event* as an event that is sudden, relatively rare, can be reasonably defined as harmful or revealing the possibility of potentially greater future harms, inflicts harms or suggests potential harms that are or could be concentrated in a definable geographical area or community of interest, and that is known to policy-makers and the public virtually simultaneously (1997, 22).

The term *potential* means that we may "know" intuitively that an event will gain a lot of attention; it is difficult to know in advance how *focal* that event will be. Will the event make a more significant and discernible difference on the agenda than competing events and issues? Will that event have any discernible influence on policy changes? We cannot know the answers *ab initio*, but we can certainly analyze the history of similar events to make preliminary guesses.

A focusing event, by definition, draws increased attention to a public issue or problem. Baumgartner and Jones (1993) note that increased attention to a policy problem is usually *negative* attention, and negative attention often yields further political debate, thereby moving issues closer to potential policy changes. To understand focusing event politics, one must consider the use of rhetoric, language, stories, metaphors, and symbols in conjunction with the event. In this "social constructionist" thread of research, students of the media explain how issues gain the attention of journalists, how journalists and their sources use symbols and stories to explain complex issues, and how news consumers respond to these issues and symbols (Edelman 1967; Edelman 1988; Stone 1989; Stone 2002; Majone 1989; Schneider and Ingram 1991). Related to this is research on problem framing in government and the mass media (Entman 1993; Burnier 1994; Lawrence 2000), which has been profitably applied to studies of focusing events (Glascock 2004; Gunter 2005; Liu, Lindquist, and Vedlitz 2011). Framing theory, discussed in some detail by Frederick W. Mayer's chapter for this volume (Chapter 6), argues that the participants in policy debate frame stories about problems to fulfill newsgathering routines designed to make story both efficient and compelling (Bennett 2003), and to motivate action by its supporters, inaction by its opponents, or both. These symbols are often promoted by a policy entrepreneur – an individual who has credibility in a given policy community because of technical expertise, political savvy, and the ability to broker deals that lead to new programs and policies (Kingdon 2003; Mintrom 1997; Mintrom and Vergari 1998).

As we have already intimated, the politics of focusing events is firmly based in studies of group coalescence and mobilization. Groups coalesce to form advocacy coalitions based on mutual interests and values. The political scientist Paul Sabatier predicts that two to four such coalitions will form in most policy domains. However, some domains prone to sudden disasters may be characterized as "policies without publics" (May 1990), in which policy-making is often the domain of technical experts, with little public mobilization around particular policy changes. These domains are characterized by one advocacy coalition, which may be quite weak (see, e.g., Birkland 1998). Thus one may usefully see focusing events as occurring on a spectrum of notice and impact. Some occurrences prompt an "internal mobilization" effort among existing group members within a relatively small and cohesive policy domain, but do not make too many waves among major media outlets, powerful interest groups, or the broader public. In these cases, focusing events may yield internal mobilization efforts to promote the change that policy elites or experts prefer (Cobb and Elder 1983). Other events may receive wall-to-wall media coverage and reshape policy debates throughout a society. Thus in the United States, the earthquake and hurricane policy domains were very much characterized by an internal mobilization model of agenda setting and policy change, while in the oil spill and nuclear power policy domains, there were discernible advocacy coalitions that were engaged in direct competition to shape policy outcomes (Birkland 1997).

One must be careful not to overstate the role of focusing events in generating new policy ideas, as opposed to the role of heightened political space for consideration of those ideas. Among the public there is widespread belief that the Three Mile Island (TMI) nuclear power plant accident in 1979 was the event that stalled the movement toward more nuclear power plants for generating electricity. But public support for greater regulatory scrutiny of nuclear power predated the TMI accident, and even before that event, the nuclear power policy domain was highly polarized (Baumgartner and Jones 1993; Birkland 1997). All the participants in the nuclear policy debate could use the accident at TMI for their own rhetorical ends: as evidence either for the "defense in depth" notion of nuclear safety, or for the idea that nuclear power contained unknown risks, that other incidents were equally or more serious than TMI, and that heretofore poorly understood systems' accidents could yield catastrophic disasters (Perrow 1999).

Apart from group efforts to expand issues, major events often reach the agenda without group promotion through media propagation of news and symbols of the event. This coverage of a sudden and shocking event makes the public aware of these events with our without efforts on the part of group leaders to induce attention. This media propagation of symbols gives less powerful groups advantages in policy debates that they may not otherwise possess. Pro-change groups are relieved of the obligation to create and interpret powerful images and symbols of the problem. Rather, groups only need repeat the already existing symbols that the media have seized upon as emblematic of the current crisis. These obvious symbols are likely to carry more emotional weight than industry or governmental assurances that policy usually works well. In 1938, for example, photographic images of fragile American children who had been poisoned by sulfanilamide triggered a strong outcry that lead to more stringent regulation and licensing for medications in the Food and Drug Act of 1938 (Carpenter and Sin 2007). Thus, media-generated symbols of health, environmental, or other crises or catastrophes are often used by groups as an important recruiting tool – thereby expanding the issue – and as a form of evidence of the need for policy change.

The politics of agenda setting, then, is not a neutral, objective, or wholly rational process. Rather, it is the result of a society, acting through its political and social institutions, building a consensus on the meanings of problems and the range of acceptable solutions. There are many possible constructions that compete with each other to tell the story of why a problem is a problem, who is benefited or harmed by the problem, whose fault it is, and how it can be solved.

A HISTORICAL INSTITUTIONAL PERSPECTIVE ON FOCUSING EVENTS

Political institutions play an important role in both providing opportunities and constraints for policy change following a focusing event (Hacker 1998). A historical institutional perspective helps explain this phenomenon since it is "grounded in the assumption that political institutions and previously enacted public policies structure the political behavior of bureaucrats, elected officials and interest groups during the policy-making process" (Béland 2005, 3). This approach requires attention to both path dependency and policy feedback, which together generate a "historical sequence" of political decisions that structures the options available in the policy debate. As Béland and Hacker (2004) have argued, "a central reality of politics is often it is the sequence and timing of an event or decision that is most crucial determinant to policy outcomes" (46).

Indeed, because "organizations are not smoothly adaptive to changing circumstances" (Jones 2001, 4), purely event-driven policy change may be quite rare, with incremental changes being filtered by pre-existing ideological or epistemological commitments, even when events compel some sort of policy action. Policy-makers also frequently respond to focusing events by borrowing from the "lesson drawing" in nearby or similar jurisdictions so as to avoid the steep information costs inherent in innovation – in other words, settling for "satisficing." Furthermore, the types of problems that governments typically address are the "hard," complex problems that lack consensus, easy solutions and are comprised of many underlying attributes and involve entrenched interests that place a high priority on particular policy outcomes. These "hard" problems also lead to path dependency as "each step in the process adds constraints, so that the history of attempts to solve the problem affect current opportunities to solve it" (Jones 2001, 76). Events are unlikely to overcome these forces unless they are sufficiently severe to overcome these constraints.

Furthermore, individuals and groups have limited information-processing capacity, and therefore no system can pay attention to all issues and ideas that are prevalent in society (Baumgartner and Jones 1993; Cobb and Elder 1983; Walker 1977). This agenda-setting process then represents a process of identifying the most prevalent problems and selecting appropriate solutions (Birkland and Lawrence 2009; Hilgartner and Bosk 1988; Lawrence and Birkland 2004). It is essentially a triage process. This process of triage is constant, and the "half-life" of many issues is likely to be quite short, as other pressing matters also intrude on the agenda and compete for elite and public attention. But while the half-life of issues may be short and yield long-term inattention, in the short term, policy actors can become overly attentive to new information out of proportion to its broader meaning about a class of social phenomena (Jones 2001). Nonetheless, events can sometimes have an immediate influence on policy, given the right circumstances. And even if they do not initially have a detectable influence on policy, they may still contribute to experience about the problems revealed by events and drive substantive policy learning. The resulting insights may then shape policy formulation down the line, after new focusing events once again widen the opportunities for policy change or after the ramifications of earlier events become clearer.

Effective study of event-driven policy change, then, must not focus solely on events in isolation: it should consider a sequence of related events within the context of a policy domain or, perhaps, a policy regime (May, Jochim, and Sapotichne 2011). As Sabatier (1993) has persuasively noted, the effective study of policy change in most important contexts can only occur over a period of a decade or more. The research question that one must address, then, in studying event influence on policy agendas, is not whether a particular event had an influence on policy change (although such a question is interesting, in context). Rather, the more fruitful question is whether one or more events in a domain, considered over a sensible time frame, had an influence on policy change, and if so, when and how. Conversely, when a series of events does not lead to policy change, though they do clearly influence policy debate, it should raise questions about the institutional and political factors stymying action.

Undertaking such research requires attention to the impact of past events on problem framing, which tends to have a significant influence on the consideration of potential policy solutions (Cohen, March, and Olsen 1972; Jones 2001; Kingdon 2003). The initial framing of any such event, of course, like the attempts to identify proper solutions to the issues that policy elites highlight, typically reflects political conflict and divergent understandings of causation. As Deborah Stone has argued,

> In politics, causal theories are neither right nor wrong, nor are they mutually exclusive. They are ideas about causation, and policy politics involves strategically portraying issues so that they fit one casual idea or another. The different sides in an issue act as if they are trying to find the "true" cause, but they are always struggling to influence which idea is selected to guide policy. Political conflicts over causal stories are therefore more than empirical claims about sequence of events. They are fights about the possibility of control and the assignment of responsibility (2002, 197).

The framing of previous events, then, generally shapes policy-makers' definitions of current events and their implications for policy options. Any given causal narrative, moreover, "is more likely to be successful if its proponents have visibility, access to the media, and prominent positions and if it accords with widespread deeply held cultural values" (Stone 2002, 203). Finally, the realities of bounded rationality offer another reason to think about the impact of multiple focusing events over the time frame of a decade or more. Decision-makers cannot know *every* outcome of *every* policy choice, and determine the "optimal path" for policy. Instead, policy-makers often engage in "satisficing" (Simon 1957) in systems characterized by incremental decision-making (Lindblom 1959; 1979) or path dependency. As a result, a series of events that seem to demonstrate a pattern has more impact than simply one event on its own.

SOME EMPIRICAL APPLICATIONS OF FOCUSING EVENT THEORY

The formulation of American airline safety and security regulation since the mid-1980s demonstrates the importance of considering policy reforms along a generational time scale, with close attention to the cumulative impact of focusing events. In this policy arena, various incidents prompted extensive learning and formulation of new ideas. Those proposals often did not immediately translate into new policies. But they laid the intellectual groundwork for the more dramatic shifts that occurred after September 11, 2001 (Cobb and Primo 2003). Prior events, such as the bombing of Pan Am 103 and the crash of TWA Flight 800 in 1996, allowed for comprehensive debate of various regulatory proposals so that when the 9/11 attacks occurred, many of these ideas were already "on the shelf" (Birkland 2004). These prior events served in a sense as "dress rehearsals" in terms of raising ideas to the agenda, which greatly enabled the sweeping changes enacted after 9/11 (Birkland 2004, 356).

In the Appendix to this chapter, we provide an extensive list of focusing events that posed questions about the American aviation safety and security domain from the 1980s through the 2000s. Although there are many events in this list, they can be categorized, for analytic purposes, into four overlapping stories:

1. Aviation *safety*, with the cause of the safety problem being an underappreciated hazard inherent to the aircraft that had not been properly mitigated (TWA 800; United 585 and US Airways 427, American 587).
2. Aviation *safety*, in which pilot or organizational error created the conditions under which the incident could occur (ValuJet 592, American 587).
3. Aviation *security*, in which criminals or terrorists interfered with the normal operations of (Pam Am 103, September 11)
4. Events that looked like terrorism, but that turned out to be category 1 events (TWA 800).

All these incidents, with the exception of two rudder deflection incidents (UA 585 and US 427), gained considerable attention in aviation safety and security circles, and led to considerable attention to particular problems in enacted legislation in the United States, as shown in Table 5.1. In most cases, the incidents also prompted at least some policy changes.

The legislation in this table corresponds with the major aviation security events of from 1988 to 2001, starting with Pan Am 103. It suggests that policy-makers were responsive to the key issues raised by investigations into Pan Am 103, such as baggage matching with passengers. The 1996 legislation similarly reflected analyses of TWA 800, although some of the legislation in that act spoke to broader concerns about aviation safety without sound knowledge of the cause of the ValuJet crash. But most interesting from the perspective of this chapter is the degree to which the September 11 attacks brought together all these disparate ideas for legislation into one much more sweeping bill that not only changed the law with respect to the proximate cause of the September 11 hijackings and cockpit intrusions, but also changed the law with respect to the already-known modes for attacking commercial airplanes. One can argue, then, that the prior events were rehearsals for September 11 in the sense that the proponents of these disparate ideas were able to hone their ideas and then were able to use the September 11 attacks as reasons

TABLE 5.1 *Key issues addressed in aviation security legislation*

Topic	Aviation Security Improvement Act of 1990	Federal Aviation Authorization Act of 1996	Federal Aviation Administration Authorization Bill	Airport Security Improvement Act of 2000	Aviation and Transportation Security Act	Homeland Security Act of 2002
Airport access control	•	•			•	
Baggage matching	•	•			•	
Background checks of employees	•	•			•	
Cargo and mail security	•	•			•	
Cockpit security					•	
Employee ID systems					•	
Explosives and explosives detection	•	•			•	•
Create or restore air marshals					•	•
Modify existing organizations	•				•	•
Create new organizations						•
Passenger profiling		•			•	
Allow pilots to carry fatal weapons						•
Allow pilots to carry non-fatal weapons					•	
Certification of screening companies		•				
Require screening personnel be US citizens					•	
Require all airport personnel be screened (including flight crews)					•	
Screeners – general issues	•				•	
Make screeners federal employees		•			•	
Provide security training to the aircrew					•	
Provide security training to pilots					•	
Total	7	8	1	1	17	5

Source: Birkland (2004).

why their ideas should be legislated. The September 11 attacks thoroughly opened up the whole range of issues for review; far more than the previous incidents, those attacks received unrelenting media coverage and a political consensus that far-reaching policy reforms were in order.

In this case, an overwhelming focusing event (9/11) provided the opportunity for policy entrepreneurs to engage in the "selling" of their ideas in the "marketplace of ideas" for solutions to various problems revealed by that focusing event as well as several earlier smaller-scale focusing events.

A RISK-BASED EXPLANATION FOR FOCUSING EVENT POLITICS

We turn now to a further discussion of the relative salience of focusing events. Birkland's definition of a "focusing event" includes the critical notion of "harm." If the event was not harmful or did not raise the specter of potential harms in the future, it would be unlikely to significantly influence agenda setting. The idea of harm relates directly to the literature on risk perception, discussed at length by Elke U. Weber in this volume (Chapter 4). Paul Slovic, through his distinction between "dread risk" and "unknown risk," has been crucial in shaping our understandings of risk perception. Dread risk is a person's "perceived lack of control, dread, catastrophic potential, fatal consequences, and the inequitable distribution of risks and benefits." Unknown risks are the "hazards judged to be unobservable, unknown, new, and delayed in their manifestation of harm" (Slovic 1987, 283). Dread risk has the greatest potential to lead to increased attention. Indeed, Slovic argues that events that relate to dread risks place considerable pressure on regulators to more strictly regulate the processes involved in the recent event; in short, people press regulators to "fix the problem." More dread risks will yield greater pressure for regulation. The concept of "dread risk" can help to explain why some types of disasters issues get more attention than others.

The matrix in Figure 5.1 arranges risk into dimensions of relative dread risks and relatively known or unknown risks. Based on psychometric research, Slovic and his collaborators have shown that many activities that carry some risk, such as driving cars or riding bicycles, elicit little dread and are well known, and so carry minimal emotional punch. Some risks that are well known but still elicit considerable dread, such as crime and warfare, carry more emotional punch. Still other risks, like food irradiation, are unknown in terms of risk but not particularly "scary."

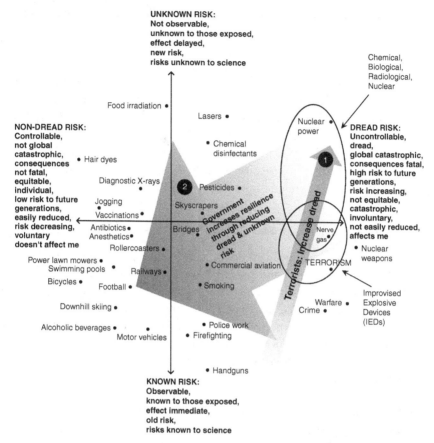

FIGURE 5.1 Risk perception matrix and the competing influences of terror
Source: Taken from Sheppard (2011). This figure was originally adapted from Paul Slovic, Baruch Fischhoff, and Sarah Lichtenstein, "Facts and Fears: Understanding Perceived Risks" (pp. 181–214) in Richard C. Schwing and Walter A. Albess, Jr. (eds.), Societal Risk Assessment: How Safe Is Safe Enough? New York/ London: Plenum Press, 1980, p. 201, Figure 5.

Figure 5.1 demonstrates that the act of terrorism increases the dread risk of an action or a hazard. Therefore, the dread risk of commercial aviation increases more following an act of terrorism than it does following an aviation safety incident. Since greater dread risk equates to greater attention to an event and a call for regulatory action, the implications of this can be seen in terms of the regulatory change following acts of terrorism. Evidence of this can be seen in Table 5.1 and how the September 11 attacks were a stronger catalyst for media coverage and policy change than previous aviation safety events.

Bringing together all the concepts we have discussed this far, we argue that a focusing event can lead either to a *narrowing* of the range of immediately acceptable solutions to a problem, or a *sudden expansion* of the types of ideas being discussed. This degree of consensus/dissensus is a function of the risk perceptions that surround the problem. The greater the "dread" risk, the more likely there will be greater attention to an event and call for regulatory action. Indeed, as Slovic notes, risks that rank higher on the "dread" dimension are usually accompanied by more intensive demands for the regulation of those risks.

THE MEANING OF FOCUSING EVENTS FOR REGULATION

As we have noted, focusing events generally reveal policy failures – which are the fodder for interest group debate, because these failures reveal the problems that policy entrepreneurs join to "solutions" in the post-event window of opportunity, whether or not that opportunity proves wide enough, or the policy entrepreneurship savvy enough, to facilitate substantive policy change. As the policy arena of American airline safety and security policy suggests, this process can generate an accumulation of event-related knowledge. We conclude by considering what this generational approach to the impact of focusing events mean for regulatory politics. Our consideration of the regulatory politics of disasters and other focusing events rests on a few assumptions. The first, and most important, is that "regulation" is *any* intervention into the social or economic system by government to achieve a particular public policy goal, such as safety or market competition. Thus, regulation can derive from any institution of government.

Regulation is not, we assume, simply the province of regulatory agencies, such as the FAA or the FDA. Of course, many other agencies do issue regulations relating to the prevention and management of disaster, but, as a matter of policy and professional doctrine, most of this regulatory activity is focused in the states. We are now beginning a project to understand whether this notion of "dread risks" can help explain the considerable regulatory and political differences between the regulation of commercial aviation safety and commercial aviation security. These areas do, of course overlap considerably. But they are also quite different classes of events. A terrorist incident is intentional; an aviation accident is, by definition, unintentional. The "causes" of the demise of an airplane due to technical faults or human error are often well understood, either before or after the investigation of the disaster. By contrast, the root causes of terrorist attacks on aviation are often not well understood, and such events are not presented in the same manner as "normal" aviation accidents are. But it is of considerable interest that before the September 11 terrorist attacks, the regulation of civil aviation security and safety was largely entrusted to the Federal Aviation Administration (FAA), a federal regulatory agency charged with, over its history, the regulation of aviation safety and the promotion of aviation as an important transport industry sector. Our research thus far reflects the idea that civil aviation safety is, like policies intended to address natural disasters, an area of "policies without publics," because of the highly technical nature of aviation safety.

This research has raised some important theoretical considerations. One of these rests on the assumption of how a focusing event works to change policies across the broad range of policy types. Public policy does not merely consist of laws and regulations. Policies also need to be implemented, and ultimately, are implemented by people. Table 5.2 shows the various levels at which public policy is codified, implemented, or both.

Policy change can be detected at levels ranging from constitutional change, which is clearly very visible to most members of a political system, all the way to subtle changes in the behavior of "street-level bureaucrats" (Lipsky 1980), whose vigilance or other behaviors may be hiding by the most recent event. A good example of this is the behavior of airport screeners in the days immediately after the September 11 attacks. These screeners became much more thorough and careful in their searches for dangerous or prohibited items in passenger luggage. Thus investigations into regulatory responses to focusing events have to be alert to the possibility of policy adjustments at all of these jurisdictional levels, and through both formal and informal channels. The origins of important shifts can come from various points on this hierarchy of regulatory authority.

This description is rather different than the assumptions that several scholars who have closely studied focusing events, including one of this chapter's authors, have made about how focusing events influence the agenda. The dominant assumption has been that focusing events have a profound influence on the legislative branch, which then enacts policy change to remedy the problems were revealed by the most recent focusing event. The regulatory agencies, in turn develop regulations to implement the will of Congress. Such a model assumes a very static bureaucracy, and assumes that the bureaucracy simply waits for direction from Congress,

TABLE 5.2 *Levels of policy codification*

Level of policy	Where codified	Accessibility of codification
Constitutional	In the federal or state constitutions	Highly visible at the federal level: the Constitution has been edited very few times. Some state constitutions are more easily amended for minor changes.
Statutory	United States Code, *Statutes at Large*	Highly visible through codification in statute law, publication in *Statutes at Large*.
Regulatory	*Federal Register, Code of Federal Regulations*	Moderately visible through the Code of Federal Regulations and the Federal Register.
Formal record of standard operating procedures	Operating Procedures Manuals	Low visibility because SOPs are often only internally published.
Patterned behavior by "street-level bureaucrats"	Not formally codified; evidence of a "policy" may be found in some agency records	Low visibility because these are behavioral changes with variations among actors
Subtle changes in cognition, in emphasis on problems, etc.	Not formally codified. Often revealed by the behavior of street-level bureaucrats themselves.	Very low visibility. Not codified, and changes in perceptions and emphases may be sublet.

Source: Birkland (2011), Table 7.1, p. 204.

with little or no capacity of its own to detect, diagnose, and seek to fix problems. A more realistic model of the influence of a focusing event on policy change, regulatory change, and other behavior of regulators themselves would account for the fact that policy-makers, at whatever level of government or responsibility, do not necessarily sit and wait passively for direction from superordinate levels of government. Indeed, regulators may have a strong incentive to be seen as being proactive problem solvers, who can address the immediate problem and persuade Congress that no statutory change is required. Regulators may wish to avoid statutory change, because the industries with which they negotiate regulations are

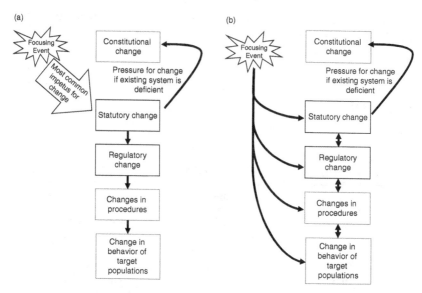

FIGURE 5.2 Two models of the influence of focusing events on policy actors and levels

likely to oppose more stringent statutory changes, and because senior regulators may fear that their influence and power would be diluted or even stripped from their agency in any statutory change. Of course, higher-level administrators in the executive branch may well make changes in the organization of regulators even before Congress acts.

Figure 5.2 depicts two models of the influence of focusing events on various levels of policy-making. The first model is the classic assumption, in which the primary influence of focusing events is transmitted through Congress to regulators. The second model depicts what we believe is a more realistic model of the influence of focusing events, which tend to alter the perceptions and behaviors of participants across the policy spectrum ranging from the most highly placed decision-makers to the street-level bureaucrats who have the obligation to carry out congressional efforts to protect the health and safety of the public. Furthermore, we assume that most street-level bureaucrats will react with greater vigilance and attempts to improve performance because most such bureaucrats take some pride in their work and take seriously their role in public service.

The case studies in this volume bear out the assumptions that drive the second model In almost all of those case studies, officials throughout the state grappled with the implications of focusing events, while policy adjustments came from various places within the government, and often outside it as well. This more dynamic model should guide future research on the influence of focusing events and regulatory politics, whether with regard to the types of crises emphasized in this volume, or in other domains, such as aviation security and safety, chemical plant accidents, or gun safety.

APPENDIX

Major American Aviation Safety and Security Incidents, 1988–2001

Event	Desired policy change	Actual change
Pan Am 103 – Lockerbie, Scotland, 1988. Bomb in luggage. (Security) http://aviation-safety.net/database/record.php?id=19881221-0	Baggage matching to passengers Explosion-resistant baggage containers in cargo hold	Aviation Safety Improvement Act of 1990: improved explosives detection, changes in intelligence gathering at Dept. of State and FBI
United 585, Colorado Springs, CO, 1991. Uncommanded rudder deflection. (Safety) http://aviation-safety.net/database/record.php?id=19910303-0	Not obvious until US Airways 427. NTSB unable to issue probable cause report for lack of data	See US Airways 427
US Airways 427, Aliquippa, PA, 1994. Uncommanded rudder deflection during landing approach. (Safety) http://aviation-safety.net/database/record.php?id=19940908-0	Correct defect in rudder control servo that allowed for uncommanded rudder deflection. NTSB recommended changes to pilot training, better flight data recorders, and improved rudder design. (Cobb and Elder 2003, 73)	FAA ordered new rudder systems, pilot training for jammed rudders. Retrofit program started in 2003, following an Airworthiness Directive issued in 2002, complete by 2008. This event helped explain the United 585 crash
ValuJet 592, Miami, FL, 1996. Fire in class D cargo hold (hold without smoke detection and fire	Require class D cargo holds to have smoke detection and fire-suppression capabilities similar to those in other	Conversion of class D to class C or E holds by March 18, 2001, per www.boeing.com/commercial/aero

(continued)

(continued)

Event	Desired policy change	Actual change
suppression), caused by improper loading of hazardous materials. (Safety) http://aviation-safety.net/database/record.php?id=19960511-0	cargo compartments (in essence, eliminate the class D type).	magazine/aero_06/textonly/s03txt.html

References

Baumgartner, Frank R., and Bryan D. Jones. 1993. *Agendas and Instability in American Politics*. Chicago: University of Chicago Press.

Béland, Daniel. 2005. "Ideas and Social Policy: An Institutionalist Perspective." *Social Policy & Administration* 39 (1): 1–18.

Béland, Daniel, and Jacob S. Hacker. 2004. "Ideas, Private Institutions and American Welfare State 'Exceptionalism': The Case of Health and Old-Age Insurance, 1915–1965." *International Journal of Social Welfare* 13 (1): 42–54. doi:10.1111/j.1369-6866.2004.00296.x.

Bennett, W. Lance. 2003. *News, The Politics of Illusion*. New York: Longman.

Birkland, Thomas A. 1997. *After Disaster: Agenda Setting, Public Policy and Focusing Events*. Washington, DC: Georgetown University Press.

———. 1998. "Focusing Events, Mobilization, and Agenda Setting." *Journal of Public Policy* 18 (1)(April): 53–74.

———. 2004. "Learning and Policy Improvement After Disaster: The Case of Aviation Security." *American Behavioral Scientist* 48 (3): 341–64.

———. 2006. *Lessons of Disaster*. Washington, DC: Georgetown University Press.

———. 2011. *An Introduction to the Policy Process: Theories, Concepts, and Models of Public Policy-Making*. Armonk, NY: M.E. Sharpe.

Birkland, Thomas A., and Regina G. Lawrence. 2002. "The Social and Political Meaning of the Exxon Valdez Oil Spill." *Spill Science & Technology Bulletin* 7 (1-2)(June): 17–22. doi:10.1016/S1353-2561(02)00049-X.

———. 2009. "Media Framing and Policy Change After Columbine." *American Behavioral Scientist* 52 (10): 1405–25. doi:10.1177/0002764209332555.

Burnier, DeLysa. 1994. "Constructing Political Reality: Language, Symbols, and Meaning in Politics: A Review Essay." *Political Research Quarterly* 47 (1): 239–53.

Carpenter, Daniel, and Gisela Sin. 2007. "Policy Tragedy and the Emergence of Regulation: The Food, Drug, and Cosmetic Act of 1938: 1." *Studies in American Political Development* 21 (2): 149.

Cobb, Roger W., and Charles D. Elder. 1983. *Participation in American Politics: The Dynamics of Agenda-Building*. Baltimore: Johns Hopkins University Press.

Cobb, Roger W., and David M. Primo. 2003. *The Plane Truth: Airline Crashes, the Media, and Transportation Policy*. Washington, DC: Brookings Institution.

Cobb, Roger W., and Marc Howard Ross. 1997. *Cultural Strategies of Agenda Denial: Avoidance, Attack, and Redefinition*. Lawrence, KS: University Press of Kansas.

Cohen, Michael D., James G. March, and Johan P. Olsen. 1972. "A Garbage Can Model of Organizational Choice." *Administrative Science Quarterly* 17 (1)(March 1): 1–25. doi:10.2307/2392088.

Edelman, Murray J. 1967. *The Symbolic Uses of Politics*. Urbana: University of Illinois Press.

   1988. *Constructing the Political Spectacle*. Chicago: University of Chicago Press.

Entman, Robert M. 1993. "Framing: Toward Classification of a Fractured Paradigm." *Journal of Communication* 42 (4): 51–59.

Glascock, Jack. 2004. "The Jasper Dragging Death: Crisis Communication and the Community Newspaper." *Communication Studies* 55 (1): 29.

Gunter, Valerie J. 2005. "News Media and Technological Risks: The Case of Pesticides After." *Sociological Quarterly* 46 (4): 671–98.

Hacker, Jacob S. 1998. "The Historical Logic of National Health Insurance: Structure and Sequence in the Development of British, Canadian, and U.S. Medical Policy." *Studies in American Political Development* 12 (01): 57–130.

Hilgartner, James, and Charles Bosk. 1988. "The Rise and Fall of Social Problems: A Public Arenas Model." *American Journal of Sociology* 94 (1): 53–78.

Jones, Bryan D. 2001. *Politics and the Architecture of Choice : Bounded Rationality and Governance*. Chicago: University of Chicago Press

Kingdon, John W. 2003. *Agendas, Alternatives, and Public Policies*, 2nd ed. Longman Classics in Political Science. New York: Longman.

Lawrence, Regina G. 2000. *The Politics of Force: Media and the Construction of Police Brutality*. Berkeley: University of California Press.

Lawrence, Regina G., and Thomas A. Birkland. 2004. "Guns, Hollywood, and School Safely: Defining the School-Shooting Problem Across Public Arenas." *Social Science Quarterly* 85 (5): 1193–1207.

Lindblom, Charles E. 1959. "The Science of 'Muddling Through'." *Public Administration Review* 19: 79–88.

———. 1979. "Still Muddling, Not Yet Through." *Public Administration Review* 39: 517–26.

Lipsky, Michael. 1980. *Street-Level Bureaucracy: Dilemmas of the Individual in Public Services*. New York: Russell Sage Foundation.

Liu, Xinsheng, Eric Lindquist, and Arnold Vedlitz. 2011. "Explaining Media and Congressional Attention to Global Climate Change, 1969–2005: An Empirical Test of Agenda-Setting Theory." *Political Research Quarterly* 64 (2): 405–19. doi:10.1177/1065912909346744.

Majone, Giandomenico. 1989. *Evidence, Argument and Persuasion in the Policy Process*. New Haven: Yale University Press.

May, Peter J. 1990. "Reconsidering Policy Design: Policies and Publics." *Journal of Public Policy* 11 (2): 187–206.

———. 1992. "Policy Learning and Failure." *Journal of Public Policy* 12 (4) (December): 331–54.

May, Peter J., Ashley E. Jochim, and Joshua Sapotichne. 2011. "Constructing Homeland Security: An Anemic Policy Regime." *Policy Studies Journal* 39 (2)(May): 285–307. doi:10.1111/j.1541-0072.2011.00408.x.

Mintrom, Michael. 1997. "Policy Entrepreneurs and the Diffusion of Innovation." *American Journal of Political Science* 41 (3)(July): 738–70.

Mintrom, Michael, and Sandra Vergari. 1998. "Policy Networks and Innovation Diffusion: The Case of State Education Reforms." *The Journal of Politics* 60 (01): 126–48. doi:10.2307/2648004.

Perrow, Charles. 1999. *Normal Accidents: Living with High-risk Technologies, with a New Afterword and a Postscript on the Y2K Problem*. Princeton, NJ: Princeton University Press.

Sabatier, Paul A. 1993. "Policy Change over a Decade or More." In *Policy Change and Learning: An Advocacy Coalition Approach*. Boulder, CO: Westview.

Sabatier, Paul A., and Hank Jenkins-Smith. 1999. "The Advocacy Coalition Framework: An Assessment." In *Theories of the Policy Process*. Boulder, CO: Westview.

Sabatier, Paul, Hank Jenkins-Smith, and Edward F. Lawlor. 1996. "Policy Change and Learning: An Advocacy Coalition Approach." Journal of Policy Analysis and Management 15 (1): 110–46.

Schattschneider, E. E. 1975. The Semisovereign People. Hinsdale, IL: The Dryden Press.

Schneider, Anne, and Helen Ingram. 1991. "The Social Construction of Target Populations: Implications for Politics and Policy." *American Political Science Review* 87(2): 334–48.

Sheppard, Ben. 2011. "Mitigating Terror and Avoidance Behavior Through the Risk Perception Matrix to Augment Resilience." *Journal of Homeland Security and Emergency Management* 8 (1) (January 24). doi:10.2202/1547-7355.1840. www.degruyter.com/view/j/jhsem.2011.8.issue-1/jhsem.2011.8.1.1840/jhsem.2011.8.1.1840.xml.

Simon, Herbert Alexander. 1957. *Models of Man: Social and Rational; Mathematical Essays on Rational Human Behavior in Society Setting*. New York: Wiley.

Slovic, P. 1987. "Perception of Risk." *Science* 236 (4799)(April 17): 280–85. doi:10.1126/science.3563507.

Stone, Deborah A. 1989. "Causal Stories and the Formation of Policy Agendas." *Political Science Quarterly* 104 (2): 281–300.

   2002. *Policy Paradox: The Art of Political Decision Making*. New York: Norton.

Walker, Jack L. 1977. "Setting the Agenda in the U.S. Senate: A Theory of Problem Selection." *British Journal of Political Science* 7: 423–45.

6

The Story of Risk

How Narratives Shape Risk Communication, Perception, and Policy

Frederick W. Mayer

The organizing question of this volume is when and how events in the world lead to recalibrations of risk perception and changes in policy. A naïve model might assume a simple analytic process: existing regulations are based on prior assessments of risk, new information about risk becomes available, and regulators update their assessments and make cost-effective policy changes. Under some circumstances, such a Bayesian process might not be too far off the mark. Viewed over a sufficiently long time period, regulatory change in some cases might approximate such a rational process. Arguably the cases of lead in gasoline, drunk driving, and smoking are examples. But in myriad other cases – including the nuclear accidents, offshore oil spills, and financial crises covered in this volume – the path from risk to policy strays far from the policy analytic ideal.

Sometimes events trigger rapid and significant policy responses, as in the case of America's reaction to the nuclear accident at Three Mile Island, Europe's to Chernobyl and Japan's to Fukushima. Indeed, responses to certain events can be greater than warranted by cool analysis: the US policy response after the 9/11 attacks might be one such example. Yet in other cases, even an event that one might presume would lead to considerable change barely makes a ripple: the utter lack of policy response to New Town and other mass shootings is one such example.

Clearly, the policy response to new information about risk is a good deal more complicated than the analytic model would suggest. When events occur, information about them may or may not be transmitted accurately by news media or other sources. When citizens and regulators receive information, they may or may not appropriately recalibrate their beliefs. And, not least, even if policy-makers and publics have reasonably accurate assessments of risks, the politics of translating beliefs into policy change is far from straightforward, and depends, among other things, on the political dynamics at play.

Many literatures can help shed light on these processes, among them economic, political, sociological, and cultural analyses of news production and consumption (Downs 1957; Bennett 1983; Hamilton 2004; Entman 2004); psychological research on the ways in which we process information (Tversky and Kahneman 1974; Slovic 2000; Loewenstein et al. 2001); and political research on agenda setting (Kingdon 1984; Nelson 1986; Downs 1972; Baumgartner and Jones 2010); political communication and framing (Edelman 1964; Goffman 1974; Entman 1993); and political mobilization (Gamson 1992; Tilly 2002; Benford and Snow 2000). But missing from the literature is a full appreciation of the role of narrative in risk communication, risk perception, and the politics of risk.

This is surprising for two reasons. First, narrative is at the core of how we make sense of and respond emotionally to our experiences generally, and it plays an important role in both motivating and scripting our actions. We are, whatever else we are, storytelling animals. And, second, risk makes a good story. Indeed, many of our stories are about risk – narratives of dangers over the horizon, cautionary tales of what is to be feared, accounts of what happens when we don't pay attention to the fate of canaries in coal mines. Indeed, one could argue that an evolutionary advantage of human narrativity is to enable us to imagine risks we have not experienced, and to help us avoid them.

This chapter develops a narrative theory of risk. It explores the role of narrative in the transmission of risk information, the processing of information into risk perception, and the impact of risk perceptions on the politics of policy-making, arguing that at each stage, stories matter. For an event to have impact, it must first become news, and the currency of news is the story. For news to shape risk perceptions, it has to alter how we imagine risk and how we feel about it, and stories are the natural vehicle for imagination and feeling. For risk perceptions to inform regulatory policy, it must pass through the filter of politics, and stories are the lifeblood of politics.

THE STORYTELLING ANIMAL

As a starting point for understanding why stories might play a significant role in risk perception and policy-making, fields usually thought of as matters for technical analysis, the first answer is that narratives are central to almost every realm of human perception and behavior. As Barbara Hardy (1977, 3) has put it, we "dream in narrative, daydream in narrative, remember, anticipate, hope, despair, believe, doubt, plan, revise, criticize, construct, gossip, learn, hate and live by narrative." To all that we could add fear, perhaps the most relevant of emotions for risk, which so often is the product of stories in mind.

As I have explored in greater depth elsewhere, narrative serves many important psychological functions (Mayer 2014). First, our memory is deeply narrativized. More than we like to believe, our memory is far from a literal record of our experience, but rather it is both highly selective and malleable. We remember stories better than other information, and the better the story, the better we remember it. Moreover, we make stories of our memory. When we are "remembering," we are not so much recalling events as reconstructing them, taking those fragments we do recall and putting them back together into a pattern that makes sense to us, into a story (Bartlett 1933). What we seem to remember best is the significance and emotional impact of the past – the point of the story – an organizing point of departure as we reconstruct the tale that makes that point, the story that "must have been" (Kotre 1996).

Second, how we perceive our experience is strongly influenced by narrative. In part, this is because we rely heavily on schema already held in memory, many of which are encoded by narrative. In common usage, to say we "understand" something comes very close to saying that we can tell a story in which it makes sense, in which what happened is the plausible consequence of circumstance and character. For example, when an unarmed black man was shot by a white policeman in Ferguson, Missouri, immediately, our minds started constructing stories to account for how such an event could have happened. Note that the story we construct will depend fundamentally on the form that we expect such stories to take. Those expectations are shaped by the more general stories we already hold in mind. For some, the tale will be about the dangers of "walking while black," a case of police overreach and prejudice; for others, perhaps, it will be a story of a policeman doing a tough job in a dangerous world.

Beyond its purely cognitive functions, narrative is intimately intertwined with emotion. Obviously, narrative has the power to evoke emotion: Who has not cried in a movie theater? (See Schelling 1987 for a lovely essay on this phenomenon.) The relationship between story and emotion goes deeper still. Forgas (2001) makes a persuasive case that one's affective state depends fundamentally on narrative framing. The same objective outcome – the death of child in a car accident, for example – can evoke completely different emotions depending on the story. If the driver who caused the accident had been drinking and was driving recklessly, we feel anger. If, on the other hand, she had been distracted by her two children fighting in the backseat when she swerved across the center line, we feel pity.

Narrative is also intimately connected to identity. In McAdams' (2001) words, we live "storied lives." In early childhood, self-awareness appears at just the moment we are able to tell a story about ourselves and can see ourselves as a character in a story (Miller et al. 1990). As we mature, our identity is further established by the autobiographical narrative we hold in mind, a narrative that includes particular vignettes but also tends to have an overall narrative form. As we script our autobiographical narrative, we seek to maintain the integrity of our character, to see ourselves as more-or-less consistent actors in the past, present, and possible future.

Finally, much of our behavior is *acting*, in the dramatic sense of acting on a stage. The metaphor of life as theater has a long pedigree, going back at least to Shakespeare. As Bruner (1990, 34) puts it, "[w]hen we enter human life, it is as if we walked on stage into a play whose enactment is already in progress – a play whose somewhat open plot determines what parts we may play and towards what denouements we may be heading." The suggestion is that when we act, we are engaging in what he calls "acts of meaning," actions that make sense for our character in the story in which we find ourselves. Similarly, MacIntyre (1984) argues that we act in ways that are intelligible because they are explicable in the context of a story we can tell ourselves and others.

Given the extent of human narrativity, it would be surprising if stories did not play an important role in how we experience and respond to risk. In the sections that follow, I argue that what is true about human experience more generally is true about risk. As storytelling animals, we communicate about risk through stories, apprehend risk through stories, and are moved to act in response to stories.

FROM EVENTS TO NEWS: THE SOCIAL CONSTRUCTION OF RISK COMMUNICATION

An accident occurs at a nuclear power plant; an oil tanker runs aground; Lehman Brothers collapses. Before there can be a change in risk perceptions, and long before any regulatory response, information about what has happened must be transformed into a meaningful form and transmitted to some audience. Occurrences must be turned into *news*, or more specifically, into news *stories*. Aside from those few who experience these events directly, we learn about what happened through the stories we hear about them (and even those who experienced the event may not know what hit them without a story). For most of us, therefore, the event *is* the story.

The connection between the raw occurrence and the news story is far from simple. Always there is considerable latitude for interpretation, many possible stories that could be told. Was Chernobyl a story about an unavoidable accident or a negligent act, a product of Soviet design flaws or evidence of the inherent dangers of nuclear energy? Was the financial crash of 2008 a story about lax regulation or investor malfeasance? Was the Gulf oil spill a story about corporate malpractice or operator error? As the next two sections of this paper will explore, the framing of the news story makes a difference, both for our perceptions of risk and for the logic of policy response.

The question is what we can say about what makes news. A starting point is to recognize that there is at best a very tenuous connection between news and information that might be most important for the public to know. As Downs (1957) first argued, there is precious little demand for information that informs the public about policy choices. The reason is simple: if the value of information to an individual citizen depends on the impact that information is likely to have on policy outcomes, we would predict little demand for policy information since few individuals have significant ability to affect policy outcomes. The marginal cost of acquiring information, if only in time spent so doing, will almost always outweigh the marginal benefits. And without demand for policy information, little is supplied.

So, then, what is covered? First, good stories make news. Aside from whatever other biases they may have, the news media have a bias for story. The local news is a series of mini-tragedies: shootings, robberies, fires, and car wrecks (Stone 1988). (Hence the aphorism, "if it bleeds, it leads.") The national news is about the prospects for war, epic battles between Congress and the president, and human tragedies from around the world. Campaigns are covered as horseraces, not policy debates (Patterson 2011). Occasional heart-warming stories about heroic rescues or other selfless acts get their play, too, but always it is a story, with a plot (a sequence of events structured with beginning, middle, and end), characters (heroes, villains, and victims), and a point, a meaning.

Risk makes a good story. Inherent in risk is the essential ingredient of drama: things may turn out well or they may turn out badly, the story may have a happy or a sad ending. But not all risks are equally dramatic. Rare and extreme risks make better stories than common and less extreme risks: the possibility of a nuclear meltdown, however remote, is more dramatic than the fact that burning coal involves greater risks for all of us who breathe the soot of power plants; a shooting at an elementary school is more dramatic than the relationship between easy access to guns and America's high suicide rate. A corollary is that complexity can be the enemy of a good story. The financial crash of 2008, for example, is much harder to encapsulate in a simple narrative (and affords fewer visuals for our video era) than the Exxon Valdez or Chernobyl accidents, which may in part account for the emergence of a less cohesive policy response.

The consequence of the bias for story, therefore, is an overrepresentation of *dramatic* risks. For example, in 2013, the American extreme distance swimmer Diana Nyad swam from Cuba to Key West, an extraordinary physical feat. The framing that the media chose to give the swim was that she had swum "without a shark cage" in "shark-infested waters" (Shiffman 2013). Objectively, however, Nyad faced minimal risk of a shark attack. The real danger was with the deadly box jellyfish, whose stings can kill. Indeed, Nyad's previous attempts had been thwarted by jellyfish stings. But account after account highlighted sharks. Why? Almost certainly because shark attacks make better stories than jellyfish stories. The movie is *Jaws*, not *Tentacles*. To say that Nyad swam through "sharked-infested waters" (the usual formulation) is to evoke those stories, and the dread that comes with them.

Second, the stories that we demand of our news are the stories we want to hear. As humans, we are particularly receptive to stories that fit familiar forms, that confirm our beliefs, and that resonate with the stories we already hold in mind. Every society has a canon of such stories that together serve to define what *we* remember, what *we* believe, and who *we* are (Hinchman and Hinchman 1977; Macintyre 1984). Narrative culture is not determinative, but it does prepare a community to be more receptive to stories that reinforce or resonate with the stories it already shares in common.

With social media, what "goes viral" seems to have less to do with the real import of events than with the eagerness of the culture to consume a certain tale. Stories in the mainstream press generally have some basis in fact, but the extent of interest in a story will still be powerfully affected by fit with our prior narratives. This phenomenon might be called the "*China syndrome*": we demand "true" stories that fit our fictions. Arguably, Three Mile Island was a huge story not just because of the real risks involved, but also, in part, because it so perfectly fit our nightmares. At the extreme, the demand for a story that "fits" the story we want to hear is so great that it almost requires no event at all to produce it. Urban legends have this characteristic, whether the classic "razor in the apple" Halloween legend (Best and Horiuchi 1985) or the widespread "vaccines cause autism" story, facts have little to do with the eagerness with which the story is consumed.

Third, what story reaches the public depends on who gets to tell the story. Although cultural fit is important in determining what stories circulate in the public arena, fit alone in not determinative. In part, this is a function of formal position. As Entman (2004) has argued in the case of news about 9/11, information tends to cascade from elites who have greater access to the media megaphone. Raw footage of the planes striking the towers required no mediation to be dramatic. But from the beginning, the form the story took, and therefore the meaning of what was happening, was powerfully shaped by those in position to interpret the event. In this case, President Bush had unique ability to shape the form of the story, which he used to frame the attacks as analogous to Pearl Harbor, requiring a "war on terror" in response.

It is worth noting, too, that the capacities of the media to tell the story matter as well. As has been well documented, there are declining numbers of reporters with experience covering finance, science, and international affairs (Pew 2008; Kennedy and Overholser 2010). As a consequence, unless there is some dramatic event that can be told as a simple story ("nuclear meltdown" fits the bill), the media may simply not have the capacity to convey the nature of the risks involved. In sum, events in the world are conveyed to us in the form of stories. Before we can form judgments about the risks events represent, narrative imperatives both filter and mold the raw occurrences into meaningful forms.

FROM NEWS TO PERCEPTION: NARRATIVE AND THE APPREHENSION OF RISK

Just as the relationship between a raw event and the social construction of news about it is far from straightforward, the relationship between a story about risk and our apprehension of that risk is far from simple. Thinking about risk is hard (as I rediscover whenever I teach decision analysis).

In an ideal analytic model, one would need to estimate accurately the probability of possible outcomes, establish the relative value of those possibilities, and calculate the expected value of the risky options. These are not easy tasks, and for rare and extreme events, the estimates and valuations are even more difficult.

It should come as little surprise then that the way we humans actually think about risk bears scant resemblance to the analytic ideal. The best-developed line of research on this is what is now generally referred to as behavioral economics. Beginning with the seminal work of Tversky and Kahneman, there has been an explosion of experimental work exploring how humans perceive risks. That literature is well summarized in this volume by Weber (Chapter 4), one of the major contributors to it. The insights of the behavioral economics literature are numerous and have far advanced our understanding of how humans actually process risk information. But largely missing from the risk perception literature is an account of the role of stories.

Risk assessment is an act of imagination. It requires anticipating what might happen. Furthermore, our judgments about rare but extreme risks cannot depend on personal experience, since most of us never experience them (fortunately!). Such risks, therefore, require even greater leaps of imagination. To a great extent, therefore, we make sense of risk, as we do much else, through the stories we construct and hold in mind.

Below, I explore how several psychological tendencies described in the behavioral economics literature can be explained, in part at least, by a narrative theory of mind.

Availability

A first explanation for distortions in risk assessment is *availability*. As Tversky and Kahneman (1974, 1127) put it, "[t]here are circumstances in which people assess the frequency of a class or the probability of an event by the ease with which instances or occurrences can be brought to mind." And although the actual frequency or probability will likely affect the ease of recall, "factors other than frequency and availability" also matter, among them the "retrievability of instances" and the "imaginability" of scenarios (Tversky and Kahneman 1974, 1127).

Both retrievability and imaginability can be well explained by how narrative operates in memory. First, the retrievability of an instance is, by definition, a function of memory. And, as discussed above, memory is highly narrativized. We remember stories better, and the better the story – that is, the more dramatic, the more emotionally arresting, the more meaningful – the better we remember it. Given this perspective on memory, we would expect that dramatic stories of risk would be relatively more represented in our memory than mundane and that we would be more prone to retrieve such stories. If it is true, therefore, that we assess risk by their salience in memory, we would overestimate storied risks and underestimate unheralded ones.

Similarly, the ease with which we can imagine risks will be bound up in the ease with which we can construct a story. As Tversky and Kahneman (1974, 1128) put it, "[t]he risk involved in an adventurous expedition, for example, is evaluated by imagining the contingencies with which the expedition is not prepared to cope. If many such difficulties are vividly portrayed, the expedition can be made to appear exceedingly dangerous, although the ease with which disasters are imagined need not reflect the actual likelihood." Note how close Tversky and Kahneman's example comes to saying that the perception of risk depends on the ease with which we can imagine a story in which some disaster occurs. And what we know about story construction is that those that draw on familiar tropes are easier to imagine than those that don't. The Diana Nyad story recounted above to illustrate the proclivity of the media to tell dramatic stories also demonstrates why it is that we fear sharks more than jellyfish when we go into the water (or riptides, the deadliest danger at the beach): it's easier construct a story about a great white attack than one involving jellyfish (or the complex interaction of waves and beach topography).

Representativeness

In Tversky and Kahneman's taxonomy, a second class of heuristics leading to distortions in risk perception is *representativeness*. The idea is that when asked to evaluate the likelihood that a welfare recipient would be black, for example, we focus on the frequency with which blacks are welfare recipients compared to the frequency with which whites are, and ignore the fact that there are many more whites than blacks in America. (More formally, we are bad Bayesians: we ignore priors and focus on the wrong conditional probability.)

Why might this be? Tversky and Kahneman offer a number of explanations. But many of them come close to the idea that we make judgments based on conformity to stereotypes rather than on statistical analysis. And what we know about stereotypes is that they are a form of schema constructed by narrative. Stereotypes of blacks as welfare recipients, librarians as shy, lumberjacks as hearty men with beards, teachers as female, and so forth, are established by the stories of our culture. When we are casting their roles in the stories we construct, we naturally turn to the familiar tropes. If we hear about a nuclear accident, our minds rush to the culturally available narratives of nuclear meltdown and widespread radiation poisoning. Lurking somewhere in the background, no doubt, are the emotive stories of Hiroshima and Nagasaki.

Framing (Prospect Theory)

Perhaps the most influential of Tversky and Kahneman's empirical findings was that logically identical risks, if described differently, will elicit different choices, a violation of a key tenet of rational choice theory. A typical scenario is the following:

Imagine that the U.S. is preparing for the outbreak of an unusual Asian disease, which is expected to kill 600 people. Two alternative programs to combat the disease have been proposed. Assume that the exact scientific estimates of the consequences of the programs are as follows:

- If Program A is adopted, 200 people will be saved.
- If Program B is adopted, there is 1/3 probability that 600 people will be saved, and 2/3 probability that no people will be saved.

A second group of respondents was given the same cover story with the following descriptions of the alternative programs.

- If Program C is adopted 400 people will die.
- If Program D is adopted there is 1/3 probability that nobody will die, and a 2/3 probability that 600 people will die (Tversky and Kahneman 1981, 453).

What Tversky and Kahneman found was an overwhelming preference for the risk averse alternative (Program A) in the first case and for the risk seeking alternative (Program D) in the second case, even though the two scenarios are logically identical. It turns out that when a situation is framed in terms of possible gains, most people are risk averse, but when individuals perceive the same situation in terms of losses, they become risk seeking.

The behavioral economics literature has demonstrated the robustness of this phenomenon, but it has not provided a full explanation for it. Narrative analysis can help explain the behavior. First, framing depends on how the story is told. Stories always begin with an initial state of affairs that then serves as an emotional reference point, or anchor, for the story that follows ("Once upon a time ... "). In the first telling, the story begins with a presumption of 600 dead; in the second, no one is yet dead. These initial conditions ensure that the first story will be about gains achieved or squandered, the second about losses accepted or avoided. Second, the initial framings then restrict the shape of the available plots. In the first scenario, which begins with 600 expected to die, the choice is between a certain story with a happy ending (200 saved!) and a chance of a story with a happy ending (even more saved) but a better-than-even chance of a story that ends in tragedy (200 could have been saved!). In the second scenario, which begins with none yet dead, the choice is between a sure tragedy (400 dead!) or a dramatic tale that could end in tragedy as well (even more dead) or a one-third chance of triumph (400 saved!).

In an expected value calculation the two scenarios are equal, but in narrative terms they are not. A narrative theory of choice might suggest that rather than weighing options in terms of expected outcomes, people evaluate them in terms of what stories might be told about their actions if they act one way or the other. In the first scenario the choice is between being the hero for sure or a better-than-even chance of being the villain. In the second, the choice is between being the villain for sure or a chance of being the hero. When we evaluate our choices we imagine our emotional states under each of the possible story outcomes and conclude: "I couldn't live with myself if I took the chance and lost in the first scenario, but I couldn't live with myself if I didn't in the second."

Illusion of Control

Our feelings about risk also appear to depend in some fundamental way on whether or not we believe that we have a measure of control. Slovic (1987) identified two types of risk for which it seems we have disproportionate fear. The first, *dread* risk, involves risks in which we have little control and that have catastrophic consequences. Nuclear reactor accidents and nerve gas score particularly high on that measure. The second is *unknown risk*, characterized by being "unobservable, unknown, new, and delayed in the manifestation of harm" (Slovic 1987, 283). Drug, chemical, and genetic engineering risks score high on this dimension. On the other end of both spectrums, riding a bicycle (actually quite a risky enterprise) scored particularly low. The implication is that we systematically overestimate the dread and unknown risks, and we underrate that of such familiar and (we believe) controllable risks as riding a bicycle or driving a car.

To some extent, this pattern can be explained by the availability heuristic already discussed. Certainly, stories about catastrophic dread risks are likely to be more salient in memory than those of car or bike accidents. But the pattern Slovic describes adds another element, that of whether or not we perceive an ability to act in ways that would reduce or avoid the risk. Almost certainly, we are overconfident. Statistically, we are just not that good at avoiding enough auto accidents to make driving safer than flying, yet many who do not fear driving fear flying.

A narrative interpretation of the anomaly might be that as we imagine facing the risk, we also imagine ourselves in the story and see ourselves taking actions that avoid the risk. We are riding a bike, a car swerves and is about to hit us, but we react quickly enough to avoid the accident. In a sense we are casting ourselves in these stories as competent and prudent, and imagining a world in which our agency determines our fate. That is why inexplicable accidents beyond our control are so shocking; they are allegories that threaten our preferred narrative. As Slovic notes, such accidents are signals that things may not be what they seem, and that our control may be illusionary. This might account for why the public appeared much more alarmed about the outbreak of Ebola than, say, the diabetes epidemic, even though the latter is a far greater risk (in the United States, at least). The story of Ebola is one in which medicine is helpless, whereas diabetes is, usually, quite treatable. In the cases discussed in this volume, perhaps the nuclear accident scenario comes closest to invoking dread risk. However small the probability, a nuclear meltdown is, by definition, out of our control.

Affect

With the introduction of "dread" risk, already we are beginning to shift away from pure cognition to the realm of affect. More can be said. Our experience of risk is intimately connected with emotion. Research on how people actually apprehend risk (as opposed to the ideal analytic model) suggests that risk perception is more a matter of feeling than of calculation, an affective rather than a cognitive state, as Weber points out in this volume. When we are swimming in the ocean and feel something bump against our legs, it is hard not to experience a sudden rush of emotion, even though logically we might know with the rational part of our brain that there is little chance that we are about to be attacked by a shark. In that moment, as in others, fear can overwhelm all rational calculation of risk.

Narrative plays its part in this. In part, this is because fear is at the root of other cognitive tendencies that we have already considered. Thus the vividness of "the fear" emotion leads to availability; the fear of failure helps to account for prospect theory; the durability of stereotypes rests in the fears that they encapsulate. But also some stories more effectively invoke fear than others because they are more easily imagined. A specter of a nuclear catastrophe, for example, will likely trigger greater fear than a financial crisis, the imagined scenario of which is both more abstract and less existentially terrifying.

To summarize, then, narrative is important in how we humans interpret the information we receive about events. Although we are capable of rational risk assessment if we put our minds to it (and experts are more likely to do so than general publics) all of us prefer "thinking fast," making use of efficient heuristics that enable us to form quick judgments about risk (Kahneman 2011). Stories play an important role here, both in shaping cognitive heuristics and, even more directly, in enabling the act of imagination that is risk perception.

FROM PERCEPTION TO POLICY: NARRATIVE AND THE POLITICS OF RISK

If anything, the road from a change in risk perception to a change in policy is even more circuitous than that from event to news or from news to perception. To be sure, in some circumstances a change in belief may be sufficient to induce a change in action. If regulators are sufficiently empowered to act and sufficiently insulated from political pressures, new understandings of risk may well translate directly into changes in policy. But likely more often, regulators operate in a political environment in which they cannot ignore the external political environment and in which policy reflects political processes.

The politics of policy-making is enormously complex, involving interests, institutions, and ideas in multiple arenas. Clearly, anything approaching a full account of the political processes through which recalibrations of risk result in policy change is beyond the scope of this chapter. My focus here, therefore, is on two specific obstacles to policy change. The first is the need to align divergent views into a shared perception of the problem; the second is the translation of that shared perception into political pressure. Both, I will argue, are greatly facilitated by the power of shared narrative.

Aligning Risk Perceptions and Policy Preferences

A first political problem is to align individual risk perceptions and policy preferences into shared perceptions and preferences. How widely shared they will need to be will depend on the degree to which politics spills into the public arena, but however narrow the circle of policy-makers may be, as long as power is shared, some degree of alignment will be necessary.

Several literatures bear on this issue. A first is the agenda-setting literature, the focus of which is the question of what issues the policy process will attend to. The presumption is that time and attention are limited, and that policy-makers can only attend to so many issues at once. Shared public narratives appear to be quite important here. Nelson (1986), for example, highlights the importance of widely circulated stories of child abuse as key to bringing that issue to the fore. Downs (1972) emphasizes the role of news stories in shaping the "issue attention cycle." Baumgartner and Jones (2010) argue for the importance of converging on common "policy image" as a necessary element for policy change, and illustrate that point with reference to how the image is framed by narratives. It seems that shared narratives help focus attention, particularly if those narratives are dramatic, with a clear depiction of the problem and an (apparently) straightforward solution. Similarly, the literature on "focusing events" literature draws on the more general literature to describe how dramatic events can shape public discourse and drive policy responses (Birkland 1998). As Birkland explores in this volume, there is a politics to focusing events involving the instrumental use of narrative by special interests and politicians.

Both the agenda-setting and focusing event literatures draw on a more general framing literature in political communication (Edelman 1964; Goffman, 1974; Entman 1993). The term "framing" here has a broader meaning than its narrower use by Tversky and Kahneman, involving associations, connotations, and logics that attend to a particular characterization of an issue. Frames orient us to an issue, often carrying both a normative evaluation and an implicit solution.

Political Mobilization

As noted above, political mobilization is not always needed for political change. But sometimes political pressure may be necessary to countervail entrenched habits or interests. It is important to recognize that a shared perception of a problem is insufficient. A shared belief that banks are too big to fail, or that easy access to semi-automatic weapons is unacceptably risky, does not guarantee collective action to pressure for change. Political mobilization is, at its heart, a matter of collective action. It is useful, therefore, to begin with reminding ourselves of the core obstacles to collective action.

Before there can be collective action, three other problems must be addressed. The first is the familiar problem of free riding. A regulatory change desired by a group is a collective good for its members. But if there are costs to taking action and if the benefits of successful action cannot be restricted to only those who act, each member has an incentive to free ride on the action of others. The second barrier, seemingly more mundane but often quite difficult, is the problem of coordinating action. Always there are many forms that political action might take. What specific policy of the many should we promote? What form should political action take? And third, even if we would prefer to participate in collective action rather than to free ride, and even if we know the script for what form action should take, we may need to be assured that others will join too. We do not want to protest alone.

Narratives, I have argued (Mayer 2014), are our "go to" tool for solving these problems. When we participate in collective action, we are participating in a shared narrative, a collective drama in which joining, marching, giving, voting, and other private actions became meaningful and satisfying acts in our autobiographical narrative. Such actions become expressions of our identity, not choices based on a calculus of interest. And because participation in collective action is its own reward, there is no temptation to free ride. Indeed, to shirk is shameful. Similarly, a shared narrative can facilitate coordination. By providing a common script to choreograph our actions, we know what form collective action should take. And, finally, because we believe that others, like ourselves, are caught up in the shared narrative, we are assured that they, too, will act.

From this perspective, much of politics is a contest of public narratives, instrumentally used by those who would frame an issue to spur political action (or to deter it). So, too, the politics of risk. Those who would spur action, either because they perceive a grave threat or because they would benefit from that action, or both, spin a narrative of common peril, in which acting can avert disaster. Conversely, those who would deter action, for either idealistic or egoistic reasons, counter with narratives in which the risk is overstated, the real danger is to act, or even, as we have seen in the case of climate change, that the advocates of action are the real villains.

Usually, general forms of these stories are already in circulation when some event happens. News of the accident at Three Mile Island arrived in the context of an already mature political battle over the future of nuclear power, in which opponents of nuclear energy were animated by a risk narrative that made opposition a heroic act. For them, the story of TMI then became an instantiation of the meta-narrative (in literary terms an example of synecdoche) they already held. It was a story waiting to happen. And for opponents of nuclear power, the story of TMI was a powerful tool for pushing the danger frame out to a wider audience (see Birkland and Warnement's Chapter 5 in this volume). In this case, the ability of those who favored continued expansion of the nuclear industry in the United States to tell a story in which that made sense were at a severe disadvantage, although there was certainly a valid case to be made that the alternative to nuclear posed the greater risk.

Narratives do not always contribute to a coherent policy response, of course. After the financial crash, in contrast, an event that might have been expected to result in a significant regulatory response, an initially powerful narrative about the risks of unregulated banks that animated the Occupy Wall Street movement and, in part, the Tea Party was countered by a narrative in which government regulation had caused the problem by forcing lenders to make risky loans or a blame-the-victim narrative in which poor credit decisions by borrowers was to blame. In this case, the consequence of the multiple stories, arguably, was that policy responses went in multiple directions. And sometimes even powerful narratives are completely blocked by counter-narratives. The horrific mass shootings in Columbine, Aurora, and New Town might have been expected to lead to significant gun regulation. Certainly, they were effective in mobilizing an element of the public. But that political energy has been effectively countered by opponents of gun control, who have been very successful in spinning counter-narratives of government overreach and threats to personal freedom.

CONCLUSION

The role of narrative in shaping the policy responses to risk shocks has been under-appreciated. Certainly there are many factors that affect those responses, as the chapters in this volume attest. But narrative plays a role in each link of the causal chain, shaping the information we receive about an event, the way in which we perceive that information, and the politics through which perceptions are translated into action. A narrative theory of risk, therefore, can help explicate why some events lead to change and others do not, and why policy change takes the sometime peculiar form that it does.

References

Bartlett, F. C. and C. Burt. 1933. Remembering: A Study in Experimental and Social Psychology. *British Journal of Educational Psychology*, 3(2): 187–92.

Baumgartner, F. R., and B. D. Jones. 2010. *Agendas and instability in American politics*. Chicago: University of Chicago Press.

Benford, R. D., and D. A. Snow. 2000. Framing Processes and Social Movements: An Overview and Assessment. *Annual review of sociology*: 611–39.

Bennet, L. 1983. *News, The Politics of Illusion*. New York: Longman.

Best, J. and G. T. Horiuchi. 1985. The Razor Blade in the Apple: The Social Construction of Urban Legends. *Social Problems*, 32(5): 488–49.

Birkland, T. A., 1998. Focusing Events, Mobilization, and Agenda Setting. *Journal of Public Policy* 18(01): 53–74.

Bruner, J. 1990. *Acts of Meaning*. Cambridge, MA: Harvard University Press.

Downs, A., 1972. Up and Down with Ecology: The Issue-Attention Cycle. *The Public Interest* (28): 38–50.

Downs, A. 1957. *An Economic Theory of Democracy*. New York: Harper.

Edelman, M.J., 1964. *The Symbolic Uses of Politics*. Urbana, IL: University of Illinois Press.

Entman, R. M., 1993. Framing: Toward Clarification of a Fractured Paradigm. *Journal of Communication*, 43(4): 51–58.

Entman, R. M., 2004. *Projections of Power: Framing News, Public Opinion, and US Foreign Policy*. Chicago: University of Chicago Press.

Forgas, J. P., 2001. *Feeling and Thinking: The Role of Affect in Social Cognition*. Cambridge, UK: Cambridge University Press.

Gamson, W. A., 1992. *Talking Politics*. Cambridge, UK: Cambridge university press.

Goffman, E., 1974. *Frame Analysis: An Essay on the Organization of Experience*. Cambridge, MA: Harvard University Press.

Hamilton, J., 2004. *All the News That's Fit to Sell: How the Market Transforms Information into News.* Princeton, NJ: Princeton University Press.

Hardy, B., 1977. Narrative as a Primary Act of Mind. In M. Meek, A. Warlow, and G. Barton (eds.), *The Cool Web: The Pattern of Children's Reading.* London: Bodley Head: 12–23.

Kahneman, D., 2011. *Thinking, Fast and Slow.* New York: Macmillan.

Kennedy, D. and G. Overholser. 2010. *Science and the Media.* Washington, DC: American Academy of Arts and Sciences. www.amacad.org/pdfs/scienceMedia.pdf.

Kingdon, J. W., 1984. *Agendas, Alternatives, and Public Policies.* Boston: Little, Brown.

Kotre, J., 1996. *White Gloves: How We Create Ourselves Through Memory.* New York: W.W. Norton and Company.

Loewenstein, G. F., E. U. Weber, C. K. Hsee, and N. Welch. 2001. Risk as Feelings. *Psychological bulletin,* 127(2): 267–86.

MacIntyre, A. C. 1984. *After Virtue: A Study in Moral Theory.* Notre Dame, IN : University of Notre Dame Press.

Mayer, F. W. 2014. *Narrative Politics: Stories and Collective Action.* Oxford: Oxford University Press.

McAdams, D. P., 2001. The Psychology of Life Stories. *Review of General Psychology,* 5(2): 100–22.

Miller, P. J., R. Potts, H. Fung, L. Hoogstra, and J. Mintz. 1990. Narrative Practices and the Social Construction of Self in Childhood. *American Ethnologist,* 17(2): 292–311.

Nelson, B. J., 1986. *Making an Issue of Child Abuse: Political Agenda Setting for Social Problems.* Chicago: University of Chicago Press.

Patterson, T. E., 2011. *Out of Order: An Incisive and Boldly Original Critique of the News Media's Domination of America's Poliical Process.* New York: Vintage Books.

Schelling, T. C., 1987. "The Mind as a Consuming Organ. In Jon Elster, (ed.) *The Multiple Self,* Cambridge, UK: Cambridge University Press: 177–98.

Shiffman, D., 2013. Swimming with Sharks: Was Diana Nyad Really at Risk of Being Bitten During Her Historic Swim. *Slate.* September 4. www.slate.com/articles/health_and_science/science/2013/09/diana_nyad_shark_swim_how_dangerous_are_the_sharks_between_cuba_and_florida.html.

Slovic, P. E. 1987. Perception of Risk. *Science* 236: 280–85.

Slovic, P. E., 2000. *The Perception of Risk.* New York: Earthscan publications.

Stone, D. A., 1988. *Policy Paradox and Political Reason*. New York: Addison-Wesley Longman.
Tilly, C., 2002. *Stories, Identities, and Political Change*. Oxford, UK: Rowman and Littlefield Publishers.
Tversky, A. and Kahneman, D., 1974. Judgment under Uncertainty: Heuristics and Biases. *Science*, 185(4157): 1124–31.
Tversky, A. and Kahneman D., 1981. The Framing of Decisions and the Psychology of Choice. *Science* 211(4481): 453–58.

PART II

CASE STUDIES ON OFFSHORE OIL SPILLS

7

From Santa Barbara to the Exxon Valdez

Policy Learning and the Emergence of a New Regime for Managing Oil Spill Risk[1]

Marc Allen Eisner

Crises have played an important role in the history of regulation of the offshore oil and gas industry. The conventional wisdom is that major changes in US policy were the direct result of significant oil spills. This chapter provides much-needed context and nuance to that conventional wisdom by examining the policy response to two events separated by 20 years: the Santa Barbara oil spill of 1969 and the Exxon Valdez spill of 1989. While both events did result in significant new environmental legislation and regulation, the nature of the responses were the product of the political times and built upon a substantial, albeit slower, evolution in environmental policy. The Santa Barbara accident resulted in the institution of a new liability-based regulatory system that satisfied the need for significant policy response during the peak of the environmental movement and appeased the market-based sensibilities of the Nixon administration. The new regime for oil spills created in the wake of the Exxon Valdez – the Oil Pollution Act of 1990 – did not spring fully formed from Congress in the wake of the accident. Rather it evolved over two decades, and reflected successive changes in oil spill policy and the policies created to manage a related problem, toxic wastes. Over the course of this period, members of the policy community expanded and clarified liability, created new mechanisms for funding oil removal and remediation, and ultimately emphasized prevention and response preparedness.

---

[1] The author would like to thank Jonathon Free and Vanessa Freije for research assistance.

7.1 POLICY COMMUNITIES AND THE ENVIRONMENT

In the past several decades, political scientists have devoted a great deal of attention to the dynamics of policy-making. This research has generated a rich literature on policy communities or subsystems, issue networks, and advocacy coalitions. While the differences between competing frameworks is beyond the scope of this paper, each recognizes that the development of public policy is shaped by the interaction of interested actors drawn from the relevant congressional committees, bureaucracies, interest groups and private sector organizations (e.g. universities, corporations) (see Sabatier 2007). Following Frank R. Baumgartner and Bryan D. Jones, the institutional structures and a shared understanding of policy (a "policy image") create a structurally induced equilibrium that conveys stability and promotes incremental change along a particular trajectory (see Baumgartner and Jones 1993). However, focusing events such as crises can disrupt this stability, creating opportunities for significant changes in policy. Thomas Birkland provides a useful definition of a focusing event as "an event that is sudden, relatively uncommon; can be reasonably defined as harmful or revealing the possibility of potentially greater future harms; has harms that are concentrated in a particular geographical area or community of interest; and that is know to policy makers and the public simultaneously" (Birkland 1998: 54) (see also Chapter 5 in this volume). Such events are important because they can change popular perceptions of risk, leading what was once an apathetic public to mobilize. Advocacy groups, previously excluded from the policy monopoly, may enter the fray, hoping to change the dominant policy image and seek significant changes in public policy. Political elites may simply hope to capitalize on the crisis to take advantage of the spike in issue salience.

Following the disruption, new actors, new institutions, and new understandings of core policy problems may once again lead to a new equilibrium, creating a context for successive incremental changes along an altered trajectory. Two qualifications are necessary at this point. First, one might suspect, given this dynamic, that policy history is best understood as qualitatively different regimes separated by crises. But more often than not, new initiatives are layered upon the old and draw heavily on the legacy of past. Second, although crises and their immediate aftermath often attract a good deal of attention, a longer-term perspective is quite useful given that policies often evolve over periods that may span a decade or longer after a focusing event has occurred (Busenberg 2001). Most certainly, this was the case in environmental policy in general and oil spill policy in particular.

The Santa Barbara oil spill of January 1969 occurred near the beginning of – and played an important role in triggering – a period of rapid and substantial change in environmental policy. The fact that it was more or less concurrent with other highly visible events – the Torrey Canyon spill (1967), the burning Cuyahoga River (1969), the Lake Eerie fish kills – and reinforced the core messages of earlier publications like Rachel Carson's *Silent Spring* contributed to its impact (see Johnson and Frickel 2011; Carson, 1962). The emerging narrative was that significant new government action was required to reverse a wide range of environmental damages (see Chapter 6 in this volume for more on narratives in the policy process). High levels of mobilization were sufficient to punctuate the equilibrium that formed around issues of pollution control in the immediate postwar decades. For many activists inspired by the New Left critique of capitalism, these events provided further evidence that the government needed to force higher levels of corporate accountability. New institutions were needed to provide greater access for citizen groups that could advocate a broader understanding of the public interest (see Eisner 1993). In the aftermath of Santa Barbara, events unfolded rapidly. On February 18, 1969, Senator Henry M. Jackson (D-WA) introduced the National Environmental Policy Act, which was signed into law on January 1, 1970. The salience of environmental protection continued to grow, as exhibited by the prominence of the environment in Richard Nixon's 1970 State of the Union Address and Earth Day on April 22, 1970. By the end of the year, Congress had passed the landmark Clean Air Act Amendments and the Nixon administration had created the Environmental Protection Agency (EPA). Two years later, Congress passed the Federal Water Pollution Control Act Amendments (Clean Water Act). The modern environmental era was well underway.

But this was only the beginning. In the next several years, environmental policy-makers turned their attention to the problems of toxic and hazardous wastes. Congress created a system of cradle-to-grave regulation for hazardous waste management (the Resource Conservation and Recovery Act of 1976). It established a new regime for assigning liability for, and funding the remediation of, toxic waste sites (the Comprehensive Environmental Response, Compensation, and Liability Act of 1980). It experimented with mandatory information disclosure on the use of toxic chemicals (the Emergency Planning and Community Right-to-Know Act of 1984). Thus, within fifteen years of Santa Barbara, environmental policy-makers designed and implemented policies that provided a far more systematic means of managing the health and environmental risks associated with various forms of pollution, many of which had been previously ignored.

As institutions were created and innovative statutes were passed, a new environmental policy subsystem emerged, uniting the EPA and state regulators, environmental committees and subcommittees in Congress, a new generation of environmental advocacy groups, and analysts drawn from think tanks and academia. Policy subsystems, as noted above, provide a stable context within which policy evolves. They also provide a context for a process of policy learning that can cut across discrete policy areas. Policy-makers borrow solutions from one policy area and apply them to others. In environmental policy, for example, participants often drew upon the lessons learned in pollution control in one medium to other mediums. They questioned whether certain policy instruments (e.g. tradable permits in acid rain) might be applicable in other areas (e.g. water quality or climate change policy). As experience accumulates, a new understanding of how to manage related problems emerges and may find an expression across multiple related policy areas.

In the present case, the deliberations over oil spills that were triggered by the Santa Barbara oil spill and reinforced by high levels of social mobilization resulted in a series of new statutes. Even if oil spill policy seemed to reach an impasse by the late 1970s, the regulatory regime for oil spills continued to inform the policy debates as Congress turned to address the related issue of toxic wastes in the aftermath of another crisis, Love Canal. Earlier decisions regarding the correct liability regime for oil spills and possible mechanisms for funding remediation found a clear expression in the Comprehensive Environmental Response, Compensation, and Liability Act (CERCLA) of 1980. In 1989, policy-makers faced a new crisis in the Exxon Valdez disaster. As with Santa Barbara, it revealed in striking terms the inadequacy of existing policies. The policy response – the Oil Pollution Act of 1990 – would borrow from CERCLA much as CERCLA had borrowed from earlier oil spill statutes. In short, as a result of two decades of deliberations stimulated by the events of 1969, members of the policy subsystem would develop a far more comprehensive system for managing the risk of oil spills, creating a prevention, response, liability, and compensation regime that increased and broadened liability; required owner/operators to develop response plans for prevention, containment, and cleanup; and funded an Oil Spill Liability Trust Fund through a tax on the petroleum industry. There is good evidence that the new policy contributed to a reduction in the number and severity of oil spills in the subsequent decades, at least until the catastrophic Deepwater Horizon spill of 2010 once again revealed the limitations of the existing regime (see Birkland and DeYoung 2011).

7.2 SANTA BARBARA AND A NEW REGIME

The Santa Barbara oil spill began on January 28, 1969, as oil began to gush from a six-inch gap between the drill and the drilling hole in Union Oil's Platform A, located some six miles off the California Coast. By February 7, up to 230,000 barrels of oil had been leaked, despoiling some 40 miles of coastline along Santa Barbara and Ventura counties. It was the single largest offshore spill thus far in the nation's history – a fact that was driven home by heavy media coverage (Graetz 2011). In an editorial entitled "Deadly Blanket of Blackness," the *Los Angeles Times* described the carnage in striking terms: "Half an inch deep and an untold hundreds of miles in length and breadth, a black blanket of crude oil was still riding the long Pacific swells, spreading death and destruction along the Southern California coastline" ("Pollution: Deadly Blanket of Blackness" 1969: G5). The *Washington Post* editorial board refused to view this event in isolation. It attributed Santa Barbara and other recent events to the "crass indifference to the consequences of technological advance in exploiting nature which is leading to the despoiling of nature. That is to say, the gains from technology seem to run only one way – to profits rather than to the preservation of a planet on which man can comfortably live" ("The Loss of a Few Birds" 1969: G5).

The industry response appeared quite callous. In a Senate hearing on February 5, Fred Hartley, President of Union Oil, stated, "I don't like to call it a disaster, because there has been no loss of human life. I am always impressed by the publicity the death of birds receives compared with that of people." He continued, "relative to death that occurs from crime in our cities, the desecration to the offshore area of Santa Barbara – although it's important and a problem we are fully devoted to – should be given a little perspective" (Quoted in: Jackson 1969: A34). The Nixon administration, in contrast, understood the potential political implications of the spill. By February 9, a newly formed advocacy group, GOO (Get Oil Out), had secured 40,000 signatures on a petition – nearly 30 percent of Santa Barbara residents – calling for an end to oil operations (Greenwood and Dye 1969). Senator Edmund Muskie (D-ME) – the Senate's chief environmentalist and, in Nixon's mind, the likely

Democratic contender in the 1972 presidential election – had arrived on site to hold informal hearings. President Nixon moved quickly to visit Santa Barbara, along with his newly confirmed Interior Secretary, Walter Hickel. As Nixon surveyed the cleanup efforts, he remarked,

It is sad that it was necessary that Santa Barbara should be the example that had to bring [the environment] to the attention of the American people. What is involved is the use of our resources of the sea and of the land in a more effective way and with more concern for preserving the beauty and the natural resources that are so important to any kind of society that we want for the future (Quoted in Farmbry 2012: 162).

Although the Santa Barbara oil spill stimulated the larger debates about environmental protection, it also forced new thinking about federal regulation of petroleum industry. When the spill occurred, the federal government had little formal authority to act. When Secretary Hickel sought to force oil companies working off the Santa Barbara coast to suspend operations, for example, he discovered that he did not have legal authority to do so unless they had violated the terms of their leases. The best he could achieve was the voluntary commitments of six oil company presidents to suspend operations – and this commitment was broken in 24 hours (Hickel 1971). Moreover, it was unclear whether Union Oil could be held liable for the damages. As the *Los Angeles Times* reported, "Under terms of the lease granted by the federal government … Union … is required only to exercise 'reasonable diligence' in drilling and producing operations" (Blake 1969). Although Union voluntarily assumed the costs of the cleanup, there was little question that the existing regulatory regime for petroleum was wholly inadequate.

To place things in context, one must step briefly into policy history. Traditionally, maritime law imposed limited liability for loss of cargo, personal injury and death, and collision damage. In 1851, Congress drew on this tradition and passed the Shipowner's Limitation of Liability Act, capping liability to the value of the vessel and its cargo as a means of promoting investment in the shipbuilding and maritime industries (Zimmermann 1999). Subsequently, the Oil Pollution Act of 1924 prohibited discharge of oil in coastal waters, imposing a fine ($500 to $2,500) and/or imprisonment (30 days to one year). The Clean Water Restoration Act (1966) built on the 1924 legislation and introduced civil liability for oil pollution. It expanded the reach of the law to prohibit discharge on shorelines and the navigable waters of the United States. Those responsible for the discharge were required to remove it immediately. Failing

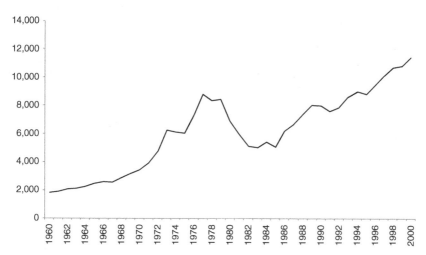

FIGURE 7.1 Petroleum imports, 1960–2000 (Thousands of Barrels per Day)
Source: US Energy Information Administration, *Annual Energy Review* (2012), Table 5.4.

this, the Interior Secretary was authorized to arrange for the removal and the responsible party was subject to a higher fine ($10,000) and reasonable removal costs incurred by the government. However, liability was capped at the lesser of $5 million or $67 per gross ton of the vessel in question. Moreover, the federal government faced a significant obstacle in assessing these penalties insofar as it had to prove that the spill resulted from gross negligence (Wang 2011).

The regime for oil spills clearly lagged behind the changes in the sources of oil for domestic use. Traditionally, oil was extracted from wells on land and oil imports were quite limited. The 1960s, however, had witnessed a significant growth both in the international transportation of oil and the exploitation of offshore petroleum resources (see Figures 7.1 and 7.2). Between 1960 and 1970, imports almost doubled, from 1.8 million barrels per day to 3.4 million barrels per day. Moreover, offshore oil production increased fivefold, from 319 thousand barrels per day to 1.6 million barrels per day. But public policy had failed to keep pace with the changes in the sources of petroleum and the risks they imposed (see Murchison 2011; US Energy Information Administration 2012).

In the immediate aftermath of the Santa Barbara spill, Interior Secretary Hickel introduced new regulations for oil companies working under federal leases. Some of these regulations were technology-based, the kinds of regulations that would be referred to subsequently as

FIGURE 7.2 Offshore petroleum production, 1960–2000 (Thousands of Barrels per Day)
Source: US Energy Information Administration, *Annual Energy Review* (2012), Table 5.2.

"command-and-control" regulations (e.g. technical specifications for well casings, blowout preventers, the inspection regime). Henceforth, any variances from these regulations would have to be approved by the Geological Survey rather than on site (see Rich 1969). The most important changes, however, came in the area of liability, departing from the prevailing approach to oil spills described above. Hickel imposed *absolute* liability without having to prove gross negligence. As Hickel explained, this decision largely reflected the practical need for an immediate response:

> Cleaning up a spill cannot wait for a court judge to decide who is liable. It has to be done before the pollution kills the wildlife and ruins the beaches. For this reason I demanded that all companies who hold drilling leases on the outer Continental Shelf accept liability for clean up even before the cause of a spill is determined. This became known in short as "absolute liability without cause." (Hickel 1971: 91)

All of this came as something of a surprise. As Governor of Alaska and Interior Secretary-designee, he had expressed concerns over the impact of water standards on industrial development (Frome 1992). Now, as Interior Secretary, he proved unresponsive to claims that absolute liability could undermine oil production and the incentives of current leaseholders, and render some independent operators uninsurable.

A New Regime for Managing Oil Spill Risk

As oil pollution policy was moving forward at the Department of the Interior, US delegates were simultaneously participating in the development of an international convention. Following the 1967 Torrey Canyon disaster, which resulted in the spill of some 32 million gallons of oil, damaging the coastlines of France and the UK, an international convention was called in Brussels to address issues of liability on an international level. The resulting International Convention on Civil Liability for Oil Pollution Damage, finalized in 1969, imposed strict liability on ship owners and required that ship owners carrying in excess of 2,000 tons of oil have insurance or exhibit the capacity to pay for damages. Assuming that the incident was not a result of negligence, liability was capped at $125 per ton of the ship's tonnage, with an aggregate cap of $140 million. Although the US delegation was actively involved in negotiating the terms of the convention, and President Nixon urged prompt ratification, in the end, the Senate refused to ratify it. The core problem for the Senate was rather straightforward. The international convention would have required the federal pre-emption of state laws, many of which imposed no cap on liability (a point we will return to later). Moreover, given the urgency of the problem in the aftermath of Santa Barbara, there was little desire to tie national policy to the protracted negotiations that were common in the development and ratification of international conventions (Zimmermann 1999).

Although the Senate refused to ratify the international convention, Congress nonetheless moved quickly to incorporate expanded liability into the Water Quality Improvement Act (WQIA) of 1970, an amendment to the Federal Water Pollution Control Act. The WQIA declared "there should be no discharges of oil ... into or upon the navigable waters of the United States, adjoining shorelines, or into or upon the waters of the contiguous zone."[2] It authorized the president to issue regulations regarding the quantities of oil deemed harmful, the methods and procedures for the prevention of discharges, and remediation.[3] Moreover, in recognition of the patchwork of state policies, it authorized the president to coordinate and direct public and private cleanup efforts, establish criteria for local and regional contingency plans, and develop a National

---

[2] 33 U.S.C. § 1321 (b)(I)

[3] In September 1970, the Department of the Interior provided a rather imprecise definition of harmful quantities as "any discharge which (a) violates any applicable water quality standard (state or federal), or (b) causes a film or sheen upon or discoloration of the surface of the water or adjoining shorelines, or causes a sludge or emulsion to be deposited." See Keener (1971).

Contingency Plan for the removal of oil (borrowing from the international response to the Torrey Canyon spill) (see Murchison 2011; Keener 1971).

The WQIA required that the owner/operator report any discharge of a harmful quantity to the US Coast Guard (USCG), or face a fine of $10,000, up to a year in prison, or both. More important, it established civil and criminal penalties for violations of the act. Assuming the spill was not the product of acts of God, acts of war, or acts/omissions of a third party with no contractual relationship with the potentially liable party, Congress authorized the federal government to recover cleanup costs within statutorily prescribed limits (e.g. $100 per gross ton or $14 million for a vessel; an $8 million liability ceiling was imposed for onshore facilities). Vessels of over 300 gross tons, moreover, were required to establish and maintain evidence of financial responsibility to meet their potential liability. The most controversial provision of the act – one which met with the opposition of insurers, the oil and shipping industries, and the Nixon administration – established absolute (i.e. strict) liability. That is, liability existed "without regard to whether any such act or omission was or was not negligent" (Kenworthy 1970: 28). If the government could prove that the spill was a product of willful negligence or willful misconduct, the owner/operator's liability was unlimited. The government could seek to recoup additional cleanup costs via tort actions, although its success could be constrained by the Shipowner's Limitation of Liability Act of 1851, which capped liability at the value of the vessel and its cargo ("Oil Spills and Cleanup Bills: Federal Recovery of Oil Spill Cleanup Costs" 1980). In a nod to federalism, the WQIA did not pre-empt state laws (Murchison 2011).

When Congress passed the Federal Water Pollution Control Act Amendments (Clean Water Act) in 1972, it incorporated the provisions of the WQIA into § 311 of the new statute and extended them to hazardous wastes. The act once again established liability limits and required that owner/operators provide proof of their financial ability to meet the cleanup costs noted above. Although the act did not include an explicit strict liability provision, the courts nonetheless interpreted it as imposing this standard (Klass 2004). Under § 311(b)(4), the EPA was required to determine "those quantities of oil and any hazardous substances the discharge of which may be harmful to the public health or welfare." The need to set reportable quantities of some 271 substances deemed hazardous resulted in enormous delays, a product of the complexity of the underlying science and legal challenges issued by the Manufacturing

Chemists Association (see Garrett 1979). Section 311(k) created a $35 million revolving fund (the 311 Fund) in the US Treasury, controlled by the USCG. Funded by appropriations and monies recovered from past liable parties, it could be used to finance future cleanup and mitigation before liable parties reimbursed the costs incurred by the government.

The Clean Water Act was amended again in 1977, further strengthening the oil spill regime. Under the amendments, jurisdiction was extended 200 miles offshore. The amendments raised the liability of vessels to $150 per gross ton and removed the $14 million cap; at the same time, it raised the liability caps for onshore facilities to $50 million, with lower liabilities for inland oil barges, which posed less of an environmental threat (Hall 1978). Importantly, the 1977 amendments went well beyond the removal of oil. Section 311(f)(4) noted that liability extended to "any costs or expenses incurred by the Federal Government or any State government in the restoration or replacement of natural resources damaged or destroyed as a result of the discharge of oil." Section 311(f)(5) provided that the president "shall act on behalf of the public as trustee of the natural resources to recover the costs of replacing or restoring such resources. Sums recovered shall be used to restore, rehabilitate, or acquire the equivalent of such natural resources by the appropriate agencies" (Quoted in: Schenke 1990: 14). While conferees reported that both chambers were committed to some sort of "superfund" legislation to cover the costs of catastrophic spills, the creation of such a fund would have to be put off for another day (Hall 1978).

A few additional pieces of legislation are worth mentioning. In 1973, Congress passed the Trans-Alaska Pipeline Authorization Act. Following the Clean Water Act, it imposed a standard of strict liability on the owners of the pipeline right-of-way and owners and operators of oil tankers carrying Alaskan oil transported through the pipeline. In addition to removal costs, liability for holder of the pipeline right-of-way was capped at $50 million per incident (additional liability could be determined under ordinary rules of negligence) and $14 million per incident for vessel owner/operators. The Act also created the Trans-Alaska Pipeline Liability Fund, which was to be maintained at $100 million and funded through a five cent per barrel fee, collected by the pipeline operator (Stone 1975).

The next year, Congress passed the Deepwater Port Act of 1974, creating a process for licensing and regulating the construction and operation of deep-water ports (i.e. structures used as terminals for the loading and unloading oil and its transfer to onshore facilities via pipelines). Once again, Congress imposed a standard of strict liability. Responsible parties were liable for oil removal. Liability for damages was limited to $250 per gross ton or $20 million for vessels, and $50 million for the licensees of deep-water ports. A newly created Deepwater Port Liability Fund would cover immediate costs and removal expenses and damages that could not be compensated by responsible parties. It was to be maintained at $100 million, funded from fees collected by port licensees from owners of oil (Pfennigstorf 1979).

The new regime for oil spills was further strengthened in 1978, when Congress passed the Outer Continental Shelf Lands Act Amendments. Under Title III of the act, owner/operators of vessels and offshore facilities were jointly and severally and strictly liable for the costs of cleanups and other economic damages. The act, once again, established different levels of liability for vessels and offshore facilities but in both cases owner/operators were responsible for removal and cleanup costs paid by the federal or state governments. The act required that owner/operators establish "evidence of financial responsibility sufficient to satisfy the maximum amount of liability" to which they could be exposed (e.g. in the case of offshore facilities, a maximum of $35 million). In recognition of the fact that cleanup costs could exceed the statutory limits, the Act also created a new Offshore Oil Pollution Compensation Fund, capitalized at $200 million and funded through a fee on oil extracted from the Outer Continental Shelf (Murchison 2011).

Congress sought to strengthen the regime in subsequent years to produce a single, comprehensive policy for oil spills. But these efforts were stymied by the same vagaries of federalism that had previously prevented ratification of the International Convention on Civil Liability for Oil Pollution. Several states had their own oil spill statutes, and the majority of these states placed no caps on liability. For states with unlimited liability, federal legislation that pre-empted state laws to impose a national liability standard would constitute a weakening of state laws. Senators representing coastal states with stricter laws were steadfast in their opposition, whereas a majority in the House of Representatives, the petroleum industry, shippers and insurers strongly supported uniform federal liability caps.

To recap, within less than a decade of the Santa Barbara oil spill, Congress had passed a series of statutes that expanded and clarified liability and required that owner/operators have the resources to meet their potential liability. Government can adopt various strategies for managing risks. In the areas of environmental and occupational health, much of the core statutes of the 1970s focused on risk reduction via command-and-control regulations. Regulators prescribed acceptable levels of emissions, imposed technological standards, and banned the production or use of various substances known to constitute an unreasonable risk to health or the environment. Certainly, the regime for oil spills had some command-and-control elements. But as David A. Moss (2002: 283) notes, these statutes "harnessed the power of risk shifting." They "purposefully and aggressively utilized liability as an environmental tool." The underlying theory was that higher levels of liability would create incentives for greater risk management on the part of industry actors, leading to a reduction in oil spills. Moreover, to fund removal and remediation, four funds were created, three of which would draw on fees from the petroleum industry. Arguably the liability approach is more pro-business than equivalent command-and-control approaches, as it gives businesses the right to choose the most cost-effective means of reducing risk. Thus, given the political demands for action in the wake of the Santa Barbara spill, it may have been more appealing to the pro-business elements of the Nixon administration.

Ultimately, the patchwork of state and federal statutes would remain in place until the next salient event: the Exxon Valdez disaster in 1989. But before turning to that event, it is worth noting that the liability-based regulatory approach adopted after Santa Barbara also set the trajectory for environmental policy in the US much more broadly. To see that we must examine the efforts to design a new regime for toxic and hazardous wastes that drew heavily on the oil spill statutes.

7.3 FROM OIL SPILLS TO TOXIC CHEMICALS

While congressional efforts to strengthen the oil spill regime bore little fruit in the years following the passage of the Outer Continental Shelf Lands Act Amendments, another incident triggered major changes in the regulation of toxic chemicals. In 1978, a series of exposés revealed a tragic series of events at Love Canal in Niagara Falls, New York. During the 1940s, the Hooker Chemical Company had disposed of some 21,000 tons of toxic waste in a clay-lined canal. Although Hooker had sold the property to Niagara in 1953 for $1 – with full disclosure of the wastes – the city subsequently permitted development on the site (including a school, a playground, and housing). By the 1970s, the toxic mix of mercury, benzene, chlorinated compounds, and dioxins contaminated the groundwater and the soil, resulting in a range of serious health problems

(Barnett 1994). The contemporaneous revelations of the "Valley of the Drums" – a site with more than 20,000 leaking barrels of hazardous wastes in Louisville, Kentucky – only increased the public attention drawn to the problem. Members of Congress clearly understood that the "nation's hazardous waste problem was made more difficult by the fact that existing statutory authority, common law authority, and funding for cleanup of existing hazardous waste sites was woefully inadequate" (Klass 2004: 928).

In 1976, Congress had passed the Resource Conservation and Recovery Act (or RCRA) to regulate hazardous wastes. It created a cradle-to-grave system of regulation that imposed stringent record-keeping and reporting requirements to document the generation, use, transportation, and disposal of wastes. Treatment, storage, and disposal facilities, moreover, were required to meet construction and performance standards promulgated by the EPA. But even if RCRA performed as intended – and there is much evidence that the demands of the new policy simply overwhelmed the analytical and financial resources of regulators – it did nothing to address the problems posed by Love Canal, the "Valley of the Drums," and other yet-undisclosed toxic waste sites that predated RCRA. In the wake of these striking revelations, Congress turned to strengthen regulations with the Comprehensive Environmental Response, Compensation, and Liability Act of 1980 (see Eisner 2007).

Although the details of CERCLA are beyond the scope of this paper, the act created a system for identifying and prioritizing hazardous waste sites, assessing liability for responsible parties, and funding remediation. Several connections between CERCLA and earlier oil spill policies are worth noting. First, CERCLA drew on the example of the National Contingency Plan that had been authorized by the WQIA to coordinate cleanups and further expanded to cover hazardous wastes under the provisions of the Clean Water Act in 1972. Under CERCLA, the National Oil and Hazardous Substances Contingency Plan was broadened once again to cover the releases from hazardous waste sites. Second, Congress drew on the examples of the recently passed oil statutes (see above) that created various funds that were financed through a fee on oil. Under CERCLA, Congress created the Superfund, which was financed through an excise tax on the petroleum and chemical industries. It would be used to cover the costs of remediation where it was impossible to identify and/or recover costs from responsible parties (e.g. because the firms in question no longer existed).

Third, CERCLA adopted the liability regime and thus the risk-shifting strategy that had been applied to oil spills. The House and Senate bills (H.R. 7020 and S. 1480) that provided the foundations for CERCLA included language expressly referring to strict, joint, and several liability. In the waning days of the lame duck Congress, opposition to the liability language threatened to stall passage. The sponsors feared that the window of opportunity for policy change could close quite abruptly once the 97th Congress was sworn in and Ronald Reagan assumed the presidency. To facilitate passage, the sponsors substituted new language in CERCLA § 101 (32), noting that the act adopted "the standard of liability which obtains under section 1321 of Title 33," that is, § 311 of the Clean Water Act. As noted above, the courts had interpreted this section of the Clean Water Act as imposing strict liability. The concession, in short, was not a genuine concession. It did nothing to weaken the liability standards in CERCLA but rather extended the existing liability standards in petroleum to hazardous waste sites more generally (Klass 2004).

Congress drew heavily on existing oil spill policy when shaping CERCLA and there was some hope of extending the new legislation to oil spills, thereby creating a unified scheme. This goal, alas, became another victim of lame duck session. During the legislative debates, a House bill covered oil spills, but the final version of the Senate bill that became CERCLA contained an exclusion for petroleum. While members of both chambers recognized this as a major defect that could complicate liability under CERCLA, there was insufficient time to forge a compromise. In the end, § 101(14) of CERCLA stated that the term hazardous substance "does not include petroleum, including crude oil ... natural gas, natural gas liquids, liquefied natural gas, or synthetic gas usable for fuel (or mixtures of natural gas and such synthetic gas)." Ironically, CERCLA would not extend the new regime to oil spills, even if its very design drew on the legislative efforts stemming from the Santa Barbara spill (Knopf 2011).

7.4 THE EXXON VALDEZ AND THE OIL POLLUTION ACT OF 1990

Since the 1970s, further progress on oil spill liability law has been thwarted by a conflict over states' rights. As noted earlier, the core question was whether federal law should pre-empt state laws. Some observers believed that the stalemate would not be broken until another devastating spill forced the issue onto the agenda (Hager 1989a). On March 24, 1989, this proposition was tested when the Exxon Valdez ran aground on Bligh Reef. The resulting spill dumped some 250,000 barrels of North Slope crude oil into Alaska's Prince William Sound, creating a slick larger than the state of Delaware and giving rise to an environmental disaster that was far greater than the Santa Barbara spill two decades earlier. Media coverage was, once again, critical in bringing images of the ecological impacts to a broader public. As one analyst observed: "The Valdez spill, with its dramatic television footage of a huge and grotesque environmental disaster was the 'Pearl Harbor' of the US environmental movement" (Randle 1991). Conservative estimates of the initial casualties included 250,000 seabirds, 144 bald eagles, 4,400 sea otters, and 20 whales, as well as significant and long-term declines in the local fisheries. Moreover, the initial human impacts included "high levels of collective trauma, social disruption, economic uncertainty, community conflict, and psychological stress." In short, the biological, commercial and social impacts were nothing short of catastrophic (Gill, Picou, and Ritchie 2011).

The accident revealed in stark terms the inadequacy of industry practices. The tanker's captain, Joseph Hazelwood, was intoxicated and asleep, leaving control of the ship to the third mate who lacked a license to pilot the ship in Prince William Sound. (Hazelwood had completed an alcohol treatment program a few years earlier, a fact that was known by Exxon.) Alyeska Pipeline Service (a consortium co-owned by Exxon, British Petroleum, Arco, Mobil, Amerada Hess, Phillips, and Unocal) was responsible for deploying a contingency plan for responding to spills. However, its response was delayed and piecemeal due to inadequate supplies of barrier booms and dispersants and the fact that the containment barge had been stripped of equipment for repairs from storm damage some two weeks earlier. Because Alyeska had disbanded its full-time response team in 1981 as a cost-cutting measure, it proved particularly difficult to assemble an emergency crew on a holiday weekend. Alyeska's problems dated back several years, with documentation of failed oil spill drills, inadequate equipment, and poor personnel training. Indeed, state records had described its spill response capacity as having "regressed to a dangerous level" (McCoy and Wells 1989). As the *Wall Street Journal* reported: "The oil companies' lack of preparedness makes a mockery of a 250-page containment plan, approved by the state, for fighting spills in Prince William Sound." Regardless of the causes of this particular wreck, "the disaster…exposed a much deeper problem: the seeming inability of the oil industry to fight major oil spills" (Wells and McCoy 1989).

The corporate response did not help matters. Exxon's chairman, Lawrence Rawl, did not comment on the event for six days, and then there was a consistent effort to deflect responsibility for the slow response away from Exxon and Alyeska (e.g. Rawl claimed that some of the blame should be assigned to environmentalists who questioned the toxicity of dispersants and to the Alaskan officials and USCG that "wouldn't give us the go-ahead to load those planes, fly those sorties, and get on with it."). Although Exxon ran full-page apologies in the newspaper, the chairman consistently argued that the company had gotten a "bad rap." When asked in a *Fortune* interview what lessons he would offer other CEOs based on the crisis, Rawl responded: "You'd better prethink which way you are going to jump from a public affairs standpoint before you have any kind of a problem. You ought to always have a public affairs plan...I just keep putting one foot in front of the other, and I'm hoping with a little bit of luck to prove to you that we're going to make this thing work out better than the greatest, most optimistic expectations ("Rawl: In Ten Years You'll See Nothing" 1989).

While Exxon's chairman may have viewed the spill through the lens of public relations, Congress drew a different lesson: the existing patchwork of policies for oil spills was simply inadequate. Although Congress had established four separate funds to cover the costs of cleanups, three were location specific (the outer continental shelf, the Louisiana Offshore Oil Loop, and the Alaskan pipeline), and the revolving fund created under § 311(k) of the Clean Water Act was grossly underfunded. Although it had been authorized at $35 million, at the time of the Exxon Valdez spill, it contained a mere $4 million. It could cover but a small fraction of the estimated $1 billion that would be required for cleanup, restoration, and economic damages. To place things in perspective, Exxon itself was spending $1 million per day on the cleanup. Even if there was dissatisfaction with Exxon's efforts, there was a valid concern that any efforts to federalize the cleanup could lead the company, in the words of USCG Commandant Paul A. Yost Jr., "to close their checkbook." If the federal government sought to replicate Exxon's efforts, the 311 revolving fund would be fully depleted in a matter of days (Hager 1989b; Hager 1989a).

As the media attention and public outrage over the Exxon Valdez mounted, Congress turned again to the issue of oil spill liability. The salience of the event – when combined with contemporaneous spills in Rhode Island, Delaware, and the Houston Ship Channel – was sufficient to break the stalemate of the past, resulting in the passage of the Oil Pollution Act (OPA) of 1990. As Congress considered liability, it was

TABLE 7.1 *State oil pollution liability regimes*

Limited Liability on Shipping Sector (vessel owner or operator) and Oil Cargo Sector (oil cargo owner)	Limited Liability on the Shipping Sector (vessel owner or operator)	Unlimited Liability on Shipping Sector (vessel owner or operator) and Oil Cargo Sector (oil cargo owner)	Unlimited Liability on the Shipping Sector (vessel owner or operator)
Florida	Delaware	Arkansas	Alabama
New Jersey	Louisiana	California	Connecticut
New York	Texas	Hawaii	Georgia
	Virginia	Maryland	Illinois
		North Carolina	Maine
		Oregon	Massachusetts
		Washington	Michigan
			Minnesota
			Mississippi
			New Hampshire
			Ohio
			Pennsylvania
			Rhode Island
			South Carolina

Source: Kim (2002)

forced to return to the issue of pre-emption and the related question of international protocols. The majority of states had oil pollution liability regimes in place and the majority of these imposed unlimited liability on shippers and in many cases, the oil cargo owners as well (see Table 7.1). For the past decade, further progress on oil spill regulation had been blocked by the inability of the House and Senate to agree on the question of pre-emption. Under the 1984 Protocols to the International Convention on Civil Liability for Oil Pollution Damage and the Protocol to the International Fund for Compensation for Oil Pollution Damages, liability would be have to be capped at $260 million per vessel (assuming that the spill was not a product of willful negligence or misconduct). Participation in these international protocols would be impossible absent federal pre-emption. In the deliberations following the Exxon Valdez, the House forwarded legislation that would pre-empt state laws – a position that was supported by the Bush administration and the shipping industry. Insurers also supported the caps, arguing that they could not provide insurance to cover unlimited liabilities and, hence, caps were required to secure functioning insurance markets (see insurance discussion in Chapter 3 of this volume). In the Senate, however, there was once again a steadfast refusal to accept legislation that would weaken state laws (or to support a protocol that would have the same effect). Senate Majority Leader George J. Mitchell (D-ME) warned that there would be no vote on any law that pre-empted state laws. Ultimately, the conferees agreed that the OPA would not interfere with state laws, even if it precluded participation in the international conventions (Cushman 1990; Kuntz 1990).

As with CERCLA, the OPA adopted "the standard of liability which obtains under section 311 of the [Clean Water Act]." In other words, responsible parties are strictly, jointly and severally liable for cleanup costs (without limit) and damages (within statutory limits). With respect to damages, the OPA increased liability well above the levels established under the Clean Water Act. For tankers above 3,000 tons, liability was capped at the greater of $1,200 per gross ton or $10 million. For vessels below this size, liability was capped at $2 million (with liability limits of $500,000 or $600 per gross ton for nonoil vessels). As in the past, liability for damages was unlimited if the spill was the product of willful negligence or willful misconduct, or if the responsible party failed to report the spill or make reasonable efforts to remove the oil. As with § 311 of the Clean Water Act and CERCLA, acts of God or war, or acts/omissions of third parties with no contractual relationship to the responsible party were recognized as defenses against liability for removal and damages (see Nichols 2010; Grumbles and Manley 1995).

Under the OPA, the range of recoverable damages was comprehensive, including (1) natural resources damages; (2) real and personal property damages including use value; (3) the loss of subsistence use of natural resources; (4) the loss of tax and other revenues; (5) the loss of profits or earning capacity; and (6) the increased costs of public services. Of these categories, damages to natural resources, loss of tax revenues and increased costs of public services were recoverable only by government (federal, state, foreign) and Indian tribes. In the remaining categories, government and private parties could both recover damages (Randle 1991).

The OPA also addressed another gap in the oil spill regime: the lack of funding sufficient to clean up a catastrophic spill. In 1986, Congress had created the Oil Spill Liability Trust Fund as part of the Omnibus Budget Reconciliation Act, building on the precedents set in the Outer Continental Shelf Lands Act Amendments (1978) and CERCLA. However, it failed to authorize its use or provide a funding mechanism.

The OPA addressed these issues. It consolidated the four existing funds into the Oil Spill Liability Trust Fund; the Offshore Oil Pollution Compensation Fund transferred $216 million in 1990, with another $335 million subsequently transferred from the Trans-Alaska Pipeline Liability Fund (National Pollution Funds Center 2008). Moreover, the OPA authorized a five-cents per barrel fee on imported and domestic oil to bring the fund up to $1 billion (the tax ceased on December 31, 1994, due to a sunset provision but was reinstated under the Energy Policy Act of 2005). Under the process established by the OPA, claimants could seek compensation from the trust fund if the responsible parties denied liability or failed to make payment within 90 days. At that point, the Attorney General could recover costs incurred from the responsible party (Meltz, Ramseur, and Pettit 2010).

In addition to establishing liability and a funding mechanism to assist in remediation, the OPA focused on preventing the recurrence of events like the Exxon Valdez. It imposed tighter regulations for licensure of merchant mariners and the manning of vessels. It required that new tankers operating in US waters have double hulls; existing single hull tankers would have to be retrofitted or retired in accordance with a statutory timetable. It also mandated that tankers participate in the USCG vessel monitoring and tracking system, to help ships avoid navigation hazards, and the use of escort vessels in Prince William Sound and Puget Sound.

More importantly, it strengthened contingency planning requirements at all levels. At the national level, the OPA required that the National Contingency Plan be expanded to include a fish and wildlife response plan and a worst-case discharge response plan. It mandated a more demanding system of contingency planning, involving a national response unit to be established by the USCG at Elizabeth City, North Carolina, to maintain an inventory of spill removal resources and equipment and to coordinate public and private responses to spills. The national response unit would also coordinate with newly created USCG strike teams and district response groups that were created for each of the ten USCG districts. The OPA strengthened the requirements that vessels and facilities have response plans, including emergency response procedures to address worst-case scenario spills. These plans had to be filed with the relevant agencies (e.g. the USCG, the EPA, or the Department of Transportation) by February 1993 and approved by February 1995. Facilities and vessels were prohibited from transporting or storing oil if they failed to meet these requirements. There was no expectation that each vessel would have the equipment and personnel on hand to implement their response plans.

Rather, they could secure these resources via contractual relationships (Ramseur 2010; Randle 1991). This provision, in turn, stimulated an increase in the self-regulatory capacity of the petroleum industry.

The financial consequences of the spill created incentives for Exxon to invest more heavily in safety and environmental performance. In the immediate aftermath of the spill, environmentalists called for a boycott of Exxon's gasoline, and some 40,000 consumers returned their Exxon charge cards. While Exxon's stock initially lost value, the impact was surprisingly minor and short-lived.[4] In the end, however, Exxon paid out more than $3.8 billion in cleanup and damages, with an additional $507 million in punitive damages (reduced from $5 billion awarded earlier by a jury). Exxon, in turn, filed suit against its insurers for their "bad faith" conduct in refusing to cover the costs of the spill. The insurers, led by Lloyd's of London, had justified their refusal on number of factors, including "Exxon's willful, wanton, reckless and/or intentional misconduct." Exxon demanded $3 billion, but ultimately settled for $300 million, a fraction of the costs it had incurred (Lenckus 1996).

Beyond Exxon, the spill was a public relations disaster for the petroleum industry as a whole. It revealed its lack of preparation for a catastrophic spill (e.g. prior to the event, the oil industry's worst-case scenario was a spill one-quarter of the size of the Exxon Valdez) (Hager 1989b). In an immediate response, the American Petroleum Institute – the industry trade association – convened a task force to consider the need for new industry procedures for preventing, containing, and cleaning up oil spills. The end result was the creation of a new Petroleum Industry Response Organization, which was subsequently chartered as the Marine Spill Response Corporation (MSRC) (see O'Reilly 1993). The MSRC was created to maintain the equipment and personnel necessary to mitigate catastrophic spills. The industry also created a new nonprofit corporation, the Marine Preservation Association, to fund the MSRC via the dues paid by member corporations. Rather than developing their own catastrophic response plans and negotiating agreements with contractors, members could simply join the association and obtain a MSRC service contract to meet the requirements of OPA.

In recognition of the fact that the resources necessary for future cleanups would need to come from the private sector, Congress supported the

---

[4] On March 24, 1989, Exxon's stock closed at $11.13 a share. It fell to a low of $10.44 on April 11, recovering its full value within three months of the spill. Surprisingly, by the end of the year, Exxon's stock closed at $12.50 a share (see White 1995).

efforts of the Petroleum Industry Response Organization (which became the MSRC). The Conference Committee for the OPA, which was conducting its work as the consortium of twenty oil companies was developing the industry response, explicitly cited it as an organization with which the USCG National Response Unit should work as it engaged in its contingency planning (Randle 1991). Over the next five years, the MSRC would devote $900 million to create five regional response centers with 23 equipment staging areas (Grumbles and Manley 1995; Wald 1991). In sum, as a result of the deliberations surrounding the OPA, the petroleum industry increased its capacity for managing risks and its efforts became partially integrated with those of the federal government. Regulation gave rise to coregulation that was embedded into a stronger scheme for liability and prevention.

7.5 THE OIL POLLUTION ACT AS A SUCCESSFUL REGIME?

Crises can stimulate high levels of political mobilization, force issues on to the agenda, and ultimately lead to rapid and significant changes in public policy. Yet, one must be cautious. As John W. Kingdon reminds us, there are always solutions waiting for problems to happen (see Kingdon 1984). When crises occur, advocates move quickly to couple their pre-existing solutions to the problems at hand before the window of opportunity closes, as it invariably will (and often quite abruptly). In so doing, they may move to premature closure, framing policies before they have an adequate understanding of the factors that caused the crisis in question and the appropriateness of competing policy instruments. There is nothing to guarantee that the resulting policy will be successful in enhancing the government's capacity to prevent new crises from emerging or to manage the attendant risks.

Ideally, the response to a crisis would incorporate what had been learned about the underlying causality. If policy-makers are successful in understanding issues of causality and select the appropriate policy instruments, they may avert future crises or, at the very least, reduce their magnitude so that they fall within acceptable parameters. The response to the Santa Barbara spill did not primarily engage issues of causality. Rather, it worked to shift risk to industry actors by adjusting the liability scheme. By imposing norms of strict liability and raising the monetary ceilings on liability, Congress hoped to create the incentives for corporations to adopt a more proactive stance and attend to issues of risk management. In the event that this failed, funds were established to facilitate oil removal and compensate those who had borne the costs of a spill. The contrast with the Exxon Valdez is clear. Congress once again increased the caps on liability and dramatically increased the resources available for remediation. But there was also a stronger emphasis on prevention (e.g. the mandate for double hulls on vessels and participation in the USCG monitoring and tracking system, and the requirements that vessels and facilities develop response plans and have the resources on hand to respond to spills at an early stage). The introduction of these preventive measures, one might argue, marked a clear departure from the regime put in place in the 1970s.

How did the new oil spill policies perform? The USCG collects data on oil spills, and this data, as presented in the Polluting Incident Compendium provides clear evidence of significant improvement following the passage of the OPA in 1990.[5] In the 1973–1990 period, an average of 11.86 million gallons of oil were spilled on an annual basis. In the 1991–2009 period, an average of 1.9 million gallons of oil were spilled annually. This reduction is particularly impressive given that this latter period includes the 8 million gallons discharged from facilities following Hurricane Katrina and another spill of 1.8 million gallons from a tank barge and a platform that had sunk as a result of Hurricane Rita, both in 2005 (US Coast Guard 2012). As Figure 7.3 shows, the sharp drop in the total volume of spills occurred following the passage of the OPA in 1990. Figure 7.4 provides a graphical representation of the number of significant spills by spill size over 1,000 gallons. What is particularly notable is the dramatic reduction in spills over 100,000 gallons that, in turn, suggest the importance of contingency planning and the rapidity of response. Indeed, between 1991 and 2009, there were only two years out of 18 when there were spills of over 1 million gallons, and as noted above, the most significant spills of 2005 were a product of hurricanes. This contrasts sharply with the pre-1991 period, when spills in excess of 1 million gallons occurred on an annual basis every year but 1977.

The data presented here supports the conclusion that the series of policy changes that stemmed from the Santa Barbara spill and, following the Exxon Valdez, culminated in the Oil Pollution Act of 1990 significantly enhanced the nation's ability to manage the risks from oil spills, in many cases, it would appear, preventing spills (e.g. through double-hulled vessels) or reducing their magnitude (e.g. through expanded contingency

---

[5] The Coast Guard began collecting data with the creation of the Pollution Incident Reporting System in 1973.

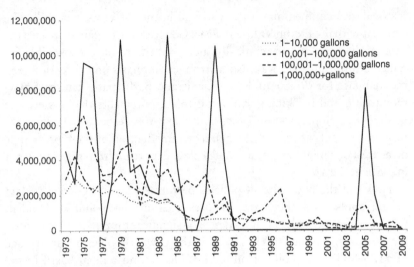

FIGURE 7.3 Total volume of spills by spill size, 1973–2009
Source: US Coast Guard, Pollution Incident Compendium (2012).

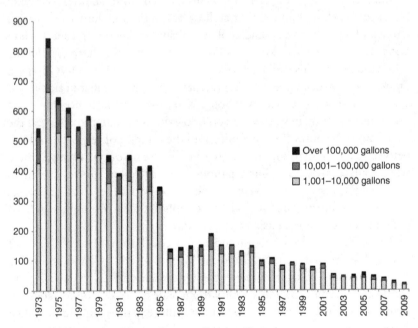

FIGURE 7.4 Number of significant spills by spill size, 1973–2009
Source: US Coast Guard, Pollution Incident Compendium (2012).

planning and response plans based on more realistic worst-case scenarios). Yet, as subsequent events would reveal, there was much work yet to be done.

On April 20, 2010, an explosion on BP's Deepwater Horizon mobile offshore drilling unit would result, ultimately, in a spill of 206.6 million gallons of oil and another 400,000 gallons of oil products. To place things in perspective, the spill from Macondo 252 would account for 86 percent of all the oil discharged between 1973 and 2010. Even though things would return to "normal" in 2011 – there were no spills above 5,000 gallons for the first time on record – the Deepwater Horizon catastrophe triggered a new round of critical reflections on the policies and practices in place for managing the risk of oil spills, the quality of regulation, conflicts of interest in the regulatory model, and the adequacy of industry response plans. Certainly, BP had a comprehensive spill plan (some 582 pages in length, with 52 pages dedicated to the Deepwater Horizon rig). But it grossly understated the worst-case scenario and was riddled with errors, identifying walruses, sea otters, sea lions and seals as "sensitive biological resources" in the Gulf and listing experts to be consulted who had long-since relocated and in one case died years before the completion of the plan (see Mohr, Pritchard, and Lush 2010). Clearly, the kind of complacency that had been so evident two decades earlier in the Exxon Valdez event had not been eliminated.

Based on the precedents set by earlier significant spills, one might have expected the Deepwater Horizon to stimulate another substantial change in the regime for oil spills. The growing technological complexity of deepwater drilling and the evidence of inadequate industry preparation suggested in stark terms that the regime created under the Oil Pollution Act of 1990 was no longer sufficient to manage the risks of catastrophic damages. According to a Gallup (2010) poll conducted in May 2010, 87 percent of the nation was following media coverage of the spill closely; fully 72 percent believed that the Deepwater spill constituted a disaster, with 37 percent agreeing that it was the worst environmental disaster in the past hundred years. With the White House and Congress under unified Democratic control and President Obama's emphasis on environmental issues and sustainable energy in the 2008 campaign, one might have anticipated the kind of rapid and significant policy change witnessed in the wake of the Exxon Valdez. Yet, the response was surprisingly inconsequential.

Perhaps the explanation can be found in the larger economic context. As the nation was slowly recovering from the deepest recession since the

Great Depression, the environment remained a second-tier issue. The Pew Research Center conducts an annual poll on what the public believes the top priorities should be for the president and Congress. In 2010, the top two priorities were strengthening the economy (83 percent) and improving the job situation (81 percent). Protecting the environment (44 percent) ranked seventeenth on a list of 21 domestic priorities. One year later, the priority attached to the economy and jobs had increased (87 and 84 percent, respectively), whereas protecting the environment had fallen to 40 percent (Economy Dominates Public's Agenda, Dim Hopes for the Future 2011). Further evidence can be found in the Gallup poll. Since 1984, Gallup had polled Americans on whether they prioritized economic growth or environmental protection. In the wake of the 2008 financial collapse, a majority of Americans prioritized economic growth for the first time. The Deepwater Horizon reversed this trend, but only temporarily. By 2011, economic growth was once again prioritized over environmental protection (54 percent to 36 percent) (Jones 2011). Similarly, while a majority (55 percent to 39 percent) prioritized environmental protection over energy development in the wake of the BP spill, by 2011, a majority once again favored energy development (Jones 2012).

As for the Deepwater Horizon, public opinion data from polls taken during the crisis was difficult to interpret. Although 75 percent of respondents in a Gallup poll believed that BP bore a great deal of blame for the leak and 81 percent rated the BP's response as poor or very poor, a majority (53 percent) also had a negative assessment of President Obama's handling of the crisis. Despite this evaluation, a majority believed that BP (68 percent) rather than the federal government (28 percent) should be in charge of the cleanup. And although the federal government had issued a temporary ban on offshore drilling in the Gulf of Mexico, a narrow majority (47 percent to 46 percent) believed the ban should be lifted and BP (49 percent to 46 percent) should be allowed to drill in the same area in the future (Gallup 2010). There was little evidence that the Deepwater Horizon stimulated a broad demand for a thoroughgoing overhaul of the oil spill regime. This, when combined with the priority attached to the economy, created few incentives for elected officials to invest in policy change that would expand the role of the federal government, particularly in a year that was dominated by the small-government rhetoric of the Tea Party and would end with hotly contested midterm elections and a new Republican majority in the House of Representatives.

There is another interpretation worth consideration. The muted policy response to the Deepwater Horizon may have been a product of policy success. Following Santa Barbara, Congress created a liability regime that shifted the risk to the industry. This regime, as refined and extended by a series of statutes culminating in the Oil Pollution Act of 1990, would ultimately combine risk-shifting, preventive measures, and a new funding mechanism for remediation. As the performance record reveals, the period since 1990 was characterized by sharp reductions in the total volume and number of significant spills. While there were exceptional events – most notably, the Deepwater Horizon – the larger record is quite positive, particularly when compared to the decades separating Santa Barbara and the Exxon Valdez. It may be the case that further advances are most difficult to achieve when the risk of disaster has already been brought within what are broadly seen as acceptable parameters.

References

Barnett, H. C. 1994. *Toxic Debts and the Superfund Dilemma*. Chapel Hill: University of North Carolina Press.
Baumgartner, F. R., and Jones, B. D. 1993. *Agendas and Stability in American Politics*. Chicago: University of Chicago Press.
Birkland, T. A. 1998. Focusing Events, Mobilization, and Agenda Setting. *Journal of Public Policy*, 18 (1), 53–74.
Birkland, T. A., and DeYoung, S. E. 2011. Emergency Response, Doctrinal Confusion, and Federalism in the Deepwater Horizon Oil Spill. *Publius*, 41 (3), 471–93.
Blake, G. 1969. Union Oil Co. May Not Be Liable for Damage Caused by Big Leak. *Los Angeles Times*, A1.
Busenberg, G. J. 2001. Learning in Organizations and Public Policy. *Journal of Public Policy*, 21 (2), 173–189.
Carson, R. 1962. *Silent Spring*. Boston: Houghton Mifflin.
Cushman Jr., J. 1990, June 29. Conferees Agree on Bill to Cover Cost of Oil Spills. *New York Times*, 1.
Economy Dominates Public's Agenda, Dim Hopes for the Future. 2011. Pew Research Center for the People and The Press.
Eisner, M. A. 1993. *Regulatory Politics in Transition*. Baltimore: Johns Hopkins University Press.
Eisner, M. A. 2007. *Governing the Environment: The Transformation of Environmental Protection*. Boulder, CO: Lynne Rienner.
Eubank, S. R. 1994. Patchwork Justice: State Unlimited Liability Laws in the Wake of the Oil Pollution Act of 1990. *Maryland Journal of International Law*, 18 (2), 149–71.

Farmbry, K. 2012. *Crisis, Disaster, and Risk: Institutional Response and Emergence*. Armonk, NY: M.E. Sharpe.

Frome, M. 1992. *Regreening the National Parks* (Vol. 31). Tucson: University of Arizona Press.

Gallup. 2010. *Oil Spill in the Gulf of Mexico*. Retrieved from www.gallup.com/poll/140978/oil-spill-gulf-mexico.aspx.

Garrett, T. L. 1979. Federal Liability for Spills of Oil and Hazardous Substances Under the Clean Water Act. *Natural Resources Lawyer*, 12 (4), 693–719.

Gill, D. A., Picou, J. S., and Ritchie, L. A. 2011. The Exxon Valdez and BP Oil Spills: A Comparison on Initial Social and Psychological Impacts. *American Behavioral Scientist*, 56 (1), 3–23.

Graetz, M. J. 2011. *The End of Energy: The Unmaking of America's Environment, Security, and Independence*. Cambridge: MIT Press.

Greenwood, L., and Dye, L. 1969. Backers Say 40,000 Petition to End Offshore Oil Drilling. *Los Angeles Times*, A1.

Grumbles, B. H., and Manley, J. M. 1995. The Oil Pollution Act of 1990: Legislation in the Wake of Crisis. *Natural Resources and Environment*, 10 (2), 35–42.

Hager, G. 1989a, May 20. Deadlock Likely to Continue. *CQ Weekly*, 1183.

Hager, G. 1989b, April 8. Spill May Halt Drilling Bill, Help Liability Efforts. *CQ Weekly*.

Hall, Jr., R. M. 1978. The Clean Water Act of 1977. *Natural Resources Lawyer*, 11 (2), 343–72.

Harris, R. A., and Milkis, S. M. 1996. *The Politics of Regulatory Change: A Tale of Two Agencies*, 2nd ed. New York: Oxford University Press.

Hickel, W. J. 1971. *Who Owns America*. Englewood Cliffs, NJ: Prentice Hall.

Jackson, R. L. 1969, February 6. Oil Firm Official Defends Drilling at Senate Hearing. *Los Angeles Times*, A1 and A34.

Johnson, E. W., and Frickel, S. 2011. Ecological Threat and the Founding of the US National Environmental Movement Organizations, 1962–1998. *Social Problems*, 58 (3), 305–29.

Jones, J. M. 2011, March 17. Americans Increasingly Prioritize Economy over Environment: Largest Margin in Favor of Economy in Nearly 30-Year History of the Trend. Retrieved from Gallup Politics: www.gallup.com/poll/146681/Americans-Increasingly-Prioritize-Economy-Environment.aspx.

Jones, J. M. 2012, March 23. Americans Split on Energy vs. Environmental Trade-Off. Retrieved from Gallup Politics: www.gallup.com/poll/153404/Americans-Split-Energy-Environment-Trade-Off.aspx.

Keener, K. C. 1971. Federal Water Pollution Legislation and Regulations with Particular Reference to the Oil Industry. *Natural Resources Lawyer*, 4 (3), 492.

Kenworthy, E. 1970, April 4. Curb on Oil Spills Is Signed by Nixon: New Law Raises Penalties and Expands Liabilities. New York Times, 28.

Kim, I. 2002. Financial Responsibility Rules under the Oil Pollution Act of 1990. *Natural Resources Journal*, 42, 591.

Kingdon, J. W. 1984. *Agendas, Alternatives, and Public Policies*. Glenview, IL: Scott, Foresman and Company.

Klass, A. B. 2004. From Reservoirs to Remediation: The Impact of CERCLA on Common Law Strict Liability Environmental Claims. *Wake Forest Law Review*, 39, 904–70.

Knopf, C. D. 2011. What's Included in the Exclusion: Understanding Superfund's Petroleum Exclusion. *Fordham Environmental Law Review*, 5 (1), 3–42.

Kuntz, P. 1990. Conferees Break Logjam. *CQ Weekly*, 2042.

Lenckus, D. 1996. Exxon Seeks More Spill Cover: Oil Giant Reaches Partial Agreement with Insurers. *Business Insurance*, 1.

McCoy, C., and Wells, K. 1989. Alaska, US Knew of Flaws in Oil-Spill Response Plans. *Wall Street Journal*, A3.

Meltz, R., Ramseur, J., and Pettit, C. 2010. Questions Regarding Liability Under the Oil Pollution Act and the Oil Spill Liability Trust Fund. *Congressional Research Service*, 5, 8.

Mohr, H., Pritchard, J., and Lush, T. 2010, June 9. BP's Gulf Oil Spill Response Plan Lists the Walrus as a Local Species. *Christian Science Monitor*.

Moss, D. A. 2002. *When All Else Fails: Government as the Ultimate Risk Manager*. Cambridge: Harvard University Press.

Murchison, K. M. 2011. Liability Under the Oil Pollution Act: Current Law and Needed Revisions. *Louisiana Law Review*, 71 (3), 917–56.

National Pollution Funds Center. 2008. *Oil Spill Liability Trust Fund, Annual Report, FY 2004-FY2008*. Washington DC: US Coast Guard, National Pollution Funds Center.

Nichols, J. E. 2010. *Oil Pollution Act of 1990(OPA): Liability of Responsible Parties*. CRS Report for Congress. Washington, DC: Congressional Research Service.

Oil Spills and Cleanup Bills: Federal Recovery of Oil Spill Cleanup Costs. 1980. *Harvard Law Review*, 93 (8), 1761–85.

O'Reilly, G. 1993. *Marine Spill Response Corporation (MSRC): How it Hopes to Fill the Oil Recovery Requirements of the Oil Pollution Act of 1990*. Paper submitted in partial fulfillment of the requirements for the degree of Master of Marine Affairs, University of Rhode Island.

Pfennigstorf, W. 1979. Environment, Damages, and Compensation. *Journal of the American Bar Foundation*, 4 (2), 347–448.

Pollution: Deadly Blanket of Blackness. 1969, February 9. *Los Angeles Times*. G5.

Ramseur, J. L. 2010. *Oil Spills in US Coastal Waters: Background, Governance, and Issues for Congress*. CRS Report for Congress, Washington, DC: Congressional Research Service.

Randle, R. V. 1991. The Oil Pollution Act of 1990: Its Provisions, Intent, and Effects. *Environmental Law Reporter*, 3 (91), 10119–35.

Rawl: In Ten Years You'll See Nothing. 1989. *Fortune*, 119 (10), 50.

Rich, S. 1969, February 19. Tight Rules Imposed on Oil Drilling. *Washington Post*, A5.

Sabatier, P. A. (ed.). 2007. *Theories of the Policy Process*. Boulder, CO: Westview Press.

Schenke, D. L. 1990. Liability for Damages Arising from an Oil Spill. *Natural Resources and the Environment*, 4 (4), 14–16.

Stone, A. G. 1975. The Trans-Alaska Pipeline and Strict Liability for Oil Pollution Damage. *Urban Law Annual*, 9 (1), 179–201.

The Loss of a Few Birds. 1969. *Washington Post*, A22.

US Coast Guard. 2012) *Polluting Incidents In and Around US Waters. A Spill/Release Compendium: 1969–2011*. Washington DC: United States Coast Guard.

US Energy Information Administration. (2012). *Annual Energy Review*.

Wald, M. 1991, September 5. Oil Spill Organization Gearing Up. *New York Times*.

Wang, H. 2011. *Civil Liability for Marine Oil Pollution Damage: A Comparative and Economic Study of the International, US and Chinese Compensation Regimes*. Alphen aan den Rijn: Kluwer Law International.

Wells, K., and McCoy, C. 1989. How Unpreparedness Turned the Alaska Spill into Ecological Debacle. *Wall Street Journal*, A1.

White, M. A. 1995. Investor Response to the Exxon Valdez Oil Spill. Working Paper, University of Virginia, McIntire School of Commerce.

Zimmermann, J. A. 1999. Inadequacies of the Oil Pollution Act of 1990: Why the United States Should Adopt the Convention on Civil Liability. *Fordham International Law Journal*, 23 (5), 1499–1539.

8

The Nordic Model of Offshore Oil Regulation

Managing Crises through a Proactive Regulator

Ole Andreas Engen and Preben H. Lindøe

8.1 INTRODUCTION

Over the last 50 years, Norway has developed a unique approach to regulation of risks in the petroleum industry. This regime is characterized by close interactions between the Norwegian government, the oil companies, the supplier industry, and the unions. Accidents in the industry have shaped the regulatory framework by shifting alliances, power relations, trust, and distrust among these various actors. Accidents have served as focusing events (see Chapter 5 in this volume) that have created policy windows for change, but also influenced the relationships between the involved actors and created new institutional settings for regulating risk and safety.

The Norwegian risk regulatory regime was created out of two major crises. In 1977 and 1980, two large accidents on the Ekofisk field in the southern part of the North Sea took place; the blowout at the Bravo oil rig and the capsizing of the oil rig Alexander L. Kielland. The first was a public eye-opener about the potential environmental consequences of offshore oil production in the North Sea, and the other was the largest industrial disaster in Norway since World War II, with 123 lives lost. Both events mobilized public opinion and put the debate about safety on the public and political agenda. The two events also created a process of self-reflection and renewed effort to improve safety within the industry (Lindøe et al. 2011). The regulatory responses to the accidents represented a paradigm shift from an old reactive regime based on prescriptive and technical requirements towards a risk-based, proactive regime with functional legal requirements. The regulatory regime after Alexander L.

Kielland in 1980 differed from the old regulatory regime in several respects, particularly with regard to partnership between public regulators and industry, supervising and fostering self-regulation by industry, the involvement of labor force and other stakeholders, and the issue of mutual trust among the parties.

This chapter addresses the impact of focusing events on the Norwegian regulatory regimes for offshore oil production. In order to understand substantive policy change in Norway, we will introduce the prevailing organizing principle of industrial relations in Scandinavian countries; namely "the Nordic Model." The principles of the Nordic model of risk regulations have thoroughly been analyzed in Lindøe et al. (2014), where it has been compared with risk regulatory regimes in the UK, the US and Australia. The main objective of the analysis in Lindøe et al. (2014) is to compare different traits of robustness of the four regimes respectively. Accordingly this chapter expands upon this robustness perspective by putting it in relation to crises and further connecting the robustness of the regime to notions of learning and change.

The chapter is organized as follows. First we will briefly introduce the Norwegian context and the basic organizing principles for the petroleum industry in Norway. Thereafter, we introduce the Nordic model where risk in the petroleum industry is regulated through a public/private partnership by combining safety management systems (self-regulation) with public/state regulation with legal binding laws and regulation. We then examine the impact of focusing events on the Norwegian petroleum regulatory risk regimes and discuss the robustness of the regime today.

8.2 THE NORWEGIAN OIL INDUSTRY

The regulatory foundations of Norway's oil industry were created in the 1970s with the establishment of Statoil, the Norwegian Petroleum Directorate (NPD), and the Ministry of Oil and Energy (MOE). From its establishment in 1972, Statoil became the main instrument of the state in developing Norwegian petroleum competence, and Statoil relied on the concession system to strengthen its dominance. Under the concession system, agreements regarding training and transfers of knowledge and technology from other companies were negotiated, and Statoil itself took the role of intermediary in delegating tasks to Norwegian industry. Industrially, the company contributed greatly to technical and organizational adaptation by utilizing traditional industrial networks and functioning as an agent that would transfer and adapt international petroleum techniques and competence.

The outlook for long-term production and positive investment returns also induced the international oil companies to be flexible in meeting the demands of public officials regarding their development strategies, choice of suppliers, and measures to transfer competence to Norwegian companies and research institutes. This form of governmental organizing for a national petroleum industry is fully compatible with what Michael Porter denotes an "infant industry policy," and it contributed to the creation of a petroleum industry cluster in Norway. Representatives of the Norwegian government considered these terms as absolutely necessary for securing the greatest possible share of oil profits for Norwegian society.

The UK developed a regulatory framework for exploitation of its North Sea oil and gas deposits that resembled that of Norway in many respects. The UK government initially implemented a petroleum tax system and a system of handing out licenses, and it established institutional bodies to oversee offshore exploration and production. But developments in the UK domestic oil industry followed a different path from that of Norway after the 1970s. The Norwegians pursued the "infant industry policy" during the 1980s, while the UK chose to deregulate and de-emphasize active governmental participation. In general, UK oil policy in the 1980s consisted of little more than a passive concession practice that relied on a tax policy that was compatible with the general economic policy of the Thatcher government (Engen 2009).

As the Norwegian oil companies and main suppliers gradually built up their competence under the shelter of a protectionist policy, the public administrative apparatus for oversight of the oil and gas industry matured. The formation of NPD in 1973 followed traditional administrative practice (Olsen 1989), and the Directorate engaged personnel with little experience in the petroleum industry, frequently resulting in an inflexible approach to regulation of the industry. This rigidity began to change in the late 1970s and 1980s, however, as NPD built up independent competence to handle complicated technical problems and complex development projects.

With respect to economic policies, Norway has been different from other "Petrostates" in two important ways. First, the prevailing institutional framework in Norway helped to better integrate the international oil industry into the existing domestic industrial structure, creating channels for technological transfer and national competence building, and laid the foundation for a Norwegian oil industry that is able to compete in the international arena. Second, the political system has, to a certain extent, managed to contain itself to restrict overspending of the public fortune by an elite and, instead, to allocate these returns over time to a broad swath of the public. The concession policies have, at least to some extent, prevented an excessive petroleum depletion rate, and the country's petroleum fund was set up to isolate petroleum rents from politicians, thereby preventing them from spending too much money for the purpose of being re-elected or for cronyism (Engen et al. 2012).

Economically speaking, the petroleum industry is the largest single contributor to the Norwegian National Account. In 2015 about 250,000 people were employed in the petroleum industry, representing about 10 percent of Norway's work force. The petroleum industry has also developed technological competence through large-scale training of engineers and research activities, with approximately 90 percent of the country's energy-related research focused on oil and gas. The industry accounts for about 15 percent of the Norwegian GDP and contributes to regional development, low unemployment, and economic and technological growth.

8.3 THE NORDIC MODEL

The term "regulatory regime" can be used to describe the overall way risk is regulated in a particular policy domain (Hood et al. 2001). A regulatory regime consists not only of rules and enforcement mechanisms, but it also includes everything from overall policy to concrete implementation, stakeholders and agencies at various levels, and all formal and informal mechanisms that keep the regime together. A certain amount of stability and durability over time will also characterize a risk regulatory regime, although dynamic, changing processes and interactions between elements are important topics in regime studies. In this chapter we define the term "robustness" of a regime to constitute the degree to which a regime can accommodate and adapt to changes in the nature of risk, the actors in the policy system, or other dynamic features of the regime. The concept of robustness is useful as a normative description of regime performance over time as discussed by Hale (2014) in his reflection on lessons learnt across risk regulatory regimes in Norway, the UK, the US and Australia.

In terms of regulation "The Nordic model" refers to a high degree of formalized industrial relations. This implies a centralized regulatory structure within the national government, but at the same time a trinity of cooperation among employers, employees, and the government concerning economic policy, exchange of information, and consultation. In Norway, this model of industrial organization also supported a national system for the collective negotiations between employers and employees and, moreover, contributed to the maturation of oil companies according to the formal and informal rules of the Norwegian setting (Engen et al. 2012). Unlike in the UK and the US, in Norway, the working conditions offshore were subjected to the same legal framework as the working conditions onshore. The Environmental Act in 1977 gave employees in Norway extended privileges in general and became a powerful instrument for offshore workers in terms of influencing security and safety regulations. A safety deputy, for instance, had the same power as the platform manager to stop the production stream if there was any suspicion of technical or organizational irregularities that could increase the risk of undesirable incidents.

"The Nordic model" refers to institutional frameworks organizing and regulating negotiations, wealth distribution, and conflict resolution. Conflicts between parties are solved through extensive laws and systems of agreements. Historically speaking, the Nordic model implied that employers supported unions and their professional activities to a certain degree. Moreover, employers have several times been forced to de-emphasize short-term profit goals to advance longer-term managerial objectives. The success of this policy may be explained by the strength of the unions in national and local political processes. From this perspective, we may say that the Nordic model has functioned as a stabilizing factor in Norwegian politics and society. It has formed and shaped the political strategies concerning how to balance a growing resource economy with other economic sectors, how to find balance between the public and private sector, and finally how to consider challenges created by the fact that petroleum is a non-renewable and exhaustible resource.

In terms of safety, "the Nordic model" is embodied in the tripartite collaboration – involving employer, employees, and the government. A common feature within the tripartite system is the in-house use of an Occupational Health and Safety Organization that offers three different collaborating structures. First, Safety Committees provide opportunities for employer and employees to meet and discuss important issues. Second, there are independent and autonomous Safety Representatives, and third, there are a number of experts on occupational health and safety who may be called upon in disputes, either as an in-house service or external consulting expertise. Hence, safety and optimal working environment is one of the cornerstones of the model.

8.4 A FRAMEWORK FOR RISK REGULATION AND RISK MANAGEMENT

Historical events and processes within the petroleum industry and decisions particularly about risk regulation goals have given rise to a distinct legal framework for governing offshore oil and gas in Norway (Lindøe and Engen 2013). The US offshore regime is typically characterized as "command-and-control," a top-down approach where regulators set a system of prescriptive rules and enforce industry compliance through inspections, fines, and other penalties. In contrast, the UK and Norway follow principles of enforced self-regulation, relying on the capability of the industry to manage their own risks according to accepted norms and standards set, and ultimately enforced, by the government (Lindøe, Baram, and Paterson 2013). The Norwegian regime goes even further in developing a tripartite system based on egalitarian values and mutual trust among involved actors including industry, employees, and government officials. Norway, promotes a symmetrical partnership between public agencies, industrial actors, and labor unions for risk governance with the asymmetric role of sanctioning industry for violation of law given to the public agencies.

Modes of risk regulation are distinguished by their focus on either compliance or risk management (Hopkins 2011). Regulations that focus on compliance are often referred to as "hard regulations" or "command-and-control," and they contain prescriptive rules with specific consequences for violations. In contrast, regulations that focus on risk management are often referred to as "soft regulations" and include approaches coined as "self-regulations," where industry develops risk regulations, rules, and procedures itself (Sinclair 1997, Short and Toffel 2010), or "meta-regulations"(Gilad 2010), where industry-implemented regulations are overseen by the government. An important question is whether these two modes of regulation represent a dichotomy or whether they are complementary (left side of Figure 8.1). Our argument follows Sinclair (1997), stating that the dichotomy is false. In practice risk regulation regimes will combine responsibilities and roles in a public-private partnership with a top-down approach composed of legal binding norms and a bottom-up approach composed of industrial standards and "best practice" as indicated in the right part of Figure 8.1.

The Nordic Model of Offshore Oil Regulation

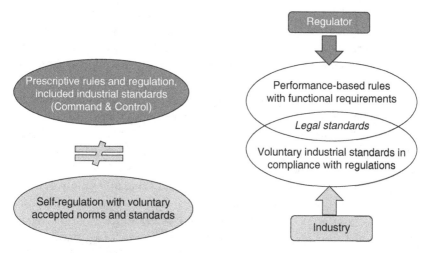

FIGURE 8.1 Two modes of risk regulations

The Norwegian risk regulatory regime is typically function based. Function-based regulation needs some form of discretionary criteria that are considered "legal standards," linking functional requirement in the law to industrial standards. The term and concept is challenging to interpret, but can be defined as words or phrases in a law that stipulates a scale or norm beyond the law, for example a particular practice, widespread attitude in the community, or other conditions that change with time. One example of practical use of a legal standard is the ALARP-principle ("as low as reasonably possible"). Another example is Sec. 1–1 in the Norwegian Working Environment Act, which states that the purpose of the law includes "a standard of welfare at all times consistent with the level of technological and social development of society."[1] The point is that the law refers to norms existing outside the law that are contextually defined and change over time as technology improves. Legal standards tie the unchanging word of law to the ever-changing implementation of the norms and ideas imbedded in that law. The use of legal standards aims to achieve an appropriate level of regulation in highly dynamic industries. It can also be seen as an expression of respect for the importance of expert knowledge to ensure the safety and quality in key areas of society in changing circumstances.

---

[1] The full text of the Norwegian Working Environment Act (Norway) is available at: www.regjeringen.no/no/tema/arbeidsliv/arbeidsmiljo-og-sikkerhet/innsikt/arbeidsmiljoloven/id447107/.

Legal standards probably safeguard the goal of safety and quality better than if they had been fully formulated in laws and regulations. The underlying measure of the legal standards is based on an understanding of the issues, terminology, and solutions that are understood in the professional and scientific community. Through stakeholder involvement in the process of developing these norms, the use of legal standards may enjoy greater legitimacy than only legal terminology and legal text. In a system of "enforced self-regulation" based on voluntary technical standards and linked to legal standards, compliance with requirements to implicit norms and legal standards becomes the responsibility of private enterprises. Requirements formulated as legal standards follow developments in technology and societal demands. But their interpretation requires continuous dialogue between the relevant actors in the sector. From legal history we may learn that the development of legislation under the common law principle requires a very active "legislative zone" in which different parties may claim their interests and independent bodies, typically courts, sort out the disputes (Kaasen 2014).

A consistent application of a function-based regulation requires a comprehensive and systematic review of how the various provisions are to be understood and how the appropriate standards should be used to meet the requirements. Procedures must stipulate relationships between laws, regulations, and technical/professional standards. For the regulatory authorities and inspectors, this can be a demanding and comprehensive system to keep up to date, and it requires that the standards keep pace with developments and new knowledge (Bieder and Bourrier 2013). There is an inherent tension between the desire to develop comprehensive guidelines representing expectations of industry best practice and the desire to require industry to continually innovate and implement any new expertise and scientific knowledge that may improve safety. Risk regulation with stakeholder involvement, tripartite collaboration, and internal control may in principle be described as a balance of power between state control and industrial degrees of freedom.

Industrial activities based on "command–and–control" and the state as external controller based on performance-based rules with functional requirements have different logics of action. Adjustment from one system towards the other represents a change in goal setting, use of legal binding norms, industrial standards, professional and legal competence, etc. Responsibilities regarding the safe conduct of activities transferred or delegated from regulators to companies provide for new cooperative approaches in the implementation of regulatory regimes (Olsen and

Lindøe 2009). Norway was the first country to make this transition, and it has now been followed by the UK and to a lesser extent the US (Bennear 2015). In Norway, like in the US and UK, this shift toward took place after major focusing events as described in the next section.

8.5 FOCUSING EVENTS AND PUBLIC POLICY IN THE NORWEGIAN CONTEXT

According to Birkland and Warnement, the "influence of focusing events on policy change, regulatory change and other behavior of regulators themselves would account for the fact that policy-makers, at whatever level government of responsibility, do not sit and wait passively for directions from subordinate levels of government. Indeed, regulators may have strong incentives to be seen as proactive problem solvers" (Birkland and Warnement, Chapter 5 in this volume). This statement seems to describe the way in which Norwegian regulatory bodies have behaved following both major and minor accidents in the petroleum industry. In this section, we will show that the adoption of the Nordic model of risk regulation was event-driven, where accidents along the Norwegian continental shelf opened up a window for policy change. This opportunity for change was, however, used by regulatory agents rather than politicians in order to change the regulatory modes for safety and demonstrate to the public that the system was more sophisticated and that the regulators had learned from the accidents. We will show how the Norwegian risk-regulating regime has undergone one substantive policy change and several additional cosmetic makeovers which together provided its character.

8.5.1 The Aftermath of Bravo and Alexander L. Kielland

On April 22, 1977, the Phillips Petroleum operated Bravo platform drilling into the Ekofisk field (also referred to as the Ekofisk Bravo or Ekofisk B platform) experienced a "blowout" or uncontrolled release of oil and gas. The blowout occurred during a "workover," which describes a situation in which the completion materials for a well are unstable and need to be replaced. During the replacement, the well pressure became uncontrolled, leading to a spill of over 200,000 barrels of oil into the North Sea. The oil spilled for seven days before the well was ultimately capped. Much of the oil evaporated (up to 30 percent by some estimates). Because of the evaporation, combined with the turbulent nature of the

North Sea and the distance from shore of the rig, there was little obvious environmental damage from the spill, despite its size (NOAA 1977).

Despite the lack of concrete environmental damage, the Bravo accident underlined the environmental risk of offshore oil production in Norway. Even though environmental perspectives had been discussed from the beginning of international oil interest in the North Sea, the Bravo accident visualized, not least through substantial international media coverage, what kind of environmental accidents could occur.

If the blowout on the Bravo platform was a wake-up call regarding the environmental risk, the capsizing of the Alexander L. Kielland platform in 1980 in which 123 lives were lost became a shocking reminder of the human risk. The Alexander L. Kielland was also operated by Phillips Petroleum in the Ekofisk field on the Norwegian continental shelf. The Alexander L. Kielland was not an active drilling or production platform, but rather served as a "flotel" or floating hotel, for the Edda platform. On March 27, 1980, in gale force winds, braces supporting one of the five legs of the Alexander L. Kielland broke. The platform then tilted 35 degrees to one side, partially submerging the rig, including some sleeping quarters. Of the 212 men on board at the time of the accident, only 89 survived (BBC, n.d.).

Following the Alexander L. Kielland accident, the newly established Norwegian Petroleum Directorate (NPD) started a process of developing new regulations, and in 1981 new rules concerning licensees' internal control were established, followed in 1985 by the Regulation of Internal Control. According to Bang and Thuestad (2014), the Alexander L. Kielland accident highlighted the need to simplify and enhance the efficiency of risk regulation on the Norwegian continental shelf. The accident underlined the importance of being able to establish and maintain a high level of safety with clear regulatory boundaries. Hence the regulatory regime that was established consisted of two elements: an organization change and the beginnings of the tripartite approach to regulation in the Nordic model.

On the organization front, a single agency – at that time the NPD – was assigned responsibility to draw up detailed regulations and to make overall safety and working environment assessments. Formal agreements with the pollution control authorities for the natural environment and with the health authorities for health-related issues were also negotiated and implemented through the NPD. This level of coordination and support between many different governmental agencies represented a new approach to administration and differed from the model previously used in Norway, which assumed that single agencies were responsible for their individual aspects of safety and working environment (Hovden 2004).

The organizational changes eventually led to the wholesale change in paradigm toward a tripartite, collaborative approach to offshore oil and gas regulation in Norway. The coordination of supervision at NPD gradually identified the need to develop a joint set of regulations based on a number of overall considerations, including safety and the working environment, the natural environment, and occupational health. Industry believed the reform represented a simplification of the regulatory process as NPD was the single point of contact for all health, safety, and environmental issues, but furthermore, the more integrated regulations also stimulated a greater degree of industry participation and collaboration in influencing risk regulation. However, the challenge became to handle collaboration issues and possible conflicts of interest across government agencies through formal procedures and agreements in order to ensure that the system functioned well in practice.

While the Alexander L. Kielland catastrophe had an impact on the formal regulatory regime, the general political consequences were more moderate. The opposition was reluctant to criticize the government, thereby obtaining political benefits. Accordingly, no minister or industrial managers were forced to leave their positions. The political turbulence that followed in the aftermath of the accident was more concerned with how the memorials should be organized, rather than channeling the responsibilities in any particular direction. However, Alexander L. Kielland became a part of the Norwegian collective narrative of the cost and risk of being an oil-producing country (Ryggvik and Smith Solbakken 1997; Ryggvik and Øye Gjerde 2014; see also Chapter 6 in this volume) and created a momentum towards new regulatory approaches.

8.5.2 Focusing and Amplifying Event: Piper Alpha

On July 6, 1988, there was a fire and explosion at the Alpha platform in the Piper oil field off the UK continental shelf. The Piper Alpha platform was owned and operated by Occidental Petroleum. The proximate cause of the initial explosion was the failure of a pressure valve on one of two gas compression pumps, but the damage was made worse by a series of additional failures. The investigation reports revealed a complex mixture of failures on all levels rooted in culture, structure, and the procedures of the operating oil company (Occidental Petroleum), but also pointed to underlying factors such as short-termed financial focus, lack of coordination by regulatory authorities, and neglect of safety issues by the British government (Paté-Cornell 1993). In the end, 167 of the 229 men onboard the Piper Alpha died, making it the deadliest accident in offshore oil history (Paté-Cornell 1993).

The accident sent shock waves in the entire international petroleum industry and created a redesign of the British regulatory regime and other petroleum regulatory regimes as well. Following the accident, the British government commissioned an independent commission to evaluate the accident. The commission produced the Cullen Report, named after Lord William Cullen of Whitekirk who chaired the commission. The 1990 Cullen Report played a major role, both through its recommendation of a specific health, safety, and environment regime and by promoting Formal Safety Assessments (FSA) called "safety cases."

The British safety regime looked to Norway when redesigning its regulatory regime. The Health and Safety Executive (HSE) requires that all operations be covered by detailed safety cases in which potential hazards, their consequences, and the methods of controlling any risks are described and explained. The overall responsibility for safety on an installation falls on the Safety Case Duty Holder, who appoints an Offshore Installations Manager (OIM) to discharge this responsibility. In the case of mobile drilling rigs, the duty holder is the drilling contractor. This is similar with the principle of internal control at the Norwegian sector which requires that responsibility is assigned to the operator. Before an operator brings a drilling rig into the UK or operates a fixed platform, they have to prepare a safety case for the HSE to approve. The Operator, or License Holder is subject to separate and additional verification requirements under the Design and Construction Regulations in the form of well examinations carried out by an independent and competent person. All parties involved have legal duties to cooperate with both the OIM and the Well Operator when the well is under construction. The Safety Case Duty Holder and the Well Operator must demonstrate how their safety management systems will operate together, who has primacy in an emergency, and who has overall responsibility. However, as argued by Paterson (2014), even though the safety case approach may remain the best option among alternatives (especially detailed prescriptions), question remains as to whether it has been implemented as well as it might.

The new trajectory of British offshore regulation after Piper Alpha seemed to confirm that the Norwegian regulatory regime was on right track. According to the official Norwegian petroleum history, Magne Ognedal (the leader of the Safety Department of NPD) was assigned a significant role in the public hearings after the accident – apparently so momentous that the Cullen Report by certain British observers later on was nicknamed "the Ognedal Report" (Ryggvik and Smith Solbakken 1997). Accordingly, the process in the aftermath of Piper Alpha illustrates how regulators play a vital and proactive role in redesigning the regulation. Even though there were no criminal charges against the companies, the impact on Scottish society and the British regulatory system was similar to the Norwegian experiences after Alexander L. Kielland. From a Norwegian perspective, we may also claim that the influence of the "Nordic model" on the British process and the direct participation of central Norwegian regulatory agents also contributed to strengthening the Norwegian system and induced further regulative developments in other industrial sectors. The general Internal Control Regulation reform on Health, Environment, and Safety, which came into force in Norway on January 1, 1992, was inspired by the positive experiences from safety regulation in the North Sea.

8.5.3 Challenging the Nordic Model: NORSOK, Snorre, Gullfaks, and Deepwater Horizon

The regulatory regime established after Alexander L. Kielland was challenged from many directions. Among the Norwegian authorities and the regulatory agents, a similar accident as Alexander L. Kielland was unthinkable. At the same time there was also an increased consciousness among the industrial actors and certain representatives from the government that the general cost level of developing petroleum fields on the Norwegian continental shelf was far too high. This cost-consciousness was exacerbated by the general fall of world oil prices in the mid-1980s (Engen 2014). In 1993 a program to improve cost-effectiveness, denoted NORSOK, was introduced. NORSOK (an abbreviation for *NORsk SOkkels Konkurranseposisjon*; in English, "the competitive situation of the Norwegian shelf") was an industrial program for development of new technologies and standards, organizational development and new contractual relations, regulations, and new initiatives for cooperation and negotiations between oil companies and their suppliers. The main objective was to reduce average costs by as much as 50 percent. The program was inspired by the similar British initiative, CRINE (Cost Reductions In a New Era). NORSOK represented a break with the infant industry policy of the 1970s and 1980s and a political shift from an active and interventionist oil policy to a more passive one that sought to link various actors rather than dictating terms to them. NORSOK introduced a process that allowed the both oil companies and main suppliers to enjoy greater freedom when choosing technological concepts, suppliers, location of bases, headquarters, etc.

The incidents and number of accidents however increased during the 1990s. According to Hovden (2003), several trends in the risk picture could be identified: First, an increase in major hazards related to problems with gas leakage, and second, increased risk and uncertainty regarding the safety level for floating installations. This was mainly related to new technology. A helicopter accident in 1998 in the Norne field in the northern part of the North Sea killed 12 people and demonstrated very clearly the risks associated with shuttle traffic to and from rigs by helicopters. There were also increased accidents related to the activities of service vessels and increased vulnerability caused by use of information and communication technologies (ICT systems), automation, and the resultant reduced staffing on the installations. However, most studies showed low and stable frequencies of occupational accidents.

The exact reasons for these incidents and accidents are not unambiguous. Some are due to lack of maintenance, but also a large part can be traced back to new technology and contractual relations. The investigation report after the Norne accident argued that tighter schedules and increased use of rotating teams increase the risk of such type of accidents. Additionally, Norne is a floating installation with reduced capacity in terms of living quarters and therefore was also dependent on shuttling people more frequently. New contractual relations, which relied on supermajors as the owner/operators, but outsourced the design and operation to smaller subcontractors, implied that the oil companies became less responsible for general HSE. The increased economic pressure combined with lack of knowledge and experiences in risk and safety management constituted new vulnerabilities and reduced robustness in the sociotechnical system. Accordingly the governance structure and institutional mechanisms of the Nordic model were put under pressure. NORSOK implied shifts in the technology and organization, and the regulatory system – as exemplified with Norne – was not able to keep up with requirements, procedures, and norms for the new technology and working environments. Thus, we see an institutional inertia where risk regulation and safety management were sacrificed on the altar of new cost efficient technological solutions (Engen 2014).

Neither the Norne accident nor other smaller accidents or incidents on the Norwegian shelf changed the regulatory regime in a substantial way.

The Nordic Model of Offshore Oil Regulation   195

There were extensive discussions concerning changes, but the main principles established in the wake of Alexander L. Kielland remained more or less stable. However, the struggle to improve cost efficiency and increased rate of accidents induced periods of conflicts between different groups of actors. The unions expressed distrust towards the oil companies, which strained the climate of collaboration concerning safety.

Just before the beginning of the new millennium, this controversy threatened to disintegrate the tripartite collaboration on HSE. The Norwegian historian Helge Ryggvik (2003) has characterized the Norwegian HSE controversy in 1999 and 2000 as the "major accident that never happened," and accordingly it can be considered as a focusing event because it induced action both from the Norwegian politicians and regulatory agents. In response to the growing tensions and controversy, in September 2000, the NPD issued a letter to the licensees on the Norwegian shelf requesting new actions to improve safety. Two regulatory bodies for collaboration were established, the Safety Forum and Working for Safety. In 2001, NPD launched the project "Risk Level on the Norwegian Shelf" in order to get a better assessment of risk governance and management. In 2001, the Ministry of Work and Administration issued the white paper "About Health, Environment and Safety in the Petroleum Sector." These efforts resulted in changes that refined the principles of tripartite collaboration by facilitating avenues for unions and industrial associations to meet regularly to discuss safety issues. The ability of the tripartite system to self-correct in response to growing tensions reveals the resilience of this regulatory regime. Such processes may be described as "boxing and dancing" where the authorities "promoted a process where converging sense making and the development of organizational structures for collaboration mutually reinforced each other" (Rosness and Forseth 2014, 336).

During the first decade after the millennium shift there was only one substantial incident on the Norwegian continental shelf, namely the Snorre A incident. Snorre A is a floating installation in the north region of the North Sea operated by Statoil. In 2004, a gas leakage was discovered and the platform was evacuated. The investigation report showed afterwards that the situation onboard the Snorre A was very grave, with successful intervention occurring only a minute away from ignition and a very serious accident. The Snorre A incident created a lot of attention in the media, but in this case there was not any serious debate whether the ground principles in the Norwegian regulation model should change. Statoil was, however, challenged both from the Norwegian Petroleum Safety Agency (PSA) and the unions to improve their procedures concerning compliance and learning. Several internal organizational programs where introduced, but these programs were, to a certain degree, overshadowed by the merger between the two biggest Norwegian oil companies, Statoil and Hydro Oil and Gas, in 2007.

The next accident happened between December 2009 and May 2010, when the Gullfaks C field, operated by Statoil, was subjected to several critical incidents with leakages of hydrocarbons from wells. The initial well trajectory was plugged during the fall 2009, and drilling in a sidetrack well started in December 2009. During the next couple of months, Statoil experienced three serious well-control incidents, the most critical one resulting in a complete loss of well control on May 19, 2010. Again the investigation showed that the incident was only minutes away from becoming a full-scale disaster. An investigation was run by an independent research institute. The strong Norwegian environmental group Bellona used the report from this independent investigation as an indication that Norway should adapt EU regulations in order to avoid major oil pollution in the future. In this case Bellona proposed a major critique not only of Statoil, but also towards PSA and thus the Norwegian regulatory regime in general.

The Gullfaks C incident and the subsequent investigation and pressure from environmental groups, triggered the Norwegian government to start an evaluation of the character of the robustness of the current regulatory regime. Two large investigations were launched: One carried out by the Norwegian Petroleum Safety Agency (PSA) and an independent one carried out by a Norwegian research institute. PSA's report aimed to clarify to what extent Statoil had met the regulatory safety requirements related to the planning and preparation process for the well. The investigation points to deficiencies with risk management and compliance with internal requirements for drill operation planning and execution. In sum, it was identified that Statoil had specific challenges related to the quality of the planning process, quality and precision in executing the processes, risk comprehension, compliance and management.

Based on the PSA report, Statoil was asked to further clarify why such an incident actually happened and to identify and carry out improvements concerning quality and resilience in the organization as a whole. Statoil interpreted this request as an instruction to accomplish an independent study, and after a procurement procedure the International Research Institute in Stavanger, Norway (IRIS) was assigned the task. The report was critical towards Statoil and created a public debate after being published, including during the Question Hour in the Parliament.

Gullfaks C happened in the international context of Deepwater Horizon, which created further momentum for review of offshore regulatory procedures through media and NGO pressure not only in Norway, but also among most European countries. After the Deepwater Horizon disaster, members of the EU Parliament questioned the existing risk regulatory approaches for offshore health and safety regulation among the member states and whether these regulatory approaches would prevent a similar incident from happening in EU-country waters. The government and industry in Norway and the UK established reviews of existing regulatory processes to assess potential lessons to be learned from the Deepwater Horizon accident and to make recommendations with regard to well control and safe offshore operations. In the UK, some groups were quick to declare that offshore regulatory standards, as exemplified by the "Safety Case Regime" were superior to those found in the Gulf of Mexico at the time of the Deepwater Horizon disaster (House of Commons 2011). Both in the UK and Norway, the reaction towards those who wanted to utilize the focusing event as an engine of regulatory change was unambiguous; in both countries, the assessments argued that the North Sea regulatory regimes were the best way of organizing safety and that any change towards more standards-based or prescriptive regulations was not required (House of Commons 2011; Petroleum Safety Authority 2011).

8.6 THE ROBUSTNESS OF THE NORWEGIAN REGIME

There is strong evidence that offshore risk regulations in Norway and the UK have been shaped by focusing events. There remain questions as to whether the regimes in these two countries are robust when confronting new challenges and incidents. Ideally, one could examine these questions empirically, by looking at accident rates or near-miss data across time under different regulatory regimes to determine which type of regime seems best at reducing risk. Unfortunately, comparable data to make such assessments are not available, and those that are available present a very mixed picture (Bennear 2015). In this section, we instead examine the threats to robustness in the Norwegian regime and the possibilities for addressing those threats.

Today, the Norwegian regime is function-based where compliance rests on a number of assumptions, not least the existence of trust between the regulators, firm, and industry-partners including employees (unions) and employers (industry associations) (Hale 2014, 422). "Trust" means that the people/organizations that interact with each other act in expected ways. To rely on one another means further that one expects that the person or organizations will, for example, use the functions-based system and framework of norms in a way that is in accordance with the rules and overall objectives that have been previously agreed upon. However, displaying trust in others makes oneself also vulnerable. The other party might act differently than expected, and opposite one's own interests and desires. In a Norwegian-style regulatory regime, robustness is fundamentally based on trust between the key players. However, trust is a vulnerable concept and may easily turn over to distrust. In a function-based and trust-based system where distrust always is a possibility, there is a vulnerability built in as a potential risk that threatens the robustness of the system. In order to make the function-based system work, there have to be mechanisms that balance trust and distrust between the actors. Such mechanisms may be governmental agents that act as mediators or facilitators. However playing such a role also assumes a certain kind of trust from the other players and power to exercise such a role.

Reducing vulnerabilities from fragile trust relationships requires power relations and the exercise of power. "Power" can be defined as a persons' ability to achieve his/her will, despite resistance from others. Government actors such as PSA may reduce vulnerability by exercising its legitimate power to enforce legally binding rules and following up with sanctions. Strong governmental agents can exercise power by limiting firm and supplier discretion and thereby reducing the scope of options available that are consistent with the legal standards. In the Norwegian safety regime there has traditionally been a balance of power and trust between the regulators and the regulated entities. This balance of power and trust has resulted in a hybrid regulatory system of limited command-and-control or prescriptive rules combined with performance-based rules and voluntary standards.

The function-based regulations give greater leeway to all actors than prescriptive rules would. By "greater leeway," we mean that both companies and the PSA have been given autonomy to decide how they will meet the HSE standards. But the regime rests on the assumption that the involved parties have a common interest in maintaining this system and that the conflicts of interest that may arise will naturally be solved without threatening the foundation of the trust between the involved parties. How much power each of the involved agents actually possesses varies depending on the character of the event. Over the last 10–15 years we have seen several examples where the parties have used their power base in a way that has reduced their trust in each other, and in some cases it has turned into distrust and blocking of cooperation. Such situations have often occurred in connection with focusing events such as Norne, Snorre A, and Gullfaks C.

The function-based character of the regulatory regime creates a large degree of autonomy for how employers and companies can design the safety practices they think are appropriate. Such autonomy can be advantageous for employers in several ways, not the least of which is that it enables them to meet safety goals in the least costly manner. Similarly, the employees have a vested interest in the regime because it gives them relatively large influence – at least formally. This is provided through the formal arenas for collaboration and has given unions more power than other regulatory regimes would provide.

The various groups' use of power to protect their (special) interests in the HSE field represents a type of politics that can be a challenge for the regime, including its robustness. This is especially true in cases where special interest groups have such large impact, ending up in a situation where objectively reasonable solutions are excluded in favor of solutions preferred by the interest group. Focusing events often induce politicization, where instead of seeking neutrality and well-reasoned decisions, the solution becomes a political battlefield. This can happen because the focusing events temporarily change the power dynamics of the actors, providing some with an opportunity to change the nature of the system. For example, after a large loss-of-life event, industry arguments for cost-effectiveness are not politically saleable, diminishing the power of the industry actors and increasing the relative power of the unions. Politicization may, in other words, disturb trust and power relations. Politicization of the Nordic regulatory model is problematic because decisions may be made on the basis of political expediency without clear theoretical or normative grounds in risk analysis and cost-benefit analysis. After a focusing event, politicization has the same effect as risk narratives. It could create and fortify changes, but it could also exclude rational actions and risk decisions.

Furthermore, the autonomy to navigate between performance-based rules and voluntary standards may develop into a complex hierarchy of norms. Overlapping legislation, legal standards, standardization, and detailed management documentation (for companies) are further challenging aspects of the Norwegian regime. Mastering such complexity requires knowledge and skills. This complexity also gives power to those who possess knowledge in handling and navigating in the system. This can naturally lead to distrust among those who cannot understand or manage the complex system and must rely on others to do so. Complexity thus creates an imbalance between power and confidence, as those with competence get increased power, while those who lack knowledge respond with distrust. Such a disparity can be further developed where there is a significant leeway to achieve regulatory objectives by applying different industry standards or other "best practices" rather than adhering to a single set of prescribed practices. How can those without detailed knowledge of the system truly know if the proposed best practices are actually satisfactory? Indeed, new studies have shown that unions in Norway are experiencing an increasing lack of knowledge causing a reduction in their confidence in the function-based system (Engen et al. 2014).

Accordingly defining the balance of power is one of the most complex questions concerning robustness/vulnerability of the Norwegian regulatory regime. The issue depends on a multitude of factors and on defining how the different roles and responsibilities should be distributed between the state, the industry and the scientific and professional stakeholders. The important factors influencing the balance of power are risk perception and risk assessments among the stakeholders, the capabilities of the regulator and regulated, the regulatory and legislative culture, and so forth (Lindøe and Engen 2013).

8.7 FINAL REMARKS

The major accidents occurring in the North Sea during the 1970s and 1980s put the issue of technical safety and integrity on the offshore installations on the political agenda. The global oil and gas industry creates a complex regulatory environment. Major operators, entrepreneurs, and subcontractors are operating with similar exploration, drilling, and production equipment and procedures. They are subject to similar industrial standards created by a network of experts within an international scientific and technical community. These activities involve sophisticated analytic methods, advanced engineering, large-scale investment, and complex projects. They must be managed appropriately to ensure that benefits are gained without incurring major accidents and other unacceptable harms to the public, the workers involved, and the human and natural environments. However, major accidents such as the Deepwater Horizon blowout and oil spill in the Gulf of Mexico in 2010, and near accidents at Snorre A and Gullfaks C in the North Sea in the last years, have demonstrated that combining productivity and safety is still a major challenge, particularly in deepwater regions and other difficult areas.

The function-based Norwegian regulatory regime based on principles of the Nordic Model has until 2015 survived; perhaps because there has not been a focusing event similar to Alexander L. Kielland. The incidents on Snorre A and Gullfaks C have, however, brought to light some weaknesses in the seemingly robust character of a function-based regulatory regime. A function-based regime demands continuous improvements and learning in order to maintain its effectiveness over time. At the same time it requires strong and competence-based regulators that are able to strike a balance between functions-based and prescriptive-based rules and also have the power to maintain the regime by serving brokers with industry and labor actors. The Nordic model of risk regulation has shown that risk regulation emphasizing legal writings and text do not have real substance without a proactive and strong regulator and empowered industrial actors working in mutual communities of practice. In determining a robust regulatory practice, many elements must be weighed toward a legal framework: the maturity of the industry, its use of standards and "best practices," and the commitment of external and internal stakeholders to utilize information about unsafe conditions to make improvements in order to make safety management systems more efficient.

References

Bang, P. and O. Thuestad (2014) Government-Enforce Self Regulation: The Norwegian Case in P. Lindøe, M. Baram, and O. Renn (eds.), *Risk Governance of Offshore Oil and Gas Operations*. Cambridge, UK: Cambridge University Press: 243–73.

Bennear, L. S. (2015) "Positive and Normative Analysis of Offshore Oil and Gas Drilling Regulations in the US, UK, and Norway," *Review of Environmental Economics and Policy*, 9(1): 2–22.

Bieder, C. and M. Bourrier (2013). *Trapping Safety into Rules. How Desirable or Avoidable Is Proseduralization?* Surrey, UK: Ashgate.

British Broadcast Company (BBC) n.d. "On This Day: 27 March 1980, North Sea Platform Collapses" available at: http://news.bbc.co.uk/onthisday/hi/dates/stories/march/27/newsid_2531000/2531091.stm.

Engen O. A (2009) The Norwegian Petroleum Innovation System A Historical Overview in J. Fagerberg, D. Mowery and B. Verspagen (eds.) *Innovation, Path Dependency and Politics*. Oxford University Press: Oxford: 179–207.

Engen O. A (2014) Emergent Risk and New Technologies in P. Lindøe, M. Baram, and O. Renn (eds.) *Risk Governance of Offshore Oil and Gas Operations*. Cambridge, UK: Cambridge University Press: 340–59.

Engen O. A., O. Langhelle, and R. Bratvold. (2012) Is Norway Really Norway in B. Schaffer and T. Ziyadov (eds.) *Beyond the Resource Curse*. University of Pennsylvania Press: Philadelphia: 259–83.

Engen, O. A., P. H. Lindøe, and K. Hansen. (2014) Power, Trust and Robustness-The Politicization of HSE in the Norwegian, Petroleum Regime, Paper at Working on Safety, Sept. 2014. Glasgow.

Gilad, S. (2010). "It Runs in the Family: Meta-Regulation and Its Siblings" *Regulation and Governance* (4): 485–506.

Hale, Andrew (2014). Advancing Robust Regulation in P. Lindøe, M. Baram, and O. Renn (eds.) *Risk Governance of Offshore Oil and Gas Operations*. Cambridge: Cambridge University Press: 403–24.

Hood, C., H. Rothstein, and R. Baldwin (2001). *The Government of Risk: Understanding Risk Regulation Regimes*. Oxford, UK: Oxford University Press.

Hopkins, A. (2011). "Risk-Management and Rule Compliance: Decision-Making in Hazardous Industries" *Safety Science* 49: 110–20.

Hovden J. (2003) The Development of New Safety Regulations in The Norwegian Oil and Gas Industry in B. Kirwan, A. R. Hale, and A. Hopkins (eds.) *Changing Regulation: Controlling Risks in Society*. Oxford: Pergamon: 57–78.

Hovden, Jan (2004). "Public Policy and Administration in a Vulnerable Society: Regulatory Reforms Initiate by a Norwegian Commission." *Journal of Risk Research* 7(6): 629–41.

House of Commons, Energy and Climate Change Committee. (2011). "UK Deepwater Drilling—Implications of the Gulf of Mexico Oil Spill " London: The Stationery Office Limited. www.publications.parliament.uk/pa/cm201011/cmselect/cmenergy/450/450i.pdf.

Kaasen. K (2014) Safety Regulation on the Norwegian Continental Shelf in P. Lindøe, M. Baram, and O. Renn (eds.) *Risk Governance of Offshore Oil and Gas Operations*. Cambridge, UK: Cambridge University Press: 103–31.

Lindøe, P. H., M. Baram, and J. Paterson (2013) Robust Offshore Risk Regulation in G. Marchant, K. Abbot, and B. Allenby (eds.) *Innovative Governance Models for Emerging Technologies*. Cheltenham: Edward Elgar Publishing Ltd: 235–53.

Lindøe, P., M. Baram, and O. Renn (2014) *Risk Governance of Offshore Oil and Gas Operations*. Cambridge, UK: Cambridge University Press.

Lindøe P. and O. A. Engen (2013) Offshore Safety Regimes – A Contested Terrain in M. Nordquist, J. N. More, A. Chircop, and R. Long (eds.)

The Regulation of Contental Shelf Development. Rethinking International Standards. Leiden, The Netherlands: Martinus Nijhoff Publishers: 195–214.

Lindøe, P., O. A Engen, and O. E. Olsen (2011). "Reponses to Accidents in Different Industrial Sectors." *Safety Science* 49: 90–97.

National Oceanic and Atmospheric Administration (NOAA) 1977. Ekofisk Bravo Oil Field Situation Report. Available at: http://incident news.noaa.gov/incident/6237/506648

Olsen J. P (1989) *Petroleum og Politikk* (Petroleum and Politics) Oslo. Tano

Olsen, O. E. and P. H. Lindøe (2009). "Risk on the Ramble: The International Transfer of Risk and Vulnerability." *Safety Science* 47: 743–55.

Paté-Cornell E. (1993) "Learning from the Piper Alpha Accident: A Postmortem Analysis of Technical and Organizational Factors" *Risk Analysis*, 32(11): 215.

Paterson J. (2014) Health and Safety Regulation on the UK Continental Shelf. Evaluation and Future Prospects in P. Lindøe, M. Baram, and O. Renn (eds.) *Risk Governance of Offshore Oil and Gas Operations.* Cambridge, UK: Cambridge University Press: 132–53.

Petroleum Safety Authority of Norway. 2011. "The Deepwater Horizon Accident–Assessments and Recommendations for the Norwegian Petroleum Industry (English Summary)." Olso: Petroleum Safety Authority, Norway. www.ptil.no/getfile.php/PDF/DwH_PSA_sum mary.pdf.

Rosness R. and U. Forseth (2014) Boxing and Dancing: Tripartite Collaboration as an Integral Part of a Regulatory Regime in P. Lindøe, M. Baram, and O. Renn (eds.) *Risk Governance of Offshore Oil and Gas Operations.* Cambridge, UK: Cambridge University Press: 309–39.

Ryggvik, H. (2003) *Fra forvitring til ny giv. Om en storulykke som aldri inntraff* (From deterioration to a new beginning. A major accident that never happened). University of Oslo, TIK nr 26.

Ryggvik, H. and K. Øye Gjerde (2014) *On Edge, Under Water* Norli. Oslo

Ryggvik, H. and M. Smith Solbakken (1997) *Norsk Oljehistorie* (Norwegian Oil History) Vol III. Oslo, Norway: Ad Notam Gyldendal.

Short, J. L. and M. W. Toffel (2010). "Making Self-Regulation More Than Merely Symbolic: The Critical Role of the Legal Environment." *Administrative Science Quaterly* 55: 361–96.

Sinclair, D. (1997). "Self-Regulation Versus Command and Control? Beyond False Dichotomies." *Law and Policy* 19(4): 529–59.

9

Reform in Real Time

Evaluating Reorganization as a Response to the Gulf Oil Spill

Christopher Carrigan

Notwithstanding the human, economic, and ecological suffering that accompanies regulatory disasters, such tragedies present opportunities for improvements to the regulatory structures designed to control the risks that prompted the failure.[1] Still, precisely because they are calamitous, such events create intense pressure for forceful action even when the set of solutions is inadequate to address the issues at hand (Kingdon 2003; Carrigan and Coglianese 2012; Chapter 5 in this volume). While outwardly dramatic, the political response in such cases can be merely symbolic, doing little to actually address the public's desire that its government better manage the associated hazards (Edelman 1967; Mayhew 1974). In these situations, apart from presenting the appearance of being responsive, the resulting actions will be of little use in effecting productive change and, worse still, can convey the false message that the underlying problems have been remedied.

In this chapter, I ask which of the two boundaries along the range of possible reactions to regulatory disaster – effective change or symbolic action – more closely describes the US response to the tragic 2010 oil spill in the Gulf of Mexico. The disaster began with the death of 11 oil rig workers after the April 2010 explosion and fire on the BP-leased Deepwater Horizon drilling rig and ended with several million barrels of oil being spilled into the Gulf. I consider how the disaster impacted public

---

[1] I am grateful to participants at the Kenan Institute for Ethics' *Recalibrating Risk* seminars held at Duke University for their insightful comments on earlier drafts. I also thank Lori Bennear for her helpful feedback as well as Mike Brooks and Stefan Roha for their excellent research assistance and valuable conversations in writing this chapter.

and political views of the appropriate balance among competing objectives: developing offshore oil and gas reserves, collecting the accompanying tax revenue from oil and gas companies, and ensuring that drilling was conducted in an environmentally responsible way. Moreover, I review the range of US government and industry responses to the tragedy before specifically linking public and political attitudes to perhaps its most dramatic permanent reform – the disbanding of the Department of the Interior's (DOI) regulator of offshore oil and gas development. Prior to its breakup one month after the onset of the spill, the Minerals Management Service (MMS) employed roughly 1,600 workers (MMS 2010d), not only to regulate offshore production but also to lease federal offshore properties and collect taxes from private oil and gas production on all government-owned property.

I first demonstrate that the spill interrupted a pronounced long-term swing in political and public attitudes toward emphasizing oil and gas production relative to ensuring that drilling was conducted safely and environmentally responsibly. Yet, while the Gulf disaster did persuade politicians and citizens to question their previous views, the effect proved transitory. Even before the well was capped, attention again shifted to finding ways to encourage exploration to control energy costs. Second, contrasting popular perception, I describe how the reorganization of government oil and gas management functions had been a source of ongoing debate well before the onset of the spill. Still, these reform proposals envisioned a very different structure that, instead of dividing MMS's missions between agencies, would have streamlined minerals management activities and potentially consolidated all of them in one government entity. The purpose of the proposals for such a radically different structure was to resolve perceived long-standing inadequacies in DOI's performance as tax collector. Third, contradicting its outwardly dramatic aura, I show that the decision to formally divide the three missions had little impact on how the personnel charged with achieving them interacted. Rather, from an operational standpoint, the reorganization appears to have done little to alter governmental oil and gas management processes.

In combination, these insights demonstrate that the dismantling of MMS in reaction to the Gulf oil disaster does not fit neatly into the category of a reform that can be expected to significantly reduce the possibility of future breakdowns. Neither is it one that in its role to placate the intense pressure for action in the wake of calamity offers no hope of promoting the public interest. Rather, the disbanding of MMS in the aftermath of the Deepwater Horizon tragedy reveals the possibility that symbolic responses can serve – either by accident or on purpose – a third objective, one which staves off meaningful action until social preferences completely recalibrate.

Particularly when a dramatic swing in popular perceptions of risk is ephemeral and diverts attention from more durable reform efforts, a truly symbolic act can limit the degree to which associated actions lock in programs that move policy away from true social preferences. This is true precisely because symbolic responses have few measurable effects. Dissolving MMS did not address long-standing concerns about the failure of DOI to collect royalties due the federal government. Even so, the reorganization also did little to impede the federal government's ability to remain politically responsive given the possibility (and resulting reality) that the public and its political representatives would shift their attention back to finding ways to reduce energy costs and increase oil and gas taxes as environmental concerns raised by the disaster began to fade.

BUILDING POLITICAL AND SOCIAL PRESSURE FOR DEVELOPMENT

In compiling its top 10 US news stories of 2010, *Time* magazine named the Gulf oil spill number one (Tharoor 2010). The disaster, which began on April 20, 2010 with an explosion aboard the Deepwater Horizon drilling rig, resulted in an estimated 4.9 million barrels of oil being dumped into the Gulf of Mexico, making it far and away the biggest oil spill in US history (National Commission 2011). In comparison, the Exxon Valdez spill in March 1989, which up to 2010 was the largest US offshore spill, deposited approximately 250,000 barrels into Alaska's Prince William Sound (Skinner and Reilly 1989; see also Chapter 7 in this volume). While no small number, the Gulf tragedy created a spill more than an order of magnitude greater than *Exxon Valdez*. By the time the well was more or less capped in mid-July 2010, the story had captivated the nation, introducing people to a litany of new terms including "top-kill" and "junk shot" as BP, scientists, and the federal government tried frantically to stop oil from gushing from the Gulf seafloor (Fountain 2010).

The accident occurred in a political climate fraught with tension among two competing narratives (see Chapter 6 in this volume). The first, captured by the political chant "drill, baby, drill," views the

US as a nation of abundant natural resources that have been used, and should continue to be used, to further economic growth and political stability. The second narrative derives from the environmental movement of the 1970s (see Chapter 7 in this volume) and sees ongoing use of fossil fuels as a key component of continued environmental degradation both locally and globally, thus arguing for significant government regulation to stem this degradation.

In the period leading up to the Gulf oil spill, political and social preferences had evolved, in a gradual but pronounced manner, toward emphasizing oil and gas production (consistent with narrative 1) over environmental and drilling safety (narrative 2). Prompted by President Bill Clinton's National Partnership for Reinventing Government, MMS began to experiment with more cooperative ways of regulating drilling and production in the early to mid-1990s, forming committees to propose gas valuation rules and settle issues connected to controversial Pacific Outer Continental Shelf (OCS) leases[2] (Cedar-Southworth 1996; MMS 1995). MMS's interest in collaboration was encouraged by a broad set of interests. Reacting to a 1993 report by a group tasked to study OCS policy that included representatives from the oil industry, coastal states, and environmental organizations, Secretary of the Interior Bruce Babbitt noted the significance of the committee's recommendation that OCS operations "be regenerated based on consensus" (MMS 1994).

While MMS's efforts to adopt a more cooperative approach were already in motion, the November 1995 Deep Water Royalty Relief Act (DWRRA) – which amended the 1953 Outer Continental Shelf Lands Act (OCSLA) – accelerated the process, promoting deep-water drilling despite the fact that the technology to support it safely was not yet available (National Commission 2011). The DWRRA waived royalty payments on western and central Gulf leases offered for sale until near the end of 2000 as long as the associated lease was in deep water (i.e. required drilling in water deeper than 200 meters) and had not produced substantial oil and gas. The act also provided royalty relief for existing leases if the company could show that it was not able to extract oil and gas from the property economically without it (DWRRA 1995).

Not surprisingly, the law had important effects on Gulf oil and gas operations. MMS noted in a later budget justification that it "triggered record-breaking lease sales in 1997 and 1998 ... and opened the door to

---

[2] The OCS includes all federally owned submerged lands in the Gulf, along the Pacific coast, and surrounding Alaska.

TABLE 9.1 *Subject matter of congressional hearings in which MMS or successor agency personnel testified by topic (1982–2012)*

Period	Evaluation	Leasing	Environment	Regulation	Revenue	Total
1982–1995	28	25	41	22	24	77
1996–2009	20	14	13	1	20	41
2010	0	4	8	13	1	14
2011	3	1	5	7	2	12
2012	0	3	1	3	0	5

Notes: The table does not include budget hearings. To categorize a hearing, its title and summary description were reviewed. Where clarification was necessary, the testimony was examined as well. For each table row, the sum of the topic count exceeds the total because hearings often involve multiple topics. Evaluation refers to identifying areas for oil and gas exploration, and leasing refers to leasing properties to oil and gas producers. The Bureau of Ocean Energy Management, the Bureau of Safety and Environmental Enforcement, and the Office of Natural Resources Revenue are the successor organizations to MMS.
Source: Proquest Congressional.

increased deepwater production" (MMS 2004, 80). In recounting the lessons from initial efforts to oversee deep-water exploration, Carolita Kallaur, MMS's Associate Director at the time, explained, "there is tremendous value from collaboration between government, industry and the scientific community in the area of research and operational requirements. This is particularly true if it is found that the operating environment is totally different from what one is used to" (Kallaur 2001).

Subsequent congressional lawmaking continued to support this policy shift. Each law Congress passed during the 15 years from 1995 through 2009 primarily focused on either encouraging offshore exploration or boosting tax collections. Beginning with the 1995 DWRRA, this list included the 1996 Federal Oil and Gas Royalty Simplification and Fairness Act, the 2005 Energy Policy Act, and the 2006 Gulf of Mexico Energy Security Act. For example, despite extending moratoria near Florida's coast, the Gulf of Mexico Energy Security Act concurrently mandated that MMS attempt to lease 8.3 million acres within one year, of which close to 70 percent had been previously off limits to drilling (Gulf of Mexico Energy Security Act 2006).

This congressional shift is further illustrated in Table 9.1. The table categorizes the subject matter of hearings, which included testimony of an MMS employee or, after its breakup, one of its successors over the period from MMS's creation through 2012. Over the 14 years from 1982 to 1995, environmental and regulatory issues generated more combined interest than evaluation and leasing issues as measured by how often they were the subject of hearings over the period. In contrast, during the 14 years from 1996 through 2009, evaluation and leasing were close to two-and-a-half times more likely to be considered than the environment and regulation.

The numbers for regulation are even more striking. From 1982 through 1995, 22 hearings included a substantial discussion of offshore regulation. In contrast, only one hearing involved an important consideration of regulatory issues over the 14 years leading up to the Gulf spill. Even that hearing was primarily focused on the Bureau of Land Management's (BLM) onshore regulatory program and included relatively little mention of MMS's offshore responsibilities (Subcommittee on Energy and Mineral Resources 1996).

Like Congress, the White House accelerated its push for greater exploration during the presidency of George W. Bush. President Bush's 2008 Memorandum for the Secretary of the Interior represented a significant break from previous policy, opening up all areas of the OCS except marine sanctuaries for exploration (Bush 2008a). The president noted, "One of the most important steps we can take to expand American oil production is to increase access to offshore exploration" (Bush 2008b). Prior to the onset of the Gulf spill, the Obama administration shared this enthusiasm for offshore drilling. Accompanying his 2010 Memorandum, which removed only Alaska's Bristol Bay from leasing consideration (Obama 2010a), President Barack Obama declared "today we're announcing the expansion of offshore oil and gas exploration" (Obama 2010d).

Shifting political priorities over the period reflected public sentiment on energy issues as well. Figure 9.1, which summarizes repeated Gallup polls asking respondents their preference for environmental protection or economic growth, reveals a growing concern for the economy, particularly beginning around 2000. In April 1990, respondents favored focusing on environmental protection even if it caused the economy to suffer by close to a four to one margin. By early 2010, those that prioritized economic growth exceeded those for the environment by 15 percentage points. This dramatic reversal is consistent with the results of a related Gallup poll which, beginning in March 2001, asked people specifically about whether energy production or the environment should be prioritized. Although environmental protection held a 16 percentage point margin in 2001, the March 2010 poll showed a similar reversal, revealing a seven percentage point advantage for energy development (Gallup 2013).

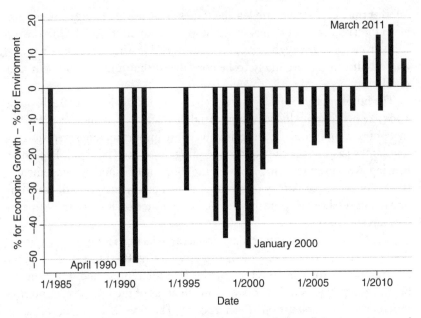

FIGURE 9.1 Gallup poll results measuring public preference for economic growth or environmental protection (1984–2012)
Notes: The percentages were compiled using all Gallup polls conducted from 1984 to 2012 that asked: "With which of these statements about the environment and the economy do you most agree – [ROTATED: protection of the environment should be given priority, even at the risk of curbing economic growth (or) economic growth should be given priority, even if the environment suffers to some extent?" % for Economic Growth – % for Environment was computed by subtracting the percent of people that preferred environmental protection from the percent that placed greater importance on economic growth.
Sources: Gallup 2010, 2012.

PUBLIC AND POLITICAL RESPONSE TO THE GULF OIL SPILL

While the oil and gas industry itself certainly supported efforts to expand production, the previous section demonstrates that politicians and the public did as well over an extended period prior to the Gulf oil spill. The dramatic images and sheer magnitude of the disaster had important short-term impacts on those preferences. As one consequence, intense media coverage in the days following the explosion raised awareness of the technological sophistication required to drill for oil and gas in deep water.

Prior to the spill, the US had few recent examples to draw upon that would demonstrate what could go wrong when drilling for offshore oil.

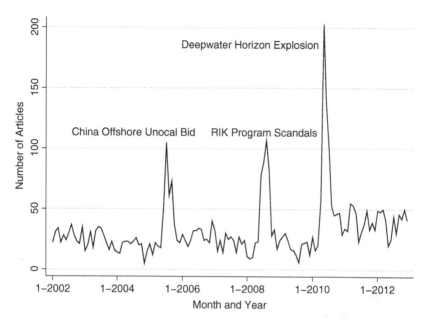

FIGURE 9.2 *New York Times* and *Washington Post* article mentions of offshore oil and gas (January 2002–December 2012)
Notes: The number of articles each month was computed as the sum of all *New York Times* and *Washington Post* articles which included the words "offshore" plus either "oil" or "gas" in them. The RIK Program refers to the Royalty in Kind Program, overseen by a subgroup within MMS's Minerals Revenue Management division, which accepted tax payments from oil and gas producers in kind.
Source: Lexis Nexis Academic.

Relative to the 10-year period that preceded it in which over 430,000 barrels were deposited in offshore waters, during the entire 35 years from 1975 through 2009, OCS activities only accounted for about 121,000 barrels spilled, under a third of the amount in 25 more years (BOEMRE 2011, see also Chapter 7 in this volume). Perhaps more poignantly, the BP disaster deposited more than 40 times more oil into the Gulf in three months than what was spilled during that same 35-year period (Carrigan 2014). Reflecting a broadly held view drawn from several decades of experience, President Obama declared less than three weeks before the Deepwater Horizon fire that "oil rigs today generally don't cause spills. They are technologically very advanced" (Obama 2010b).

In addition to presenting an instance intensely at odds with prior history, the Gulf oil spill simply focused attention on offshore oil and gas drilling. Figure 9.2 shows monthly the number of articles from the

New York Times and *Washington Post* that included the words "offshore" and either "oil" or "gas" over the period from January 2002 through December 2012. As the figure demonstrates, the period between 2002 and the beginning of 2010 demonstrated a consistent level of interest in oil and gas with the exception of two spikes. The first corresponded to coverage of the political controversy associated with China National Offshore Oil Corporation's effort to buy Unocal. The second spike, which occurred during the second half of 2008, was in response to a DOI Office of the Inspector General (OIG) communication released in September 2008 (Devaney 2008).

The memorandum and associated investigative reports principally focused on the unethical actions during the first half of the 2000s of members of the Royalty in Kind (RIK) Program, a group of roughly 50 employees in MMS's Minerals Revenue Management (hereafter, Revenue Management) division (MMS 2007). The reports documented the large volume of industry gifts – including invitations to parties, trips, and event tickets – that nine of the 19 accused workers received, in addition to the sexual relationships that two of these employees had with industry contacts (OIG 2008c). During the investigations, OIG uncovered evidence that at least one RIK employee held another job that was not disclosed (OIG 2008b) and that three senior Revenue Management employees ignored "post-employment restrictions," organizing opportunities for two retired employees to provide consulting services to MMS (Devaney 2008; OIG 2008a).

Still, the attention resulting from RIK Program employees' indiscretions paled in comparison to the surge in news coverage as the events unfolded with the onset of the spill in April 2010. Relative to March, when 20 articles mentioning offshore and oil or gas appeared in the *New York Times* and *Washington Post*, by May, coverage had increased by 10 times to over 200 articles. However, almost as dramatic as the spike in interest, Figure 9.2 also shows how quickly attention waned in the months following the capping of the well in July 2010. While the well was still releasing oil into the Gulf during June and the first half of July, the number of articles had already declined to a level that approximated the coverage when the OIG communication was released in 2008. By August 2010, reporting on offshore oil and gas in the *New York Times* and *Washington Post* had settled back to a much lower level that continued to exceed, but began to approach, what had been the long-term trend between 2002 and 2010.

Not only was the dramatic increase in newspaper coverage of offshore operations short-lived, but so was the relative emphasis that the public placed on environmental protection relative to economic growth and energy development during the crisis. Recalling Figure 9.1, the Gallup poll conducted in May 2010 while oil was still pouring into the Gulf did demonstrate a measurable increase in respondents who viewed environmental protection as more important than economic growth. Yet, the poll conducted in March 2011 – less than a year after the initial explosion – showed that attitudes toward energy development and environmental protection had already largely returned to their longer-term trends. Relative to the poll conducted in March 2010 when respondents emphasizing economic growth exceeded those for the environment by 15 percentage points, by March 2011, the spread had actually widened to 18 percentage points despite the Gulf disaster in between.

Gallup polls asking individuals to specifically prioritize energy production or the environment show a similar rebound. A month after the onset of the spill, concern for the environment outpaced development by 16 percentage points, which represented exactly the same spread as the first poll in 2001. By March of the next year, 50 percent of respondents agreed that development was a greater priority while 41 percent opted for the environment, a nine-point difference, which represented the largest in the poll's history. Although the spread shrank in subsequent polls, even in April 2013, respondents still showed a slight preference for energy development relative to environmental protection (Gallup 2013).

Congressional activity reveals analogous patterns to those found for the news media and in public opinion polls. As displayed in Table 9.1, the number of hearings that considered offshore regulation and the environment substantially increased directly after the onset of the spill. In fact, the volume of hearings on regulatory issues in 2010 was greater than the total that considered regulation in the 20 previous years combined. But similar to media coverage and public opinion, interest quickly tapered off. The total number of hearings declined slightly from 14 to 12 in 2011 and decreased more quickly to 5 in 2012, as did the number that specifically focused on regulatory and environmental issues. By 2012, the volume and makeup of hearings looked remarkably similar to patterns of congressional oversight prior to the spill.

This same pattern can even be seen in the Obama administration's approach to conducting offshore lease sales following the spill. Citing a focus on ensuring drilling and production were conducted safely, Secretary Salazar announced in December 2010 a much more cautious plan for leasing federal offshore property to oil and gas companies than the strategy released in March, just prior to the Deepwater Horizon explosion. In modifying its strategy, DOI removed the eastern Gulf of Mexico and mid- and south Atlantic from consideration for development through 2017, and Alaska's Chukchi and Beaufort Seas until at least 2012. The revised schedule also called for delayed sales in the western and central Gulf until at least late 2011. In announcing the changes, Secretary Salazar noted, "Our revised leasing strategy lays out a careful, responsible path for meeting our nation's energy needs while protecting our oceans and coastal communities" (Office of the Secretary of the Interior 2010a).

Still, by early 2012, President Obama's "all-of-the-above" energy strategy had renewed the federal government's interest in offshore leasing. The president's statement during his January 2012 State of the Union address that, "tonight, I'm directing my administration to open more than 75 percent of our potential offshore oil and gas resources" (Obama 2012) spearheaded a series of DOI announcements regarding future leasing activity. Just two days later, the Department announced a June 2012 sale of "all available unleased areas in the Central Planning Area offshore Louisiana, Mississippi and Alabama" (Office of the Secretary of the Interior 2012c), which would eventually attract $1.7 billion in high bids on close to 39 million acres (Office of the Secretary of the Interior 2012d). In two separate March announcements, DOI asked for industry's interest in potentially including the coast of south-central Alaska in the 2012 to 2017 leasing plan while also indicating its plans to assess the available oil and gas reserves in the mid- and south Atlantic (Office of the Secretary of the Interior 2012a; 2012e). By June 2012, the administration had announced that its final leasing program for 2012 to 2017 had made "all areas with the highest-known resource potential – including frontier areas in the Alaska Artic – available for oil and gas leasing in order to further reduce America's dependence on foreign oil" (Office of the Secretary of the Interior 2012b). Contrasting the much more cautious approach to development in late 2010, Secretary Salazar noted, "Put simply, this program opens the vast majority of known offshore oil and gas resources for development over the next five years ... President Obama has made clear his commitment to expanding responsible domestic oil and gas production in America as part of this all-of-the-above energy strategy" (Office of the Secretary of the Interior 2012b).

REFORMING OIL AND GAS OPERATIONS AFTER THE SPILL

Despite the fleeting political and public reaction to the Gulf spill, as summarized in Table 9.2, the disaster did spawn reforms within both government and the industry, particularly during the period when collective attention fixated on the spill. Just over three weeks after the initial explosion, on May 14, 2010, President Obama announced his intention to place a temporary moratorium on offshore drilling. By the end of that month, DOI had released a directive outlawing drilling of new wells in waters greater than 500 feet and mandating those operators currently drilling in deep water to stop and secure the associated wells (MMS 2010c). The directive further stipulated that applications to drill would not be considered for six months with the idea that the moratorium would give a key presidentially created committee – the National Commission on the BP Deepwater Horizon Oil Spill and Offshore Drilling – time to complete its investigation (DOI 2010a). The order was subsequently successfully challenged in a federal district court, which characterized the moratorium as overly expansive in granting a preliminary injunction (Feldman 2010b). Although the ruling was upheld by an appeals court, DOI quickly replaced that moratorium with another one in the same spirit (Salazar 2010a). Because the revised suspension focused on drilling that relied on the types of technologies utilized to prevent blowouts in deep water, it had a very similar effect as the overturned broader suspension.

Under pressure from the oil and gas industry as well as the affected states, DOI lifted the suspension approximately a month and half before its original estimate (Salazar 2010b). Yet, it did so while imposing a series of new requirements, including two directives and an interim final rule. Although shallow-water applications were approved (albeit at a reduced rate) during the period beginning with the initial moratorium, the additional measures effectively meant that drilling in deep water only resumed with the first permit approval at the end of February 2011 (Broder and Krauss 2011). The more stringent requirements had effectively extended the suspension for over four more months, even prompting the aforementioned district court in New Orleans to order the first agency created to replace MMS prior to its permanent breakup, the Bureau of Ocean Energy Management, Regulation, and Enforcement (BOEMRE), to respond to drilling applications (Feldman 2011).

Central to the new requirements were two Notices to Lessees (NTLs), issued within 10 days of each other in June 2010, that implemented

TABLE 9.2 *Reforms enacted in response to Gulf oil spill (May 2010–May 2013)*

Action	Date	Summary
Drilling Safety		
Interim Final Rule	October 2010	Imposed requirements for drilling including those to maintain both bore integrity and blowout preventer
Final Rule	August 2012	Affirmed interim rule and formalized API's Recommended Practice (RP) 65 mandating process for isolating potential flow zones
Workplace Safety		
Workplace Safety Final Rule	October 2010	Codified API's RP 75 requiring operators to prepare plans to identify, address, and manage safety and environmental hazards
SEMS II Final Rule	April 2013	Empowered any and all industry personnel to halt operations due to dangerous activity
Moratorium		
Initial	May 2010	Suspended pending, approved, and current offshore drilling of new deep-water wells in Gulf and Pacific for six months
Revised	July 2010	Restricted moratorium to drilling operations utilizing types of blowout preventers used in deep water
Lifted	October 2010	Repealed moratorium but stipulated that operators comply with NTLs and Drilling Safety Interim Final Rule before resuming
Directives		
Safety NTL	June 2010	Implemented recommendations from DOI's May 2010 "Safety Measures Report" including CEO certification of operations
Environmental NTL	June 2010	Required operator exploration plans submitted to BOEM to include blowout scenario and description of steps taken to prevent blowout
Safety NTL Judgment	October 2010	District court in New Orleans declares NTL unlawful because violated APA. Vacated officially by BSEE in April 2012

(continued)

TABLE 9.2 *(continued)*

Action	Date	Summary
Institutions		
Disbanding of MMS	May 2010	Divided MMS into three DOI agencies, each separately focused on tax collection, offshore leasing, and offshore oversight
Center for Offshore Safety	March 2011	Organized to promote effective practices for operators and to focus on implementing API's RP 75
Ocean Energy Safety Institute	May 2013	Created to spur offshore safety research and identification of "Best Available and Safest Technology"

Notes: The reforms listed include those specifically enacted and not those contemplated but not implemented. API refers to the American Petroleum Institute. SEMS refers to a Safety and Environmental Management System. An NTL is a Notice to Lessees and Operators. APA refers to the Administrative Procedure Act of 1946.

a series of DOI recommendations made in response to President Obama's directive for Secretary Salazar to conduct a review of the incident soon after the initial explosion (DOI 2010b). The first notice, referred to as the Safety NTL, applied to both shallow-water operations and deep-water production platforms, as well as those drilling activities suspended by the moratorium (MMS 2010a). The Safety NTL mandated an increased focus on testing and inspecting drilling and blowout prevention equipment as well as ensuring that well design met stricter standards. Moreover, the NTL required CEO certification of the associated company's operations as well as independent verification and additional reporting to MMS before the end of the month in which the notice was issued to avoid the possibility of having the associated operations shut down.

The second, referred to as the Environmental NTL, centered on improving responsiveness to a blowout like the one that prompted the Gulf spill (MMS 2010b). Specifically, for each new or pending exploration or development plan, the directive required that the company present a more detailed assessment of the worst-case scenario with respect to the amount of oil that could be spilled as part of the activity. In addition, the operator needed to describe the resources and strategies available to minimize and respond quickly to such a situation.

Roughly four and a half months later, because MMS had not followed the notice-and-comment procedures outlined in the 1946 Administrative

Procedure Act (APA) in issuing the directive, the Safety NTL was declared to be "of no lawful force or effect" by the same court which overturned the initial moratorium (Feldman 2010a). Regardless, by that time, MMS's initial replacement, BOEMRE, had already issued an interim final rule covering many of the aspects of the NTL (BOEMRE 2010a). First announced on the last day of September, the Drilling Safety Rule became effective upon its publication in the *Federal Register* on October 14, 2010. Building from a report issued by Secretary Salazar to President Obama recommending ways to make offshore operations safer, the rule strengthened requirements surrounding procedures for readying and maintaining wells during drilling. Moreover, it mandated increased testing of blowout preventer components designed to stop oil from escaping in an accident and further required that a combination of professional engineers and independent experts audit operator well design and control equipment. The final version, which replaced the interim rule two years later, largely affirmed while also clarified its predecessor. In addition, it extended the requirements to include operations associated with preparing wells for production as well as closing them when they were no longer producing (BSEE 2012).

One day after its interim Drilling Safety Rule was published, BOEMRE's Workplace Safety Rule, addressing human error in oil and gas operations, appeared in the *Federal Register* (BOEMRE 2010b). The rule made mandatory existing American Petroleum Institute (API) recommended practices surrounding the creation of a Safety and Environmental Management System (SEMS), requiring oil and gas companies to develop plans to identify, respond to, and minimize ongoing offshore risks as they arose. This more proactive regulatory approach had its origins in collaborations between MMS and API dating back to the early 1990s, with MMS's preliminary rulemaking efforts commencing in 2006 (MMS 2006). Still, only the spill provided the impetus to make SEMS mandatory. In addition to annual reporting requirements beginning in March 2011, operators were required to have their SEMS plans in place by November 2011.

DOI consequently supplemented the Workplace Safety Rule through further rulemaking activity. Promulgated in April 2013, SEMS II mandated that, by June of the following year, operators implement additional procedures and plans through which employees could assist in identifying hazards, reporting violations, and stopping activities which posed serious safety and pollution risks (BSEE 2013b). The added requirements also obligated operators to both define exactly who had authority for a project at all phases and use independent and accredited providers to perform required SEMS audits.

Beyond new regulatory requirements, the spill also prompted the creation of a multitude of commissions to study various aspects of the disaster. In addition to the teams directly involved in the response such as those led by National Incident Commander Thad Allen and Energy Secretary Steven Chu (Obama 2010c), soon after the explosion, multiple committees were tasked with identifying the causes of the disaster. These included the aforementioned National Committee created by President Obama as well as a National Academy of Engineering committee and the DOI-led Outer Continental Shelf Safety Oversight Board, both commissioned by Secretary Salazar (Office of the Secretary of the Interior 2010b; 2010c).

Yet, the catastrophe also provided the impetus for more permanent organizational responses, both within industry and the federal government. To respond to an immediate need to develop technologies capable of containing spills of the unprecedented size of the Deepwater Horizon disaster, oil and gas operators in the Gulf formed joint ventures to produce the needed containment equipment. For example, the Marine Well Containment Company, formed by ExxonMobil, Chevron, ConocoPhillips, and Shell, was designed as an independent firm to both develop a spill containment system and make it available to member companies to encourage DOI to allow deep-water drilling to resume in the Gulf (Marine Well Containment Company 2014). In fact, based on its ability to tap into the well-capping system designed by another consortium called the Helix Well Containment Group, Noble Energy was able to secure the first deepwater permit in late February 2011 following the moratorium (Broder and Krauss 2011).

A month later, API announced the oil and gas industry's intent to fund a safety institute modeled after existing organizations such as the nuclear industry's Institute for Nuclear Power Operations (Porter 2011). Named the Center for Offshore Safety and positioned within API, the institute was to initially focus its efforts on the development of the safety management systems required through the aforementioned Workplace Safety Rule. After officially opening its doors in October 2011, the Center also began work to certify auditors to conduct the independent audits of these systems as would be eventually required by SEMS II (Dlouhy 2012).

By May 2013, DOI had announced that it was creating a safety institute of its own, with the goal of "help[ing] federal regulators keep pace with new processes employed by the industry" (BSEE 2013a). While initially proposed by Secretary Salazar in late 2010 (Wald 2010), the creation of the Ocean Energy Safety Institute was further endorsed by the Offshore Energy Safety Advisory Committee, a permanent federal advisory board established in January 2011 to advise the secretary on offshore safety issues (Office of the Secretary of the Interior 2011). While not given regulatory authority, the Institute, housed at Texas A&M, was intended to both be a source of independent advice for government agencies and train offshore government employees on the latest offshore technologies (BSEE 2013a).

DISBANDING THE MINERALS MANAGEMENT SERVICE

Still, the most outwardly dramatic of these organizational responses was the decision to disband MMS. As President Obama explained at a press conference a little over a month after the onset of the spill, following the Inspector General communication describing the aforementioned unethical activities in the RIK group, "Secretary Salazar immediately took steps to clean up that corruption. But this oil spill has made clear that more reforms are needed ... That's why we've decided to separate the people who permit the drilling from those who regulate and ensure the safety of the drilling" (Obama 2010c). The restructuring was more fully described in Secretary Salazar's Order 3299, which was released eight days prior to President Obama's news conference. In the Order, Secretary Salazar outlined the creation of three separate organizations from MMS – the Bureau of Ocean Energy Management (BOEM), the Bureau of Safety and Environmental Enforcement (BSEE), and the Office of Natural Resources Revenue (ONRR) (Salazar 2010c). BOEM would inherit Offshore Energy's leasing and resource evaluation functions, BSEE would assume regulatory oversight, and the Revenue Management group would become known as ONRR.

Given the difficulties in separating offshore leasing from regulatory oversight, as an interim step, MMS was replaced by BOEMRE. By early October 2010 when ONRR began operations, BOEMRE still managed all of the remaining functions of the former MMS. The desire to complete the reorganization, which would take almost a year and a half to finish, as well as to strengthen offshore oversight brought with it a resultant need to increase DOI's resources. Just three weeks after the onset of the spill, the Obama administration requested (and Congress approved) $29 million in emergency funds (Salazar 2010d). Four months later, the Obama administration submitted a supplemental request to amend DOI's 2011 budget (Obama 2010e). The request provided for $91 million in additional resources, of which $25 million would be funded through additional inspection fees. The additional $66 million in budget authority represented a 36 percent increase over the original request. Its purpose was "to address safety and environmental concerns highlighted by the Deepwater Horizon oil spill" and to divide BOEMRE "based on the premise that the missions within BOEMRE – including Outer Continental Shelf management, safety, and environmental oversight and enforcement, and revenue collection – need to be clearly defined and distinct from each other" (Obama 2010e).

By the end of 2011, the US had joined other major offshore oil countries in the OECD, particularly the UK and Norway, in separating the leasing and safety components of offshore oil regulation (Bennear 2015). Given that other OECD countries had separated these functions much earlier in response to similar concerns about conflicts of interests, one may wonder why the US waited until a major oil spill to make this change. But the history of MMS reveals a far more nuanced dynamic between the leasing and safety branches of MMS that had, heretofore, not been viewed as problematic. To understand the turnaround in perspective one must go back 28 years to the founding of MMS.

DRIVING OIL AND GAS ROYALTY COLLECTION THROUGH REORGANIZATION

Interestingly, the decision to combine the leasing, safety and environmental oversight, and tax collection missions to form MMS some 28 years earlier was borne from a different – but also dramatic – crisis. Over a period of several decades, Congress and other commentators remained persistent in their criticism of DOI as the federal government's oil and gas tax collector. By the late 1950s, the US Geological Survey (USGS) – the agency within DOI authorized in 1926 to collect royalties associated with energy and minerals extraction on government-owned land – was the object of intense scrutiny by the Government Accounting Office (later renamed the Government Accountability Office or GAO) and the OIG (MMS 1995). An independent panel known as the Linowes Commission, formed in 1981 to study royalty collection, accused USGS of costing the government hundreds of millions each year in lost tax revenue. Beyond habitually collecting too little in royalties, USGS proved incapable of combating efforts by oil companies to remove oil from fields without declaring it to avoid paying taxes (Linowes Commission 1982).

In the view of many, USGS's organizational design contributed to its failure. Because tax collection was independently conducted in each of its 11 regional offices, the function never gained traction in the "scientifically oriented" agency (Linowes Commission 1982). Hoping to spur the creation of a strong, independent minerals revenue collector, the Linowes Commission recommended that the function be split from USGS and located at a new agency that would be staffed with finance professionals working with a centralized accounting system (Linowes Commission 1982).

Reacting to this perception that USGS had relegated minerals revenue collection to a secondary status within the agency, Secretary of the Interior James Watt established MMS in 1982 by moving all oil and gas revenue collection functions from USGS to the new agency. He further transitioned all offshore leasing responsibilities, which had been formerly split between USGS and BLM, to MMS and shifted these same functions for onshore leases to BLM (Hogue 2010). The effect was that BLM assumed all responsibilities connected to onshore development, leasing, and regulation, and the newly created MMS was tasked with all duties for offshore oil and gas operations as well as revenue collection for all leases.

Secretary Watt's reorganization plan was broadly supported. Merging offshore functions into one agency was both consistent with the Linowes Commission's recommendations (Durant 1992) and advocated by GAO. In fact, GAO actively lobbied for greater consolidation (Socolar 1982). The House Appropriations Committee concurred, suggesting in its bill, "The bulk of the appropriation ... is associated with the ... evaluation of resources, regulations, and activities associated with Federal and Indian lands. These are functions formerly divided between the Geological Survey and the Bureau of Land Management. That division of function often caused problems of neglect, duplication, and turf wars. The Committee agrees with the consolidation" (Committee on Appropriations 1982).

As described by the House Appropriations Committee, the view at the time was that the prior structure – in which BLM administered offshore lease sales while USGS managed offshore lease management and revenue collection – was responsible for the issues that plagued DOI. These included functional disputes, application backlogs, and the "underpayment and inadequate collection of royalties owed to the United States" (Committee on Appropriations 1982). For most commentators, any concern surrounding the formation of MMS was not about the combination of leasing, revenue, and safety in one agency, but rather why that agency had not also been assigned BLM's onshore minerals management functions (Durant 1992).

From the outset, MMS was designed specifically to implement its two core functions, tax collection and offshore management, based on the view that the prior structure had failed. It is thus not surprising that MMS's organizational design reversed what had preceded it. Rather than assimilate tax collection with its other functions as USGS had, personnel and operations associated with MMS's Royalty Management (later renamed Minerals Revenue Management) activity were consolidated and housed in Lakewood, Colorado, separate and distant from offshore management which was centered in the Gulf region. The goal was to "provide efficiency and economies of scale in the financial and data collection process" (MMS 1993).

At the same time, to combat the "problems of neglect, duplication, and turf wars" (Committee on Appropriations 1982) that plagued DOI's offshore program when it was divided between USGS and BLM, MMS joined the associated sub-activities into one tightly knit program. Outer Continental Shelf Lands, eventually renamed Offshore Energy and Minerals Management (hereafter, Offshore Energy), combined resource evaluation, leasing, and regulation. Corresponding roughly to their timing in the process of developing offshore lands, the overlap was nevertheless substantial. For example, although studies to identify the location of oil and gas reserves were primarily intended to aid leasing decisions, these efforts also supported "regulatory personnel to ensure that discoveries [were] developed and produced in accordance with the goals and priorities of the OCSLA" (MMS 2004, 108). Because offshore oil and gas drilling primarily occurred in the Gulf of Mexico, the personnel associated with these functions were stationed in New Orleans or one of MMS's other offices near the Gulf.

MMS's organization design – which separated revenue collection from offshore energy management at the same time it integrated the elements of the offshore program – maintained much the same structure over its entire lifespan (Carrigan 2014). Just prior to its breakup in 2010, the vast majority of MMS's scientists, engineers, and inspection personnel, as members of the Offshore Energy program, were still located in Louisiana. In contrast, MMS's Colorado operations, which housed Revenue Management, were overwhelmingly staffed with tax and business professionals. In implementing the calls for "top quality financial managers" (Socolar 1982) and a more integrated offshore program, MMS had developed into an organization whose core functions were not only separated geographically but also through the competencies of those charged with carrying out the two programs.

EARLY EFFORTS TO SPUR REVENUE COLLECTION

Although MMS's creation did provide DOI with a respite from the criticism it received for its revenue collection difficulties, the effect proved temporary. When MMS testified in April 1985 before the House Committee on Government Operations, it was to scrutinize Revenue Management's performance in collecting and disseminating royalties associated with oil and gas production on Indian lands (Subcommittee of the Committee on Government Operations 1985). In addition to missing payments or making them for incorrect amounts, a congressional inquiry had unearthed multiple examples where Revenue Management had not responded to requests by the Bureau of Indian Affairs (BIA) to audit individual Indian accounts, a task MMS was required to perform. In response to questions about delays in responding to specific inquiries, Revenue Management revealed that it was "an obvious case of something 'falling through the cracks'" (Subcommittee of the Committee on Government Operations 1985).

The 1985 inquiry marked the beginning of a pattern of congressional oversight predicated on the idea that Revenue Management needed to do better. The quantity of hearings focused on oil and gas tax collection did not stray noticeably from those focused on offshore operations. Table 9.1 reveals that at least during the first half of MMS's existence, regulatory and environmental issues commanded significant attention in Congress. Still, the tone of these inquiries was different. Many of the hearings were prompted by the 1989 Exxon Valdez oil spill, which was not within the purview of Offshore Energy's statutory authority. While the group became intimately involved in the cleanup effort, garnered appropriations to conduct spill research, and received authority to write associated rules (Committee on Energy and Natural Resources 1989; MMS 1990, 1991), Offshore Energy's inclusion in these hearings was not because of any failure of drilling regulation.

At the same time Offshore Energy was testifying in the aftermath of Exxon Valdez, Revenue Management was again reporting to Congress on the problems it was having collecting royalties for Indian tribes (Special Committee on Investigations of the Select Committee on Indian Affairs 1989). Similarly, one year earlier, to open a hearing to review audits of MMS's revenue collection efforts, subcommittee chairman John Melcher noted, "To date, action by the Department falls far short of adequately carrying out the requirements of [the] law. Today, we are going to hear attempts to explain why" (Subcommittee on Mineral Resources Development and Production 1988).

Further evidence for Congress' dissatisfaction with Revenue Management is also displayed through GAO reports covering MMS's tenure as tax collector. In the first four years of its existence, nine GAO inquiries focused on offshore management issues while only three centered on royalty collection and one on both missions. In contrast, from 1986 through 2009, seven reports centered on offshore energy management relative to 34 on revenue management and eight on both. Moreover, even a cursory look at the titles confirms GAO's dissatisfaction, including a 2007 report titled "Royalties Collection: Ongoing Problems with Interior's Efforts to Ensure A Fair Return for Taxpayers Require Attention."

Revenue Management's issues were numerous, but an important source of many of the problems was the initial design of MMS, which allowed the accounting and auditing function to build expertise but impeded its ability to collect needed data from BLM and Offshore Energy. The geographical separation and the divergent backgrounds of operations personnel in Revenue Management and Offshore Energy suggest the difficulties the groups had in collaborating were not necessarily surprising. Multiple reports document these concerns (see, e.g., Subcommittee on Royalty Management 2007; GAO 2008). For example, in a December 2007 communication, a committee commissioned by DOI on the heels of an OIG investigation described the coordination problems created by having BLM and BIA as well as Offshore Energy and Revenue Management all involved in the royalty collection process. In addition to documenting how BLM's manual process for sharing onshore production data with MMS was inefficient, the report went on to note how similar issues existed even within MMS. The committee concluded, "Increased sharing of electronic information between BLM and MRM [Revenue Management], as well as between OMM [Offshore Energy] and MRM, would dramatically increase the consistency of Federal lease status and production information across these agencies" (Subcommittee on Royalty Management 2007).

REFORMING REVENUE COLLECTION IN THE WAKE OF SCANDAL

Coupled with the ongoing operational problems within Revenue Management, the previously described 2008 OIG investigative reports describing the scandalous activities in the RIK group only intensified

Congress' desire to reform MMS during the first decade of the 2000s. As described in Figure 9.3, prior to the explosion on Deepwater Horizon, various bills were introduced in Congress with the purpose of restructuring MMS and the broader organization surrounding it. Each had the goal of addressing, through design, MMS's revenue collection problems and the associated coordination issues between MMS and BLM in particular. No fewer than eight bills were introduced between November 2005 and April 2009 that proposed to rename MMS as the "National Ocean Resources and Royalty Service" to emphasize the agency's roles in collecting oil and gas revenue and managing offshore development (see, e.g., Deep Ocean Energy Resources Act of 2008 2008). For those bills that sought to reform MMS by redesigning or repositioning it in the federal government prior to the onset of the spill, none proposed breaking up MMS into its component parts.

The Clean, Affordable and Reliable Energy Act as well as the Minerals Management Service Reform Act, both introduced in 2009, sought to increase congressional control over MMS by either making it an independent agency or reforming the appointment process so that the secretary of the Interior could not directly name MMS's director without congressional approval. While also reacting to the RIK scandal, a second set of proposals attempted to directly address the coordination issues that were plaguing Revenue Management's ongoing operations as well. In September 2008, Representative Joe Barton introduced a bill "to improve interagency coordination and cooperation in the processing of Federal permits for production of domestic oil and gas resources" (US House of Representatives 7032 2008). By creating an agency called the Office of the Federal Oil and Gas Permit Coordinator, the reorganization would have both "promote[d] process streamlining" and eliminated "duplication of effort" by tasking the office to more formally coordinate BLM and MMS activities.

Addressing the same ongoing issues that were plaguing MMS royalty collection, the Consolidated Land, Energy, and Aquatic Resources (CLEAR) Act, introduced in the House, went even further. Instead of simply creating a federal coordinator, the CLEAR Act sought to combine BLM's Oil and Gas Management program with MMS. In a hearing before the House Committee on Natural Resources, the bill's sponsor Representative Nick Rahall explained:

This bill would establish the Office of Federal Energy and Minerals Leasing, combining the energy development work currently split between the MMS and the Bureau of Land Management. Having one agency doing the leasing and one agency collecting the money is inefficient, unnecessary, complex, and potentially costs the American people millions in lost royalties (Committee on Natural Resources 2009, 3).

By joining the BLM's onshore minerals management functions with the Revenue Management and Offshore Energy groups at MMS, the new agency would be better positioned to overcome the long history of shortcomings in royalty collection. Interestingly, the proposed reorganization mirrored GAO's recommendation to consolidate oil and gas functions when MMS was originally created in 1982. Like Senator Barton's proposal to create a federal oil and gas coordinator, the CLEAR Act was specifically focused on improving coordination and information flow to Revenue Management – a long-standing impediment to the success of the federal government's oil and gas management program.

EXAMINING THE DISBANDING OF MMS

On the surface, the decision to disband MMS appeared to be a dramatic step away from the congressional organizational reform efforts that preceded the disaster which had generally sought to consolidate government oil and gas management functions. For example, in contrast to the proposal in the 2009 version of the CLEAR Act which sought to unite onshore and offshore functions, the administration's decision – enacted not through legislation but through Secretary Salazar's Order 3299 – actually appeared to further divide these administrative functions. While the narrative describing MMS's conflicts of interest was largely absent from discussions of its issues prior to the spill, the story was largely supported by Congress and a broader set of commentators after the onset of the disaster (see, e.g., Flournoy et al. 2010; Honigsberg 2011).

By initially structuring the agency such that it was tasked to collect revenue – and given that revenue could not be collected without production – the logic behind the conflict narrative rested on the idea that MMS's original design impeded its ability to regulate effectively since doing so would require limiting oil and gas production. A similar story described the discord that existed within the Offshore Energy group. In its role as government offshore leasing agent, MMS would have an incentive to promote oil and gas development. Just as tax collection allegedly hindered regulation by encouraging development, having MMS oversee that development process directly further weakened MMS's desire to be an effective regulator, thereby creating the conditions for a disaster like the Gulf spill.

In addition to the broad agreement among commentators that MMS needed to be divided, as Figure 9.3 demonstrates, the intent of congressional proposals to formalize the structure of oil and gas operations through organic legislation also clearly shifted after the spill. Of the six bills initiated after its onset, only one, the Oil Spill Prevention Act, intended to keep the revenue collection, leasing, and regulatory oversight functions in one organization. For example, although both the Outer Continental Shelf Reform Act and the Clean Energy Jobs and Oil Company Accountability Act sought to place limits on the number of organizations that could be responsible for various aspects of the oil and gas management process, each still supported dividing offshore management and separating it from revenue collection. The bills backed the administration's stated rationale for the imposed reform as well. The Clean Energy Jobs and Oil Company Accountability Act, introduced by Senator Harry Reid, provided that, "the Secretary [of the Interior] shall ensure, to the maximum extent practicable, that any potential organizational conflicts of interest related to leasing, revenue creation, environmental protection, and safety are eliminated" (Clean Energy Jobs and Oil Company Accountability Act 2010).

While the disbanding of MMS appeared on the surface to both diverge substantially from pre-spill proposals as well as institutionalize dramatic changes to the government oil and gas infrastructure, in its implementation, the reorganization did much less. Secretary Salazar's report to Congress two months after announcing the breakup described his implementation plan (DOI 2010a). Highlighting the existing division between Revenue Management and Offshore Energy, the plan noted, "The Office of Natural Resources Revenue can be transitioned most quickly ... with the transfer of the largely intact Minerals Revenue Management function" (DOI 2010a). In contrast, the report documented the complexity associated with dividing Offshore Energy into BOEM and BSEE. It indicated that the "two Bureaus will be created from a single bureau in which functions and process are tightly interconnected, making the separation complicated and demanding" (DOI 2010a). Salazar's plan called for a protracted implementation schedule that resulted in the formal separation of the offshore management and regulatory groups almost a year and a half after the process began. Furthermore, the recognition remained that ongoing "close program coordination" was necessary between the two organizations to "maintain a functioning and effective process" (DOI 2010a).

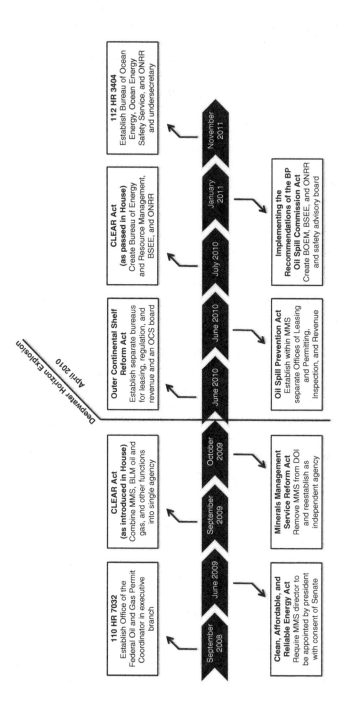

FIGURE 9.3 Timeline of select congressional bills proposing reorganization of MMS and related functions (2008–2012)
Notes: Timeline tracks bills that proposed to reorganize the functions of MMS, sought to move it, or introduced new organizational arrangements to interact with it. All relevant bills are listed except those that included a proposal which duplicates another bill that preceded it. In that case, only the first bill that mentioned the proposal is included.
Source: Searches in Proquest Congressional database of bills.

Salazar's report underscores the division between Offshore Energy and Revenue Management. The degree to which the Offshore Energy programs relied on each other clearly contrasted the independence maintained by the Revenue Management group, an independence embedded in MMS's creation. The fact that Revenue Management was a freestanding unit within MMS would allow it to transition within two and a half months of the initial implementation plan, a deadline that DOI easily met. In addition to further demonstrating the organizational divide within MMS, the actual implementation also reveals how little actually changed operationally with respect to oil and gas revenue collection through the reorganization. Numerous reports along with a plethora of hearings suggested that the two groups were already operating as separate entities. In simply formalizing the separation and renaming the tax collection function, the reorganization did little to change how the organizations operated.

Moreover, a GAO report released in July 2012, recapping the separation of Offshore Energy's leasing and regulatory functions, showed that – like the creation of ONRR from Revenue Management – little had changed with the formation of the two offshore energy bureaus (GAO 2012). In addition to continuing to use the same IT system, BOEM and BSEE each retained their headquarters in the same New Orleans office that had housed them both prior to the split. Employees of the newly formed agencies indicated their intention to continue to collaboratively manage the federal offshore process. As one senior official highlighted, "the split would not 'put up a wall' between the two bureaus" and "that staff would be able to 'walk down the hall' to discuss and resolve issues with colleagues in both bureaus" (GAO 2012: 27). The report further indicated, "Interior officials ... stated that the initial reorganization will not significantly change the bureau's work processes" (GAO 2012: 27). In orchestrating the reorganization effort, a taskforce identified no less than 49 interdependencies to ensure that information and processes would flow between the organizations. Thus, GAO's description of how the reorganization was implemented clearly supported Salazar's hope in his 2010 plan that the development and regulation missions would continue to work closely with each other (DOI 2010a).

In the end, Salazar's Order 3299 simply formalized a division between Revenue Management and Offshore Energy that already existed and had impeded MMS's ability to effectively audit oil and gas company tax remittances from the outset. At the same time, the process of allocating the regulatory and development functions to BSEE and BOEM included a plethora of provisions that would allow them to continue as work as closely as they had while they were both part of the Offshore Energy group at MMS.

DIVIDING MMS AS PURPOSEFUL SYMBOLIC REFORM

In many ways, the breakup of MMS presents a classic example of symbolic political action. As described, alternative reorganizational proposals, embodied in bills at various stages in Congress, existed prior to the spill. Each addressed long-standing coordination failures both between MMS and BLM and within MMS itself, failures that were instrumental in explaining DOI's difficulties in collecting offshore and onshore taxes. By making MMS an independent agency, adding additional organizational layers to encourage collaboration, or locating additional functions at the agency, each of these reforms proposed that MMS retain its existing roles while potentially garnering new ones. However, because it would seem hard to argue that any agency should receive more authority if it was perceived to act recklessly with what it already had, none of these actions was politically palatable as a response to the Gulf disaster.

In contrast, a reorganization that disbanded MMS conveyed the message that strong action had been taken, a particularly important aim given the criticism that the Obama administration was receiving for its lack urgency in reacting to the spill. Similar reforms had been enacted with success in other countries (Bennear 2015; GAO 2012). One senior DOI official noted, "Separating resource management from the safety and environmental functions had been a best practice used by some European nations such as Norway" (GAO 2012: 25). Thus, the decision to separate MMS into its component parts was swift, outwardly dramatic and responsive, and easily defended.

Even so, in implementation, the reform itself did little to change existing practices, retaining both the strengths and limitations of the organization of oil and gas functions that characterized the structure of MMS as it existed prior to the Gulf disaster. Although it did create independent agencies with separate budgets, the process by which the government leases offshore lands, regulates those same leases, and collects revenue is much the same as it was before MMS's breakup. MMS's revenue collection and offshore management missions were already effectively divided prior to the split. Moreover, in creating BSEE and BOEM, substantial effort went into ensuring these agencies could operate much as they did when they collectively comprised the Offshore Energy group at MMS.

Still, underneath the symbolism was a useful side-effect, either intended or not, that made the reform more than simply a political response which met the need to respond with policy actions that were available and acceptable (Baumgartner and Jones 1993; Kingdon 2003). Because it did not change the underlying infrastructure in any dramatic way, the reorganization also did not close the door to the reforms, which were intended to remove impediments to how DOI managed revenue collection, that had been in the works prior to the spill. Congressman Rahall's statement to open the July 2009 hearing to discuss the CLEAR Act underscored how deeply many felt reform was needed to correct deficiencies in royalty collection and leasing processes. He indicated, "Just this week, three – count them, three – new GAO reports detailing major flaws in the Federal oil and leasing program are being released. The reports add significantly to the massive body of investigative work done over the past 25 years calling into question the management of the entire Federal oil and gas program" (Committee on Natural Resources 2009).

Many of the bills introduced in the House and Senate after the onset of the disaster supported Secretary Salazar's decision to break up MMS. Still, in some cases, they incorporated subtle but important differences that reflected a persistent congressional interest in oil and gas revenue reform. The CLEAR Act amendment that passed the House in July 2010 presents one example. Although the amendment represented an effort to affirm Secretary Salazar's administrative action through statutory action, the act also sought to combine offshore and onshore regulatory functions in one agency and offshore and onshore development functions in another (Consolidated Land, Energy, and Aquatic Resources Act 2010).

Despite the dramatic and vivid images of oil-soaked birds and tar balls washing onto Gulf beaches during the summer of 2010, the evidence presented in this chapter has shown that political priorities as well as social views have largely returned to what they were before the Gulf disaster. In this way, the tragedy shares a feature of many others that have come before it (Birkland 2006). Patterns of congressional oversight, administration leasing decisions, media coverage, and shifts in public opinion polls all demonstrate that not only did the Gulf oil spill hold people's interest for only a short time, the long-term trend reflecting a growing preference for energy security and economic growth over environmental protection was interrupted only temporarily. Unlike the persistent attention on revenue collection, renewed interest in regulatory oversight and environmental stewardship was only fleeting.

Given the ephemeral nature of the push to take environmental safety more seriously, symbolic action – such as the decision to enact a reorganization that did little to shift the infrastructure in a direction that could not be easily reversed – offered value. In presenting the impression that action was being taken swiftly and dramatically, the disbanding of MMS served its purpose and, yet, provided the opportunity to revisit ways to improve revenue management and promote US energy independence when the status quo returned. And this may not be a bad outcome. Although tragedies like the Gulf oil spill are potentially debilitating, dramatic reform enacted during those moments can have costs (Carrigan and Coglianese 2012). Moving too far one way or the other in response to a dramatic, but temporary, reshaping of attitudes toward risk is not always the best option.

A reorganization like what was recommended by various commentators which would have spread MMS's missions among different federal departments (e.g. Flournoy et al. 2010) would likely have made it more difficult to resurrect the reform efforts proposed before the spill. In a hearing to examine the Outer Continental Shelf Reform Act of 2010, Michael Bromwich, who was tapped by President Obama to oversee MMS's dissolution, acknowledged the potential perils of creating additional organizational units. In response to an inquiry by Senator Lisa Murkowski questioning the usefulness of further fragmenting oversight of oil and gas operations when the Gulf spill had provided evidence of the difficulties of having multiple players involved, Bromwich responded, "I agree with you and understand the reluctance to believe that creating yet more pieces is a cure-all" (Committee on Energy and Natural Resources 2010). As the analysis has demonstrated, organizational decisions can have real consequences and, furthermore, involve real trade-offs. By enacting reform that did not truly respond to the temporary public uproar and preserved the opportunity to consolidate activities in one federal oil and gas agency, the reform was able to achieve an end while costing very little.

If the intense congressional focus on the perceived shortcomings of DOI's revenue collection program over at least the last 60 years is any indicator, ensuring that the government collects all federal oil and gas taxes due is one of its top priorities. From this perspective, the largely symbolic disbanding of MMS in the wake of the Gulf oil spill did more than simply reaffirm the Obama administration's control over the failure. Rather, it demonstrated that, in the wake of disaster, true symbolic action can serve an important purpose that extends beyond ensuring the political survival of those forced to act when such regulatory disasters occur. Through the very act of doing nothing, unlike true reform, symbolic reform eliminates the possibility that the action is only responding to an emotional but fleeting public display, a possibility which is not unusual in disaster. Particularly when truly responsive action undercuts other actions more reflective of longer-term preferences, only symbolic reform can limit the damage. As this analysis has demonstrated, while the Gulf tragedy fueled intense demand for regulatory reform and greater environmental accountability, this pressure quickly receded as concerns about energy and economic growth returned to the fore. Such actions like the DOI's organizational response, which serve as placeholders, can delay real action until the right action – guided by both introspection and a focus on more deep-rooted public preferences – is clearly understood.

References

Baumgartner F. R., and Jones B. D. 1993. *Agendas and Instability in American Politics*. Chicago: University of Chicago Press.

Bennear L. S. 2015. Positive and Normative Analysis of Offshore Oil and Gas Drilling Regulations in the US, UK, and Norway. *Review of Environmental Economics and Policy* 9(1): 2–22.

Birkland T. A. 2006. *Lessons of Disaster*. Washington, DC: Georgetown University Press.

Broder J. M., and Krauss C. 2011. Oil Drilling to Resume in the Gulf's Deep Waters. *New York Times*, March 1, B1.

Bureau of Ocean Energy Management, Regulation and Enforcement (BOEMRE). 2010a. *Oil and Gas and Sulphur Operations in The Outer Continental Shelf—Increased Safety Measures for Energy Development on the Outer Continental Shelf*. RIN 1010-AD68, Washington, DC: US Department of the Interior.

Bureau of Ocean Energy Management, Regulation and Enforcement (BOEMRE). 2010b. *Oil and Gas and Sulphur Operations in the Outer Continental Shelf—Safety and Environmental Management Systems*. RIN 1010-AD15, Washington, DC: US Department of the Interior.

Bureau of Ocean Energy Management, Regulation and Enforcement (BOEMRE). 2011. *All Petroleum Spills ≥ Barrel from OCS Oil and*

Gas Activities by Size Category and Year, 1964 to 2009. www.boemre.gov.

Bureau of Safety and Environmental Enforcement (BSEE). 2012. *Oil and Gas and Sulphur Operations on the Outer Continental Shelf—Increased Safety Measures for Energy Development on the Outer Continental Shelf.* RIN 1014-AA02, Washington, DC: US Department of the Interior.

Bureau of Safety and Environmental Enforcement (BSEE). 2013a. *BSEE to Establish Ocean Energy Safety Institute.* Press Release, Washington, DC: US Department of the Interior.

Bureau of Safety and Environmental Enforcement (BSEE). 2013b. *Oil and Gas and Sulphur Operations in the Outer Continental Shelf—Revisions to Safety and Environmental Management Systems.* RIN 1014-AA04, Washington, DC: US Department of the Interior.

Bush G. W. 2008a. *Memorandum on Modification of the Withdrawal of Areas of the United States Outer Continental Shelf from Leasing Disposition.* July 14. The American Presidency Project. www.presidency.ucsb.edu/ws/?pid=77685.

Bush G. W. 2008b. *Remarks on Energy.* July 14. The American Presidency Project. www.presidency.ucsb.edu/ws/index.php?pid=77683.

Carrigan C. 2014. Captured by Disaster? Reinterpreting Regulatory Behavior in the Shadow of the Gulf Oil Spill. In *Preventing Regulatory Capture: Special Interest Influence and How to Limit It*, eds. D. Carpenter and D. A. Moss. New York: Cambridge University Press.

Carrigan C., and Coglianese C. 2012. Oversight in Hindsight: Assessing the US Regulatory System in the Wake of Calamity. In *Regulatory Breakdown: The Crisis of Confidence in US Regulation*, ed. C Coglianese. Philadelphia: University of Pennsylvania Press.

Cedar-Southworth D. 1996. Regulatory Reform—It's Working at MMS. *MMS Today* Summer: 4–5.

Clean Energy Jobs and Oil Company Accountability Act. 2010. 111 S. 3663.

Committee on Appropriations. 1982. *Department of the Interior and Related Agencies Appropriations Bill, 1983.* House Report 97–942, Washington, DC: US House of Representatives.

Committee on Energy and Natural Resources, US Senate. 1989. *Gulf of Mexico Oil Spill Prevention and Response Act.* Hearing, 101st Congress.

Committee on Natural Resources, US House of Representatives. 2009. H.R. 3534, *"The Consolidated Land, Energy, and Aquatic Resources Act of 2009" (Parts 1 and 2).* Hearing, 111th Congress.

Committee on Natural Resources, US House of Representatives. 2010. *Discussion Draft, Amendment in the Nature of a Substitute to H.R. 3534, Dated June 22, 2010 (5:25 P.M.).* Hearing, 111th Congress.

Consolidated Land, Energy, and Aquatic Resources Act of 2010. 2010. 111 H.R. 3534.

Deep Ocean Energy Resources Act of 2008. 2008. 110 H.R. 6108.

Department of the Interior (DOI). 2010a. *Implementation Report: Reorganization of the Minerals Management Service.* Washington, DC: US Department of the Interior.

Department of the Interior (DOI). 2010b. *Increased Safety Measures for Energy Development on the Outer Continental Shelf.* Washington, DC: US Department of the Interior.

Devaney E. E. 2008. *OIG Investigations of MMS Employees.* Memorandum, Office of the Inspector General, Washington, DC: US Department of the Interior.

Dlouhy, J. A. 2012. Tougher Offshore Scrutiny? Not Yet. *Fuel Fix*, December 13. http://fuelfix.com/blog/2012/12/13/tougher-offshore-scrutiny-not-yet/.

Durant R. F. 1992. *The Administrative Presidency Revisited: Public Lands, the BLM, and the Reagan Revolution.* Albany, NY: State University of New York Press.

Edelman M. 1967. *The Symbolic Uses of Politics.* Urbana, IL: University of Illinois Press.

Feldman M. L. C. 2010a. *Ensco Offshore Co. versus Kenneth Lee Salazar, et al.* US District Court, Eastern District of Louisiana. No. 10–1941, Section "F".

Feldman M. L. C. 2010b. *Hornbeck Offshore Services, L.L.C. et al. versus Kenneth Lee "Ken" Salazar et al.* US District Court, Eastern District of Louisiana. No. 10–1663, Section "F".

Feldman M. L. C. 2011. *Ensco Offshore Co., et al. versus Kenneth Lee "Ken" Salazar, et al.* US District Court, Eastern District of Louisiana. No. 10–1941. Section "F".

Flournoy A., Andreen W., Bratspies R., Doremus H., Flatt V., Glicksman R., Mintz J., Rohlf D., Sinden A., Steinzor R., Tomain J., Zellmer S., and Goodwin J. 2010. *Regulatory Blowout: How Regulatory Failures Made the BP Disaster Possible, and How the System Can Be Fixed to Avoid a Recurrence.* Center for Progressive Reform White Paper 1007, Washington, DC: Center for Progressive Reform.

Fountain H. 2010. 'Junk Shot' Is Next Step for Leaking Gulf of Mexico Well. *New York Times*, May 10. www.nytimes.com/2010/05/15/us/15junk.html.

Gallup. 2010. *Energy Environment*, www.gallup.com/poll/137888/Energy-Environment.aspx.

Gallup. 2012. *Americans Still Prioritizing Economic Growth Over Environment*, www.gallup.com/poll/153515/Americans-Prioritize-Economic-Growth-Environment.aspx.

Gallup. 2013. *Americans Still Divided on Energy-Environment Trade-Off*. April 13, www.gallup.com/poll/161729/americans-divided-energy-environment-trade-off.aspx.

Government Accountability Office (GAO). 2008. *Mineral Revenues: Data Management Problems and Reliance on Self-Reported Data for Compliance Efforts Put MMS Royalty Collections at Risk*. GAO-08-893R.

Government Accountability Office (GAO). 2012. *Oil and Gas Management: Interior's Reorganization Complete, But Challenges Remain in Implementing New Requirements*. GAO-12-423.

Gulf of Mexico Energy Security Act. 2006. Public Law No. 109–432.

Hogue H. B. 2010. Reorganization of the Minerals Management Service in the Aftermath of the Deepwater Horizon Oil Spill. *CRS Report for Congress, 7–5700*, Washington, DC: Congressional Research Service.

Honigsberg P. J. 2011. Conflict of Interest That Led to the Gulf Oil Disaster. *Environmental Law Reporter*. http://elr.info/sites/default/files/articles/41.10414.pdf.

Kallaur C. 2001. *The Deepwater Gulf of Mexico – Lessons Learned*. Proceedings: Institute of Petroleum's International Conference on Deepwater Exploration and Production in Association with OGP. Washington, DC: Bureau of Ocean Energy Management, Regulation and Enforcement.

Kingdon J. W. 2003. *Agendas, Alternatives, and Public Policies*. 2nd ed. New York: Longman.

Linowes Commission (Commission on Fiscal Accountability of the Nation's Energy Resources). 1982. *Fiscal Accountability of the Nation's Energy Resources*. Washington, DC: Commission on Fiscal Accountability of the Nation's Energy Resources.

Marine Well Containment Company 2014. *MWCC Overview*. March, www.marinewellcontainment.com.

Mayhew D. R. 1974. *Congress: The Electoral Connection*. New Haven/London: Yale University Press.

Minerals Management Service (MMS). 1993. *United States Department of the Interior Budget Justifications, F.Y. 1994*. Washington, DC: US Department of the Interior.

Minerals Management Service (MMS). 1994. *OCS Legislative Group Wades into Controversy; Finds Consensus Elusive But Possible*. Press Release, No. 40065, Sep. 21.

Minerals Management Service (MMS). 1995. *United States Department of the Interior Budget Justifications, F.Y. 1996*. Washington, DC: US Department of the Interior.

Minerals Management Service (MMS). 2004. *United States Department of the Interior Budget Justification, F.Y. 2005*. Washington, DC: US Department of the Interior.

Minerals Management Service (MMS). 2006. *Oil and Gas and Sulphur in the Outer Continental Shelf (OCS)—Safety and Environmental Management Systems*. RIN 1010-AD15, Washington, DC: US Department of the Interior.

Minerals Management Service (MMS). 2007. *United States Department of the Interior Budget Justification, F.Y. 2008*. Washington, DC: US Department of the Interior.

Minerals Management Service (MMS). 2010a. *Increased Safety Measures for Energy Development on the OCS*. NTL No. 2010-N05. Washington, DC: US Department of the Interior.

Minerals Management Service (MMS). 2010b. *Information Requirements for Exploration Plans, Development and Production Plans, and Development Operations Coordination Documents on the OCS*. NTL No. 2010-N06. Washington, DC: US Department of the Interior.

Minerals Management Service (MMS). 2010c. *Notice to Lessees and Operators of Federal Oil and Gas Leases in the Outer Continental Shelf Regions of the Gulf of Mexico and the Pacific to Implement the Directive to Impose a Moratorium on All Drilling of Deepwater Wells*. NTL No. 2010-N04. Washington, DC: US Department of the Interior.

Minerals Management Service (MMS). 2010d. *United States Department of the Interior Budget Justification, F.Y. 2011*. Washington, DC: US Department of the Interior.

National Commission on the BP Deepwater Horizon Oil Spill and Offshore Drilling. 2011. *Deep Water: The Gulf Oil Disaster and the Future of Offshore Drilling*. Report to the President, Washington, DC: National Commission on the BP Deepwater Horizon Oil Spill and Offshore Drilling.

Obama B. 2010a. *Memorandum on Withdrawal of Certain Areas of the United States Outer Continental Shelf from Leasing Disposition*. March 31. The American Presidency Project. www.presidency.ucsb.edu/ws/index.php?pid=87708.

Obama B. 2010b. *Remarks by the President in a Discussion on Jobs and the Economy in Charlotte, North Carolina*. Washington, DC: The White House, Office of the Press Secretary.

Obama B. 2010c. *Remarks by the President on the Gulf Oil Spill.* Washington, DC: The White House, Office of the Press Secretary.

Obama B. 2010d. *Remarks of President Obama on Energy Security.* March 31. The American Presidency Project. www.presidency.ucsb.edu/ws/index.php?pid=89980.

Obama B. 2010e. *Requests for FY 2011 Budget Amendments: Communication from the President of the United States.* September 22. House Document 111-144, Washington, DC: US House of Representatives.

Obama B. 2012. *Remarks by the President in State of the Union Address.* January 24. Washington, DC: The White House, Office of the Press Secretary.

Office of the Inspector General (OIG). 2008a. *Investigative Report: Federal Business Solutions Contracts.* Office of the Inspector General, Washington, DC: US Department of the Interior.

Office of the Inspector General (OIG). 2008b. *Investigative Report: Gregory W. Smith.* Office of the Inspector General, Washington, DC: US Department of the Interior.

Office of the Inspector General (OIG). 2008c. *Investigative Report: MMS Oil Marketing Group – Lakewood.* Office of the Inspector General, Washington, DC: US Department of the Interior.

Office of the Secretary of the Interior. 2010a. *Salazar Announces Revised OCS Leasing Program: Key Modifications Based on Ongoing Reforms, Unparalleled Safety and Environmental Standards, and Rigorous Scientific Review.* Press Release, Office of the Secretary, Washington, DC: US Department of the Interior.

Office of the Secretary of the Interior. 2010b. *Salazar Launches Full Review of Offshore Drilling Safety Issues During Visit to Oil Spill Command Centers on Gulf Coast.* Press Release, Office of the Secretary, Washington, DC: US Department of the Interior.

Office of the Secretary of the Interior. 2010c. *Salazar Launches Safety and Environmental Protection Reforms to Toughen Oversight of Offshore Oil and Gas Operations.* Press Release, Office of the Secretary, Washington, DC: US Department of the Interior.

Office of the Secretary of the Interior. 2011. *BOEMRE Calls for Nominations for Offshore Energy Safety Advisory Committee.* Press Release, Office of the Secretary, Washington, DC: US Department of the Interior.

Office of the Secretary of the Interior. 2012a. *Interior Department Invites Industry Interest in Potential Oil and Gas Lease Sale in Alaska's Cook Inlet.* Press Release, Office of the Secretary, Washington, DC: US Department of the Interior.

Office of the Secretary of the Interior. 2012b. *Interior Finalizes Plan to Make All Highest-Resource Areas in the US Offshore Available for Oil and Gas Leasing: Next Five-Year Strategy Includes Frontier Areas in the Alaska Artic.* Press Release, Office of the Secretary, Washington, DC: US Department of the Interior.

Office of the Secretary of the Interior. 2012c. *Obama Administration Announces Proposed Central Gulf of Mexico Oil and Gas Lease Sale: Sale Will Make Nearly 38 Million Acres Available as Part of the President's Blueprint for a Secure Energy Future.* Press Release, Office of the Secretary, Washington, DC: US Department of the Interior.

Office of the Secretary of the Interior. 2012d. *Obama Administration Holds 39 Million Acre Lease Sale in Central Gulf of Mexico: Salazar Announces Over $1.7 Billion in High Bids for Over 2.4 Million Acres.* Press Release, Office of the Secretary, Washington, DC: US Department of the Interior.

Office of the Secretary of the Interior. 2012e. *Secretary Salazar, Director Beaudreau Announce Next Steps for Potential Energy Development in the Mid- and South Atlantic.* Press Release, Office of the Secretary, Washington, DC: US Department of the Interior.

Outer Continental Shelf Deep Water Royalty Relief Act (DWRRA). 1995. Public Law No. 104–58.

Porter R. 2011. *API: Board of Directors Approves Industry Center for Offshore Safety.* March 17. American Petroleum Institute, www.api.org.

Salazar K. 2010a. *Decision Memorandum Regarding the Suspension of Certain Offshore Permitting and Drilling Activities on the Outer Continental Shelf.* July 12. Secretary of the Interior, Washington, DC: US Department of the Interior.

Salazar K. 2010b. *Decision Memorandum: Termination of the Suspension of Certain Offshore Permitting and Drilling Activities on the Outer Continental Shelf.* October 12. Secretary of the Interior, Washington, DC: US Department of the Interior.

Salazar K. 2010c. *Establishment of the Bureau of Ocean Management, the Bureau of Safety and Environmental Enforcement, and the Office of Natural Resources Revenue.* Order No. 3299, Secretary of the Interior, Washington, DC: US Department of the Interior.

Salazar K. 2010d. *Statement of Ken Salazar, Secretary of the Interior before the Senate Appropriations Subcommittee on Interior, Environment, and Related Agencies on the Continuing Reform of the Outer Continental Shelf Program.* Washington, DC: US Senate.

Skinner S. K. and Reilly W. K. 1989. *The Exxon Valdez Oil Spill.* Washington, DC: The National Response Team.

Special Committee on Investigations of the Select Committee on Indian Affairs, US Senate. 1989. *Federal Government's Relationship with American Indians*. Hearing, 100th Congress.

Socolar M. J. 1982. *Statement of Milton J. Socolar, Special Assistant to the Comptroller General, before the Chairman, Subcommittee on Interior of the House Committee on Appropriations on the Report of the Commission on Fiscal Accountability of the Nation's Energy Resources*. Washington, DC: US House of Representatives.

Subcommittee of the Committee on Government Operations, US House of Representatives. 1985. *Problems Associated with the Department of the Interior's Distribution of Oil and Gas Royalty Payments to Indians*. Hearing, 99th Congress.

Subcommittee on Energy and Mineral Resources, Committee on Resources, US House of Representatives. 1996. *BLM Oil and Gas*. Hearing, 104th Congress.

Subcommittee on Mineral Resources Development and Production, Committee on Energy and Natural Resources, US Senate. 1988. *Royalty Management Program*. Hearing, 100th Congress.

Subcommittee on Royalty Management, US Department of the Interior. 2007. *Report to the Royalty Policy Committee: Mineral Revenue Collection from Federal and Indian Lands and the Outer Continental Shelf*. Office of Policy Analysis, Washington, DC: US Department of the Interior.

Tharoor I. 2010. The BP Oil Spill: The Top 10 of Everything of 2010. *Time Magazine*, December 10. http://content.time.com/time/specials/packages/article/0,28804,2035319_2035315,00.html.

Wald M. L. 2010. A Collaborative Effort to Prevent the Next Spill. *New York Times*, November 2. https://green.blogs.nytimes.com/2010/11/02/a-collaborative-effort-to-prevent-the-next-spill/?_r=0.

US House of Representatives. 2008. 110 H.R. 7032.

PART III

CASE STUDIES ON NUCLEAR ACCIDENTS

10

Recalibrating Risks of Nuclear Power

Reactions to Three Mile Island, Chernobyl, and Fukushima

Elisabeth Paté-Cornell

THREE LEVELS OF REACTION TO NUCLEAR ACCIDENTS: THE PUBLIC, THE REGULATOR, AND THE INDUSTRY

Three nuclear power plant accidents have deeply affected the public and the nuclear industry worldwide: Three Mile Island in Pennsylvania in 1979, Chernobyl in Ukraine in 1986, and Fukushima Daiichi in Japan in 2011. The perception of the risk by the public at the time of each accident was shaped by its direct consequences, fears of delayed effects of nuclear radiations, the media coverage, the political climate, and the general understanding of different reactor types at the time. The public was thus influenced to some degree by the distance, both geographic and cultural, from the accident site, and by political and economic situations.

Formal responses to these accidents came from the regulatory agencies – in the US, the Nuclear Regulatory Commission (USNRC) – and from the industry itself. Their reactions depended in part on the circumstances of the accident, but also on the type of reactor and the organizations involved. The Chernobyl reactors, for instance, were graphite-moderated RMBKs (*Reaktor Bolshoy Moshchnosti Kanalniy* or "High Power Channel-type Reactor") of Soviet design that were not in use in the US. The industry and the regulatory agencies thus responded by addressing the technical and organizational problems that these accidents had revealed based on the lessons learned. In each case, most governments issued new regulations and required a review of existing systems and upgrades of the new ones. For example, the USNRC, after the Three Mile Island accident (TMI), required technical and operational improvements, and also some levels of probabilistic risk assessment (PRA) to complement the deterministic bases of regulation and to set priorities. In addition, the US nuclear power industry created a self-regulatory body, the Institute for Nuclear Power Operations (INPO) (described further below), while the World Association of Nuclear Operators (WANO) provided a forum for international sharing of operations information.

This chapter describes first the common factors and the differences that underlined these three major nuclear accidents. For each of them, it then provides a brief description of the accident sequence and the reactions of the public, the industry, and different governments. There is no intent here to cover the response from the whole world, but mostly from the US and from countries where the measures that were adopted reflected or influenced the global industry.

A "recalibration" (reassessment) of the risk occurred in the wake of each of these events. It varied – and in some cases widely – across countries according to local technologies and circumstances, and to the lessons learned in different geographic, economic and political situations. "Recalibration" in this context can be defined in two different ways: changes of perceptions of the same known facts and new knowledge about system failures.

In some instances, the perception changed even if the risk itself had not, since neither the actual hazard nor the information about the system had changed. This was the case for instance in Germany following the Fukushima accident. It was caused by a tsunami that could not happen in Northern Europe, but nonetheless it intensified the opposition of environmental groups (and a large part of the German public) to nuclear power plants in Germany. They demanded the decommissioning of nuclear reactors. Chancellor Merkel appointed an Ethics Committee that recommended the total shutdown of German nuclear power plants by 2020 (Office of Chancellor Merkel 2011) under the assumption that nuclear energy would be replaced by renewable, non-polluting energy sources, even if the transition required temporarily burning more hydrocarbons.

Recalibration can also happen with the realization that a phenomenon that had not been seriously envisioned could happen, or was more likely than deemed before. In the case of TMI, a hydrogen bubble may have formed in the dome of the pressure vessel, but without significant explosion in the absence of oxygen (USNRC 1979). After Chernobyl, there was a realization that the risk itself had changed because the operators could engage in dangerous behaviors – way beyond classic human errors – that had not been envisioned: operators had performed an extremely hazardous test in disobedience to an explicit order not to do it. This was also the case to some degree at Fukushima when it became clear, based on existing data that had been deliberately overlooked, that the risk of large tsunamis in that part of Japan was much higher than the intuitive estimates of the decision-makers who had chosen the plant site.

In those instances where the accident changed the understanding of potential reactor failures, a Bayesian updating of the probabilities of accident sequences and their consequences provides a quantitative tool to recalibrate the risk based on what was known before and on the new information about accident sequences and their root causes. These data may involve new possible events and/or new probabilities of known accident sequences and their consequences. The new data can also include the ability – or not – of different organizations, operators, governments, and public authorities to handle serious failures, manage the situation, and control the damage after the fact. After Fukushima, for instance, the consequences were worse than what was expected assuming efficient response, and there was serious disappointment about the performance of both the operator (Tokyo Electric Production Company; TEPCO) and the public authorities. In addition, new information about the consequences of an accident (health effects, environmental impact, and social disruption) also affects the immediate readjustment of the risk.

One of the emphases of this chapter is on the role of PRA and damage assessment short of perfect information in post-accident recalibration (USNRC 1975; Kaplan and Garrick 1981; Paté-Cornell 2009). Perception often becomes reality, and the argument here is that a modicum of reality should be injected into perception, both of the (uncertain) damage after an accident and of the risks looking forward to support better decisions later. After the fact, it is often difficult, as discussed further, to assess the losses and the number of casualties because many of the consequences of exposure to ionizing radiations are delayed, confounding factors may intervene, and the dose-response relationships in human beings are still poorly known (UNSCEAR 2008). Therefore, meaningful quantification of the consequences of these accidents may be difficult at first. The numbers can range from severe underestimation by counting only the immediate losses, to gross overestimations of long-term unknown effects based on conservative assumptions about dose-response relationships. One can then simply decide to let the existing process – political, regulatory, judicial, public opinions – dictate changes in design and operation criteria, or choices of energy sources. The updating may have to be improved later as further information becomes available and the costs become clearer.

Once reliable information is gathered, Bayesian methods can be helpful in providing new (posterior) probabilities of different accident sequences, based on prior knowledge and on the likelihood of new events (Jaynes 2003). One caveat of course, is that these probabilistic assessments – like any others unless there are sufficient statistical data – may be subjected to the biases that have been underlined in the psychology literature (for instance, by Tversky and Kahneman 1974). Optimistic or pessimistic self-serving estimates and a natural tendency to keep doing what one has always done and promote one's prior beliefs can distort these estimates and defeat their purposes.

On the whole, there is a serious risk of not doing a risk analysis for external loads. This point is illustrated further by the case of the Fukushima nuclear power plant, where the tsunami design criterion was clearly not based on the whole historical record, the argument being that the database was too old to be relevant. Tsunamis, obviously, cannot be pinpointed in time, but probabilistic hazard analysis, generally based on record and other information, is meant to represent these uncertainties in a given time window. These probabilistic estimates – both of loads and systems' capacities – are an essential part of the information basis needed to support rational choices of design criteria for external events such as earthquakes and tsunamis.

One thus hopes that the information about risks is properly gathered and used for the design and operation of the 60 new nuclear reactors currently under construction in the world, including 20 in China, 7 in Russia (some of which will replace existing units), and 5 in India (World Nuclear Association 2017). The risk level at each site depends on the reactors' location and design, the local practices and politics, and the state of the economy; but nuclear accidents are not local, and they have worldwide repercussions. The risks involved and the way in which they are assessed and managed anywhere is thus of global concern. The World Association of Nuclear Operators offers a forum for exchange of operations information, but each country is responsible for its own management and regulatory regime. Whether or not to do a serious site-specific risk analysis and how to use the results, therefore, depends on the willingness and ability of each country to do so. A major global challenge is thus to support the different governments and their nuclear power industries – if they are willing to accept that support – in an effort to improve their safety based on the lessons of past nuclear accidents elsewhere. No amount of (mostly Western) anti-nuclear arguments is likely to change drastically the current programs of nuclear reactor development in many countries, given a growing need for energy, the pollution of burning hydrocarbon, and the high costs (for the moment) and intermittent operation of alternative renewable sources of energy (International Energy Agency 2012).

THREE DIFFERENT TYPES OF NUCLEAR REACTORS AND THREE DIFFERENT TYPES OF ACCIDENTS

These three accidents (Three Mile Island, Chernobyl, and Fukushima) involved three different types of reactors and three different accident sequences. Three Mile Island (Reactor 2) was a pressurized water reactor (PWR), located in Pennsylvania, constructed by Babcock and Wilcox. Like all US nuclear reactors, it was protected by a secondary containment structure. The accident occurred because a valve was stuck open in the primary system of reactor cooling, and the poor design of the control panel did not allow the operators to identify the problem immediately. It took several days to control the situation; there were severe miscommunication problems and widespread, deeply seated fears, but no known casualties.

The Chernobyl plant was of a different type. It was a graphite-moderated reactor of Soviet design without secondary containment. One problem with that system is the imbalance caused by a positive void coefficient, which means a positive feedback loop that makes things worse in case of loss of coolant (more steam, therefore more heat, etc.). The immediate cause of the accident was an unauthorized (and it turned out, quasi-criminal) experiment that involved shutting down the cooling system to try prove the reactor's safety. Of the site's 600 workers, 2 died within hours of the accident, 28 in the first four months after the event, and another 106 received doses high enough to cause acute radiation sickness (USNRC 2013a). The World Health Organization estimated that the accident would cause a number of birth defects and later shorten the lifespan of about 4,000 people through different forms of cancer under specific hypotheses of human response to different doses of radiations (WHO 2006)

Fukushima Daiichi was a pressurized water reactor of General Electric design. What triggered the accident was a tsunami caused by a large subduction-plate earthquake off the Sanriku coast in March 2011. The effect of the tsunami on the plant was the result of the choice of a low 5.7 meters tsunami design criterion (slightly increased later). An analysis of historic data published shortly after the accident showed that chances were about 5 percent that a tsunami at the site could exceed

10 meters in a 30-year period (Epstein 2011). The tsunami itself killed about 24,000 people. It is unclear – as it was at TMI – if the radiation from the plant claimed any additional victims, but the potential radiation exposure definitely added anxieties and tragic social disruptions to the huge damage and loss of lives caused by the tsunami.

Each of these accidents and the reactions that followed are described further. Again, the issues addressed here are the worldwide reactions of the public, the regulators and the nuclear power industry to these disasters. Changes were made in the design and operations of nuclear power plants and questions were raised about choices of sources of electric production. A corollary question is thus how these disasters could have been avoided by proper risk assessment and management, including design decisions and emergency planning.

ORGANIZATIONAL CHANGES SINCE THESE ACCIDENTS AND IMPROVEMENT OF DECISION SUPPORT

What Changed after Each of These Accidents: Better Knowledge and New Organizations

Each of these accidents led to regulatory and institutional responses intended to improve reactor safety. Clearly, of course, the industry is still vulnerable to failures triggered both by internal and external events, including human errors. Several things, however, have changed since the TMI and Chernobyl accidents and decades later the Fukushima power plant failures. First, the technology has matured worldwide. Few graphite-moderated RMBKs are left in operation (10 at this time, all in Russia), and those that were kept were retrofitted, for instance by adding secondary containment protection. Another example of technical improvement is the shift from analog to more reliable digital controls, for instance in the control of feed-water injection in the secondary circuits.

Perhaps more importantly, there is greater exchange of information about design, construction and operations through new procedures and organizations. A major organizational change in the US has been the creation of INPO, the Institute of Nuclear Power Operations (Rees 1996; Reilly 2013) after the Three Mile Island accident. It is an industry organization approved by USNRC for self-regulation, training, and accreditation. INPO sets safety objectives, criteria and guidelines. It conducts regular evaluations – with increasingly stringent requirements – of all nuclear power plants and communicates the results to the nuclear power industry. Its ratings influence the insurance rates of nuclear power plants. More importantly, perhaps, it provides a forum for candid exchange of information regarding operations and safety. The confidentiality of INPO reports is thus critical to its effectiveness. The USNRC is ultimately in charge of ensuring that the safety of nuclear reactors meets all required standards.

Because of the success of INPO, other industries (e.g. the oil industry) have considered reproducing the model. It is not clear, however, that it can be generalized, because the effectiveness of that institution is based on the national scope of its mission and on the relative homogeneity (with limited competition) of the US nuclear power industry.[1] On a global scale, the World Association of Nuclear Operators, mentioned earlier, provides a forum and mechanisms for sharing lessons learned across countries, but without the same powers as INPO because safety criteria vary widely around the world according to the local economy, politics and competences.

Altogether, these organizational changes have been clearly beneficial. INPO has received increasing support from the USNRC, and the two organizations cooperate to improve the overall safety of US nuclear plants.

Increased Use of Probabilistic Risk Assessments

PRAs have been performed early in the history of the US nuclear power industry. The method was developed in the 1960s and 1970s, when there was little experience with the operation of nuclear reactors beyond that of the US Navy and when the Atomic Energy Commission (later the USNRC) was concerned by the potentially disastrous effects of release of radioactive material in the atmosphere (USNRC 1975). The problem was in computing the probability of such a disaster, given that there was not sufficient experience to derive the answer from existing statistics. PRAs are increasingly required in the US, especially, and with a different focus after each accident, to recalibrate the risk and to improve the systems in a cost-effective way based on lessons learned. In the process of design and operation of nuclear reactors, PRA adds to deterministic methods some

---

[1] Whether or not the INPO model can be extended to other sectors such as the oil industry to decrease their risk in operations is debatable but was suggested in the wake of the BP Deepwater Horizon / Macondo well accident in 2010. See Carrigan's discussion in Chapter 9, this volume.

information about the relative chances and consequences of different types of potential system failures and accident sequences. In any case, the results of PRAs are not meant to replace deterministic knowledge of physical phenomena, but to complement that information.

The US Nuclear Regulatory Commission (USNRC) required early that a PRA be performed for all US nuclear power plants. Generic PRAs have been developed for most reactors in the Western world. They can be performed at several levels: level 1 to estimate a generic core damage frequency (CDF) and level 2 to estimate the frequency of different amounts of radioactive release (source term). The challenge is to extend these analyses to a site-specific level 3 in order to account for local characteristics: population, weather patterns, and external events such as earthquakes or tsunamis. Whereas PRA is common in the US, it is not necessarily performed in many reactor sites elsewhere, as was the case in Fukushima.

THE THREE ACCIDENTS AND THE REACTIONS TO EACH OF THEM

The reactions to the three accidents were driven in part by the differences among the three types of reactors involved and by the location of the accident.

Three Mile Island, 1979

As mentioned earlier, Three Mile Island was a pressurized water reactor in Pennsylvania that was constructed by Babcock and Wilcox. The fundamental causes of the 1979 accident were a combination of component malfunction, bad information about the system state, and poor design of the control panels. A partial meltdown occurred, a small amount of radioactive gases and iodine was released in the atmosphere, and some coolant release in the Susquehanna River was authorized. In the end, there were no known victims (Kemeny 1979), but the fear of radiation at that time was intense, and the cleanup costs were in the order of one billion dollars.

The TMI accident had an enormous impact on the American public and was a shock to the nuclear power industry at a time when the technology was taking off, but was poorly known and often considered too complex. Intense fears of nuclear radiations were fueled further by media reports and by a movie (*The China Syndrome*) that happened to be released at the same time. On top of real uncertainties, there was also a series of miscommunications and disseminations of wrong information in a political climate in which there was little incentive for a rational discussion of what was happening. A wide evacuation was ordered, and President Carter toured the installation in radiation protective gear. Some of these reactions can be attributed to lack of knowledge, but also a political desire to reflect public fears at a time when the atomic bombing in Japan at the end of World War II was still fresh in memories.

The media provided scary coverage of the accident, calling it "a nuclear nightmare" and emphasizing the likeness to the movie *The China Syndrome*, in which the industry tries to cover up the problems. On March 30, a few days after the accident, Walter Cronkite stated that "The horror tonight is that it could get much worse. It is not an atomic explosion that is feared, the experts say that is impossible, but the specter was raised of perhaps the next most serious kind of nuclear catastrophe – a massive release of radioactivity" (Nimmo and Combs 1989: 65). On March 30, the *Washington Post* immediately claimed that the "future of nuclear energy is clouded" (Smith 1979).

The reaction in many circles, including many in academia, was to try to ban nuclear energy. For instance, Perrow (1984) argued that in complex systems, in which he included aircraft and dams, one should expect "normal accidents." He states that tight couplings in these systems make them incomprehensible to their operators, and therefore, in his opinion, all nuclear power plants should be shut down. His views on that topic influenced a large group of social scientists and engineers (e.g. the Union of Concerned Scientists) to begin focusing on the issue of complexity (see also LaPorte 1996).

In the United States, the Three Mile Island accident arguably had an enormous impact on the number of reactors, both active and in the planning stage. Figure 10.1 shows these numbers, following both the TMI and the Chernobyl accidents, until 2005. The *New York Times* in 2009 quotes an industry executive regarding the lasting effects of TMI: "The credibility of an industry was lost," Bruce Williams, a vice president of Exelon Nuclear, which now owns the Three Mile Island station, told a Pennsylvania newspaper in 2004" (Behr 2009).

The reaction to TMI, both in government and industry, was immediate and drastic, with the cancellation of 51 nuclear reactor projects, effectively stopping the growth of the US nuclear program.[2] As early as 1975, the

---

[2] There are about 100 nuclear reactors in operation in the US at this time.

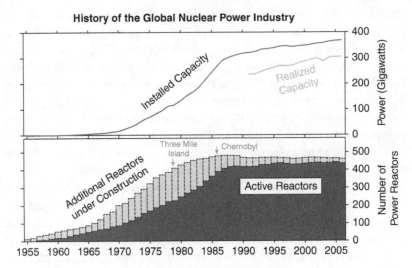

FIGURE 10.1 History of the global nuclear power industry
Source: https://commons.wikimedia.org/wiki/File:Nuclear_Power_History.png

US had already decided by law to stop the reprocessing of spent fuel in the US[3] and to keep the wastes onsite in pools[4] for reasons relating to waste storage and non-proliferation, but the number of nuclear power plants had still been planned to increase. According to the USNRC, "TMI permanently changed both the nuclear industry and the NRC" (USNRC 2013a) Plant designs were modified, operator training was improved, emergency preparedness was updated and many other changes were made to improve the safety of nuclear facilities (USNRC 2013a). The emphasis was on human performance, human factors, safety culture, operator training, emergency preparedness, and increasing plant inspections to identify early precursors of potential failures. The USNRC also stepped in to require the upgrading of a number of subsystems such as the auxiliary feed-water system. Soon after the TMI accident, with USNRC support, the industry-based Institute for Nuclear Power Operations (INPO) was created (USNRC 2013a).

The "recalibration" of nuclear risks following TMI was thus hugely negative against a political background of anti-nuclear sentiments in large

---

[3] Other countries such as France and China have made a different choice, since 95 percent of the energy is still contained in the spent fuel and can be reused for power generation.
[4] The problem, of course, is now to ensure the safety of these pools and adequate storage capacity at a time where the Yucca Mountain permanent storage site program has been shut down.

segments of the population. There were great uncertainties: confusion at the time about the unfolding of the events, the political climate, and the media contributed to a large increase in risk perception, and the accident reinforced existing opposition in the public to nuclear power technology. But the measures taken after TMI, both from the USNRC and INPO, were effective in improving the safety of US nuclear operations. Along with rising oil prices and government research programs, TMI also arguably helped accelerate investments in the development of renewable energies, solar and wind in particular, with mixed success in the following decades. The main response to TMI in the 1980s was thus to increase the use of existing sources of energy such as hydrocarbons, with corresponding increases in pollution and their consequences for health and effects on climate.

Chernobyl

The accident that happened at Chernobyl in Ukraine in 1986 was the worst in the history of nuclear power worldwide. As described earlier, the reactors were Russian-type RMBKs, graphite-moderated, without secondary containment. The accident started with an inside explosion, the graphite caught fire and was further inundated, and large amounts of radiations were dispersed in the atmosphere. Some of the firefighters and employees who tried immediately to put down the fire died shortly after of radiation poisoning. A large area was evacuated – although several hundred people refused to move.

The public reactions were most intense in the former Soviet Union and its neighboring European countries. Radiations reached the ground in remote areas as clouds and rain transported the radioactive material released. Increases in the rate of atmospheric radioactivity were felt in large areas of Europe, and the contamination of agricultural products was a real threat. The number of casualties beyond the immediate deaths is still unknown (with estimates ranging anywhere from about 200 to 985,000). Some reasonable toxicological studies put the number of people whose lives might have been – or may still be – shortened by exposure to Chernobyl radiations at around 4,000 (WHO 2006).

As mentioned earlier, two main factors contributed to the occurrence of the accident and influenced greatly world perceptions. First, the design of these reactors created an imbalance such that once the accident started, things got worse. Second, the operators of the plant had decided to perform a forbidden experiment in which they shut down the core cooling system. The flow of steam to the turbines was shut off, thus disconnecting the protection systems. There was a power spike causing a rapid increase in steam pressure, a sequence of explosions severing some coolant lines, a steam explosion of the reactors and the plant, and a huge fire with graphite dispersion, which was fought by helicopters. The plant had no secondary containment to prevent the release of radioactive material after the explosion. The total number of victims is impossible to track down with certainty because the dispersion of the population – among several confounding factors – made an epidemiological study unfeasible. But the consequences of the accident involved more than human casualties, birth defects, and environmental damage. A large zone around the plant and the whole town of Pripyat were evacuated, and 135,000 people were displaced while a few hundred simply refused to move. The accident thus caused terrible anxieties and the degradation of the country's social fabric.

The reactions of the public worldwide were intense but variable in their consequences. This time, there was a realization of the facts (and not only the risks) of a massive explosion in a nuclear reactor. There was also widespread skepticism about loss reports: the Soviet numbers were short-term estimates, thus considered gross underestimations of the actual casualty number, while others were clearly wild overestimations of the losses. The Soviets were slow to respond, to evacuate the danger zones, and to acknowledge the number of casualties. They ended up enclosing the plant in a concrete sarcophagus, and they closed most other RMBKs under intense international pressure.[5] The Europeans were alarmed by what they saw as a direct threat to their safety and that of their agricultural products. In the US, the public response was perhaps less intense than the reaction to TMI, because there were no graphite-moderated plants in the US, and Ukraine was far enough away to avoid the radioactive cloud. But Chernobyl clearly caused an increased awareness of nuclear reactor hazards, the dangers of poorly trained operators, and the risks of human errors or downright criminal behavior.

The reaction of the US Nuclear Regulatory Commission was first to determine the facts, then to assess their implications for regulating US nuclear power plants and conducting longer-term studies. They were less intense than after the TMI accidents since, again, the US does not have graphite-moderated reactors and the Chernobyl accident was a clear case of violation of safety rules. There were some changes in US reactor design, construction, and maintenance to ensure availability of backup safety

---

[5] The 10 remaining ones were upgraded, for instance protected by a secondary containment.

systems, but mostly reviews of procedures and controls, in times of normal operations as well as emergencies. There was particular emphasis on the competence and the motivation of both management and staff.

The NRC also decided on an increase in the use of PRA to support "risk-informed regulations" on all regulatory matters, and to complement deterministic methods and defense-in-depth. Three areas were of particular interest: the nuclear reactors themselves, wastes, and materials safety. One objective was to implement safety goals with appropriate considerations of uncertainties, but also to reduce excessive conservatisms where they were of questionable value. In other terms, again, the use of PRA was to be increased as a complement but not a substitute to deterministic methods to improve safety and regulation effectiveness.

The "recalibration" of the risk following Chernobyl was thus of a different nature from what followed TMI. The Chernobyl accident was seen as another piece of evidence of the catastrophic risks of failures in nuclear power plant. As after TMI, the initial reaction after Chernobyl was a vast overestimation of the losses and a large increase in risk perception, but there were lesser technical implications than if the accident had involved a Western-type PWR or boiling water reactor (BWR). There was, however, a new assessment of the risk of absurd behaviors and actions.

As shown in Figure 10.1, the Chernobyl accident, of course, did not encourage the construction of nuclear reactors. Construction remained flat. Things were going to evolve (slowly) in the following decades. The sentiment in some quarters that the industry was doomed by TMI and Chernobyl seemed to fade as the industry gained experience and nuclear safety appeared to improve. By the 2000s, growing concerns about fossil fuel pollution and climate change, amidst the developing world's growing energy needs, seemed to invite the nuclear power industry to slowly experience a sort of rebirth. The number of planned nuclear reactors in the world began to increase again. This "nuclear revival" lasted until the conjunction of the tsunami-caused accident at Fukushima Daiichi and a decline in natural gas prices, which both made nuclear power less attractive than the earlier momentum had suggested.

Fukushima Daiichi

As described earlier, the Fukushima Daiichi nuclear power plant involved six boiling water reactors designed by General Electric and operated by Tokyo Electric Production Company (TEPCO). What triggered the 2011 accident was the massive tsunami caused by a large subduction-plate earthquake (9 on the Richter scale) with its epicenter at sea in the Sanriku area. It triggered, about 45 minutes later, a tidal wave about 14 meters high, which devastated the region, killing about 24,000 people and hitting the Fukushima Daiichi nuclear power plant. The tsunami flooded the control room and the generators, causing a loss of all electric power and all related safety features (INPO 2011). Furthermore, the debris that it carried blocked the undersea water inlets, compounding the loss of cooling problem. There was a critical failure to flood the reactors in time because from a distance, the authorities were slow to respond, leading to a partial reactor core meltdown and a sequence of explosions.

The basic problems were thus a bad choice of tsunami design criterion (initially, 5.7 meters), the clogging of undersea water coolant inlets by debris from the tsunami, and a failure of the cooling system. The design choices were made by central governmental authorities without sufficient consideration of other professional advice. The management of emergency operations was poor given the distance and communications problems between the TEPCO headquarters and the plant. The hierarchical structure of the company and of the government regulators compounded the problems of siting and design decisions, emergency planning and its implementation.

The 2011 tsunami was one of the most disastrous that Japan has ever experienced. The nuclear accident was one of its consequences, but there appear to have been few victims of radiation, if any. Luckily, the western winds blew the radioactive material over the Pacific Ocean, and the maximum exposure, beyond the operators onsite, was minimal. Most of the consequences came from the terrible effects of the tsunami, but also from the social disruption caused by the massive evacuation of large areas, both because of the tsunami devastation and the fear of population exposure to excessive radiation levels. The elderly in particular, were badly affected by this uprooting. About 40 people were injured. Only as time passes, might it be possible to assess potential increases in the rate of thyroid cancers and birth defects, but based on exposure and dose information, any additional risk seems low (UNSCEAR 2012).

The fundamental technical cause of the reactor accident was an unfortunate choice of tsunami criterion (the plant sustained the earthquake itself with minor if any problems; it was the tsunami that damaged it). There was a long history of earthquakes along the Sanriku coast and of tsunamis caused by these earthquakes starting as early as 869 AD.

Further, it had been found in a study published after the accident that the probability of exceeding in 30 years waves of 10 meters high, was in the order of 5 percent. Yet, TEPCO chose a criterion of 5.7 meters, based on the 1960 tsunami, which was caused by an earthquake in Chile. The decision was made to ignore "old" data, which were deemed unreliable in spite of the fact that stones had been erected on the hillside at the time, marking the upper reach of the past tsunamis and warning people not to build between there and the ocean. Apparently, no formal risk analysis was performed by the company (it was deemed too complicated) when, in fact, the data were there and the load analysis relatively simple. Furthermore, estimating the capacity of the structure to withstand such a wave was also simple: if the generators and the control room were flooded, the cooling of the plant could no longer function and core damage or melt would occur with a very high probability. The accident was initially deemed "an act of God" (a statement that was later toned down), which illustrates the risk of no site-specific risk analysis when making these choices. If anything, the Fukushima accident showed that accounting for local conditions in design and operations was essential to the safety of reactors.

In that instance, as in those of the TMI and Chernobyl accidents, there was an enormous worldwide public reaction. This was, after all, the case of a modern reactor, of a type used in the Western world, in a country known for its management capabilities. Furthermore, the impact of the accident on the world opinion was due in part to confusion, including in the press and in academia, between the number of victims of the tsunami and the number of additional potential victims of the nuclear reactor accident.[6] But the public reaction was driven in large part by a massive loss of confidence in the Japanese authorities. They were far away from the accident site and slow to respond. A large zone was evacuated for fear of ionized radiations, well beyond that damaged by the tsunami, with an increase in the catastrophic destabilization of the area and its social fabric. The closing of some nuclear power plants also caused a drastic reduction in the Japanese electricity-generation capacity, at a time when the economy and the demand for energy did not leave Japan with much choice but to rely, temporarily at least, on coal-burning plants, hence increasing air pollution.

---

[6] There were even some claims, including one by a German scholar, that the reactor accident had caused "24,000 casualties by ionizing radiations," confusing the deaths caused by the tsunamis with the possible effects of the reactor core melt.

The Japanese people were devastated by the physical, economic, and psychological consequences of the accident. They began doubting seriously the competence of their government decision makers and of their electric companies' managers. The evacuation of a large zone, the destabilization of the area and its social coherence (even though it was mostly due to the tsunami), and a drastic reduction of the Japanese electricity-generation capacity all contributed to that loss of confidence. The Japanese government itself acknowledged its shortcomings and later criticized the notion that the event was "unforeseen," clearly stating that it should have been anticipated (*New York Times* 2011). The decision at first was to close Japanese nuclear plants, before it became clear that it was not economically feasible. Which ones will be kept open, or reopened and when, is still unclear.

In the rest of the world, the repercussions varied. First, there was confusion in the news and in other writings, about the enormous losses due to the tsunami and the incremental damage due to the nuclear reactor accident. Even though there has been no recorded casualty from exposure to the radiation released, some people began to believe that the core melt at the nuclear plant had caused thousands of deaths.

In the US, the public reacted strongly to the cumulative effects of the tsunami and the nuclear accident, sometimes indiscriminately. Some argued that the unpredictability of nature (earthquakes in particular) makes nuclear power plants too risky to be kept in operation. The reaction was driven in large part by the realization that accidents could happen in advanced countries where one might have expected utility companies and the government to manage an accident much better than they did at Fukushima. It became clear that this accident could and should have been prevented even if the tsunami could not have been, that TEPCO had been negligent, and that the opacity of its decision process had contributed greatly to the disaster (e.g. *New York Times* 2012). The question therefore became, in part, whether the US government could do better than the Japanese did under similar circumstances.

Governments abroad reacted with a mix of political and technical changes, generally starting with a review and "stress tests" of all existing and new plants. There were wide variations in these reactions given economic and environmental constraints. Following the recommendation of the Ethics Commission appointed by Chancellor Merkel, Germany, which relied on nuclear energy for about 20 percent of its electric production, decided to close its nuclear power plants by 2020, based in part on popular risk perceptions if not actual change in the risk itself (Office of

Chancellor Merkel 2011).[7] By contrast, France, which relies on nuclear energy for about 75 percent of its electric production, decided at that time not to phase out its nuclear power, with the exception of some aging plants such as Fessenheim, which were to be shut down anyway.

In fact, in spite of the accident, countries that included Sweden, France, and Spain, actually decided to increase the capacity of their nuclear plants. Other countries, including China, Russia, and India, are building new nuclear plants at an accelerated rate. China, which now has 36 nuclear reactors, is currently building another 20.[8] India has 21 plants and is constructing 5 more, with over 20 further reactors planned. A total of 60 reactors are being built worldwide at this time (including, for instance, one in the Emirates) with many more in the planning stage (World Nuclear Association 2017).

Globally, the world finds itself facing immediate trade-offs between the risk of release of nuclear emission in a nuclear power plant accident (and from stored spent fuel), growing energy needs, and the environmental risks of atmospheric pollution by emissions of CO_2 and suspended particulates with their consequences on health and global climate change. To allow sufficient production of electricity, developing countries have opted for the construction of a large number of inexpensive and simple coal and oil burning plants (e.g. in China), but also for the construction of nuclear reactors. In some countries such as Saudi Arabia, the additional economic argument for nuclear plants was that selling oil on the world market at the current price was more profitable than consuming it at home.[9]

What changes the economic game in the US is the rapid development of natural gas production, especially shale gas in large reserves that can be exploited by hydraulic fracturing of rocks ("fracking"). It is an attractive option, even though there remain some questions about the resulting pollution of water and air,[10] because it is relatively inexpensive ($4 per million

---

[7] The immediate result was the reopening of old coal plants or the continuing use of others that were to be closed, thus increasing air pollution and its health effects, as well as greenhouse gases.

[8] China, where air pollution is most severe, decided to complete the construction of its 1000MW reactors, to update their new design, initially with the help of foreign companies (e.g. ACP 1000), then transitioning to rely on Chinese companies, and to locate their plants mostly on its Eastern coast, away from the inland seismic regions.

[9] It is not impossible also that some countries, like Iran, may be developing civilian nuclear power programs to generate enriched nuclear fuel for military purposes.

[10] Burning natural gas releases some amount of CO_2, probably about half that of burning coal for the same energy output. And fracking may release methane (CH_4), a potent greenhouse gas.

BTUs). In that respect, solar and wind energy will be clearly optimal when their costs are low enough, storage capabilities are sufficient, and transmission lines exist to distribute electricity from darker to sunnier (or windy to calm) places in an economic way; but the world is not there yet (IEA 2013).

In the US, where nuclear power accounts for about 30 percent of electric production, the reaction of the energy industry to the Fukushima accident was to close older nuclear plants both because of safety concerns and because they are no longer economically competitive with low-cost natural gas plants; subsequently, only four new nuclear reactors are currently under construction. There will thus probably be a slowdown of the US nuclear program (in fact, a reduction of the total number of nuclear power plants), for economic reasons as much as safety concerns. The problem is that at least for a while, there will be an increase in CO_2 release in the atmosphere (both from coal and gas burning) from the US, as there is in Germany.

From the perspective of the US Nuclear Regulatory Commission, there is a new emphasis post Fukushima on local risks and risk analysis, with the requirement of a level 3, site-specific PRA, which industry finds expensive and for which there are few experts (for instance in the seismic risk domain). Yet Fukushima showed that these studies are essential. Some of them involve load combination, because earthquakes, for instance, can cause tsunamis on some coasts and landslides elsewhere. Most of the increased regulatory efforts, however, focused on emergency preparedness and response (USNRC 2011; 2013c).

The Risks of No Risk Analysis

One of the main lessons of the Fukushima nuclear accident, at the time when a large number of reactors are under construction in wide variety of sites, is the importance of a site-specific risk analysis. Failure to anticipate risks and to quantify the consequences of some design decisions can obviously be catastrophic. Immediately following the Fukushima accident, an important probabilistic study of tsunami hazards was finally published by the Ninokata laboratory of the Tokyo Institute of Technology (Epstein 2011). It provides an essential lesson in looking forward at the time of design and construction. This study is a Bayesian analysis of the history of subduction-plate earthquakes along the Sanriku and Sendai coasts (see Table 10.1) based on readily available data.

The stochastic recurrence of these earthquakes does not look stationary: the intervals between earthquakes appear to decrease over time, that is, the process seems "non-ergodic." Epstein thus did a Bayesian analysis of

TABLE 10.1 *Partial history of subduction-plate earthquakes along the Sanriku and Sendai coasts*

Year	Magnitude	Interval in years
869	8.6	
1611	8.1	742
1793	8.2	182
1896	8.5	103
1933	8.1	37
1960	8.5	27

Source: Epstein (2011).

the tsunami risk under two possible hypotheses: a stationary (ergodic) process and a non-stationary one. He considered tsunamis greater than 8 meters starting with flat priors and including 6 events in 1,171 years. The result of the ergodic model was a probability of about 1/730 per year or about 4.3 percent in 30 years of exceeding that height. The result of the non-ergodic model was about 1/430 per year or about 7 percent in 30 years. Given that such tsunamis were virtually certain to flood the generators and control rooms, most likely followed by radioactive release, the resulting probability of failure (chances that the load exceeds the capacity) for the Fukushima-Daiichi reactors were at odds with the regulatory safety goals in Japan. For large early release factors, the stated Japanese safety goal is to not exceed 10^{-6} (1 in 1 million) per year (INPO 2011), much smaller than both calculated risks. The fundamental cause of this discrepancy between the reality supported by the data and the optimistic safety statements seems to be an organizational, cultural, and economic problem, with little latitude for experts at the technical level to question decisions made by higher management.

The risk analysis methods for nuclear power plants have improved over time but the industry finds them expensive and complex. After Fukushima, however, it seems that given the possibility of strong local loads, site-specific risk analyses are essential to account for local conditions in plant designs. Looking forward, this may be the case of tsunami risks in China, where a number of nuclear power plants are constructed in seismic, coastal areas such as the coast of the South China Sea.[11] Given the

---

[11] Four AP 1000 nuclear reactors are currently built at the Sanmen site. The risk of tsunamis on the coast of that area comes for the hazard of large (Richter magnitude 9) megathrust earthquakes with 20-meter waves, originating from the ridge in the Eastern China sea,

global effects of nuclear accidents, such risk analysis might support decisions that will affect large parts of the world.

RECALIBRATION OF RISK ACROSS THESE THREE ACCIDENTS

All three nuclear accidents (Three Mile Island, Chernobyl, and Fukushima) have been a shock to the public worldwide. Yet, they seem to have triggered different reactions and risk "recalibrations" because they were of fundamentally different natures, occurred in different economic situations and at different stages of experience in nuclear power plant's operations. There was a lack of understanding and of clarity at TMI, a terrible experiment and a bad design at Chernobyl, and a seriously inadequate choice of tsunami design criterion combined with poor damage control and disaster management at Fukushima Daiichi. But in the 32 years that have elapsed between TMI and Fukushima, many things have changed.

First, there was an increase of the level of experience in the design and operation of nuclear reactors. At the time of TMI in 1979, there were great uncertainties, about the general vulnerability of these plants, about what was going on during the accident and about its actual and potential health effects. At the time of Chernobyl in 1986, more had been learned in the West about boiling water reactors and pressurized water reactors. The Ukrainian plant was of a different type (carbon-moderated). The accident, while shocking to the public, especially in Europe, had a lesser effect than TMI on the industry worldwide, even though it caused more casualties than TMI, because the Chernobyl reactor design and decision error were deemed so unusual. The Fukushima accident occurred in 2011. There was more experience with nuclear operations and a greater demand for non-polluting energy than 25 years earlier. Therefore, the measures that followed involved a general upgrade of plants and improvements of damage control processes, but the global plans for the construction of new plants worldwide were generally unchanged, with the exceptions of Germany, Switzerland, and Japan.

Second, between Chernobyl and Fukushima, the economic circumstances have changed with the decrease in costs of other energy sources

---

between Japan and Taipei, with a recurrence frequency in the order of 1/000 per year (as shown by GoogleEarth based on NOAA data). The elevation of the plant above the sea level appeared to visiting experts to be in the order of 8 meters. There are some breakwaters, but it is unclear whether a site-specific tsunami risk analysis including both the benefits and possible weaknesses of "defense in depth" has been done for that facility (Epstein 2014).

and new perceptions of the environmental and health effects of burning hydrocarbons. So even though, in all three cases, there were strong immediate negative public reactions, the effects of these accidents on the global nuclear power industry were different given the changes in the economic realities, the knowledge and experience accumulated in four decades, and the concerns for the environmental effects of burning coal and gas.[12]

The use of PRA may help, not in "predicting" accidents, but in the assessment and communication of what is known and what is uncertain, decision-making about the array of risks, and the hazard of external local loads in particular. The tool can be used in siting plants, deciding whether or not the risk is tolerable, setting priorities among safety measures given the site, and making reactors less vulnerable, especially in the 60 plants currently under construction. For these new ones in particular, exchange of information and experience within the industry worldwide, before and after an accident, is essential. Industry-based organizations may not be able to implement global safety standards that depend on the economic and political situation of each country, but they may play an important role in providing a forum for that exchange of knowledge.

Altogether, the reactions and risk "recalibration" in the US and elsewhere to these three nuclear accidents are not inconsistent with Bayesian reasoning in its logic. At the time of TMI, there was relatively little knowledge about the nuclear technology and great fear of the consequences. Therefore, the reaction to the "signal" of partial meltdown was extreme because the information about the accident swamped prior knowledge. At the time of Chernobyl, more experience had been gathered. There was a strong reaction to the reality of an actual explosion and radiation dissemination, but little change in nuclear programs, which in effect remained flat. Decades later, Fukushima was a reminder of the nuclear failure risks even from known reactors in advanced countries, but also of the importance of external events on the risks of an accident. It triggered a review of all existing plants and of the technology of new ones and the shutdown of a few old reactors (with the exception of Germany, Switzerland, and Japan). But globally, it did not have a major effect on nuclear development plans worldwide.

Various countries in the world, given their needs and development levels, now have to make choices of electricity production sources

---

[12] It is not impossible also that some countries, like Iran, may be developing civilian nuclear power programs to generate enriched nuclear fuel for military purposes.

accounting for the risks of a nuclear reactor accident, the economic realities of energy needs, and the environmental effects of burning hydrocarbons with consequences for health risks and those of global climate change. For the moment most countries rely in large part on classic sources (including hydropower when feasible) and to some degree nuclear power. Success in the development of renewable, non-polluting energy sources at acceptable economic costs will be critical in the long run for a stable solution of the energy dilemmas that the world now faces.

References

Behr, P. (2009) "Three Mile Island Still Haunts U.S. Nuclear Industry." *New York Times.* www.nytimes.com/gwire/2009/03/27/27green wire-three-mile-island-still-haunts-us-reactor-indu-10327.html?pagewanted=all.

Epstein, W. (2011). "A Probabilistic Risk Assessment Practitioner Looks at the Great East Japan Earthquake and Tsunami" A Ninokata Laboratory white paper, Tokyo Institute of Technology, April 2011.

Epstein W. (2014). Personal communication.

Jaynes, E. T. (2003) *Probability Theory: The Logic of Science.* Cambridge, UK: Cambridge University Press.

Institute of Nuclear Power Operations (2011). *Special Report on the Nuclear Accident at the Fukushima-Daiichi Nuclear Power Station*, Atlanta, GA: INPO,

International Energy Agency (2012). *Key World Energy Statistics.* Paris, France: OECD.

International Energy Agency (2013). *Annual Report.* Paris, France: OECD.

Kaplan S. and B. J. Garrick (1981). "On the Quantitative Definition of Risk." *Risk Analysis*, 1(1): 11–27.

Kemeny J. G. (1979). *Report of the President's Commission on the Three Mile Island Accident: the Need for Change*, Washington, DC: The Commission.

LaPorte, T. R. (1996). "High Reliability Organizations: Unlikely, Demanding and At Risk." *Journal of Contingencies and Crisis Management.* 4(2): 60–71.

Office of Chancellor Merkel (2011). "Germany's Energy Transition" Report of the Ethics Commission for a Safe Energy Supply, Berlin, Germany.

Paté-Cornell, M. E. (2009). "Probabilistic Risk Assessment" in *The Wiley Encyclopedia of Operations Research and Management Science*, J. J. Cochran (Editor-in-Chief), New York, NY: Wiley Pub.

Perrow, C. (1984). *Normal Accidents: Living with High Risk Technologies*, Princeton, NJ: Princeton University Press.
Rees, J. V. (1996). *Hostages of each other: the transformation of nuclear safety since Three Mile Island*, Chicago, IL: University of Chicago Press.
Reilly, W. (2013). "Valuing Safety Even When the Markets Do Not Notice" in *Energy and Security*, 2nd edition, D. L. Goldwyn and J. Kalicki (eds.), Baltimore, MD: Johns Hopkins University Press: 107–20.
New York Times (December 26, 2011). "Japan Panel Cites Failures in Tsunami," Op-ed by Hiroko Tabushi. https://mobile.nytimes.com/2011/12/27/world/asia/report-condemns-japans-response-to-nuclear-accident.html.
New York Times (March 9, 2012). "Fukushima Could Have Been Prevented," Op-ed by James M. Acton. www.nytimes.com/2012/03/10/opinion/fukushima-could-have-been-prevented.html.
Nimmo, D. D. and J. E. Combs (1989). *Nightly Horrors: Crisis Coverage by Television Nightly News*. Knoxville, TN: University of Tennessee Press.
Smith, J. P. (March 30, 1979). "Pa. Reactor Mishap Called Worst in U.S. History; Accident Clouds Future Of Nuclear Power in U.S.; Future of Nuclear Power Is Clouded" *Washington Post.*: A1.
Tversky, A. and D. Kahneman. (1974). "Judgment under Uncertainty: Heuristics and Biases" *Science* 185 (4175): 1124–31.
UNSCEAR (United Nations Scientific Committee on the Effects of Atomic Radiation) (2008). *Sources and Effects of Ionizing Radiations*. United Nations, Geneva Switzerland.
UNSCEAR (United Nations Scientific Committee on the Effects of Atomic Radiation) (2012). *Interim Findings of Fukushima-Daiichi Assessment Presented at the Annual Meeting of UNSCEAR. United Nations Information Service, Geneva Switzerland.*
USNRC, Rasmussen, Norman C. et al. (1975). *Reactor Safety Study: An assessment of accidental risks in U.S. Commercial Nuclear Power Plants.* NUREG-75/014 (Wash-1400). Washington DC: Nuclear Regulatory Commission.
USNRC (1979). Annual Report (NUREG-0690).
USNRC (2011). "Recommendations for Enhancing Reactor Safety in the 21st Century: The Near-Term Task Force Review of Insights from the Fukushima Daiichi Accident," Washington DC. www.nrc.gov/docs/ML1118/ML111861807.pdf.
USNRC (2013a). "Backgrounder on the Chernobyl Accident," www.nrc.gov/reading-rm/doc-collections/fact-sheets/chernobyl-bg.html.
USNRC (2013b). "Backgrounder on the Three Mile Island Accident" (NUREG-0558). www.nrc.gov/reading-rm/doc-collections/fact-sheets/3mile-isle.html.

USNRC (2013c). "What Are the Lessons Learned from Fukushima." www.nrc.gov/reactors/operating/ops-experience/japan-dashboard/priorities.html.

World Health Organization (2006). *Health Effects of the Chernobyl Accident and Special Health Care Programmes*, Report of the UN Chernobyl Forum, B. Bennett, M. Repacholi and Z. Carr (eds.), Geneva, Switzerland: United Nations.

World Nuclear Association (2017). Plans for New Reactors Worldwide. www.world-nuclear.org/information-library/current-and-future-generation/plans-for-new-reactors-worldwide.aspx.

11

Nuclear Accidents and Policy Responses in Europe

Comparing the Cases of France and Germany

Kristian Krieger, Ortwin Renn, M. Brooke Rogers, and Ragnar Löfstedt

This [Fukushima] accident is raising a certain number of questions throughout the world concerning the safety of nuclear facilities and energy choices. France chose nuclear energy, which continues to be an essential component of her energy independence and the fight against greenhouse gases. This choice went hand in hand with an unwavering commitment to ensure a very high level of safety for our nuclear facilities. [...] I remain convinced that we made the right choices.

    Former French President Nicolas Sarkozy

In Fukushima, we have had to recognise that even in technologically highly developed countries such as Japan the risks of nuclear energy cannot be controlled safely.

    German Chancellor Angela Merkel

The phase out of nuclear energy will be undertaken until 2022 and is irreversible.

    German Premier Minister of Bavaria Horst Seehofer

This paper discusses the varying effects of major nuclear incidents on nuclear energy policies in different countries. The statements presented above offer a good illustration of the question this chapter sets out to answer: What explains the different responses to major nuclear incidents in European countries? More specifically, why does Nicolas Sarkozy underline the French commitment to nuclear energy while Angela Merkel and Horst Seehofer undertake a complete U-turn in their nuclear energy policy for Germany in the aftermath of the 2011 Fukushima accident in Japan?

On first inspection, the difference in responses to Fukushima is particularly puzzling for the cases of Germany and France: neither France nor

Germany have experienced any major nuclear incidents on their territory. Both countries were – at the time of the Fukushima disaster – governed by center-right governments with a pro-nuclear stance. Both countries have strong and advanced nuclear industries with an interest in developing and exporting nuclear technologies. Both countries are members of the European Union and are thus subject to the EU regulations of the internal energy market – in addition to being liberal democracies and thus subject to public opinion pressures. Both countries – especially Germany – are at very low risk of a tsunami of the kind that triggered the Fukushima accident. When all these factors point to convergence in policy responses to an event such as Fukushima, how can we explain the sharp contrast in post-Fukushima nuclear policies? In fact, as we will show in this chapter, the responses to the earlier major accident in Chernobyl in 1986 also displayed significant variation (while policy responses to the Three Mile Island [TMI] incident in 1979 were less pronounced in both countries and led to similar policy responses). The differences in responses to major nuclear incidents between the two countries also highlights that policy responses are not always shaped by the particular technical nature of the incident, but rather can be driven by the political associations that each of these incidents triggered (Peters and Slovic 1996).

This chapter argues that institutional structures, the degree of centralism versus decentralism in political governance, and broad ideas of the role of the state in economy and society, are essential for understanding variation in the evolution of nuclear energy policies in the aftermath of significant nuclear incidents. Before delving more deeply into the institutionalist analysis of German and French nuclear energy policies, it is important to point to the limitations of placing institutions in the forefront of the analysis: First, there are non-institutional factors, in particular the (evolution of the) economics/relative costs of nuclear energy, that make a difference and might influence institutional responses. However, this paper takes a neo-institutionalist perspective that reflects the significant role of political and economic actors in the nuclear debate who often reflect economic circumstances but also symbolic associations that show hardly any reference to factual differences between the two countries. A case in point is the fact that even "non-institutionalist" factors such as economics are embedded and shaped by discourse about its meaning in a political debate. Costs were hardly considered in the German debate about nuclear energy because all parties assured the public that safety was the prime objective and economic considerations were not allowed to compromise any safety standard. In France, however, the case for nuclear was made partially on the argument that it provides inexpensive and reliable energy to the public.

Second, France and Germany experienced a number of near misses and minor accidents throughout their nuclear history[1]. These minor incidents were highly amplified in the German media, leaving the impression that this technology is riven with failures. In contrast, the French press emphasized the ingenuity of French engineering, demonstrating that the French technical elite was able to master even unlikely incidents in its nuclear operations (Rucht 1990).

Third, many policies have evolved between disasters without being necessarily driven by the effects of major disasters (e.g. the liberalization of energy markets in the 1990s was driven by a wider adoption of neoliberal policies among European policy elites). These institutional changes were not linked to incidents but had major effects on policy arenas in both countries.

That notwithstanding, the focus of this chapter is on the impact of disasters and major incidents because of the significant role scholars assign to these events for triggering policy reflections and changes. They are often regarded as "critical junctures" in the institutionalist literature (Pierson 2000) or as "windows of opportunity" in the aftermath of disasters in the policy change literature (Kingdon 1984).

Finally, this chapter does not analyse in detail whether and, if so, how the distinctive nature of incidents (e.g. that Chernobyl happened to a Soviet-style reactor while Fukushima occurred in one of the most advanced nuclear countries) affected policy and regulatory approaches (on this topic, see Chapter 10 by Paté-Cornell, this volume). This is because even though the nature of incidents differed strongly, the policy responses in France (TMI, Chernobyl, and Fukushima) and Germany (Chernobyl and Fukushima) turned out to be relatively similar in their absorption of the technical differences of each of these incidents. It was not the technical details that can explain the differences between the two countries in their response to the three mentioned incidents, but rather the structure and implications of the following national discourse, pointing to the importance of institutions and symbolic politics.

Institutions filter the effects of nuclear incidents and subsequent changes in public opinion and social mobilization. Moreover, these factors shed light not only on the differences between countries, but also in the same country

---

[1] E.g. since 2000, 33 "anomalies" (INES scale 1, out of 7) and three incidents (INES scale 2) have occurred in Germany, but no serious incidents (INES scale 3) have occurred.

after incidents at different points in time. Ideas concerning the role of the state in the economy and society held by the political elites shape the degree of involvement of the state in the energy industry. This in turn filters the impact of public scrutiny and economic pressures on decision-makers, utilities, and other corporate stakeholders in the nuclear energy sector. While the existence of particular institutional structures and ideas does not rule out change in policies, existing (institutionalized) ideas and institutions shape and can be expected to continue shaping response to incidents.

The paper will show that the institutionally complex polity of Germany with a large number of heterogeneous players and veto points has shaped the evolution of Germany's nuclear energy policies. The government's role in the energy industry has – in line with the ideas of Germany's social market economy – traditionally been hands-off, although interventions have become more frequent since the 1990s in order to promote renewable energy. In France, authorities/agencies in the energy policy field, as well as state-dominated energy corporations, formed a tight community that withstood outside pressures for change even after major disasters. The centralized state of France – following a *dirigiste* tradition of state intervention in the economy – assumed a strong role in the energy industry in order to promote industrial competitiveness and national energy independence after World War II. It is also important to note that the development of nuclear energy was pursued for military ends. The industrial-military complex is not only far better organized in France than in Germany, its role for national security and prestige is also much less contested than in post-war Germany with a strong public sentiment against military interventions.

This chapter develops this argument as follows: Section 11.1 provides a snapshot of the nuclear policy evolution in France and Germany, with particular reference to the responses to the three major incidents of Three Mile Island, Chernobyl, and Fukushima. Section 11.2 engages with the explanation of the variation in policy responses. In the policy fields related to high-risk technologies, democratic governments are frequently argued to be following the lead of the public opinion in their policy responses. However, public opinion in the two countries broadly points into the same direction, making it necessary to seek other factors that vary. The two countries have distinctive institutional structures in which nuclear policy-makers are embedded, as well as the different ideas held by the political elites concerning the role of the state and energy policies. Section 11.3 concludes the paper by putting the two cases into a broader international context, as well as the wider scholarly debate about the changing relations between the public, experts, and policy-makers in relation to large-scale technologies.

11.1 A SHORT HISTORY OF NUCLEAR ENERGY IN FRANCE AND GERMANY

After World War II, the primary economic goal of governments in Europe was reconstruction and industrial development. To this end, policy-makers sought to ensure that energy would be available reliably and at low prices. At that time, this meant – for the supply of electricity – the support of the coal industries. However, the power of the two nuclear bombs that the United States dropped on Japan inspired not only the minds of military decision-makers but also those of decision-makers concerned with the supply of energy.

Policy-makers in France and Germany were not immune to the attractions of nuclear energy. In France, the interest was driven on the one hand by military interests. The first two reactors in Marcoule – built under the first nuclear energy plan of the government (1947–1952) – were designed for military purposes. On the other hand, French interest in nuclear development was triggered by its energy dependence and lack of indigenous energy resources. Figure 11.1 demonstrates current levels of dependence from the nuclear industry in terms of energy production.

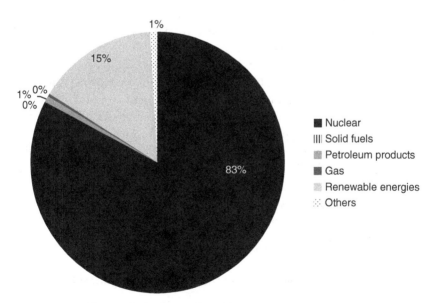

FIGURE 11.1 France, fuel mix, total production, 2010
Source: www.iea.org/stats/

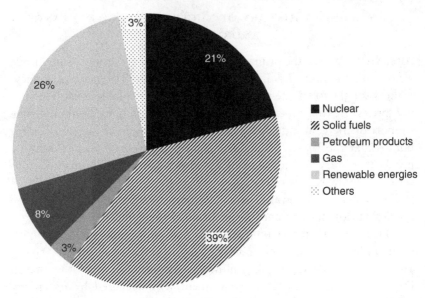

FIGURE 11.2 Germany, fuel mix, total production, 2010
Source: www.iea.org/stats/

In Germany, the interest in nuclear energy among energy policy-makers was initially not central to energy policies. Whenever the government decided to intervene into the energy sector at all, it did so in favor of coal. This reflects the strength of coal interest in Germany's post-war politics underpinned by their strong economic clout in the largest state of North Rhine-Westphalia. However, nuclear energy was viewed favorably by the government – and a dedicated Ministry of Atomic Affairs was established in 1955. The specific drivers of the nuclear program in France and elsewhere – the military use of nuclear material and the energy dependency – were not central here: After World War II, Germany's political elite did not pursue nuclear weapons – in line with the preferences of the Allies. Significant coal reserves mitigated the challenge of energy dependence. Figure 11.2 illustrates the continued central role of coal in Germany's energy generation.

11.1.2 The Development of the Nuclear Sector: Some Data

While the motivation to develop the nuclear sector is distinctive in the two countries, the subsequent development of the sectors is similar until the

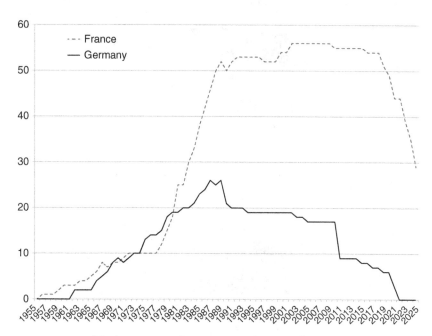

FIGURE 11.3 Nuclear reactors in operation, France and Germany, 1955–2025

Chernobyl accident in 1986 occurs. Figure 11.3 shows the number of civilian nuclear reactors in operation from the early 1950s to estimated numbers in 2025. Note that the French decline after 2020 is based on the expected reactor lifetimes of about 40 years, which may be extended or replaced by new reactors. However, there is currently only one new reactor under construction in France, expected to go online by 2016. The decline in Germany after 2011 is due to a political decision to phase out nuclear energy.

Figure 11.3 shows how the number of nuclear reactors gradually increased until the mid-1970s. This is followed by a phase of rapid expansion until 1986–1988, the years in the aftermath of Chernobyl. Afterwards, the development paths diverge in the two countries, with a gradual decline, followed by the phase out in the 2020s in Germany and a gradual increase and stability in the numbers until the 2020s in France. The development in nuclear installations is also reflected in the fuel mix for total energy consumption in the two countries where the variation in expansion, decline, and stability of nuclear energy in the mix can be observed.

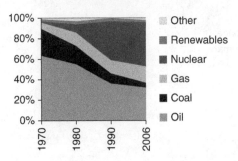

FIGURE 11.4 French fuel mix, total energy consumption, 1970–2006
Source: www.iea.org/stats/

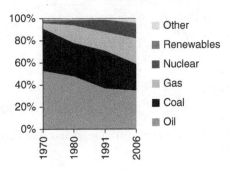

FIGURE 11.5 German fuel mix, total energy consumption, 1970–2006
Source: www.iea.org/stats/

This raises the question of why Chernobyl and Fukushima had such a distinctive impact on the development of the nuclear sector in the two countries. The impact is strongly shaped by nuclear energy policies.

11.1.3 The Evolution of Nuclear Policies: An Overview

The following sections discuss two aspects of nuclear energy policy in the two countries, namely the actors that emerged to govern nuclear energy over the course of time, as well as the policies, regulations and political dynamics between the actors.

11.1.3.1 *From the Early Days of Nuclear Energy to the Late 1970s*

Before engaging with the actors and policies found in France and Germany, it is important to understand that EU actors and policies had a rather limited impact on national energy policies in this phase. This is surprising because two of the founding pillars were concerned with energy policies. The European Coal and Steel Community (ECSC) was established in 1952 with the Treaty of Paris. As coal accounted for 80 percent of primary energy use in the early 1950s and conflicts over coal reserves contributed to hostilities between France and Germany, the architects of the ECSC foresaw a key role for the institution in the European integration process. Although the High Authority of the ECSC was formally provided with supranational powers, it failed to act against the interests of major European governments (e.g. Germany's and France's import barriers to protect domestic coal in the late 1950s). Hence, Milward concluded that "the High Authority...[was] but a powerful international committee within which separate national representatives urged for separate national policies" (Milward 1992: 117).

As a second pillar of a European energy policy, the founding fathers of the European integration process, primarily the French Jean Monnet, established 1957 the European Atomic Agency (Euratom). Euratom was to ensure that there was enough energy to form a stable basis for economic growth and that a European nuclear energy sector was established (Deubner 1979). Although France, the most advanced country in Europe in terms of nuclear research, took a special interest in Euratom, the institution and the planned European nuclear market did not take off. France hoped for a European nuclear policy that would create a common European nuclear market, a supply mechanism for uranium and a European enrichment plant. However, the United States could supply uranium more cheaply and offered to do so to Germany. As a result, the Euratom integration scheme was dubbed "stillborn" by scholars (Deubner 1979), the US became the main supplier of uranium, and France developed its own national nuclear sector. Despite Euratom's funding of joint research projects, Italy and Germany intensified their national research efforts on nuclear energy, as they wanted to prevent a further dominance of France in the nuclear sector.

In the absence of supranational actors and regulations, it is the national actors and policies that shaped the nuclear energy development in France and Germany. In France, a set of central government actors was situated at the heart of the nuclear sector, from the Ministry of Industry and the central Planning Commission of the French government to the government's nuclear research and development authority CEA (*Commissariat a l'energie atomique*) and the state-owned electricity utility EDF (*Électricité de France*). The CEA in particular became the central architect of the nuclear energy development until the 1990s, thanks to their expertise, manpower, and ownership of/involvement with other key actors of the nuclear industry (e.g. the nuclear plant construction company Framatome) (Rucht 1994). Organizations that were established in the nuclear field in 1970s were integrated or closely linked to the existing, pro-nuclear state-dominated actors: for instance, COGEMA has been set up as a CEA-owned uranium mining company; the safety regulator for nuclear plants, SCSIN, established in 1973, was supervised by the Ministry of Industry.

Beyond this integrated core of nuclear development and policy-making in the first three decades after World War II, the 1970s saw the emergence of Green Parties and anti-nuclear movement organizations, reflecting a growing concern about environmental issues. In response to protests organized by local and national movement organizations, the government established a Ministry of Environment and set up the committees for public information and communication (*Haut Comite de l'environment* and the *Conseil supérieur de la sûreté et de l'information nucléaires* [CSSIN]) (Rucht 1994).

While Germany also established a set of dedicated state actors, their involvement in nuclear energy development was much more limited. At the Federal level, the pro-nuclear Ministry of Atomic Affairs was founded in 1955 (Rüdig 2000). It was supported by an expert committee, the German Atomic Commission, from 1956 on (until it was replaced by a new specialist commission, the SSK, in 1974). This Commission comprised representatives from the Federal administration, the sciences, the nuclear industry and trade unions. The nuclear industry in Germany primarily meant Siemens and AEG both involved in reactor design and construction, as well as utilities. While the two companies closely coordinated their activities (resulting in the so-called *Kraftwerksunion* (KWU) in 1969) – the federal state did not assume majority ownership of utilities and nuclear construction companies. Since the early 1970s, local protest groups, the so-called *Buergerinitiativen*, proliferated and engaged in protests against the siting of nuclear power plants (Hatch 1986).

In terms of policies and regulations, the first three decades after World War II were shaped by pro-nuclear policies (see Table 11.2). In the 1950s and 1960s, policies revolved around developing and/or selecting suitable reactor designs, as the initial installations of research reactors (e.g. the two research reactors in Germany in 1957) and the development of country-specific designs (e.g. the French GGR reactor, the German high-temperature reactor developed in the national nuclear laboratory in Juelich) show. The

TABLE 11.1 *Key nuclear policy actors established between 1945 and 1979*

	France	Germany
Key Actors	1945 CEA *Commissariat a l'energie atomique* is established for developing nuclear weapons – Central institution for nuclear expertise in France 1952 EDF *Électricité de France* is established as part of the nationalization of energy production and to develop civilian nuclear power generation 1955 PEON Commission to assess the costs of nuclear energy development and comprising high-ranking officials from CEA, EDF, le Plan, the Ministries of Industry, Environment and Finance, and the nuclear industry 1971 Ministry for the Protection of Nature and the Environment In response to increasing environmental awareness, the MOE was established 1958 Framatome A company for nuclear power plant construction, Framatome, is established 1973 SCSIN & CSSIN The Central Authority for the Safety of Nuclear Installation and the Advisory Commission for Public Information were established 1976 COGEMA State-owned uranium mining company was established Mid- to late 1970s: Green Parties were established Three Mile Island, March 1979	1955 Ministry of Atomic Affairs Establishment of a ministry dedicated to the development and supervision of nuclear energy 1956 German Atomic Commission Advisory Commission to Ministry comprising representatives of science, industry, trade unions, and politics 1972 Nuclear Safety Standards Commission (KTA) Elaborates technical specifications and standard for the safety regulation and comprises members of operators, producers, regulators/authorities, experts, and others 1974 Commission on Radiological Protection (SSK) SSK replaces the Atomic Commission in advising the responsible ministry (then Ministry of Interior) on nuclear matters

development of nuclear energy in France was driven by specific nuclear energy five-year plans (that existed alongside general energy development plans) whilst Germany's nuclear development program was founded on the 1960 German Atomic Law.

TABLE 11.2 *Selected nuclear policies and regulations between 1945 and 1979*

	France	Germany
Policies & Regulations	1950: Energy plans Planning of energy supply/industry development through government commences through four-year plans 1952: Nuclear energy plans First five-year plan for the development of the nuclear industry for military and civilian purposes 1974/75: Messmer Plan "*Tout electricité, tout nucléaire*" program A program to increase nuclear share in meeting total energy needs from 3% (1973) to 25% (by 1985)	1960 German Atomic Law Defines licensing and operational requirements, as well as state responsibilities and penalties for non-compliance, including provisions that power plants that run orderly cannot be shut down easily 1973 oil crisis shock and federal energy programs From a hands-off approach to energy policy (along with support for coal), the government became more interventionist to develop a low-cost secure energy supply, including energy efficiency measures and an expansion of nuclear
	Three Mile Island, March 1979	

Nuclear programs in Germany and France experienced a major boost after the first oil crisis. The oil crisis from 1973 exposed the two countries' vulnerability to supply and price shocks. In France, Prime Minister Pierre Messmer announced a plan in 1974 to massively expand nuclear energy with the aim to generate all electricity from nuclear. In Germany, the government moved from its relatively limited involvement in the development of energy markets into designing more concrete energy industry development plans; the planning focused on energy efficiency measures but also included a commitment to nuclear energy as key source of electricity.

In response to the official expansion policies, the localized resistance against individual sitings of nuclear power plants turned national in the mid- to late 1970s (Hatch 1986). The protests culminated in large-scale demonstrations and occupations of nuclear sites: In 1975, the local campaign against the siting of nuclear power plant in Wyhl, Germany, turned violent and led to a deep gulf between pro-nuclear and anti-nuclear interests in Germany. Even the subsequently government-initiated *Buergerdialog Kernenergie* (Dialogue with Citizens on Nuclear Energy) did little to mitigate the opposition to nuclear energy in Germany. In France, a similarly important event was the large-scale protests in Malville against the siting of the fast breeder Superphénix. In the course of the protests, one protester was killed.

11.1.3.2 From Three Mile Island to Chernobyl

Intriguingly, the impact of the TMI accident in Harrisburg, Pennsylvania, in the United States on nuclear policies in France and Germany was fairly limited. In France, there was a brief realignment of the trade unions and the French Socialist Party (led by the presidential candidate François Mitterrand) with the environmental groups by signing an anti-nuclear petition calling for "an alternative energy policy." Moreover, once in office in 1981, Mitterrand stopped construction on five power plants and reduced the number of approved plants for 1982 from nine to six. However, Mitterrand did not question the overall commitment to nuclear energy and – as Figure 11.1 demonstrates – the number of nuclear power plants increased steeply during his presidency.

In Germany, the Social Democrats were in government as TMI happened. Chancellor Helmut Schmidt came out in opposition to abandoning nuclear power. He concluded that "the withholding of nuclear technology could mean that the benefits of nuclear will be lost for the future" (Corey 1979, 59). In fact, the Schmidt government was the first to undertake a so-called convoy licensing of three nuclear power plants of the same design type in 1982. While Germany's political and economic elites remained committed to their pro-nuclear stance, the Green Party was founded in 1980 drawing on the anti-nuclear and other social movements for their support (Renn 1995). While initially not exactly being an establishment party, the Green Party entered the German Federal Parliament (*Bundestag*) as early as 1983 and joined the SPD in government in the state of Hesse in 1985.

11.1.3.3 From Chernobyl to Fukushima

While TMI had a limited impact on actor constellations and policies in both countries, the Chernobyl accident had a significant impact on German nuclear policies and policy-making – in contrast to its impact in France.

In France, the government's response to the accident was to keep tight control over information on Chernobyl, as well as to emphasize that the Chernobyl reactor type is not in operation in France. The French nuclear program therefore continued its expansive course initially at a rapid pace (between 1986 and 1991, nine additional reactors went online) but slowed down in the 1990s. This slow down was partly driven by a tightened safety regime. However, non-safety or Chernobyl-related changes in nuclear policies also need to be taken into account.

Important adjustments in the organizational settings of the French nuclear energy policy began in the course of the 1990s only, in fact towards the late 1990s, and only some of them are related to the safety issues raised by the Chernobyl disaster. Safety-related changes include that the safety regulator IPSN became the joint responsibility of the Ministries of Industry and Environment, that it was separated from CEA in 2001 and finally – in 2006 – reorganized with extended powers into ANS. The dominant CEA also had to give up some of its control over the management of nuclear waste (a new agency called ANDRA was established independent of CEA in 1991) and the expert authority IRSN (established in 2002).

Other important changes in the organizational settings were concerned with competition in energy markets and the structure of the energy industry. A case in point is the creation of CRE, the French energy regulator, in 2000. CRE's remit includes issues such as market/network access and pricing.

These changes towards liberalized energy markets have to be seen in a European context. Since the early days of ECSC and EURATOM, the role of European institutions in energy policies changed in the mid-1980s with the advent of the Single European Act (SEA). The Single European Act, a response to the strong economic performance of the US and Japan and driven by economic and policy elites, launched the EU internal market, a program based on the widely shared political attitude emphasizing deregulation in order to achieve efficiently functioning markets (Hatch 1986). The SEA was adopted in 1987, based on a 1985 white paper by the European Commission. Although energy policy was not included in the initial inception of the Single EU market, the concept of a common market, with its emphasis on competition and transparency, constituted a contestation of the national energy markets, often – as seen in France – shaped by their national energy monopolies. Energy markets were officially added to the SEA in 1988. Against significant resistance by energy companies, particularly the monopolistic entities in various European markets, work started by the European Commission's development of a package of directives, initially on open access to gas and electricity supply. Due to this resistance and the intergovernmental European Council's reluctance to adopt the energy policy of the supranational Commission, directives of each of the packages were accepted gradually, with the EU directives on gas and electricity only adopted in 1998 and 1996 respectively. These policies were particularly challenging for the French energy sector, where the largest utility, EDF, held a market share of 90 percent in the electricity market (compared to Germany where the largest utility held 34 percent).

In Germany, the Chernobyl accident had a significant impact on nuclear policies. Most notably, a new ministry was formed (the Ministry for Environment and Reactor Safety) that took over the responsibility for nuclear energy generation and safety from the Ministries of Interior and Industry. The shift in emphasis from expanding energy generation to the safety of nuclear power plants is also reflected in the establishment of the Federal Office for Radiological Protection in 1989. However, the most important changes occurred in the realm of policies and politics. It is these changes that explain why the last reactor went online as long ago as 1989.

TABLE 11.3 *Key nuclear policy actors established between 1986 and 2011*

	Chernobyl, April 1986	
	France	Germany
Key Actors	1991: ANDRA independent ANDRA, authority in charge of managing nuclear waste, becomes independent of CEA 2000: Establishment of Energy Regulator (CRE) Following the 1994 reports and EU regulations, CRE was established 2002: IRSN established An independent expert authority IRSN is established in the field of nuclear power, independent of CEA and jointly supervised by Ministries of Health, Defence Industry and Environment. 2006: Safety Regulator ANS established ANS succeeds the SCSIN (Central Authority for the Safety of Nuclear Installations) but obtains extended powers	1986: The establishment of a Ministry of Environment (BMU) Taking over from the Ministry of Interior, the BMU also assumed responsibility for nuclear reactor safety and waste 1989: Federal Office for Radiation Protection (BfS) Federal authority to protect citizens and environment from radiation from nuclear reactors, waste, and other matter 2005: *Bundesnetzagentur* established as German energy regulator

There are two key *policy* developments in the 1990s in France that explain why the nuclear expansion slowed down. On the one hand, competition in the energy sector was promoted. This is a result from supranational policy pressures in the form of the (de-)regulation and liberalization of the internal energy market by the EU (to be discussed below) and parallel domestic developments in the form of the 1994 Souviron and 1993 Mandil reports (DGEMP 1993; French Government 1994). The competitive and financial pressures on state-owned companies such as EDF that resulted from the market liberalization and privatization programs across Europe aggravated the adverse financial impact of the overcapacity and overproduction that EDF generated as a result of the massive expansion of nuclear energy since the 1970s.

At the same time, the Green Party became one of the junior partners in the French government in 1997. They successfully applied pressure on the PS to tighten the safety regime for nuclear energy, as well as ending the construction of the Superphénix fast breeder.

The policy dynamics set off by Chernobyl in Germany were much more damaging to the nuclear "project." Most fundamentally, the Social Democratic Party (SPD) – at that time in opposition – decided in 1986 to adopt an anti-nuclear stance, culminating in the request for a phase out within 10 years (Kern et al. 2003). This had major repercussions because states with a center-left government (or a government involving the Green Party such as Hesse) now adopted policies aiming for a phase out. While the German Atomic Law puts the Federal government in charge of nuclear legislation, state authorities are responsible for the licensing and supervision of the operations of nuclear power plants. As a result, the center-left governments, for example, Rhineland-Palatine, Hesse, Lower Saxony, and Schleswig-Holstein, at that time were able to block any expansion plans of the federal government. At the same time, however, the center-right government at the federal level (Christian Democratic Union, CDU) remained committed to nuclear energy. This standoff between the Federal and Länder governments effectively blocked any changes in nuclear energy development until the federal elections of 1998 resulted in a Red-Green government (the SDP in a coalition with the Greens).

The Red-Green government emphasized sustainable development as underlying principle of their energy policy, resulting in the three broad goals of energy supply security; competition and economic efficiency; and environmental compatibility. In terms of the desired fuel mix, the government sought to promote renewable energy, continue the use of coal, and phase out nuclear energy. The phase out, however, constituted a major political and legal challenge: unless – as the Atomic Law requires – a nuclear plant does not operate in an orderly manner, the closure of a plant would create a risk for significant compensation demands by the utilities. The coalition agreement therefore aimed to amend the Atomic Law and enter negotiations with the utilities about a consensual phase out. However, these changes also turned out to be tricky because a phase out and closure – along with the ending of nuclear waste transports – would potentially violate long-term agreements on waste reprocessing and management with the UK's BNFL and France's COGEMA.

In the end, the government started negotiating with the utilities without a revised Atomic Law in place. The government and utilities reached an agreement in 2002, the so-called *Atomkonsens*, which foresaw a phase-out within 32 years. A key element of the result was however that the amount of electricity for each reactor was fixed (favorably for the utilities) and that these amounts could be transferred from one reactor to the other, thereby extending the lifetime of individual reactors beyond 32 years.

After the CDU was returned to power and led the government again in 2005, the decision to phase out nuclear energy was again under scrutiny. Industry and utility companies urged the new government to change the

TABLE 11.4 *Selected nuclear policies and regulations between 1986 and 2011*

	France	Germany
Policies & regulation	1994: Souviron report Tighter supervision for EDF 1994: Mandil report Creation of a regulatory authority to ensure greater competition on energy markets 1997–2002: Center-left coalition involving Green Party Closure of fast breeder; independence of IRSN	1986: Social Democratic Party abandons nuclear SDP passed a resolution to exit nuclear within 10 years 1998: SDP-Green Party Federal government Phase out becomes official Federal policy 2002: Atomic Consensus between government and industry Amount of energy to be produced by each reactor was fixed (even though they could be transferred) 2008: Center-right-liberal government at Federal level Phase out with extended time limits (in exchange for higher spent fuel tax on utilities)

timing of the phase out and to use nuclear energy as a longer "bridge" between the fossil and the renewable era of energy production. It took the new government until its re-election in 2008 before it actually proposed a new law extending the time limits for the phase out and providing more flexibility for energy utilities (Renn and Dreyer 2013). While the government was strongly criticized for the reversal of the phase out, the center-right government only changed course again after the Fukushima Daiichi accident occurred in March 2011.

11.1.3.4 After Fukushima

France's center-right government – as the quote at the very beginning of the paper shows – remained committed to nuclear energy even in light of the Fukushima Daiichi disaster. The response to Fukushima included two interesting aspects, first a policy aspect (the French reactor safety stress test) and second a change in the position on nuclear power by the Socialist Party, which soon captured the French presidency and was thus able to start a package of nuclear regulatory reforms.

The French government – like the German – decided to conduct an audit of its own (stress test) on all its reactors (Joskow and Parsons 2012). This was partly driven at the European level. In March 2011, the European Union decided to issue stress tests on all 143 reactors in its 27 countries (EC 2011). These audits were conducted to investigate the tolerability of all nuclear structures against all extraordinary events, including earthquakes, flooding, and all other initiating events potentially leading to multiple losses of safety functions (EC 2011). The results of the tests were published in 2012. The major lesson drawn from the test was that European reactors were safe in principle but needed substantial improvements. The European Commission urged each country with nuclear power installations to conduct a more detailed test in order to identify weak points and design measures to improve safety.

In France, the French Nuclear Safety Authority (ANS) was charged with carrying out safety assessments of all nuclear facilities in the country. The ANS stated that France's 58 nuclear reactors have a sufficient level of safety, and therefore, none of them should be closed (ANS 2011). The ANS also stated that these reactors' continued generation requires increasing their robustness in the face of situations beyond the safety margins they already have. After the audit was completed the government announced that it would introduce substantial investments to improve safety at existing sites. Some of the oldest sites are supposed to be refurbished and others will be updated. The demanded improvements of the plants have been estimated to cost more than €10 billion.

Developments confirm this concern with safety: in July 2011, EDF revised the completion date of the Flamanville reactor (Generation III European Pressurized Reactor), to 2016, with an additional cost of €6 billion; the delay stems in part from the need to carry out new safety tests (WEC 2012). In September 2011, EDF awarded contracts for the replacement of steam generators for the 13-MW-class fleet. These include 20 reactors on eight sites. Three months later, a contract was put in place to upgrade monitoring and control systems. Despite its continuous support for nuclear power, the then-government of Nicolas Sarkozy also introduced plans to boost energy efficiency and to increase the share of renewable energy. A government report issued near the election date, Energy 2050 (Pecebois and Mandil 2012), emphasized the country's policy to support efforts of increased energy efficiency and the development of renewables along with the plans to extend the life of the country's existing nuclear fleets. This combination was heralded as the best way to meet energy challenges in the future.

During the presidential campaign, the Socialist candidate François Hollande announced its intention to close 24 nuclear reactors in France and to limit French dependence on nuclear power by investing more heavily in renewable energy sources (Joskow and Parsons 2012). However, since his inauguration as the president, Hollande has not met his campaign promises. He only announced that the oldest reactor in Fessenheim (close to the German border) will be shut down in the near future. In fact, in autumn 2013, the French government announced that it would extend the maximum lifespan of nuclear reactors from 40 to 50 years (le JDD 2013). In the past, members of the Socialist party have supported the country's commitment to nuclear power partly for the reason that France has no domestic energy sources, and partly for the reason that the energy sector in France secures high employment and reasonable energy prices.

In light of growing criticism of the nuclear policies in particular from environmental groups, the Socialist government issued a new plan for safety improvements. The regulator ordered EDF (the national electricity company) to install additional safety equipment and procedures that include diesel generators for backup power, rapid response teams, and bunker-like control rooms. EDF must ensure core equipment can withstand earthquakes 50 percent more powerful than what has probably occurred over the past 20,000 years. The company also must explain to the regulators within due time how it plans to strengthen the structure of spent fuel pools, the ANS said. In response to these regulatory requirements, EDF has said it plans to spend about €55 billion ($75 billion) through 2025 to improve safety and extend the lives of its existing 58 nuclear reactors that provided 73 percent of French power production in 2012 (Patel 2014).

Overall, nuclear policies in France were not reversed or dramatically altered after Fukushima. Even if it is not clear of how the new socialist government will respond to new demands from environmental groups to follow Germany in its policy to phase out nuclear power, it will likely be a continuation of nuclear power production but with stricter safety standards, increased audits and reviews, and more governmental supervision. It is, however, very unlikely that France will build new reactors as it had planned and promised in the past. The most probable scenario is that some of the existing plants will be replaced so that the share of nuclear energy will remain more or less stable in the near future.

In Germany, the disaster of Fukushima fell into a phase in which the center-right government was under significant pressure by opposition and civil society organizations as a result of their decision to reverse the nuclear phase-out decision issued by the previous Red-Green government. Amidst the financial crisis, Chancellor Merkel had completed an agreement with the large power companies to raise the spent fuel tax in exchange for postponing the phase out of nuclear energy. Although the government denied any link between the two measures, the German public and specifically all the media were convinced that this agreement was a deal made behind closed doors. The opposition parties blamed the government for compromising public safety in exchange for revenues. Many of the environmental groups rallied against the decision to postpone the phase out and had organized powerful demonstrations against the government's plans (Buchholz 2011). This pressure was further accentuated by the loss of the traditionally conservative state of Baden-Württemberg in the state-level election a few days after the Fukushima disaster to a Green-Red coalition, installing the first state premier of the Green Party.

The response to Fukushima therefore saw a radical departure from the center-right federal government's pre-existing policies. First, it included an organizational innovation in the field of energy policy, namely the establishment of an ad hoc Ethics Committee to map the future energy policy. What is interesting about this committee is that it departs from the traditional German model of specialist/sectoral committee in its composition. Rather than energy or nuclear experts and industry, the commission was composed of elder statesmen from all political parties, functionaries of the major scientific organizations in Germany, social scientists and philosophers of ethics, the two major religious groups, Catholics and Protestants, and, as usual in the corporatist regulatory style of Germany, representatives of the corporate sector (not from the energy industry though) and the unions. Moreover, the committee's sessions, including the hearing of energy experts, were public and broadcast live on the Internet.

Second, in terms of policies, the government decided to shut down seven of its oldest nuclear units and not to re-open one unit that was out of operation during that time on March 15, 2011 (Renn and Dreyer 2013). In addition, they requested the German nuclear safety commission (SSK) to conduct a stress test on all the remaining 11 nuclear units in Germany. Similar to France, the SSK did not detect any major weakness in German reactors, and a high degree of resilience against events that went beyond the design accidents used for licensing these reactors. However, they also confirmed that the older reactors would be vulnerable to large earthquakes and all reactors to terrorist attacks (Bruhns and Keilhacker 2011).

Moreover, the government endorsed the recommendation of the ethics committee. The ethics committee's recommendations included the phase-out of nuclear energy within a 10-year period, as well as the promotion of energy efficiency and the installation of renewable energy sources. The committee also recommended that the government would establish an auditing committee to make sure that the energy transition would run smoothly and an energy public forum to boost acceptance for the new energy policies (Ethik-Kommission 2011; Nachhaltigkeitsrat 2012). In June 2011, all parties represented in the German *Bundestag* voted in favor of the new energy transition law (however, there were a few representatives who voted against the law or abstained from voting). The law mandated a phase out of all the remaining nuclear power plants by 2022. In addition, the new law included provisions to reduce the share of fossil fuel from over 80 percent in 2011 to 20 percent in 2050. Energy efficiency was to be increased by 40 percent compared to the average efficiency rates of 1990. The reactors that were shut down immediately after the accident were not reopened and remained closed (Renn 2011).

In the time between June 2011 and the end of 2012, the new energy transition law already had major effects on energy supply and consumption. The share of renewable energy in energy production increased dramatically from 17 to 23 percent and the share of nuclear fell from 23 to 16 percent. Moreover, Siemens, the construction and engineering conglomerate that had built all 17 of Germany's nuclear power plants, announced in September 2011 that it would stop building nuclear power plants anywhere in the world and also dropped plans to work with the Russian Rosatom to build new plants (WEC 2012).

It seems very unlikely that even under the constraints of recent energy price hikes and problems with energy security due to volatile energy consumption (*Economist* 2014; Wassermann und Renn 2013), Germany will

TABLE 11.5 *Nuclear actors, policies and regulations after 2011*

		Fukushima Daiichi, March 2012
Actors		2011: Establishment of an Ethics Committee Government tasked commission with developing a roadmap for future energy policies in Germany. Commission comprised of members of politics, churches, civil society and the sciences but no specialists in energy/nuclear
Policies & regulations	2011 ANS stress test of nuclear installations Demands to increase robustness on the safety margins 2012 Government report "Energy 2050" Emphasis on renewable and energy efficiency 2012 Presidential Campaign: Hollande's anti-nuclear platform Promise to close 24 reactors but so far limited action on these promises At the end 2012 ANS requires additional safety measures and EDF announces to invest €55 billion for improving safety and security of existing plants	March 2011: Immediate federal response Shut down of seven old reactor units; one reactor not in use at that time remained closed; stress test on the remaining reactors June 2011: All-party consensus in Parliament in favor of ethics committee's recommendations The recommendations include phase out within 10 years, promotion of renewable and energy efficiency Est. 2013/2014: Pending court cases utilities versus government Utilities seek financial compensation from government for policy reversal

reconsider its decision to phase out nuclear energy. In terms of installed capacity, nuclear energy has already been surpassed by renewable energy. There is little doubt that the former share of 23 percent nuclear energy can eventually be replaced by the growing renewable energy sector. Second, the consensual agreement to phase out nuclear energy, as recommended by the ethics committee as well as the overwhelming majority vote in the parliament, demonstrated the unity of commitment to the energy transition from industry to the major players in science and technology and to representatives of the environmental groups.

In sum, this overview of the evolution of nuclear energy in Germany and France revealed three important issues: First, a nuclear accident can have a different impact on policies in different countries. Second, change in the evolution of policies is not exclusively driven by accidents and safety challenges. A major element is the construction and diffusion of associations and meanings generated in a process of discourse among and between the major societal actors. Third, accidents at different points of time can have different impacts in the same country. This raises the question of factors that can explain the evolution of nuclear energy policies in the two countries. How can we explain this variable impact of disasters on policy change?

11.2 EXPLAINING DIVERGING RESPONSES TO THE NUCLEAR ACCIDENTS

There are a number of factors that are commonly associated with changing policies and institutions. In the area of risk and technology policy, one of the important lines of argument explores the negative public perceptions of large-scale technologies and their impact on institutions in charge of these technologies (Beck 1992; Slovic 1987; Slovic et al. 1981). Negative perceptions of risks associated with nuclear energy, given its potentially catastrophic consequences, have also contributed to the mobilization of local citizens against the siting of nuclear power plants and waste repositories in both countries. But can public perceptions of risk and nuclear energy technology explain the variations identified in the previous section?

11.2.1 Public Perceptions of Nuclear Energy

Table 11.6 provides an overview of internationally comparable opinion polls that asked representative samples of the German and French

TABLE 11.6 *Internationally comparable opinion polls on nuclear energy, 1982–2011*

	France		Germany	
	...is worthwhile (Agree)	...involves unacceptable risks (Agree)	...is worthwhile (Agree)	...involves unacceptable risks (Agree)
The development of nuclear power plants to produce electricity....				
EB17 Spring 1982	60%	35%		
EB22 Autumn 1984	62%	34%	44%	39%
			54%	37%
	Three Mile Island, March 1979			
	Chernobyl, April 1986			
EB26 Autumn 1986	47%	49%	33%	61%
EB28 Autumn 1987	51%	45%	36%	57%
EB31A Spring 1989	45%	51%	36%	57%
The development of nuclear power plants to produce electricity....	...is worthwhile (Agree)	...involves unacceptable risks and *should be abandoned* (Agree)	...is worthwhile (Agree)	...involves unacceptable risks and *should be abandoned* (Agree)
EB35 Spring 1991	27%	18%	21%	51%
EB46 Autumn 1996	19%	19%	12%	52%
	Energy policy preference: Continue nuclear development	Energy policy preference: Abandoning nuclear energy	Energy policy preference: Continue nuclear development	Energy policy preference: Abandon nuclear energy
Ipsos-L'express Spring 1999	18%	22%	5%	28%
	The risks of nuclear outweigh the benefits. (Agree)			
EB271 Autumn 2006	56%		51%	
EB324 Autumn 2010	53%		52%	
	Fukushima Daiichi, March 2011			
	Support/oppose nuclear way of producing electricity			
Ipsos/Globaladvisor Spring 2011	34%/67%		21%/79%	

population about the risks and benefits associated with nuclear energy. Unfortunately, there are no opinion polls available for the period of time before the TMI accident.

The TMI accident seemed to have a limited impact on the largely positive view on nuclear energy within the two countries. This is puzzling given the large-scale protests seen in both countries towards the end of the 1970s. In both countries, a larger number of survey participants perceive more benefits than risks in developing nuclear energy while a smaller proportion associated nuclear energy generation with unacceptable risks.

This positive attitude pattern experienced a major drawback in both countries after the Chernobyl disaster. In Germany, all opinion polls taken after Chernobyl show a majority of participants rating nuclear risks unacceptable or at least more significant than the benefits gained by nuclear energy. A majority expressed the opinion that Germany should not pursue this line of energy generation any longer. In France, the image turned to more ambivalence but not outright opposition. The shares of pro- and anti-nuclear opinions are often of similar size. However, out of the eight comparable opinion surveys taken between 1986 and 2011, five polls show that the percentage of anti-nuclear opinions exceeded the percentage of pro-nuclear opinions in most polls, but the two camps were close to each other. In the time period shortly before Fukushima, two opinion polls from 2006 and 2010 respectively show a stronger opposition to nuclear energy than in Germany. Opinion polls after Fukushima display a strong opposition to nuclear power in both countries, with four out of five Germans and two out of three French opposing the nuclear way of producing electricity.

In short, given the broadly similar shifts in public opinion in the two countries, public risk perceptions cannot explain the distinctive nuclear energy policies in France and Germany.

11.2.2 Ideas, Institutions and the Differential Impact of Disasters and Public Opinion

This section explores two interconnected factors, ideas and institutions that explain the different consequences of disasters and shifting public opinion in France and Germany.

11.2.2.1 *France*

French post-war energy policy evolved in a broader context of strong state interventionism into the French economy. As Peter Hall remarks, the leaders of France were eager to declare a military victory but anxious about the prospect of economic defeat in a more open world economy. A third of the nation's labor force was still employed in agriculture, and many of its firms were too small to compete on the world stage. Accordingly, French officials decided to break with the past and modernize the economy from above (Hall 2001: 49).

The envisaged renaissance of the French economy revolved around the idea of state-driven development of high-tech sectors with larger, competitive companies at their center. This can be understood as a response to the perceptions of the political elites that the previous industrial structure of smaller, family-owned companies was a cause of the defenselessness of France in World War II (Hecht 1998).

As a consequence, in 1946, the post-war government set up the *Commissariat general du Plan* which develops national plans for reconstructing the economy. These general plans were complemented by sectoral indicative plans. The plan achievement was furthermore aided by funds that were channelled by state-owned banks to the identified priority sectors, as well as para-public institutes to undertake R&D for the sector in question. In addition, these planning activities were complemented by the nationalization of producers in certain sectors.

Such a centralized planning approach reflected the institutional structure and technocratic governance culture of the French state. Power is concentrated in Paris in the executive branch, which is populated by political, economic, and bureaucratic elites that share a similar educational and social background. The structural setup ensured a high degree of autonomy of decision-makers from societal pressures while the technocratic culture underpinned faith in the capacities and the need of the state to plan the economy and a high capacity to implement even unpopular decisions.

Energy policy-making was a key field of interest to French policy-makers, driven by past conflicts with Germany over coal resources, the awareness of a lack of indigenous energy resources, as well as the importance of reliable and inexpensive energy to rebuild the country and its industrial base. Since uranium was the only fuel resources that was found abundantly in France and its overseas territories, and nuclear had the additional advantage of offering military uses, the emphasis on economic development revolving around high technology and the desire to develop capability to project military and economic strength in a more open global economy rendered the development of nuclear energy an appealing option to French policy-makers. French policy-makers felt vindicated for their strong emphasis on developing indigenous energy sources when the 1973 oil crisis demonstrated the economic consequences of import dependencies in the energy field.

Energy policy-making therefore became an area of strong state-driven planning and significant state ownership. The energy sector was subject to five-year plans and saw the nationalization of supply companies of all fuels, for example, Charbonnage de France (coal), Gaz de France (gas), Total (petroleum products), and Électricité de France (EDF) (electricity). The nuclear program was also subject to planning. In fact, EDF was set up through the second nuclear plan to complement the primarily military research and development authority CEA.

One of the important consequences of the state involvement in the energy sector of France is the creation of a cohesive policy community around the promotion of nuclear energy (Baumgartner 1989; Delmas and Heiman 2001). The CEA concentrated all expertise under a veil of (military) secrecy and thus became the central actor in a well-insulated policy network around nuclear development. CEA was joined by Framatome (monopoly rights over construction of reactors initially designed by CEA) and EDF (established as a utility with the aim to incorporate nuclear into the national grid). These three organizations controlled most of the nuclear sector development between 1946 and the 2000s. Their dominance and insulation was a result of their technocratic expertise and a wider policy context that values and promotes the role of technocratic experts. A French regulator, interviewed for a research project in 2005, describes the technocratic management of the energy sector as follows:

In France, the elite is coming from the technical high school, then move on to the Polytechnique, then, if you are among the first, the crème de la crème, then you belong to the group that is responsible for the energy system. They are a small elite driving the scenes, but they are altogether linked and they are obliged to succeed. (...) People have a great opinion of the knowledge or the capacity of them. (...) However, the problem is that we are having a problem with control. (...) If you say there is a big problem, they say "Don't worry about that (...) We will speak about that with a glass of champagne.

Political oversight or external scrutiny of the nuclear expert community has been limited: the wider political elites assumed a largely passive role that can be illustrated by the fact that the major expansion plan by Messmer 1974/75 was not even debated in the French Parliament (Hatch 1986). Another case in point is the centralized fast-track planning procedure for nuclear power plants that reduced publicity and veto points for outside actors in the development of the nuclear energy project (Hatch 1986). One French regulator interviewed in 2005 described interactions with civil society organizations as follows: "Before 2000, few things happened. More or less all energy policy relevant institutions consider themselves as a fortress against an army of evil and ignorant people outside".

Given this insulation, and the wide acceptance of the competence and leadership of technocrats in the nuclear energy development in France, as well as its wider fit with the vision of a high-tech France, the resilience of the nuclear energy program becomes more understandable in spite of disasters, changes in government, and shifts in public opinion. It is important to note, though, that the pressure on the nuclear policy community has increased significantly: For instance, in the 1990s, economic and EU regulatory pressures, as well as the involvement of the Green Party in the government, interfered with the intended further development of nuclear energy (in particular the European Pressurized Reactor). In the 2000s, the development of other indigenous fuel sources, that is, renewables, became part of the government's central energy policy objectives.

That notwithstanding, the French response to Fukushima was – at least in parts – benign for the nuclear industry: First, nuclear power stations in France continue to operate with continued support of the major parties. Second, the new safety standards require major investment and refurbishment of existing nuclear sites. At the same time, however, the new regulations also increased operating costs.

11.2.2.2 *Germany*
In Germany, the ideational and institutional set-up was different: Nuclear armament was ruled out not only by an initial ban by the Allies (even though they installed their own nuclear weapons on German soil later), but also by the vast majority of German political elites and population. At the same time, issues of energy dependence were – in view of the coal reserves – not of key importance, while the recovery and industrial development in Germany benefited from the "institutional inheritance from late industrialization that included strong industry unions, well-developed employers' associations, collaborative institutions for skill formation, and a Bismarckian welfare regime" (Hall 2001, 46).

The development of nuclear energy in Germany can therefore mostly be understood as a project initially driven by the interests of technocrats, industry, and some policy-makers (such as the first Atomic Affairs Minister Franz-Josef Strauss) in the dominant parties. The pro-nuclear forces created a web of institutions such as the Atomic Commission and the Federal Ministry for Atomic Affairs that viewed nuclear favorably.

In contrast to France, however, this emerging policy community was embedded in a wider institutional context that weakened or prevented its insulation from political and external oversight and scrutiny. In response to the experience of the Nazi dictatorship and the weakness of democracy before 1933, the German political system was re-designed in a particular manner. Power concentration was to be prevented through different sets of checks and balances, primarily a federalist structure with powerful states (*Länder*) involved in governance at both the federal and state level; a strong and independent judiciary; and a constitution whose basic rights were inalterable. Stability and social cohesion was ensured through corporatist structures that would facilitate cooperation between labor and capital, and an election system that strengthened the position of large parties. At the same time, this system led to fragmented decision making through the federalist system with strong *Länder* competencies.

It is these checks and balances in the system that effectively stopped the nuclear expansion after the 1986 Chernobyl disaster and slowed it down in some instances even before that accident. Opponents from outside the policy-making system were able to directly appeal to courts on multiple claims (as done by the anti-nuclear movement from the very beginning of their activities, e.g. in Wyhl), obstructing the government's power to implement controversial decisions swiftly. In addition, indirect pressures can be applied through (alliances with) powerful intermediary organizations (such as trade unions, local and regional party organizations) and through frequent elections on local, state- and Federal level (e.g. pressures on the government of Lower Saxony in the context of the Gorleben nuclear waste repositories decisions). The election systems on the federal and state levels that favor coalition governments enabled the anti-nuclear Green Party to become junior partners with the SPD in several state governments and, in 1998, on the Federal level.

The center-right government (CDU) therefore faced a major challenge when Fukushima happened in 2011: the main opposition parties SPD and Green Party were strongly anti-nuclear; the overall public opinion has viewed nuclear energy predominantly negatively since the Chernobyl 1986 disaster; and the anti-nuclear movement in Germany had been remobilized as a result of the reversal of the 2002 nuclear phase out by the center-right government (Buchholz 2011).

Key factors in shaping policy responses to these nuclear accidents have been the symbolic associations that key actors have assigned to each event and its meaning in the national context of opinion-forming and policy-making discourses. Symbols are powerful carriers of emotions and behavioral triggers, they are normally independent of distance (Fukushima was far away from Germany, but still Germans purchased more iodine tablets than the Japanese population) and only marginally related to technical questions such as reactor design, safety provisions or emergency preparedness. While the French public associated with its nuclear program a sense of national pride and technical superiority, to which the accidents in Japan and Russia/Ukraine posed hardly any threat, the German public associated with its nuclear reactors a fear of uncontrollable risks that were only undertaken for reasons of energy security and profit seeking of the utilities. Any incident, in particular in those countries that also have invested heavily in risk reduction, poses a major threat to the nuclear industry when it demonstrates the uncontrollable nature of this technology and precludes any form of trade-offs or compensation. Interestingly, the majority of the German Ethics Committee supported the statement that the risks of nuclear energy are so overwhelming that they cannot be compensated by any extension of expected benefits (Ethics Committee 2011). But even those who believed that the risk-benefit trade-offs were ethically acceptable shared the belief that there were less dangerous alternatives available so that the goal of energy security could also be met by alternative means. Lastly, the German nuclear debate in 2011 offered a strong argument to pro-industrial, conservative stakeholders (normally inclined to embrace nuclear power): they were sold on the idea of a fast phase out by pointing out that Germany had become an industrial leader in renewable energy technology but lost its leadership in nuclear technology. So the argument was crafted that Germany's global competitiveness and technical worldwide leadership would be supported by concentrating all efforts onto renewable energy technologies. Nuclear was seen as the past, and renewable as the new future opportunity for Germany's export-driven economy. In contrast, the French public found it reassuring that the accidents had occurred outside of France, which reinforced their belief that their reactors were better controlled and better built than the ones abroad. Furthermore, the alternative of renewable energy or coal did not seem very attractive and promising – particularly when Germany was ahead in both areas having sufficient amount of local coal and having invested more money and R&D in renewable energy technology.

In sum, while public opinion and expert analysis are both important drivers of energy politics in liberal democracies like France and Germany, it is important to take into account the institutional structures that filter public opinion to decision-makers, and include the emerging meaning associated with different policy options in the course of public debate and discourse. These two drivers can strongly influence the impact of disasters on policies, as the cases of France and Germany's nuclear sectors and policies illustrate.

11.3 CONCLUSIONS

This paper focused on presenting and explaining the varying political choices regarding the development and regulation of nuclear energy in France and Germany. The paper provides a number of interesting lessons for researchers who seek to understand the impact of accidents on nuclear energy.

First of all, it is important to note that not all accidents have political repercussions. In the case discussed here, the TMI accident had very limited effects in both country cases, while Chernobyl and Fukushima had important political repercussions, mostly in Germany. There are a number of reasons for this limited impact, including institutional factors such as the insulation and coherence of the political, technocratic and economic elites in support of nuclear energy, notably in France. Most important, however, has been the assignment of meaning to the circumstances and implications of each of these events in the public discourse.

Second, not all policies and policy change can be explained in reference to disasters. Again, this has not been at the heart of this paper's discussion, but the emergence of economic pressures in the 1990s has significantly affected nuclear energy development, involved actors, and overall institutional settings in the nuclear energy field. If state aid and guarantees to the nuclear industry are withdrawn and private investors need to be found, then these investors would be concerned about issues such as stability of political support for nuclear energy, and about possible follow-on costs such as nuclear waste management, liability insurance and decommissioning, tightening safety regulations, and more. Once German industry started to believe that there was more to gain from investing in renewable energy technology and sustainable energy systems the investment into nuclear technology lost its attractiveness.

Third, and this is at the heart of the paper, the responses to disasters differ between countries. Understanding these differences requires a closer look at the institutional settings, the dominant discourse topics and broader political ideas that underpin policy-making. In fact, given the military origins and government-driven initiation and promotion of nuclear energy, an institutionalist viewpoint can be argued to be particularly valuable when analysing the impact of disasters on nuclear policies. In Germany, the fragmentation of power (e.g. between the Länder and federal level; between the government and the judicial system; and through an election system favoring coalition governments) and a lack of a widely shared motivating idea that makes nuclear energy development a key national project made decision-makers more vulnerable to outside pressures and the nuclear policy community more prone to disintegration (e.g. the switch in the position of the SPD after Chernobyl). In France, a strong commitment to large-scale technology development combined with an institutional insulation of the nuclear policy community from outside pressures and internal scrutiny, have mitigated the impact of adverse public opinion on the nuclear energy development. Most important might be the major association of nuclear energy with national achievement and technical superiority in France, both of which had little repercussions in the German population.

These are important insights into the role of disasters and accidents in triggering/shaping policy change. They potentially help us develop a more nuanced understanding of the responses to nuclear accidents across countries but also of other accidents caused by large-scale technologies. Universalist takes on risk and society suggests that societies have moved from deep trust in the positive value in high technology as well as those that manage these technologies on our behalf (a culture of experts) to a much more sceptical view of technologies (Giddens 1999; Beck 1992; Löfstedt 2005) – and that this is reflected in the emergence and international adoption of policy principles such as the precautionary principle and of political actors such as NGOs and Green Parties (Ewald 2002). While these general shifts can indeed be observed (Renn 2008, 80ff.), it is important to take a closer look at the politics (and economics) of nuclear in individual countries.

The response to Fukushima across Europe is as diverse as we could expect from countries with such different nuclear histories and institutional settings. For instance, in Italy, a moratorium on the construction of nuclear plants for one year was adopted by the government. In the Netherlands, the government noted that the lessons learned from Fukushima will be taken into account in the definition of requirements for a new nuclear plant to be built in 2015. In Slovakia, the ongoing construction of two new reactors will be continued albeit with updated safety requirements for earthquake resilience. In Russia, President Putin asked for a review of the ambitious nuclear expansion plans. In Switzerland, the authorization process for three new nuclear power plants was frozen by the government. Similar to Germany, Switzerland decided to phase out nuclear energy but adopted a more comfortable time frame to accomplish this goal.

Both the German and French cases are peculiar. Germany's radical departure from nuclear after, having been a major producer of nuclear energy and of associated technology, is as interesting as the continued heavy reliance on nuclear in France. Many countries observe Germany's path with interest – given the potential trade-offs that 'safety' and nuclear power may have for energy security (reliable supply without nuclear base load and in the face of import dependence from Russia), climate change (coal is expected to replace at least some of the nuclear supply) and affordability of the national energy transition project *Energiewende* (some argue that once nuclear plants operate and the initial investment is undertaken/written off, they produce power relatively cheaply). The French case is also unique because of the country's strong dependence on nuclear energy for its energy production, unmatched anywhere else in the world.

However, all of these policy objectives – energy security, affordability, environmental quality and safety – are spelled out across Europe as objectives in energy policies. To what extent the German or French path allow the achievement of these goals remains to be seen. Disasters and incidents – one could therefore conclude – have at least the potential to stimulate policy innovation and experimentation – strongly shaped by institutional, discursive, and ideational conditions.

References

Ahearn, J. F. and Birlhofer, A. (2011): *Nuclear Power*. In J. B. Wiener, M. D. Rogers, J. K. Hammit and P. H. Sand (eds.): *The Reality of Precaution. Comparing Risk Regulation in the United States and Europe.* Earthscan: London, pp. 121–41.

ASN (2011): Complementary Safety Assessments of the French Nuclear Power Plants (European "Stress Tests"). French Nuclear Safety Authority (ASN), December.

Barke, R. P. and Jenking-Smith, H. C. (1993): Politics and Scientific Expertise: Scientists, Risk Perception, and Nuclear Waste Policy. *Risk Communication*, 13 (4): 425–39.

Bastide, S., Moatti, J.-P., Pages, J. -P. and Fagnani, F. (1989): Risk Perception and the Social Acceptability of Technologies: The French Case. *Risk Analysis* 9: 215–23.

Baumgartner, F. R. (1989): Independent and Politicized Policy Communities: Education and Nuclear Energy in France and in the United States. *Governance*, 2 (1): 42–66.

Beck U. (1992): *Risk Society – Towards a New Modernity*. London: Sage.

Bella, D. A., Mosher, C. D. and Calvo, S. N. (1988): Technocracy and Trust: Nuclear Waste Controversy. *Journal of Professional Issues in Engineering*, 114: 27–39.

Bosch, S. and Peyke, G. (2011): Gegenwind für die Erneuerbaren – Räumliche Neuorientierung der Wind-, Solar- und Bioenergie vor dem Hintergrund einer verringerten Akzeptanz sowie zunehmender Flächennutzungskonflikte im ländlichen Raum. *Raumforschung und Raumordnung*, 69 (2): 105–18.

Bruhns, H. and Keilhacker, M. (2011) "Energiewende" Wohin führt der Weg? *Aus Politik und Zeitgeschichte*, 46–47: 22–29.

Buchholz, W. (2011): Energiepolitische Implikationen einer Energiewende. Ifo-TUM Symposium zur Energiewende in Deutschland. Manuskript.

Corey, G. R. (1979): A Brief Review of the Accident at Three Mile Island, *IAEA Bulletin*, 21(5): 54–59.

Delmas, M. and Heiman, B. (2001): Government Credible Commitment to the French and American Nuclear Power Industries. *Journal of Policy Analysis and Management*, 20 (3): 433–456.

Deubner C. (1979): The Expansion of West-German Capital and the Founding of Euratom. *International Organisation*, 33: 203–28.

DGEMP (1993): *Reforme de l'organisation electrique et gaziere francaise (the Mandil report)*. Paris: DGEMP.

EC (2011): Communication from the Commission to the Council and the European Parliament on the Interim Report on the Comprehensive Risk and Safety Assessments ("Stress Tests") of Nuclear Power Plants in the European Union. COM(2011) 784. http://ec.europa.eu/energy/nuclear/safety/doc/com_2011_0784.pdf.

Economist (2014): Sunny, windy, costly and dirty. The Economist, 18 January 2014. www.economist.com/news/europe/21594336-germanys-new-super-minister-energy-and-economy-has-his-work-cut-out-sunny-windy-costly.

Ewald F. (2002): The Return of Descartes's Malicious Demon: An Outline of a Philosophy of Precaution. In T. Baker and J. Simon (eds.): *Embracing Risk: The Changing Culture of Insurance and Responsibility*. Chicago: The University of Chicago Press, pp. 273–302.

Ethik-Kommission (2011): *Deutschlands Energiewende – Ein Gemeinschaftswerk für die Zukunft. Bericht der Ethik-Kommission "Sichere Energieversorgung" an die Bundesregierung Deutschland*. Berlin: Kanzleramt.

French Government (1994): *Debat national Energie & Environnement-Rapport de Synthese (the Souviron report)*. Paris: French Government.
Giddens A. (1999): Risk and Responsibility. *Modern Law Review*, 62: 1–10.
Hall, P. A. (2001): The Evolution of Varieties of Capitalism In B. Hancké, M. Rhodes, and M. Thatcher (eds.): *Beyond Varieties of Capitalism: Conflict, Contradictions, and Complementarities in the European Economy*. Oxford, UK: Oxford University Press: 39–85.
Hatch M. T. (1986): *Politics and Nuclear Power – Energy Policy in Western Europe*. Lexington: The University Press of Kentucky.
Hecht G. (1998): *The Radiance of France: Nuclear Power and National Identity after World War II*. Cambridge: MIT Press.
Jones C., and Ladurie. E. L. R. (1999): *The Cambridge Illustrated History of France*. Cambridge,UK: Cambridge University Press.
Joskow, P. L. and Parsons, J. E. (2012): *The Future of Nuclear Power After Fukushima*. MIT CEEPR, Working Paper 2012–001. Cambridge, MA: MIT Press.
Kern, K., Koenen, S., and Löffelsend, T. (2003): *Die Umweltpolitik der rot-grünen Koalition: Strategien zwischen nationaler Pfadabhängigkeit und globaler Politikkonvergenz*. Discussion Paper. Abteilung Zivilgesellschaft und transnationale Netzwerke des Forschungsschwerpunkts Zivilgesellschaft, Konflikte und Demokratie des Wissenschaftszentrums Berlin für Sozialforschung, No. SP IV 2003–103. Berlin: WZB.
Kingdon, J. (1984): *Agendas, Alternatives, and Public Policies*. New York: Longman.
Le JDD (2013): L'État va prolonger le nucléaire de dix ans. www.lejdd.fr/Economie/L-Etat-va-prolonger-le-nucleaire-de-dix-ans-633771
Löfstedt, R. (2005): *Risk Management in Post-Trust Societies*. London: Earthscan.
Milward, A. (1992): *The European Rescue of the Nation-State*. Berkeley: University of California Press.
Nachhaltigkeitsbeirat Baden-Württemberg (Sustainability Council of Baden-Württemberg) (2012): *Energiewende. Implikationen für Baden-Württemberg*. Stuttgart: NBBW.
Patel, T. (2014): French Regulator Orders EDF to Make Additional Atomic Safeguards. http://sortirdunucleaire.org/French-Regulator-Orders-EDF-to-Make-Additional.
Pecebois, J. and Mandil, C. (2012): Energies 2050. Policy Brief 263. http://archives.strategie.gouv.fr/cas/en/content/energy-2050-policy-brief-263-february-2012.html.

Peters, E. and Slovic, P. (1996): The Role of Affect and Worldviews as Orienting Dispositions in the Perception and Acceptance of Nuclear Power. *Applied Social Psychology*, 26 (16): 1427–53

Pierson, P. (2000): Increasing returns, path dependence, and the study of politics. *American Political Science Review*. 94: 251–67.

Renn, O. (1995): Perzeption, Akzeptanz und Akzeptabilität der Kernenergie. In H. Michaelis and H. Salander (eds.): *Handbuch Kernenergie. Kompendium der Energiewirtschaft und Energiepolitik*. Frankfurt am Main: VVEW-Publisher, pp. 752–76.

Renn, O. (2008): *Risk Governance. Coping with Uncertainty in a Complex World*. London: Earthscan.

Renn, O. (2011): Energiesicherung. Zwischen Systemanforderungen und Akzeptanz. *Transfer*, 12: 20–22

Renn, O. und Dreyer, M. (2013): Risiken der Energiewende: Möglichkeiten der Risikosteuerung mithilfe eines Risk-Governance-Ansatzes. *DIW Vierteljahreshefte zur Wirtschaftsforschung*, 82 (3): 29–44.

Rucht, D (1990): Campaigns, skirmishes and battles: anti-nuclear movements in the USA, France and West Germany. *Organization & Environment*, 4: 193–222

Rucht D. (1994): The Anti-nuclear Power Movement and the State in France. In H. Flam (ed.): *States and Anti-Nuclear Movements*. Edinburgh: Edinburgh University Press, pp. 129–62.

Rüdig W. (2000): Phasing Out Nuclear Energy in Germany. *German Politics*, 9: 43–80.

Slovic P. (1987): Perception of Risk. *Science*, 236: 280–85.

Slovic P., Fischhoff B. and Lichtenstein S. (1981): Perceived Risk: Psychological Factors and Social Implications. *Proceedings of the Royal Society*. London A376: 17–34.

Wasserman, S. and Renn, O. (2013): Offene Fragen der Energiewende: Aufbau und Design von Kapazitätsmärkten. *GAIA, Ökologische Perspektiven für Wissenschaft und Gesellschaft*, 22 (4): 237–41.

WEC (2012): World Energy Perspective: Nuclear Energy One Year After Fukushima. World Energy Council, London. www.worldenergy.org/documents/world_energy_perspective__nuclear_energy_one_year_after_fukushima_world_energy_council_march_2012_1.pdf.

12

Public Attitudes and Institutional Changes in Japan following Nuclear Accidents

Atsuo Kishimoto

12.1 INTRODUCTION

The 2011 accident at the Fukushima Daiichi Nuclear Power Plant, which was operated by the Tokyo Electric Power Company (TEPCO), was triggered by an earthquake off the Pacific coast of Tohoku. The accident changed everyone's risk perception of nuclear power, including experts, politicians, and the general public, and it also brought about the tightening of Japan's safety standards, the establishment of a new regulatory body, and the transformation of national energy policy. Looking into the past, nuclear power governance in Japan has developed as nuclear accidents occurred. This reactive approach, where incidents and accidents caused a change in the risk perception of the general public, and were subsequently followed by changes in policy, is seen in several areas of Japanese nuclear safety policy.

This chapter discusses the ways in which significant nuclear accidents affected the risk perceptions and attitudes of the general public and experts and how, in turn, these perceptions and attitudes affected nuclear policies and organizational restructuring. Shiroyama (2010) divides nuclear safety regulatory history into three terms: the first term dating from 1959 to 1978, the second term dating from 1979 to 1999, and the third term dating from 1999 onward. To update this classification, the fourth term would begin in March 2011 when the accident took place at the Fukushima Daiichi Nuclear Power Plant.

The objectives of this chapter are to clarify the development of Japan's nuclear safety regulatory framework as a response to impacts of significant nuclear accidents, three of which took place in Japan and two outside of Japan. Section 12.2 provides a summary of major findings. Section 12.3 gives an overview of Japan's nuclear policy up to March 11, 2011, including the geographical locations of nuclear power plants, the structure of the nuclear power industry, and the enactment of the Atomic Energy Basic Act. Section 12.4 describes the 1974 accident on the nuclear-powered ship the *Mutsu*, the 1979 Three Mile Island (TMI) nuclear power plant accident, the Chernobyl accident of 1986, and the 1999 criticality accident at JCO Co., Ltd. Section 12.5 looks at the first decade of the 2000s, when no major nuclear disasters occurred. Section 12.6 focuses on the 2011 accident at the Fukushima Daiichi Nuclear Power Plant and its repercussions. Section 12.7 discusses the insights, limitations, and challenges of this study.

After the earthquake and the Fukushima Daiichi accident in 2011, concerns over the safety of power plants led to the sequential shutdown of all 54 nuclear power plants for inspections. When the Hokkaido Electric Power Company shut down Japan's last active nuclear reactor (Tomari Power Station, Unit 3) in May 2012 for such an inspection, Japan was without nuclear energy for the first time in 42 years. Due to fear of an electricity shortage, TEPCO restarted two reactors (Unit 3 and 4) at the Oi Nuclear Power Plant in the summer of 2012. These two reactors ceased operations in September 2013, however, because the power plants are required to undergo periodic inspections every 13 months. Thus, after the Fukushima accident, all of Japan's civilian nuclear power plants were shut down for at least several years.

12.2 MAJOR FINDINGS

12.2.1 Event Model

A simplified event model is tested in this chapter. First, an accident happens. If the impact of the accident is large enough, it leads to a change in risk perception among experts. Then, an investigation on possible regulatory gaps or deficits is initiated and regulatory changes are proposed and realized. Large-scale accidents can also change the risk perception among the general public, which directly leads to policy change. This paper focuses on institutional changes. Figure 12.1 shows a policy cycle under this simplified event model that resulted in institutional change.

This paper tests this model using the case of nuclear power in Japan. Among five accidents, three triggered institutional changes and two did

FIGURE 12.1 Simplified event model

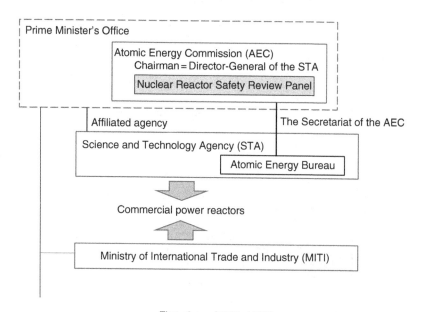

FIGURE 12.2 Transitions in the administrative systems that govern nuclear power

not. The administration system regulating nuclear power went through three significant changes since 1955 (Figure 12.2). In Figure 12.2, gray areas indicate organizations overseeing safety issues. The transition from the first phase to second phase was triggered by the nuclear-powered ship *Mutsu* accident (Section 12.4.1). The JCO accident triggered the transition from the second phase to the third phase (Section 12.4.4). Finally, the Fukushima Daiichi nuclear accident forced the change from the third phase to the fourth phase (Section 12.6.1).

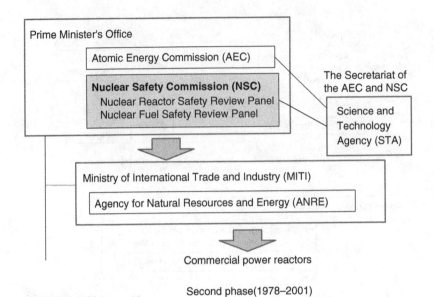

Second phase (1978–2001)

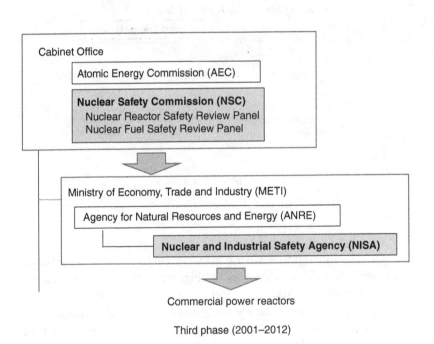

Commercial power reactors

Third phase (2001–2012)

FIGURE 12.2 (cont.)

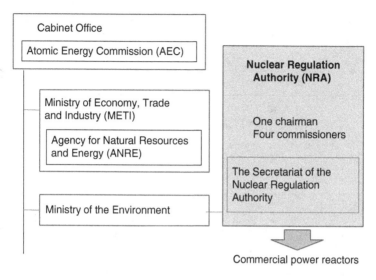

Fourth phase (2012–Present)

FIGURE 12.2 (cont.)

12.2.2 Public Opinion and the Five Major Accidents

The significant nuclear events that took place outside of Japan – namely, the TMI accident in 1979 and the Chernobyl accident in 1986 – did not lead to a shift in domestic regulatory policies nor to a realignment of governance, although they did influence public perception. The weaker reaction to these two events by the regulators in Japan explains its delay in preparing countermeasures for severe nuclear accidents when compared with the United States (US) and the European Union (EU). In contrast to the lack of response to major international nuclear crises, a few key domestic events dramatically shaped institutional changes in Japan – in particular, the nuclear-powered ship *Mutsu* accident in 1974, the criticality accident at JCO at Tokaimura in 1999, and then, the nuclear power plant disaster at the Fukushima Daiichi Nuclear Power Plant in 2011.

The impacts of nuclear accidents on public opinion differ from the impacts on expert opinion. Public opinion was shaped by a different set of three major nuclear accidents studied. The radiation leakage from *Mutsu*, as the first nuclear accident in Japan, threw cold water on the "nuclear dream" in its early stage of development. Protest movements began to take place at the planned site locations in the 1970s. After 1986,

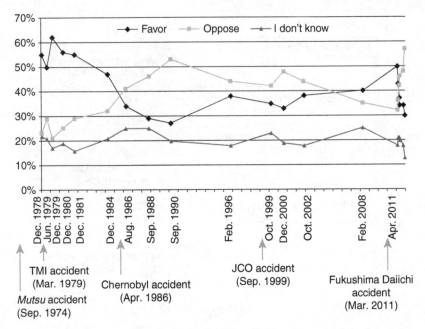

FIGURE 12.3 Time series variations of public opinion surveys and the five major nuclear accidents
Source: Public opinion surveys by the Asahi Newspaper Co.

for the first time, the proportion of respondents who opposed nuclear power exceeded that of those favoring nuclear power in the aftermath of the Chernobyl accident. Although the proportion of respondents favoring nuclear power surpassed that opposing it in the mid-2000s, the proportion of respondents who opposed nuclear power greatly exceeded that of those favoring it after the Fukushima Daiichi nuclear accident in 2011. Both the TMI accident in 1979 and the JCO accident in 1999 seem to have had little or only a temporary impact on public opinion regarding nuclear power.

12.2.3 Summary Table

The impacts of the five nuclear accidents on public opinion and regulatory policies are summarized in Table 12.1. This shows that regulatory policies were only influenced by the three domestic accidents, while public opinion was influenced primarily by the *Mutsu*, Chernobyl, and Fukushima accidents. Japanese public opinion seems to have responded to some nuclear

TABLE 12.1 *Impacts of the five accidents on public opinion and regulatory policies in Japan*

Accidents	Time	Level on the International Nuclear Events Scale (INES)	Impact on Public Opinion	Impact on Regulatory Policy
Mutsu	1974	NA (Domestic)	Significant	Trigger of the second phase
TMI	1979	5 (Foreign)	Temporary	No
Chernobyl	1986	7 (Foreign)	Significant	No
JCO	1999	4 (Domestic)	Temporary	Trigger of the third phase
Fukushima Daiichi	2011	7 (Domestic)	Very significant	Trigger of the fourth phase

accidents more than others, but Japanese regulatory policy did not always respond to public opinion. An inference from this summary table is that foreign accidents tended to be neglected by experts because they were seen as not directly applicable to Japanese technology or regulation, and therefore the lessons from those foreign accidents were not taken seriously by Japanese experts. Arguably, this inattention to foreign accidents helped lead to the subsequent domestic accidents, including the Fukushima Daiichi accident.

12.3 NUCLEAR POWER GENERATION IN JAPAN PRIOR TO MARCH 11, 2011

12.3.1 Background Information on Nuclear Energy in Japan

Nuclear research in Japan was banned shortly after World War II by the Allied powers. However, the ban was lifted in 1952, because the peace treaty signed in San Francisco did not contain terms prohibiting nuclear research. Then, research on the utilization of nuclear power was promoted politically under the rhetoric of "atoms for peace." Nuclear power generation was first introduced when the post-war years of recovery transformed into a high economic growth period. Nuclear energy was needed to help meet the growing electricity demand which was thought at the time would continue to increase more than 10 percent annually. The introduction of nuclear power also served to diversify energy sources and reduced Japan's dependence on the importation of fossil fuels. The concern about the dependence on oil became reality during the oil crisis of the 1970s. Following this, Japan began to promote nuclear power, considering it "quasi-domestically produced energy." Japan enacted three Power Source Development Laws in 1974 to help encourage the establishment of nuclear power plants: the Act on Tax for Promotion of Power-Resources Development, the Act on Special Accounts, and the Act on the Development of Areas Adjacent to Electric Power Generating Facilities. In 2010, the Special Account for Energy resulted in a tax on the promotion of power resources development of 375 Japanese yen per 1000 kWh, which was then funneled to local communities, towns, cities, and prefectures where nuclear power plants were located.

The 1980s experienced a downturn in nuclear power use worldwide due to the TMI and the Chernobyl accidents, stabilization of energy prices, and the rise of the environmental movement. The Japanese government, however, adhered to promoting nuclear power, from the viewpoint of energy security, and, later, reducing emissions of carbon dioxides. Then, in the following decade, nuclear energy was globally re-evaluated.

The Japanese Ministry of Economy, Trade, and Industry (METI) drafted the Basic Energy Plan in 2003 based on the Basic Act on Energy Policy, which passed in the Japanese legislature in 2002. The second revision of the Basic Energy Plan in 2010 referred to nuclear power and renewable energy as a "zero-emission power supply." It set as a goal that the ratio of the "zero-emission power supply" in an electric source, which amounted to 34 percent at that time, would be over 50 percent in 2020, and 70 percent in 2030. This ratio was to be achieved by establishing at least 14 additional nuclear power plants and improving the utilization capacity to 90 percent. Figure 12.4 shows the transition in the power supply share in Japan.

By 2011, Japan had 54 nuclear power plants operating across the country before the Fukushima Daiichi accident. TEPCO installed boiling water reactors (BWR), while the Kansai Electric Power Company (KEPCO) installed pressurized water reactors (PWR). Every year or two, electric companies conduct regular inspections of nuclear power plants.

The manufacturing of nuclear reactors has been occupied by an oligopoly of three heavy electric machinery companies, Hitachi, Toshiba, and Mitsubishi Heavy Industries. The company that builds the nuclear power plant undertakes its maintenance and, if needed, the extension of additional facilities. Mitsubishi Heavy Industries is affiliated with Westinghouse Electric (WH) and adopted PWR, while Hitachi and Toshiba are affiliated with General Electric and adopted BWR.

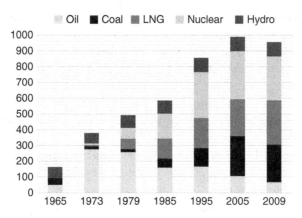

FIGURE 12.4 Transition in the power supply share in Japan prior to 2011 (billion kWh)

Toshiba obtained technical information and knowledge to manufacture a PWR-type plant, as it bought WH in 2006. Mitsubishi Heavy Industries strengthened its alliance with the French company Areva. Units 1 and 4 of the Fukushima Daiichi Nuclear Power Station were manufactured by Hitachi, while Units 2, 3, 5, and 6 were manufactured by Toshiba.

12.3.2 The Beginning of Nuclear Energy Administration

The history of the public administration of nuclear safety in Japan dates back to December 1955 when the Atomic Energy Fundamental Act (AEFA) and the Atomic Energy Commission (AEC) Establishment Act were enacted. There was no mention of "safety" in the AEFA until the first major revision in July 1978. The Act on the Regulation of Nuclear Source Material, Nuclear Fuel Material, and Reactors (hereafter, Nuclear Reactor Regulation Law) was enacted to regulate the safety of nuclear power plants in June 1957. The prime minister had legislative authority to give permission regarding nuclear energy operation, although the director-general of the Science and Technology Agency (STA) who assisted the prime minister, had substantive regulatory authority. The director-general of the STA was appointed as the chairperson of the AEC (Figure 12.2). The AEC, in practice, functioned as a policy-making body, because its opinions were generally accepted by the government, even though the AEC's formal role was that of consultative to the legislature.

Under this regulatory structure in the early years, an electric company would apply for permission to build a nuclear reactor and would submit a written application to the prime minister. The prime minister consulted the AEC on the relevance of the application to the provisions of the Nuclear Reactor Regulation Law. The AEC gave instructions to the Nuclear Reactor Safety Review Panel, which was set up within the AEC to review safety issues. The panel reported to the AEC whether or not the application met the provisions, and the AEC then returned the report to the prime minister. With regard to construction permits issued by the prime minister, they required the consent of the minister of the Ministry of International Trade and Industry (MITI), as reactors were regulated by the previously passed Electricity Business Act. In addition, regulatory approval was exempted from application of the Nuclear Reactors Regulation Law after reactor construction was permitted, and regulatory approval was based on the Electricity Business Act. There was already some criticism brewing against this interwoven governance structure, particularly for handling safety issues at this time (Tajima 1974).

The first commercial nuclear power plant in Japan launched its operations in July 1966 at the Tokai Power Station of the Japan Atomic Power Company, located in Ibaraki Prefecture. The type of reactor was a gas-cooled reactor (GCR), which had been developed in the UK but was then improved with the addition of an earthquake-resistant design introduced by Japan. In 1970, in Fukui Prefecture, the first BWR Tsuruga-1 Power Plant and the first PWR Mihama-1 Power Plant began their commercial operations. In the following year, 1971, Unit 1 of the Fukushima Daiichi Nuclear Power Plant initiated operations.

In April 1970, the AEC approved and legitimized the report by the Nuclear Reactor Safety Review Panel titled "Evaluation Guidance in Safety Design of Light Water Reactors," which was revised in 1978. In the guidelines, the design of nuclear reactors was required "to tolerate the most severe natural hazards that were predicted on the basis of those natural hazards placed in the record." This provision was interpreted as requiring very low or zero tolerance. This was backed by the spirit of the authorities and the business community who hesitated to consider the potential for severe accidents.

Meanwhile, a national survey of 3,000 people was carried out by the Prime Minister's Office in 1968 and 1969 (Prime Minister's Office 1968; 1969). As shown in Figure 12.5, only 3.2 percent in 1968 and 4.6 percent in 1969 of the general public opposed the utilization of nuclear materials intended for "peaceful purposes." As one can see by the titles of the questionnaires, the peaceful purposes of nuclear power development were emphasized.

12.4 PUBLIC AND INSTITUTIONAL REACTIONS TO THE FOUR ACCIDENTS PRIOR TO MARCH 11, 2011

12.4.1 The Nuclear-Powered Ship *Mutsu* Accident: Japan's First Nuclear Accident and the Public's Reaction

Japan experienced its first domestic nuclear "accident" in 1974. A launching ceremony for the first and last nuclear-powered ship, the *Mutsu*, was held in 1969 at a dock along the Tokyo Bay. The Crown Prince and Princess, and Prime Minister Sato attended the ceremony. In 1974, the *Mutsu* left the port of Ohminato in Aomori Prefecture for its first power ascension test on August 26 and attained a state of criticality on August 28. The leaking of radiation, however, occurred when the crew increased the reactor power to 1.4 percent of full capacity on September 1. The media sensationalized the event, highlighting the leak in radiation although it was a minor leak and there was no significant radiation exposure. As the Aomori Prefecture opposed the return of the *Mutsu* to the port of Ohminato, the *Mutsu* remained offshore for approximately one month until a compromise was reached by the central government, Aomori Prefecture, the town of Mutsu, and the Aomori Fishery Cooperation. They signed an agreement that the *Mutsu* would come back to its home harbor.

A national survey conducted by the Prime Minister's Office in 1976 on attitudes regarding the advancement of nuclear power revealed that 50 percent of respondents considered it positively, 15 percent replied negatively, and the rest responded "I do not know" (Prime Minister's Office 1976). The 15 percent of respondents expressing opposition to nuclear power development was much higher than the 3 percent or 5 percent prior to the *Mutsu* accident (Figure 12.5). Two different national surveys conducted by national newspapers showed that 12 percent and 23 percent of respondents disapproved of nuclear power development (Shibata and Tomokiyo 1999). The results of these polls are summarized in Figure 12.5. Overall, the proportion of negative attitudes in the 1970s toward the generation of nuclear power seemed to have increased since the 1960s, although the number of those who supported the generation of nuclear power continued to surpass the number of those who opposed it. The radiation leakage from the *Mutsu* seemed to stir public confidence regarding nuclear power generation. Protest movements began to take place frequently in the 1970s in residential areas near installment sites for nuclear power plants. The first lawsuit against

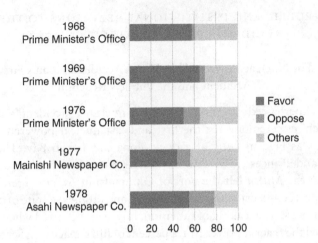

FIGURE 12.5 Public opinion before and after the *Mutsu* accident

the national government demanding the repeal of approval for the establishment of nuclear power plants in the town of Ikata in Ehime Prefecture was filed in 1973.

In the end of October 1974, the Mutsu Radiation Leak Investigation Commission, a board of investigation inquiring into the *Mutsu* accident, was launched by the leadership of the prime minister, and the investigation report was issued in May 1975. The report pointed out four problem areas: policy, organization, technology, and contracts. In the area of policy, the failure of the safety review was referred to in that the Nuclear Reactor Safety Review Panel was not attended by experts of the radiation interception design, and all panel members were part time. It also pointed out that the safety checks were only conducted in written form. The investigation report claimed that "the panel members tended to produce passable conclusions, blurring the demarcation between their responsibility for the outcomes and their roles. There seemed to remain engineering and technical gaps between the examinations and the real designs."

With the backdrop of the *Mutsu* accident, public distrust toward nuclear policy increased. In response, the Japanese government established a private advisory committee under the prime minister chaired by Hiromi Arisawa in February 1975 to re-examine the structure of nuclear-related administrative institutions. The committee submitted an opinion to the Prime Minister that following year. The opinion of the Arisawa Committee can be summarized by the following three goals.

- To establish a Nuclear Safety Commission (NSC): The functions related to nuclear safety should be separate from the AEC and moved to a newly established committee, the NSC, which "double-checks" safety reviews carried out by the competent administrative agencies. The NSC is not to regulate electric power suppliers directly. The legal position of the NSC is the same as that of the AEC; it is a consultation institution.
- To implement consistent safety regulations: Safety regulations should be consistently applied along each type of reactor to make clear the responsibility of each administrative agency. Commercial reactors will be regulated by the Minister of the MITI, commercial nuclear-powered ships will be regulated by the Minister of Transport, and test reactors will be regulated by the prime minister. The responsibility of the competent administrative agencies needed to be clarified, although regulatory power was dispersed. This also meant that the role of promotion still coexisted with that of oversight within the same agencies.
- To hold public hearings and a symposium: In order to mitigate concerns of the public and to promote understanding of nuclear power, when commercial reactors are installed, the MITI should first hold a public hearing before a decision is made on the Electric Power Development Basic Plan in the Electric Power Development Coordination Council. The NSC was to hold a second public hearing after they performed a "double check" of the safety review documents submitted by the MITI.

Based on these recommendations of the Arisawa Committee, both the Atomic Energy Fundamental Act and the Nuclear Reactor Regulation Law were amended in 1978. The NSC became established in the Prime Minister's Office by separating it from the AEC in October 1978 (Figure 12.2). Just prior to the establishment of the NSC, the AEC published the "Regulatory Guide for Reviewing Safety Design of Light Water Nuclear Power Reactor Facilities" in June 1978, which was the product of a full revision of the former guidelines established by the Nuclear Reactor Safety Review Panel in 1970. It required that the facilities be designed to withstand specific earthquake levels determined by past earthquake records and onsite field investigations in and around the region. Natural hazards to be considered other than earthquakes included floods, tsunamis, wind (or typhoons), snow, landslides, etc. However, no detailed guidelines for natural hazards other than earthquakes have been made.

The focus of these guidelines was to prevent all accidents, not to alleviate damage from potential accidents. In other words, the guidelines sought zero-risk tolerance for nuclear facilities. They also required the facilities to be designed to manage a short-term station blackout; however, the guidelines also stated that it was not necessary to consider the possibility of losing highly reliable power sources during a short-term station blackout. This means that a long-term station blackout was outside the scope of the modeled scenario.

The detailed requirements regarding earthquakes were described in the "Regulatory Guide for Reviewing Seismic Design of Nuclear Power Reactor Facilities" published in the same year, which was revised in September 1981 in response to the revision of the Building Standards Act. There was no mention of tsunamis that might be accompanied by earthquakes. The following procedures were established for newly constructed reactors and expansions:

- Electric power suppliers should examine active faults around the site of the proposed nuclear facilities. The assessed active faults should include those that date back at least 50,000 years.
- Seismic design was determined by seismic motion standards for a plant design that was calculated by empirical equations for active faults.
- The seismic motion standards were assessed with the assumption that an M6.5 earthquake could occur even if no active fault was found around the site.

As to the existing nuclear reactors, the Agency for Natural Resources and Energy (ANRE) at MITI requested electric power suppliers through the Federation of Electric Power Companies of Japan to voluntarily back-check and report the results in 1992, 11 years since the publication of the guide.

12.4.2 Three Mile Island (TMI): The Accident and Public Opinion

A loss of coolant accident occurred at the TMI nuclear power station in the United States on March 28, 1979. A partial nuclear meltdown was caused by ceasing water injection into the reactor core due to an operator's mistake, which then automatically activated the emergency core cooling system. This accident was classified as a Level 5 on the International Nuclear Event Scale (INES). The lesson from the TMI accident for global regulatory bodies was that planning for the potential of severe accidents in nuclear power plants caused by

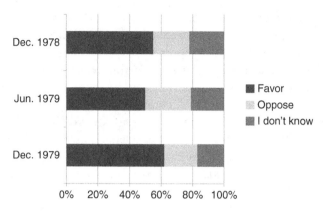

FIGURE 12.6 Public opinion on nuclear power before and after the TMI accident

multiple failures was needed. It encouraged global regulatory bodies to start considering the use of probabilistic risk assessment (PRA) methods in the risk assessment of the facilities (see also, Paté-Cornell in Chapter 10 of this volume). In the meantime, the TMI accident activated the protest movement against nuclear power around the world, while the proponents highlighted that there were no deaths associated with this accident.

A national survey was carried out three months and nine months after the TMI accident by Asahi Newspaper Co. shown in Figure 12.6 (Shibata and Tomokiyo 1999). In comparison with the results of the national survey conducted prior to the TMI accident, the proportion of respondents who answered favorably to nuclear power initially decreased slightly from 55 percent to 50 percent, but then increased to 62 percent nine months after the accident. The proportion of respondents that answered negatively slightly increased from 23 percent to 29 percent, but soon decreased to 21 percent. Three months after the accident, 67 percent of respondents answered in the affirmative to the question "do you think that a nuclear accident that involves the evacuation of residents will occur in Japan?" Half of the respondents, however, favored further development of the generation of nuclear power. The TMI accident seemed to cause only a temporary impact on public opinion regarding nuclear power in Japan. This can be explained partially by the effect of the second oil crisis that occurred in 1979, which showed the imminent dangers of an oil shortage and may have made nuclear power more acceptable.

Administrative Response to the TMI Accident

A special committee was established to investigate the TMI accident in the NSC, and it released its first report in May 1979 and its second report in September 1979. The latter listed 52 lessons from the accident. Among them, nine items were categorized as safety standards, four items as safety reviewing, seven items as safety design, 10 items as operations, 10 items as disaster prevention actions, and 12 items as safety research. Using methods such as PRA to better prepare for accidents caused by multiple factors were classified as "safety research," not as regulatory actions. This reflected a vertically divided administrative demarcation between the Science and Technology Agency (STA) holding jurisdiction over research activities and the Agency for Natural Resources and Energy (ANRE) within MITI holding jurisdiction over regulations.

Three factors can explain the reason why Japan failed to consider and/or incorporate the lessons from the TMI accident (Nishiwaki 2011):

- Several administrative lawsuits seeking the revocation of the installation licenses were filed against the government at that time. The ANRE/MITI hesitated to address and incorporate issues beyond design-based accidents into the regulatory system, because it was the central player in granting licenses.
- The TMI accident occurred shortly after the reorganization of the nuclear regulatory system in response to the *Mutsu* accident. Implementing the post-*Mutsu* accident regulatory reforms had priority over addressing new challenges.
- The STA, which lost authority to approve installation licenses but was eager to engage in nuclear power generation, attempted to retain supervision over safety research. It became more difficult to bridge the gap between safety research and regulatory actions, as the chasm between the MITI and the STA gradually widened.

The Tsuruga Accident

Shortly after TMI, a radiation leak occurred in April 1981 at the Tsuruga Nuclear Power Station of the Japan Atomic Power Co. (This accident is not included with the five major accidents summarized in Table 12.1, because it was considerably smaller than the others.) Public distrust grew, even though only a small amount of radiation leaked, because it was discovered that the electric company had attempted to cover up the leak. This public distrust was reflected in the outcomes of two national surveys conducted by the Prime Minister's Office, the first conducted shortly after

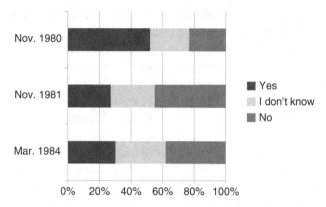

FIGURE 12.7 Public perception of safety measures before and after the radiation leakage from the Tsuruga Nuclear Power Station

the leak and the second conducted two and a half years later. The proportion of respondents who answered in the affirmative to the question "do you think appropriate safety measures are adopted at the nuclear power stations and their associated facilities?" decreased significantly shortly after the leak and remained the same in the latter survey (Prime Minister's Office 1980; 1981; 1984) (Figure 12.7).

However, it did not have a significant impact on attitudes toward the promotion of nuclear power. Another national survey conducted by the Asahi Newspaper Co. in December 1981, showed that the proportion of respondents who favored nuclear power decreased only slightly from 56 percent in the previous year to 55 percent, and the proportion of respondents who opposed nuclear power lightly increased from 25 percent in the previous year to 29 percent (Figure 12.3 above; Figure 12.7 below). In focusing on female respondents only, however, the proportion of the respondents opposing nuclear power slightly surpassed that in favor in 1984, even before the Chernobyl accident.

12.4.3 The Chernobyl Accident

Under the circumstances of distrust regarding the safety of nuclear power plants, a more severe accident occurred at Unit 4 of the Chernobyl Nuclear Power Station in the Ukraine on April 26, 1986. After a meltdown occurred, the reactor exploded, and its radioactive fallout polluted vast areas of the Soviet Union, including Ukraine, Belarus, and Russia. As it occurred during the cold war era, the first news of the accident came from Northern Europe. This accident was rated a Level 7 (major accident) according to the INES.

In a national survey conducted by the Asahi Newspaper Co. in August 1986, four months after the accident, the proportion of respondents who opposed nuclear power exceeded that of those in favor for the first time (Figure 12.3). This situation continued until the mid-2000s. A national survey conducted by the Prime Minister's Office in August 1987, however, showed that only 7 percent of respondents preferred to reduce the number of nuclear power plants or abandon nuclear power, although 93 percent of respondents had some knowledge of the Chernobyl accident (Prime Minister's Office 1987). These results suggest that the Chernobyl accident had a negative impact on public perception in Japan, but did not reach the level of an anti-nuclear power movement in Japanese society.

In response to the Chernobyl accident, the ANRE/MITI approved a nuclear safety plan called "Safety 21" in August 1986, which included the following items as planned countermeasures against severe accidents (Nishiwaki 2011):

- To conduct research and development for the prevention of human errors;
- To conduct analytical research on the behavior of reactors in the event of severe accidents;
- To develop an impact prediction system for emergency accidents; and
- To improve operation manuals for emergencies.

The second item implied that MITI began to address safety research, which thus far had been exclusively addressed by the STA, paving the way for using the results of safety research in regulatory activities.

On the other hand, the NSC created a special committee and released its first report on the investigation of the Chernobyl accident in September 1986 and a final report in May 1987. It concluded by stating:

The results of the investigation for the present safety measures suggested that it is not necessary to revise the current safety regulations and practices in response to the Chernobyl accident, because the safety of nuclear facilities in Japan is ensured due to the diligent efforts at each stage of their design, construction, operation, etc. and there is no need to change the current framework for preparing for natural disasters, because the framework has been established according to the intrinsic characteristics of the facilities (Togo, 1987).

Then, in July 1987, the NSC established a subcommittee under the Special Committee for Nuclear Safety Standards and ordered the subcommittee to consider ways to assess and manage severe accidents. Later, in March 1992, the NSC published a report on accident management as preparation for severe accidents based on the subcommittee report (NSC 1992). The NSC first insisted that the probability of the occurrence of severe accidents was determined to be sufficiently low because the safety of nuclear facilities in Japan was guaranteed under the current safety regulations. In this context, accident management was naturally regarded as extra measures to lower the already sufficiently low risk. In retrospect, their conclusions as stated above were too optimistic.

The different attitude of MITI from that of the NSC – a stronger regulatory response to Chernobyl by MITI than by the NSC – implied a lack of cooperation between these two administrative agencies in charge of nuclear safety. In July 1992, however, MITI also changed its policy of introducing regulations for severe accident measures, and ordered electric utility companies themselves to consider how to prepare for severe accidents by the end of 1993. This meant that severe accident management relied on voluntary measures by the electric companies rather than legally binding regulations. The completion of this voluntary accident management system was extended to 2002. The 1990s in Japan were more or less uneventful with respect to nuclear energy and there was a decline in the motivation for preparing for severe accidents. Activities related to safety research on severe accidents also tended to decrease around that same time.

12.4.4 The JCO Accident

The accident that triggered more extensive institutional changes in Japan was not the above-mentioned significant worldwide accidents at TMI and Chernobyl, but the JCO criticality accident that occurred in Tokaimura, Ibaraki Prefecture on September 30, 1999. The accident happened within a nuclear fuel processing facility of the JCO, which was a subsidiary of Sumitomo Metal Mining Co., Ltd. In processing nuclear fuels, a uranium solution reached a state of criticality leading to a nuclear chain reaction. Two workers at the plant were killed, one was seriously injured, and more than 600 residents were exposed to radiation. The accident was caused by violations of national safety procedures. This accident was classified as Level 4 according to the INES. Unlike nuclear reactor operators, nuclear fuel processers were not required to be regularly inspected under the Nuclear Reactor Regulation Law.

A public opinion survey conducted by the Socioeconomic Productivity Headquarters (currently the Japan Productivity Center) showed an acute change in perception among the public shortly after the JCO accident, but that change more or less disappeared one or two years later (Socioeconomic Productivity Headquarters 2002). The proportion of respondents that considered nuclear power unnecessary increased from 14.5 percent prior to the JCO accident to 21.9 percent shortly after the accident. The ratio, however, soon returned to 16.1 percent one year later. The proportion of respondents answering that nuclear power was not safe increased from 43.1 percent to 56.6 percent shortly after the accident, but it returned to 47.1 percent one year later, and was 40.0 percent two years later. Another public opinion survey conducted by the Institute of Nuclear Safety System, Inc. showed that the proportion of respondents that expressed a high degree of concern about the possibility of nuclear accidents increased from 27 percent in 1998 to 36 percent two months after the JCO accident (Kitada 2006). The ratio soon returned to the pre-accident level according to a survey one year later by the same institute. The respondents were also asked about the possibility of a Chernobyl-like accident occurring in Japan. The proportion of respondents who thought this type of accident would happen in Japan in the near future increased from 47 percent in 1998 to 61 percent two months after the JCO accident. The ratio, however, decreased gradually since then and fell below the pre-accident level in 2003.

In response to the JCO criticality accident, the NSC established an accident investigation committee and released a document titled "Immediate Measures for Ensuring Safety of Nuclear Power Generation" on November 11, based on an urgent proposal published by the committee. The final report of the committee was issued at the end of 1999. At the beginning of the section that included actions to be taken, it was pointed out that there was a significant lack of awareness of the concept of "risk."

A lack of awareness of the risk of criticality was the underlying cause of the criticality accident. Correct risk awareness must be regarded as the starting point of all efforts to ensure safety. All organizations and individuals concerned with nuclear power must maintain risk awareness while playing their respective roles. In order to have risk awareness become deeply embedded in society, we must change our attitude from the belief in the "safety myth" and the notion of "absolute safety" to, the notion of "risk-based assessment of safety" (NSC 1999: 34).

The following two reforms were implemented directly after the accident. One was to enact another law entitled the Act on Special Measures Concerning Nuclear Emergency Preparedness, which became effective on December 17, 1999. In this act, an "Article 10 warning" means detection of radiation exposure, which suggests that emergencies are occurring or are imminent. An "Article 15 warning" means a high probability of damage to the reactors such as a loss of all power or a loss of coolant, in which case the Prime Minister issues a declaration of a nuclear emergency. (These warnings were applied at the outset of the Fukushima crisis in 2011 for the first time.) The other reform was to institutionalize periodic inspections of nuclear fuel facilities, such as the JCO processing plant, after approval of installation. A tentative set of guidelines was published by the NSC in June 2000, and then the Nuclear Reactor Regulation Law was revised to introduce an inspection system in compliance with the safety rules.

Several broader institutional reform proposals were also advanced after the JCO criticality accident. The result was a reorganization of governmental ministers and agencies in January 2001. The following significant changes were made to the nuclear safety regulations:

- The regulatory authority for nuclear reactors and nuclear fuel facilities at the stages of commercial development and research and development (R&D) was centralized under the Minister of Economy, Trade, and Industry (METI), formerly known as MITI, while commercial nuclear-powered ships were regulated by the Ministry of Land, Infrastructure, and Transportation, formerly the Ministry of Transport. The Ministry of Education, Culture, Sports, Science, and Technology (MEXT), formerly the STA, only had authority to regulate test reactors.
- The NSC was relocated under the Cabinet Office and gained its own secretariat. This restructuring provided greater independence, because the administrative status of the Cabinet Office is higher than that of the other ministers and agencies.
- The Nuclear and Industrial Safety Agency (NISA) was newly established within the METI as "a special organization of the ANRE" for laws and regulations. NISA had an extent of independence within the METI, because NISA had a centralized responsibility for safety regulations associated with nuclear power operations.

As a result, a new "double-check" regime was established. From its unique standpoint, the NSC's role has been set to oversee the results of the primary safety check done by the NISA within the METI on the basis of the Nuclear Reactor Regulation Law. The need for safety checks by the NSC has been shown in various regulatory guides for reviewing safety design. As for the role of the NSC, two different viewpoints were noted in Japan (Shiroyama 2010). One viewpoint was that the NSC was unnecessary because the "double-check" function had lost its substance. The other was that the NSC needed more independence to ensure the safety of the nuclear industry. When the form of reorganization of the governmental ministers and agencies was discussed in November 1999, the idea of giving the NSC more independence had some support from the Administrative Reform Task Force of the ruling Liberal Democratic Party (Kitayama 2008).

In the late 1990s, fraud and some accidents (other than the JCO criticality accident) occurred in Japan, such as a sodium leak at the Monju fast breeder reactor in 1995, a fire and explosion in the Bituminization Demonstration Facility at the Tokai Reprocessing Plant in 1997, and falsified testing data by a manufacturer of mixed plutonium-uranium oxide (MOX) fuel in 1999. To the NSC these failures reflected incorrect risk perceptions among nuclear experts and failures in ethics and the NSC pushed toward finding a way to recover the public's trust. The ANRE/METI established a task group on nuclear safety in December 2000 to deliberate on the issue of "how future nuclear safety should be ensured in response to the changes of environment in recent years" and published a report in June 2001 (METI 2001). The premise of the discussion was that it was difficult to prevent fraud and accidents under the existing approach to nuclear safety even if nuclear power plants in Japan were equipped with sufficient safety facilities. The focus was, therefore, placed on to the question of how to create a "safety culture" in the nuclear industry.

12.5 PUBLIC OPINION AND POLICY DEVELOPMENT IN THE 2000S

12.5.1 Public Opinion in the 2000s

A string of cover-up scandals at nuclear facilities was reported in the 2000s, starting with the discovery that TEPCO had failed to report the government about shroud damage, which had been found in the voluntary inspection of its nuclear power plants in August 2002. Two years later, another accident occurred at Unit 3 in the Mihama Nuclear Power Station of KEPCO in August 2004. High-temperature vapor erupted due to bursting pipes in the secondary cooling system. Five workers died as a result of burns covering their bodies. This series of events immediately lowered public acceptance and trust regarding nuclear power. Several opinion surveys show, however, that the impacts directly after the events quickly returned to the pre-event level (Kitada 2003; 2006).

The results of a national survey from the mid-2000s regarding nuclear power by the Asahi Newspaper Co. showed that the ratio of people who favored nuclear power returned to the position exceeding that of those opposing nuclear power for the first time in approximately 20 years (Figure 12.3). As for the necessity of nuclear power, since October 2006, the proportion of respondents who answered that nuclear power was necessary for Japan largely exceeded that of respondents who answered that nuclear power was unnecessary, and the proportion continued to increase. As for safety, the proportion of respondents who answered that nuclear power was safe continued to increase after January 2006, and finally, in August 2008, exceeded that of respondents who answered that nuclear power was unsafe for the first time in 12 years. Since 2006, the proportion of respondents who were in favor of promoting nuclear power exceeded that of respondents who would like to stop or abolish nuclear power. The consecutive national survey carried out by the Atomic Energy Society of Japan also shows that the proportion of respondents who answered that nuclear power was "necessary" or was "maybe necessary" for Japan increased from 68 percent in January 2007 to 77 percent in September 2010, half a year before the Fukushima Daiichi accident (JAERO 2013). In general, the latter half of the 2000s appeared to be a time when public opinion in favor of nuclear power had increased. This trend continued until the Fukushima Daiichi nuclear accident. It can be said that the Fukushima accident occurred during the time of the highest public support for nuclear power in 25 years.

12.5.2 Countermeasures against Earthquakes and Tsunamis

The regulatory guide for reviewing the seismic design of nuclear power reactor facilities was originally issued in 1978 and revised in 1981. The "design seismic motion" was derived from a limited amount of empirical data. Some experts expressed concerns of underestimating the design seismic motions after the 1995 Hanshin-Awaji (Kobe) Earthquake, although it did not cause damage to any nuclear power reactors. A working group on the renewal of regulatory guidance was set up under the NSC based on these concerns. Seismic motion exceeding the design seismic motions was experienced for the first time during the 2005 earthquake off Miyagi Prefecture (M7.2).

As for tsunamis, TEPCO stated that a margin of safety of 2.5 meters was guaranteed assuming that the height above sea level of the lowest cooling seawater pump was 5.6 meters and the highest tidal level was 3.122 meters. This was based on records from the 1960 Chile earthquake and was submitted in the construction permit application for Unit 1 of the Fukushima Daiichi Nuclear Power Station in 1964. Responding to the tsunami damage, including more than 200 deaths caused by the 1993 earthquake of the southwest coast of Hokkaido (M7.8), the ANRE/MITI instructed the Federation of Electric Power Companies of Japan (FEPC) to assess the tsunami safety of nuclear power stations. TEPCO, however, remained firm on the conclusion that the highest tidal level was that of the Chile earthquake in 1960, although it considered 13 tsunami events accompanied by earthquakes since 1611. On the other hand, Tohoku Electric Power Co., Inc. prepared a construction permit application for Unit 1 of the Onagawa Nuclear Power Station in 1970, setting the height of the site to be 15 meters, although the highest tidal level recorded in past literature was about 3 meters. After that, a geological investigation and numerical simulation revealed that the largest tsunami was caused by the 1611 Keicho Sanriku earthquake. The highest tidal level in the site was changed to 9.1 meters. The applications for Units 3 and 4 were prepared based on that information. In the meantime, a guidance report published jointly by seven governmental ministries and agencies in 1998 introduced a new concept called "the assumed largest scale tsunami," which considers possible tsunamis in addition to those that have actually occurred. However, specific methodologies for assuming those possible tsunamis were not included in the report. The Japan Society of Civil Engineers (JSCE), therefore, launched a new group called the Tsunami Assessment Working Group and published a report in 2002 (JSCE 2002) detailing the methodologies for determining the height of tsunamis accompanying large earthquakes. This report indicated that the height of "design basis tsunami" should be "based on the seismic knowledge regarding recorded occurrences of earthquakes," which means that it turned back to the previous methodology of basing the height on the largest recorded tsunamis. TEPCO calculated the height of the largest tsunami at Fukushima Daiichi Nuclear Power Station to be 5.7 meters applying the criteria from the JSCE report. The cooling seawater pump was raised by 0.2 meters. The past tsunami events considered in the assessment were limited to those that occurred during the last 400 years.

A governmental institution, the Headquarters for Earthquake Research Promotion, which was established in response to the 1995 Hanshin-Aawaji (Kobe) earthquake, announced a report entitled "Long-Term Estimation of Seismic Activities from Sanriku-oki to Boso-oki" in July 2002. In this report, it was predicted that a class M8 earthquake accompanying a tsunami would occur with a probability of approximately 20 percent within 30 years along the Japan Trench, including the offshore area near the Fukushima Daiichi Nuclear Power Station. TEPCO simulated the possible impacts of that earthquake and obtained results that the site of the Fukushima Daiichi Nuclear Power Station would suffer a tsunami up to 15.7 meters. TEPCO, however, did not take any countermeasures at that time. Instead, TEPCO water-sealed its cooling seawater pumps in 2009, as the assumed height of the largest tsunami was changed to 6.1 meters on the basis of the methodology proposed by the JSCE.

The regulatory guidelines for reviewing seismic design were revised in September 2006 after receiving public comment. The basic policy of the renewed guidelines stated that nuclear facilities should be built where safety functions cannot be damaged by seismic motion that is assumed to occur only infrequently and where it will have a significant impact on the facilities (NRC 2006: 2). In particular, the following guidelines were emphasized:

- The standard seismic motion for the class of facilities that is the most important for seismic safety designs is set by assuming both "earthquake ground motion formulated with a hypocenter specified for each site" and "earthquake ground motion with no specific hypocenter" (NRC 2006: 4)
- Multiple earthquakes should be selected for "earthquake ground motion formulated with a hypocenter specified for each site." The definition of active faults was extended to include the faults whose seismic activities occurring after the Late Pleistocene (more than 120,000–130,000 years ago) cannot be denied. The previous guidelines limited its consideration to seismic activity that occurred 50,000 years ago (NRC 2006: 4).
- For the first time, tsunamis were referred to as accompanying events of earthquakes. The guidelines stated that "nuclear facilities should be built so that the safety functions cannot be damaged by a tsunami, which is appropriately assumed to occur only infrequently."

The responses to the guidelines, however, varied among facilities, because the guidelines did not mention either the methodologies to determine the design-basis tsunami, or an appropriate tsunami-proof design for the facilities (NRC 2006: 14).
- The existence of "residual risks," if actual seismic motion were to exceed the assumed design seismic motion, was pointed out. The guidelines stated that every effort should be made to reduce the residual risks to the lowest level reasonably achievable (NRC 2006: 2).

When the 2005 Miyagi earthquake (M7.2) occurred and Units 1 to 3 of the Onagawa Nuclear Power Station were automatically shut down, seismic motion exceeding the design seismic motion was recorded during certain wave periods, although there was no damage caused by the seismic motion. In light of this, while the revised regulatory guidelines were applied to newly built nuclear facilities, the NSC asked NISA also to back-check the seismic safety of existing nuclear power facilities in accordance with the new regulatory guidelines, even though the new guidelines were not legally binding on existing facilities. NISA, then, asked nuclear operators to make implementation plans to back-check for earthquake resistance.

Some of the observed seismic motion exceeded the design seismic motion in accordance with the old regulatory guidelines at Shiga Nuclear Power Station during the Noto Peninsula earthquake (M6.9) in 2007. When the Chuetsu offshore earthquake (M6.8) occurred in 2007, four operating units at the Kashiwazaki-Kariha Nuclear Power Station were automatically shut down and successfully led to cold shutdown conditions, despite the fact that observed seismic motion was much larger than the assumed motion at the time that all seven units were designed. NISA encouraged nuclear operators to complete the back-checks as soon as possible.

As for the Fukushima Daiichi Nuclear Power Station, TEPCO submitted an interim report of seismic back-checks for Unit 5 in March 2008, which stated that seismic safety was ensured by the assumption that the design seismic motion was to be 600 Gal. NISA confirmed the validity of TEPCO's assessment in November 2009, although it was noted that the scope of the assessment was too narrow to be an effective back-check (NAIIC 2012). TEPCO also submitted interim reports of the remaining units in April 2009. Since then, little progress was reported until the large-scale earthquake and tsunami hit the facilities in 2011. After the accident, it was revealed that the deadline to submit the reports was postponed to January 2016 internally, although TEPCO officially stated that the deadline was June 2009. It is also reported that TEPCO realized that a number of seismic strengthening projects should have been done to comply with the new regulatory guidelines, and NISA recognized the urgency for seismic back-checks with seismic strengthening projects (NAIIC 2012).

12.5.3 Severe Accident Countermeasures and PRA

While, the TMI accident in 1979 prompted regulatory agencies in many countries to tackle probabilistic risk assessment (PRA) and severe accident countermeasures, this did not occur in Japan. The basic idea of these reforms was to reduce the probability of a core damage accident as much as possible on the premise that core damage accidents occur by exceeding design specifications or from multiple failures. In doing so, it was necessary to set safety goals for the facilities before accidents occurred, and to develop response manuals for after accidents occur.

Accident Management (AM) signifies taking both measures to prevent events beyond the scope of the assumed design scaling up to severe accidents (Phase I) and measures to mitigate the impacts as far as possible when severe accidents occur (Phase II). In Japan, however, AM was treated as a countermeasure "to reduce already sufficiently small risks." AM was, therefore, not introduced as a legally binding rule, but as a voluntary activity of the electricity industry. Then, the ANRE/METI requested electric companies to carry out PRA for their reactors. Potential severe accidents were limited only to those caused by internal events. Those caused by external events were never considered, even after the 1995 Hanshin-Awaji (Kobe) earthquake. The NISA, established in 2001 within the METI, reported to the NSC that voluntary AM measures were prepared in 2002. In 2004, NISA published the results of PRA that quantified the core damage frequency (CDF) (per reactor, per year) and the containment failure frequency (CFF) (per reactor, per year) before and after AM completion. Both CDF and CFF were below 10^{-6} (1 in 1 million) in all reactors.

Parallel to PRA, the NSC initiated discussions on safety goals. An interim report on safety goals was published in 2003. With regard to quantitative goals, it proposed that "an average acute mortality risk of additional radiation exposure caused by nuclear accidents should not exceed one in a million per year for an individual living near the site boundary" and "an average (chronic) mortality risk of additional radiation exposure caused by nuclear accidents should not exceed one in a million per year for an individual living a certain distance from the nuclear facilities." Subsequently, the NISA began discussions on how to use risk information obtained by conducting PRA, and published a report describing the basic idea in May 2005.

These activities, however, have not been incorporated into the regulatory system, because the current safety regulations, including "deterministic" safety assessments in every step, were believed to adequately ensure safety as long as they conform to the laws and regulations. As a result, countermeasures to tackle "residual risks" have never been positioned as regulatory requirements, although it was clearly referenced in the revised 2006 regulatory guidelines.

12.6 REACTIONS TO THE FUKUSHIMA DAIICHI NUCLEAR ACCIDENT

12.6.1 The Accident and Public Opinion

The earthquake that occurred off the Pacific coast of Tohoku on March 11, 2011, immediately killed approximately 20,000 people in numerous towns. When that earthquake struck, Units 1, 2, and 3 of the Fukushima Daiichi Nuclear Power Station were in operation, while Units 4, 5, and 6 were under suspension due to a periodic inspection. Emergency shutdown systems were working successfully at the three units in operation. All units lost their external power supply, however, because the seismic motion damaged the electricity transmission and distribution facilities, which halted all transmission of electricity. All power sources except for one air-cooled emergency diesel generator in Unit 6 were lost as a result of the tsunami that occurred shortly after the earthquake. Following the loss of power, there were hydrogen explosions at Units 1, 3, and 4 and core damage to Units 1, 2, and 3.

The factors that directly contributed to the accident were as follows. First, updating the existing reactors to comply with the 2006 revised regulatory guide, that is, the backfitting, had been postponed, as mentioned in Section 12.5.2. Second, although the tsunami risk was recognized, the government failed to address tsunami risks based on the longer-term historical assessment published by the Headquarters for Earthquake Research Promotion, as discussed in Section 12.5.2. Third, the preparation of severe accident countermeasures focused on internal causes within the reactor and did not consider the potential for external events, such as earthquakes and tsunamis. Fourth, the cooling system back-up diesel generators at Fukushima were located at the same elevation (near sea level) as the reactor, rather than located higher up the hillside where the tsunami risk would have been lower.

In all of these factors, an underlying cause was the problem of regulatory capture. The following aspects were noted in various accident investigation reports (e.g. NAIIC 2012):

- Electrical power suppliers have become involved in drafting safety regulations because of their strong ties with regulatory bodies.
- Regulatory agencies have both regulatory sections and promotion sections within the same agency, METI.
- As regulators and the regulated share the same interests in avoiding the shutdown of existing reactors, they have attempted to ignore opinions that question the safety of existing reactors and the validity of past regulations.

Certainly, the Fukushima accident increased the risk perception of the general public in Japan. The proportion of respondents favoring nuclear power, however, still exceeded those opposing it in opinion surveys conducted shortly after the accident in March and April 2011, according to the Asahi Newspaper Co. (Figure 12.8) and the Yomiuri Newspaper Co.[1] According to a telephone survey conducted by the Asahi Newspaper Co. from April 16 to 17, 2011, the proportion of respondents favoring nuclear power was 50 percent and that opposing nuclear power was 32 percent. Approximately half of the respondents favored the status quo as opposed to the question asking them how to deal with nuclear power plants in Japan. According to the telephone survey conducted by Yomiuri Newspaper from April 1 to 3, 2011, 10 percent agreed with an increase in the number of nuclear power plants,

---

[1] The results of the telephone survey carried out by the Pew Research Center in April and May 2011 also showed that 8 percent favored an increase in the number of nuclear power plants, 46 percent supported the status quo, and 44 percent supported a decrease in the number or the abolishment of nuclear power plants (Pew Research Center 2012). In the meantime, WiN Gallup International, for instance, conducted an opinion survey with 34,000 people in 47 countries on March 21 to April 10, 2011. The survey of Japan included 1,000 people who were asked to answer via the Internet between April 5 and April 8. It showed that before the earthquake and the accident, 62 percent of respondents saw nuclear power generation favorably, while 28 percent viewed it unfavorably. At the time of the survey, however, 39 percent responded with a favorable attitude, while 47 percent responded with an unfavorable attitude (WiN Gallup International 2012).

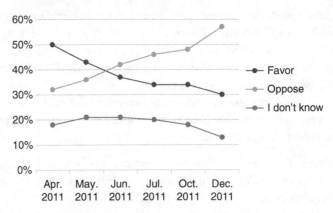

FIGURE 12.8 Change in public opinion shortly after the Fukushima accident
Source: Asahi Newspaper Co.
(This figure can be read in conjunction with Figure 12.3)

48 percent favored the status quo, 29 percent supported a decrease in the number of nuclear power plants, and only 12 percent favored abolishing nuclear power plants.

There appear to be several factors explaining why the proportion of respondents favoring nuclear power exceeded that opposing it in April, one month after the accident. The high ratio supporting nuclear power in the latter half of the 2000s might leave some inertial influence on public opinion. In addition, the following circumstances seemed to contribute to the support for the generation of nuclear power.

- The public was not aware of the fact that radioactive materials polluted widespread areas or that their effects would continue for a long time.
- At that time, the accident was understood to have been a natural disaster rather than a human-made disaster.
- The potential for an electricity shortage was an urgent matter at that time.

In opinion surveys after May 2011, it became clear that a majority of the public opposed nuclear power, and this trend has continued.[2]

---

[2] In a telephone survey conducted by Asahi Newspaper Co. on June 11 and 12, the proportion of respondents opposing nuclear power (46 percent) exceeded the proportion favoring it (37 percent) for the first time since the mid-2000s. Just a month earlier, in May 2011, the proportion opposing (36 percent) was smaller than the proportion favoring (43 percent). Since June 2011, the trend toward opposition has continued. In surveys conducted

12.6.2 Launch of the Nuclear Regulation Authority (NRA)

The Japanese government classified lessons from the accident into five categories, in a report submitted to the International Atomic Energy Agency (IAEA) in June 2011 (Nuclear Emergency Response Headquarters 2011). In category 4, which pointed out the need to reinforce the safety infrastructure, it stated that NISA should be separate from the METI and the regulatory structure of nuclear power should be re-examined including the NSC and associated agencies (NERH 2011: 40). The cabinet approved the Basic Policy on Institutional Reform of Nuclear Safety Regulation in August 2011. According to this policy, the government submitted proposed legislation that moved NISA from the METI to an affiliated agency of the Ministry of the Environment (MOE) and created a nuclear safety council as an advisory committee at the end of January 2012.

Meanwhile, the then-opposition Liberal Democratic Party launched a project team on nuclear regulatory institutions in December 2011 and initiated a counteroffer placing more weight on the independence of the new institution. As a result of discussions, the then-opposition parties submitted a legislative proposal in April 2012. This proposal was enacted as the Act for Establishment of the Nuclear Regulation Authority in June 2012. The Nuclear Regulation Authority (NRA) was established as an administrative committee. Both the NSC within the Cabinet Office and NISA within the METI were abolished, and their functions were integrated into the newly established NRA. Functions dispersed in various departments and agencies were integrated in the NRA, including nuclear safety regulations, nuclear materials security, safeguards for non-proliferation, radiation monitoring, and the regulation of radioactive isotopes.

by Yomiuri Newspaper Co., 44 percent of respondents preferred to reduce the number of nuclear power plants and 15 percent hoped to abolish all nuclear power plants. That is, 59 percent of respondents opposed nuclear power. After that, the ratio of opposition to support gradually increased. The NHK Broadcasting Culture Research Institute conducted an ongoing telephone survey of public opinion beginning in June 2011. Only 1–3 percent preferred to increase the number of nuclear power plants. While 20–30 percent of respondents supported the status quo, 40–50 percent preferred to reduce, and 10–30 percent insisted on abolishing nuclear power. This did not change in the two years following the accident (NHK Broadcasting Culture Research Institute 2013). The Nippon Research Center carried out public opinion surveys in May, July, and September 2011 and March 2012 using the door-to-door placement method. They asked 1,200 people one question about using nuclear power as an energy source. The proportion of opponents increased from 45 percent in May 2011 to 57 percent in March 2012, while the proportion of supporters continued to decrease from 33 percent in May 2011 to 24 percent in March 2012 (Nippon Research Center 2012).

The NRA was officially launched on September 19, 2012, as an organ of the MOE. The NRA consists of a chairperson and four other full-time commissioners. The Secretariat of the NRA has a staff of approximately 1,000, after merging with the Japan Nuclear Energy Safety Organization (JNES).

12.6.3 New Safety Standards

The NRA established four teams to develop new regulations for nuclear power plants. A "draft outline" of new safety standards was submitted for public comment from February 7 to February 28, 2013. The teams examined the comments, and then a new draft of safety standards was developed into 49 documents, which were again submitted for public comment from April 11 to May 10, 2013. The new safety standards were approved on June 19, and published on July 8, 2013. The difference in scope between the old and the new standards is shown in Figure 12.9. At last, severe accident countermeasures and retrofitting of existing facilities using the most current information were legally required under amendments to the Nuclear Reactor Regulation Law. For the first time, the new safety standards included standards for responding to severe accidents. The scope of "external events to be taken into account" and the level of

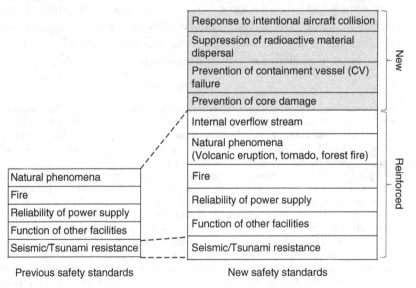

FIGURE 12.9 Scope of the new requirements compared with former requirements

the "design basis for these external events" were discussed by one of the task teams. External events now consist of both natural disasters and human-made events. Regulatory guidelines for natural disasters were developed for volcanic eruptions, tornados, and forest fires, as well as for earthquakes and tsunamis. External human-made events were divided into incidental events, such as aircraft crashes and dam failures, and intended events, such as terrorist attacks.

The design-basis tsunami, used in tsunami resistance designs, uses the tsunami that caused the greatest amount of damage to the facility among the possible tsunami scenarios and is based on the wave source model considering occurrence factors and diffusion simulation. The scale of the design-basis tsunami generally surpasses the largest recorded tsunami. Then, based on a probabilistic tsunami hazard assessment, the probability of exceeding the water level of the design-basis tsunami will be derived for the assessment site. When the NISA discussed an acceptable level of that probability, it was proposed that the frequency of exceeding the design-basis tsunami should be kept below 10^{-5}/year in order to keep the frequency of core damage below 10^{-4}/year. The safety goal, however, was not included in the new safety regulations.

As for earthquakes, the most important facilities were to be constructed on ground where active faults were not exposed. As in the prior regulations, "active faults" were defined as faults with activity records beginning 120,000–130,000 years ago. In addition, however, if needed, activity records beginning 400,000 years would be investigated. Approximately 2,000 active faults have been identified so far in Japan, and it is, therefore, generally difficult to avoid "active faults" when building nuclear facilities.

Electric companies are required to assess and take into account volcanos within a 160-kilometer radius. The 160-kilometer radius was determined by referencing the longest recorded distance of pyroclastic flows. Among the existing volcanos, the volcanos that were active after the Holocene (about 10,000 years ago) are regarded as those that may become active in the near future. The number of "active" volcanos in Japan is 110. Where the frequency of damage to nuclear power plants from volcanic events is not sufficiently small, the construction of nuclear power plants will not be permitted. It is, however, controversial that the NRA seems to have adopted a premise that large-scale volcanic eruptions, such as caldera-forming eruptions, must have some historical precursor event, unlike earthquakes.

After adoption of the new safety regulations, the NRA initiated compatibility reviews for nuclear power plants, responding to applications from electric power companies. Eleven companies applied to receive a compatibility review of 15 nuclear power plants. The NRA made determinations on the applications, declaring that the reactors at Units 1 and 2 of the Sendai Nuclear Station and the reactors from Units 3 and 4 of the Takahama Nuclear Station met the new safety standards in September 2014 and in February 2015, respectively.

12.6.4 Impacts on Energy Supply and Policy

On May 6, 2011, via the Minister of the METI, then-Prime Minister Kan requested Chubu Electric Power Co. to halt its operations at Hamaoka Nuclear Power Station in Sizuoka Prefecture, which consisted of three operating units. Chubu Electric Power decided to halt operations of Units 4 and 5 on May 9, even though the request did not have legally binding authority. The assumed height of the tsunami before the Fukushima Daiichi accident had been set at 8 meters, which was derived from the scenario of an earthquake in the Nankai Trough in the Pacific Ocean on the basis of past geological records and the results of simulations, and it was considered to be protected by a 10–15 meter sand dune. Five days after the Fukushima Daiichi accident, Chubu announced a plan to build a tsunami defense wall 12 meters high and 1.2 kilometers long to be completed within two or three years. They revised the height from 12 meters to 15 meters one month later in order to relieve the concerns of residents. Then, in July they decided to increase the height again and to build a tsunami wall 18 meters high and 1.6 kilo meters long, on the basis of the observed height of the tsunami at the Fukushima Daiichi site, which was 15 meters. Finally, they decided to add a 4-meter iron sheet to the 18-meter-high wall under construction, responding to a revised scenario of a possible Nankai Trough earthquake that could cause a 19-meter-high tsunami at the site of the Hamaoka Nuclear Power Plant, which corresponded to a 21.4 meter run-up height. This tsunami defense wall is almost completed as of 2015.

Shortly after the Fukushima accident in 2011, all 54 civilian nuclear power plants in Japan were shut down. As of April 2015, electric companies have not been approved to resume nuclear power plant operations after regular inspections. As a result, the market share of thermal fossil fuel power sources (coal, oil, and natural gas) increased to 88.3 percent in 2013, while it had been approximately 60 percent before the 2011 Fukushima accident.

On May 17, 2011, the Cabinet put forth the "Guidance on Policy Promotion: For the Revitalization of Japan" (Cabinet Decision 2011), which ordered a newly established cabinet office called the Energy and Environment Council, chaired by the State Minister for National Strategy, to formulate a new "innovative strategy for energy and the environment" (Cabinet Decision 2011: 5). The then-Prime Minister expressed an intent to undo the Basic Energy Plan that had been revised in the previous year. The Council published an "Interim Summary" in the second meeting at the end of July 2011. This document presented three basic ideas: 1) to reduce the dependence on nuclear energy; 2) to shift toward a distributed generation system; and 3) to move toward a national discussion. This new "Basic Policy" was published after the Council's fifth meeting at the end of December 2011. Based on the "Basic Policy," the Council requested that by the next spring the then-AEC would develop options for a nuclear energy policy, the Advisory Committee on Energy and Natural Resources in the METI would develop options for an energy combination, and the Central Environment Council in the MOE would develop options for a global warming policy. By referencing these three results, the government determined the "Options for Energy and the Environment," by preparing three options for the future market share of nuclear power: (1) a 0 percent scenario; (2) a 15 percent scenario; and (3) a 20–25 percent scenario, as of 2030 (Energy and Environment Council Government of Japan 2012a). A "National Debate" based on these three options was implemented in July and August 2012. It consisted of public hearing sessions, solicitation of public comments, and a Deliberative Polling event. At the same time, mass media carried out various national opinion surveys.

Responding to the high public support for the 0 percent scenario, the government, at that time led by the Democratic Party of Japan (DPJ), finally published the strategy (Energy and Environment Council Government of Japan 2012b), which presented the following three guiding principles in order to achieve the goal of zero dependence on nuclear power by the 2030s:

- To strictly apply the stipulated rules regarding the 40-year limit on nuclear power plant operations;
- To restart the operation of nuclear power plants only if the newly created NRA provides safety assurance; and
- To refrain from planning the construction of new nuclear power plants.

Before this strategy could be implemented, the House of Representatives (*Shugiin*) election took place in December 2012, and the

Liberal Democratic Party (LDP) achieved a sweeping victory, securing 294 seats and taking the government from the DPJ, which kept only 57 seats. This was the first general election since the Fukushima Daiichi accident in March 2011. The LDP decided to rescind the strategy published by the DPJ in 2012. In a policy address in February 2013, new Prime Minister Abe stated clearly that the new government planned to restart nuclear power plants as soon as they passed the safety reviews by the NRA. Six months later, in July 2013, the House of Councilors (*Sangiin*) election took place. The LDP again achieved a sweeping victory. The revised Basic Energy Plan was adopted in April 2014 and nuclear power was positioned as an "important base-load power source," although the desired proportion of each power source was not stated.

Both elections resulted in major victories for the LDP and its coalition partner, New Komeito, who also expressed intent to restart nuclear power stations, while at the same time, public opinion surveys continued to show strong opposition to nuclear power. This contrast seems strange, but might be explained in two ways. One is the effect of the voting system. The voting system of the Upper House election consists of 73 seats elected in 47 electoral zones and 48 seats elected by proportional representation. The share of the vote of the LDP and its partner New Komeito was 48.9 percent, which means that half of the voters supported various other parties against nuclear power. Figure 12.10 shows the shift in the

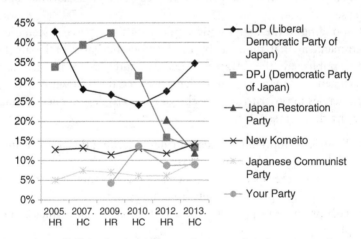

FIGURE 12.10 Shift in the share of votes in proportional representation
HR stands for House of Representatives (*Shugiin*), HC stands for House of Councilors (*Sangiin*)

party's share of several votes in proportional representation. A second explanation is that approximately half of the public is still ambivalent about nuclear power. The majority of people seem to oppose restarting nuclear power plants because the proportion of respondents opposing the restart was 58 percent and that favoring a restart was only 28 percent in a telephone survey conducted by the Asahi Newspaper Co. just before the House of Councilors election. In a telephone survey carried out by the NHK in March 2013, however, 45.5 percent of respondents selected the "no opinion" option, while 35.5 percent opposed the restart and 15.9 percent supported it, although the former Asahi survey had only two response alternatives (NHK Broadcasting Culture Research Institute 2013). Furthermore, a gradual rise in home electricity bills may have influenced public attitudes to favor nuclear power in the near future.

12.7 DISCUSSION

This chapter has focused primarily on untangling the complex relationship of events, public perceptions, and regulatory changes in the context of nuclear power, based on the event model. This section discusses some limitations and challenges of the analysis, as the main findings were summarized above in Section 12.2. A major finding is that institutional and regulatory changes in Japan were triggered primarily by three domestic accidents, not by other international accidents such as TMI and Chernobyl. Public perception in Japan seems to have been influenced by the Chernobyl accident in the Ukraine, as well as by the *Mutsu* accident and the Fukushima accident in Japan, but official changes in institutions and regulatory policy appear to have followed domestic accidents rather than foreign accidents.

Other factors may also have influenced both public opinion and regulatory policy. In light of the complexity of the real world, we must use caution when interpreting cause and effect. An ongoing poll conducted by the Asahi Newspaper Co. was adopted as an indicator of public opinion on nuclear power generation for more than 30 years, because the same questions have been repeated over time in this poll. It is well known that responses are sensitive to the framing of questions, such as the number of options identified. The response to the question that asks whether nuclear power should be promoted or abandoned differs substantially if it identifies two options "Yes" and "No" versus adding a third option, "I do not know." It is also well known that there can be a significant gender difference regarding preferences for emerging technologies, which is

TABLE 12.2 *Accidents and investigatory commissions and their reports*

Accidents	Commissions	Reports
Mutsu	Mutsu Radiation Leak Investigation Commission (in the Prime Minister's Office)	A report was released 9 months later.
TMI	Special commission on the nuclear power accident in the US (in the NSC)	The first report was released 2 months later. The second report was released 6 months later.
Chernobyl	Special commission on the nuclear power accident in the Soviet Union (in the NSC)	The first report was released 5 months later. The final report was released 1 year later.
JCO	Accident investigation board on criticality accident at uranium processing plant (in the NSC)	The first report was released 1.5 months later. The final report was released 3 months later.
Fukushima Daiichi	Fukushima Nuclear Accident Independent Investigation Commission (by the National Diet of Japan)	The report was released over a year later, on July 5, 2012.
	Investigation Committee on the Accident at Fukushima Nuclear Power Stations of Tokyo Electric Power Company (by the Cabinet Office and the Government of Japan)	The interim report was released 9 months later, on Dec. 12, 2011. The final report was released over a year later, on July 23, 2012.
	Independent Investigation Commission on the Fukushima Daiichi Nuclear Accident (by the Rebuild Japan Initiative Foundation)	The report was released on almost a year later, Feb. 28, 2012.
	Fukushima Nuclear Accidents Investigation Committee (by TEPCO)	The interim report was released 9 months later, on Dec. 2, 2011. The final report was released over a year later, on June 20, 2012.

mentioned in Section 12.4.2. Males tend to favor nuclear power technology more than females at all times.

A noteworthy feature of the regulatory response to nuclear accidents in Japan has been the repeated resort to institutional reorganization. After each major (domestic) nuclear power accident in Japan, the regulatory agencies overseeing nuclear power were replaced, combined, or elevated. From the AEC and STA, to NISA and METI, to the new NRA and MOE, Japan has attempted several different institutional models for nuclear power policy. A recurring challenge has been managing the overlapping functions of both promoting nuclear power and also regulating the safety of nuclear power; this overlap may have concentrated expertise but may also have weakened safety regulation. The new NRA is now located in the Ministry of the Environment (MOE) rather than in the Ministry of Economy Trade and Industry (METI), which may give the NRA greater independence; the results for safety remain to be seen. Meanwhile, Japan has also gradually tightened its safety standards, such as by moving from historical experience to PRA, by requiring planning of countermeasures for severe accidents, and by adding (larger) earthquakes, tsunamis, and terrorist attacks to the set of scenarios for which planning is required. After the 2011 Fukushima accident, Japan also shut down all of its 54 nuclear power plants and has yet to restart them, leading in part to greater reliance on fossil fuels (with greater emissions of air pollution and greenhouse gases).

Details of each accident, and efforts to determine the causes, were not fully investigated in this chapter. Accident investigations and reports based on these details, however, may play a role in the event cycle model leading to regulatory changes, including institutional changes (see Balleisen et al., Chapter 17, this volume). Table 12.2 summarizes the investigatory commissions and their reports in Japan for each of the five accidents discussed in this paper. As for the Fukushima Daiichi accident, four accident investigation reports were released in 2012 on the background and responses to the accident, the causes of the accident, and some lessons for policy (NAIIC 2012; Rebuild Japan Initiative Foundation 2012; Cabinet Office, Government of Japan 2012; Tokyo Electric Power Company, Inc. 2012). These investigatory bodies were temporary and have already been terminated, although new evidence continues to come to light following these publications.

References

Cabinet Decision. (2011). The Guideline on Policy Promotion For the Revitalization of Japan. http://japan.kantei.go.jp/topics/2011/20110517_guideline_1.pdf.

Cabinet Office, Government of Japan (2012). Final Report. Investigation Committee on the Accident at Fukushima Nuclear Power Stations of

Tokyo Electric Power Company. www.cas.go.jp/jp/seisaku/icanps/eng/final-report.html.
Energy and Environment Council Government of Japan (2012a). Options for Energy and the Environment. www.cas.go.jp/jp/seisaku/npu/policy09/sentakushi/pdf/Report_English.pdf.
Energy and Environment Council Government of Japan (2012b). Innovative Strategy for Energy and the Environment. www.un.org/esa/socdev/egms/docs/2012/greenjobs/enablingenvironment.pdf.
JAERO (2013). Public Opinion Survey on the unitization of Nuclear Energy in 2012 Fiscal Year. Japan Atomic Energy Relations Organization.
JSCE (2002). Tsunami Assessment Techniques for Nuclear Power Stations. http://committees.jsce.or.jp/ceofnp/system/files/JSCE_Tsunami_060519.pdf.
Kitada, A. (2003). Impact of TEPCO Scandals on Public Opinion toward Nuclear Power Generation; The Third Periodic Survey. *INSS Journal* 10, 44–62 (in Japanese).
Kitada, A. (2006). Longitudinal Survey of Public Opinion on Nuclear Power Generation; The Result of a Survey Conducted One Year after the Mihama Unit 3 Accident. *INSS Journal*, 303–01 (in Japanese).
Kitayama, T. (2008). Innovation of Nuclear Oversight System, in Mabuchi, M. and Kitayama, T., eds. *Policy Process during the Political Realignment.* Jigakusha (in Japanese).
METI (2001). Securing the Safety Basis of Nuclear Power Generations. A Report of the Task Group on Nuclear Safety (in Japanese).
NAIIC (2012). National Diet of Japan Fukushima Nuclear Accident Independent Investigation Commission's Final Report. http://warp.da.ndl.go.jp/info:ndljp/pid/3856371/naiic.go.jp/en/.
NHK Broadcasting Culture Research Institute (2013). Opinion Survey on Nuclear Power and Energy Issues (March 2013) (in Japanese).
Nippon Research Center (2012). National Opinion Survey about Time-Series Change of Attitude toward Nuclear Power after the Great East Japan Earthquake and Attitude toward Saving Electricity and Life (in Japanese).
Nishiwaki, Y. (2011). Transition of Countermeasures against Severe Accidents in Japan-When Did Nuclear Regulation Go Wrong? *Nuclear Eye* 57 (9): 37–40 and 57(10): 40–45 (in Japanese).
Nuclear Emergency Response Headquarters (2011). Report of Japanese Government to the IAEA Ministerial Conference on Nuclear Safety – The Accident at TEPCO's Fukushima Nuclear Power Stations. http://flexrisk.boku.ac.at/zitate/full_report.pdf .

Nuclear Safety Commission (1992). Accident Management as a Preparation for Severe Accidents at Commercial Light-Water Nuclear Reactor Facilities (in Japanese).
Nuclear Safety Commission (1999). *A Summary of the Report of The Criticality Accident Investigation Committee.* www-bcf.usc.edu/~meshkati/tefall99/NSC.pdf (Provisional Translation from the Japanese).
Nuclear Safety Commission (NRC) (2006). *Regulatory Guide for Reviewing Seismic Design of Nuclear Power Reactor Facilities.* www.nrc.gov/docs/ML0803/ML080310851.pdf.
Pew Research Center (2012). Japanese Wary of Nuclear Energy, Disaster "Weakened" Nation, Survey Report. www.pewglobal.org/2012/06/05/japanese-wary-of-nuclear-energy/.
Prime Minister's Office (1968). National Opinion Survey on the Utilization of Nuclear Power for Peaceful Purposes (in Japanese).
Prime Minister's Office (1969). National Opinion Survey on the Utilization of Nuclear Power for Peaceful Purposes (in Japanese).
Prime Minister's Office (1976). National Opinion Survey on Science and Technology, and Nuclear Power (in Japanese).
Prime Minister's Office (1980). National Opinion Survey on Energy Conservation (in Japanese).
Prime Minister's Office (1981). National Opinion Survey on Energy Conservation (in Japanese).
Prime Minister's Office (1984). National Opinion Survey on Nuclear Energy (in Japanese).
Prime Minister's Office (1987). National Opinion Survey on Nuclear Energy (in Japanese).
Rebuild Japan Initiative Foundation (2012). Investigation and Verification Report. Independent Investigation Commission on the Fukushima Daiichi Nuclear Accident (in Japanese).
Shibata, T. and Tomokiyo, H. (1999). Public Opinion on Nuclear Energy – Transition of Nuclear Perception Observed from Public Opinion Surveys. Energy Review Center Publications (in Japanese).
Shiroyama, H. (2010). Current situation and challenges of the Nuclear Safety Commission. *The Jurist.* 1399: 44–52 (in Japanese).
Socioeconomic Productivity Headquarters (2002). Annual Report in 2001 (in Japanese).
Tajima, E. (1974). Who Does Control the Nuclear Administration? *Economist* 52(38): 42–45 (in Japanese).
Togo, Y. (1987). Summary of Report on Chernobyl Accident Prepared by Special Committee in Japan Nuclear Safety Commission. *Journal of the Atomic Energy Society of Japan.* 29(11): 976–81 (in Japanese).

Tokyo Electric Power Company, Inc. (2012). Fukushima Nuclear Accident Analysis Report. Fukushima Nuclear Accidents Investigation Committee. www.tepco.co.jp/en/press/corp-com/release/2012/120563 8_1870.html.

WiN Gallup International (2012). Japan Earthquakes Jolts Global Views on Nuclear Energy. www.nrc.co.jp/report/pdf/110420_2.pdf.

Abbreviations

AEC: Atomic Energy Commission (1956–)
AM: Accident Management
ANRE: Agency for Natural Resources and Energy (within MITI or METI) (1973–)
BWR: boiling water reactors
CDF: core damage frequency
CFF: containment failure frequency
DPJ: The Democratic Party of Japan
FEPC: Federation of Electric Power Companies of Japan
INES: International Nuclear and Radiological Event Scale
JSCE: Japan Society of Civil Engineers
KEPCO: Kansai Electric Power Co. Inc.
LDP: The Liberal Democratic Party
METI: Ministry of Economy, Trade and Industry (2001–)
MEXT: Ministry of Education, Culture, Sports, Science and Technology (2001–)
MITI: Ministry of International Trade and Industry (2001–)
MOE: Ministry of the Environment (2001–)
MOX: mixed plutonium-uranium oxide
NISA: Nuclear and Industrial Safety Agency (within MITI and METI) (2001–2012)
NRA: Nuclear Regulation Agency (within MOE) (2012–)
NSC: Nuclear Safety Commission (1978–2012)
PRA: probabilistic risk assessment
PWR: pressurized water reactors
STA: Science and Technology Agency (1956–2001)
TEPCO: Tokyo Electric Power Co. Inc.
TMI: Three Mile Island

PART IV

CASE STUDIES OF FINANCIAL CRISES

13

Regulatory Responses to the Financial Crises of the Great Depression

Britain, France, and the United States

Youssef Cassis

The regulatory responses to the financial crises of the Great Depression greatly varied between Britain, France, and the United States, the world's leading financial powers in the interwar years – from extremely severe measures in the United States to belated and somewhat milder ones in France, and no regulation at all in Britain, at any rate before the end of the World War II. This is in sharp contrast with the measures that followed the financial debacle of 2007–2008 – without being uniform, they have nonetheless been far more homogenous. How to account for such differences, whether in rhythm, intensity, and actual policy? Everywhere, "lessons" were drawn from the Great Depression, but were they the same lessons?

As can be seen in the essays presented in this volume, regulatory responses to violent crises in general, and financial crises in particular, are the result of a combination of factors – the nature and violence of the shock, its perception, the weight of public opinion, the demand for change, the role of experts, the power struggle between interest groups, the very process of regulatory change.

In the case of the Great Depression, the severity of the financial crises, which in some cases could consist in a succession of shocks of varying degrees of intensity, played of course a decisive role. Equally important was the depth and length of the downturn and, as far as perception was concerned, the connections established between financial crisis and economic crisis. Other factors were also at work: the history of financial crises and the way past crises were handled and debated; the existing regulatory framework, or its non-existence; the economic power, social status, and political influence of the financial elites and the relative antagonism of public opinion towards big business and high finance; the frequency of financial scandals; and the political context within which the financial crises occurred and the regulatory responses were framed.[1]

This chapter intends to take into account most if not all these factors in order to understand the regulatory responses to financial crises in Britain, France and the United States during the Great Depression. The chapter is divided in four parts: the first considers the financial and economic crises in the three countries, the second looks at how they were perceived, the third discusses the regulatory measures that followed, and the fourth concludes. The approach is, as much as possible, comparative. The various factors accounting for regulatory responses are integrated into the text but are more specifically discussed in part three.

I

The Great Depression of the 1930s – sometimes simply referred to as 1929 – remains the most serious economic crisis in modern history and the yardstick by which other downturns have been measured since then. Is it a repeat of 1929? The question has been raised time and again: in 1974, for instance, with the onset of "stagflation" following the first oil shock; in 1982, when Mexico defaulted on its sovereign debt; or in 1987, when the New York Stock Exchange's fall in one day, on October 19 (over 22 percent), was actually higher than in October 1929. And of course, 2008 has been repeatedly compared with 1929 and the ensuing "Great Recession" unanimously judged the most severe contraction to happen since then.

However, on account of its duration and its effects, the Great Depression was on a far grander scale than any other crisis. It consisted of four interrelated shockwaves: the Wall Street crash of October 1929, a series of banking crises occurring over a period of five years, the collapse of the world's monetary order, and an economic slump of dramatic proportions. World industrial output dropped by 36 percent between 1929 and 1932, world trade by 25 percent in volume and 48 percent in value, the price of manufactured goods fell by 26 percent, and that of raw materials by 56 percent.[2] Between 1930 and 1933, the average unemployment rate in manufacturing industry reached 28.4 percent in the United

---

[1] Elke U. Weber highlights the "Social Amplification of Risk" – that is, the social, institutional, and cultural processes that play a role in risk perception (Chapter 4, this volume).
[2] Statistics from the League of Nations cited in Feinstein, Termin, and Toniolo (1997).

States and 34.2 percent in Germany, the two countries most affected by the depression (Bairoch 1997).

Narrowly defined, 1929 was a stock market crash on Wall Street, not a banking crisis. The New York Stock Exchange soared after 1925, partly as a result of the euphoria of the 1920s, but largely because of the progress of the American economy and particularly the rationalization of production and the introduction of new management methods. The top performing shares of the period, such as Radio Corporation of America or General Motors, reflected the companies' profitability and growth prospects. However, a bubble was clearly forming from 1928, especially in the new technologies (Rappoport and White 1994). Public confidence started to be shaken in the summer of 1929 by increasingly clear signs of an impending recession, and share prices started to fall in early October. One thing then led to another. The brokers were overwhelmed by the volume of sales, there were more frequent margin calls, and the ticker tape fell further and further behind the transactions. Without constant information on the share price levels, traders lost track of their positions and panic gripped the Stock Exchange in October 1929 on Thursday the 24th, Monday the 28th, and Tuesday the 29th, after the attempts by the market's main players to stabilize share prices (White 1990). The New York banks, supported by the Federal Reserve, increased their loans and managed to prevent a lack of liquidity and a series of bank failures. Having lost 30 percent since its peak in August, the New York Stock Exchange stabilized in the first few months of 1930, as did the production and employment indices.

The "Great Depression" was not primarily caused by the Wall Street Crash of October 1929. Its main causes lay in the international monetary system of the time, the gold standard, which was a fixed exchange-rate system, and in the monetary policies that this system generated. As Barry Eichengreen has clearly shown (see, e.g., Eichengreen 1992), the first symptoms of the crisis were seen as early as 1928, and were mainly the result of the Federal Reserve's decision to raise interest rates from spring 1928 in order to curb the wave of speculation on the New York Stock Exchange. That decision led to the interruption of America's capital exports and forced the central banks of countries dependent on that capital to adjust their balance of payments, to adopt restrictive policies to defend their currency's parity, and become more deeply entrenched in the crisis. This was typically the case with Germany, as well as other non-European countries such as Australia and Argentina. In the United States itself, the economic downturn reflected in the fall in industrial output in August 1929 became a deep depression, not because of the Wall

Street Crash, though this did not help, of course, but because of the restrictive policy adopted by the Federal Reserve from 1930 and, in particular, 1931 (Eichengreen 1992).

The banking crises of the thirties happened once the economic crisis was well under way, between 1931 and 1934. Unlike the debacle of 2008, they were more caused by, rather than a cause of, the economic crisis, even though they helped to exacerbate the situation. Moreover, these banking crises did not break up simultaneously in all countries – despite the global nature of the Great Depression. Some, like the United Kingdom, escaped them almost completely. In others such as France, the crisis was protracted, but never very acute. Not surprisingly, the most serious banking crises occurred in the countries where the economic depression was most severe, the United States and Germany.

The first banking crisis to hit a major economy took place in the United States, in the autumn of 1930: 256 banks failed in November (with 180 million deposits) and 352 in December (352 million) (Friedman and Schwartz 1963). However, it appears to have been a regional rather than a national crisis, without incidence of panic. About 40 percent of bank failures took place in the St. Louis District, mainly as a result of the collapse of one bank, Caldwell & Co., of Nashville, Tennessee, the largest investment bank in the South. In terms of deposits of failed banks, 45 percent were in two Districts, St. Louis and New York (Wicker 1996). The share of New York, where only six banks failed, was due to the failure of the Bank of United States, a fairly large bank, with 160 million deposits,[3] which collapsed on December 11. A poorly managed bank, as was the Caldwell bank, the Bank of the United States had grown somewhat recklessly in the 1920s and had been insolvent for several months. A plan to rescue it through a merger with four other banks, organized by the Clearing House and the Federal Reserve of New York, came to no avail and, being insolvent, the bank was allowed to fail. Intervention by the Fed prevented panic from spreading to the New York money market, and the impact of the crisis remained limited, though confidence was never fully restored (Wicker 1996; White 1984).

The year 1931 was laden with banking crises during the Great Depression, especially in Europe, but also in the United States.[4] Austria

---

[3] By comparison, the deposits of National City Bank, the United States' largest bank, reached 1500 million in 1930.

[4] See Bordo and James (2009) for a comparative discussion of 1929 and 1931 in connection with 2008.

was the first casualty with the failure of the Credit-Anstalt, in Vienna, in May 1931 (Schubert 1991). The run on Austria was followed by a run on Germany. Panic erupted on July 13, 1931 when the Darmstädter und National Bank (Danat Bank), one of Germany's three largest banks, did not open its doors, and government intervention just managed to recover the situation (Born 1967; James 1984; Balderston 1994; Schnabel 2004).

From Germany, the crisis moved to Britain. But it was a very different type of financial crisis – a currency, not a banking crisis, which alleviated rather than aggravated the economic crisis. It was not the city's big banks that were in danger. They held out well and, unlike in other countries, no major British bank had to close its doors. The merchant banks, on the other hand, were hit harder: with the contraction in international trade, the value of bills of exchange on the London discount market fell from £365 million to £134 million between 1929 and 1933 (Accominotti 2012). The introduction of exchange control in Germany dealt them an even more severe blow. Negotiations with the Weimar government led in September 1931 to an agreement stipulating that interest on German commercial debts would be paid in convertible currencies but the principal would be frozen. Other agreements would follow until August 1939, setting up mechanisms to reimburse a part of the acceptances in Reichsmarks; but the bills would not really be repaid before the 1950s. As in 1914, the most vulnerable banking houses were those that were the most involved in financing German trade and that found themselves with sizeable sums of unconvertible acceptances: by 1936, £9.3 million for Kleinworts – three times the firm's capital – and £4.9 million for Schroders.[5] While these two large houses extricated themselves by drawing on their reserves and by borrowing from the Westminster Bank, smaller ones, which had in fact never fully recovered from the effects of the World War I, like Fredk. Huth & Co. and Goschen & Cunliffe, disappeared from the scene.

But Britain's real crisis was that of the pound sterling, undermining for good the foundations that had underpinned its dominant position for more than a century (Sayers 1976; Cairncross and Eichengreen 1983; Kunz 1987; Wiliamson 1992). Pressure on the pound had been almost constant since its return to the gold standard in 1925; but it intensified from 1929, owing to a loss of confidence in a British currency generally considered overvalued, with the Bank of England having to face

---

[5] This situation brought about appreciable losses that reached £1.5 million for Schroders in 1931 and, probably, £3.3 million for Kleinworts in 1935 (Roberts 1992, 244–249).

increasingly frequent withdrawals of gold. The situation worsened during the summer of 1931 under the triple effect of the German crisis, the gap between foreign short-term claims on London and the British gold and currency reserves, and the extent of the budget deficit. The crisis took on a political dimension when the Labour government resigned on August 24, faced with the need to reduce its budget deficit in order to be able to obtain two loans, totaling £80 million, from the United States and France. A government of national unity, still chaired by the Labour Prime Minister, Ramsay MacDonald was formed, but the announcement of a new budget on September 10 did not soothe anxieties. Withdrawals of funds continued, and on September 21, 1931, the British government suspended the pound's gold convertibility, which henceforth floated in relation to the main currencies. It was immediately followed by twenty-five countries, mainly from the British Empire, Eastern Europe, Scandinavia, as well as key trading partners like Portugal and Argentina. Britain's abandonment of the gold standard marked the end of an era – the long nineteenth-century world economic order centered on Britain and the pound sterling. But its effects were highly beneficial to the British economy, as the Bank of England, relieved from its duty to defend the pound's parity, could embark on a policy of cheap money. Britain was the first country to emerge from the depression, unlike the countries that remained on gold.

The effects of the pound's exit from gold were rapidly felt in the United States. A second banking crisis had already hit the country between April and August 1931, with 563 banks failures, totaling 497 million deposits. However, it had remained fairly localized, with four federal districts accounting for about three-quarters of the casualties: Chicago, in the first place (one-third), as well as Minneapolis, Cleveland, and Kansas City (Wicker 1996). The third crisis, in September and October, was more serious – 817 banks failed, with deposits amounting to 747 million – and of a more nationwide character, despite the high share of three districts (Chicago, Philadelphia and Cleveland) and the absence of crisis in New York. Pressure on the dollar and gold losses – more than 369 million between September 21 and October 8 – led the Federal Reserve to raise its discount rate from 1.5 to 2.5 percent on that date, two and a half weeks following Britain's departure from gold, and to 3.5 percent one week later. The suspension of the pound's convertibility appears to have aggravated the banking crisis in the United States – the number of banks' failures increased significantly in October – mainly because, coming on top of the Austrian and German crises, it further undermined a badly shaken confidence in the financial system. However, the situation eased up at the end the year, as confidence was this time boosted by the announcement by President Hoover, on October 13, of the formation of the National Credit Corporation, whose objective was to rediscount banks' liquid but frozen assets (Wicker 1996).

Banking crises in the United States did not end in 1931. The number of bank failures remained comparatively high in 1932, without, however, the outbreak of a major panic. The open-market policy of the Federal Reserve and the loans by the Reconstruction Finance Corporation – a government-owned agency created in January 1932 to replace the National Credit Corporation, conceived as a banks' voluntary association – eased the situation. However, this action backfired in January 1933 when a decision by Congress required the Reconstruction Finance Corporation to publish a list of all the loans it had made the previous year. Such loans were seen as a sign of weakness and led to bank runs. The crisis was met with moratoria and bank holidays starting, on a small scale, in Nevada in November 1932, followed by Iowa and Louisiana. But it was the decision by the governor of Michigan to declare, on February 14, 1933, a state-wide bank holiday that sparked a panic, which, in less than three weeks, completely paralyzed the America banking system (Kennedy 1973). Moratoria and bank holidays accentuated the drain on banks in other states, not least in New York, where leading banks were dangerously dragged into the crisis. National City Bank lost 33 percent of its correspondents' deposits and 12.5 percent of its domestic deposits in February 1933 (Van B. Cleveland and Huertas 1985). Roosevelt's uncertainty about a possible devaluation of the dollar during his long interregnum – he was elected on November 8, 1932 – exacerbated the situation. By the time of his inauguration on March 4, 1933, a total of 48 states had imposed banking restrictions, with banks being closed in 33 of them. On March 6, Roosevelt declared a national bank holiday; at his first press conference on the 8th, he stated clearly that the dollar would remain on the gold standard – it was eventually devalued on April 19 – and on the 9th, Congress passed the Emergency Banking Act, which provided the legal basis for reopening the banks – a series of measures which restored confidence. One half of the country's banks, with 90 percent of total banking resources, were judged capable of reopening for business on March 15, 1933, and their soundness guaranteed by the government; another 45 percent were to be reorganized before being licensed; and the remaining 5 percent (about 1,000 banks) would have to close permanently (Wicker 1996).

In total, more than 10,000 banks, out of nearly 25,000 in existence in 1929, closed their doors until 1933.

The banking crisis was protracted but less acute in France, the other financial power of the day. There was no real panic, despite the collapse of two big banks and a score of small regional and local banks. Altogether, 670 banks failed between October 1929 and September 1937 (Laufenburger 1940). Most of them were small, often family-owned local and regional banks. They played a dynamic role in the French economy during the 1920s, especially through their support of local industry. However, unlike the big deposit banks, which remained highly liquid throughout the period, they had overcommitted themselves during the years of prosperity, becoming extremely vulnerable in times of crisis. Their difficulties were aggravated by the attitude of the local branches of the Banque de France, as some of them dramatically tightened the conditions at which they were prepared to rediscount bills. Amongst the big banks, the Banque Nationale de Crédit, the country's fourth largest bank, suspended payments in 1931: it had lost about 20 percent of its deposits between January and August when a run, starting in October, brought the loss to three quarters by the end of the year (Bonin 2002a).[6] In order to prevent a panic, depositors benefited from a state guarantee. The bank was built up again the following year, with help from the state, under the name of Banque Nationale pour le Commerce et l'Industrie. The Banque de l'Union Parisienne, the second largest investment bank, shaken by the crises in Germany and central Europe, experienced serious difficulties that brought it to the brink of bankruptcy in 1932; but it was saved by the joint intervention of the Banque de France and the main Parisian banks (Bonin 2002a).

Despite their global character, with contagious effects spreading across the globe, the financial crises of the Great Depression retained clear national characteristics, determined as much by domestic as by international conditions.

II

The "never again" feeling is a good indicator of the perception of a financial crisis – of its severity, the risks of relapse and how to prevent it. Perhaps surprisingly, this "never again" feeling has rarely been present

---

[6] From about 3,700 million francs in August 1931, its deposits had fallen to under 1,000 million by the end of the year.

in financial crises in advanced economies since the late nineteenth century, probably because very few of them had really devastating effects – whether in economic, political, or human terms (Cassis 2011). The exception is of course the Great Depression, on account of both the strength of the feeling and its durability.

The Great Depression has been analyzed time and again by economists and historians. As Ben Bernanke put it: "To understand the Great Depression is the Holy Grail of macroeconomics" (Bernanke 2000). This status of the Great Depression as a one of the defining moments in the world's history somewhat blurs the distinction between the perception of contemporaries and that of ensuing generations. The financial regulations inherited from the Great Depression go beyond the immediate measures taken as a response to the banking crises, whatever their significance, and include the war and even the postwar years. The same goes with the analyses and interpretations of the crisis. Not only has the Great Depression become the reference against which subsequent crises have been judged, it has also altered the way subsequent crises have been dealt with – not least thanks to a far better understanding of economic and financial crises, from Keynes onwards.

When considering how the financial crises of the Great Depression were perceived by contemporaries, two points should be borne in mind. First, these crises were only part of a much broader phenomenon, however intricate their links with the dire state of the economy. A first, general reaction was the sense of the sheer enormity of the depression[7] – the worldwide dimension of the crisis, the human misery of mass unemployment, the apparent collapse of capitalism, and also the rise of political extremism and growing international instability. The second is that, whatever the global character of the slump, it was felt very unevenly across countries, and analyses greatly differed as to the causes, nature, and impact of the financial crises.

These differences were not only due to the severity of the financial crises, already discussed above, but also to their effect on the "real" economy. The United States was hardest hit, with a level of GDP in 1938 still 12 percent lower than in 1929, whereas it had grown by

---

[7] In this volume, for example, Elke U. Weber (Chapter 4) and Frederick W. Mayer (Chapter 6) each detail the extent to which rare and extreme events impact risk perceptions and, potentially, policy. Whereas Weber's focus is the impact of such events on risk perceptions, especially among the general public, Meyer details the ways in which the media and policy entrepreneurs can use dramatic events as part of a narrative designed to achieve specific policy aims.

10 percent in Britain. GDP per head fell by more than 30 percent in the United States between 1929 and 1933. In Britain, it had already reached its lowest point in 1931, having fallen by 6.5 percent before the economy started to recover following the exit from the gold standard. France was in between, with GDP in 1938 still below (by 4 percent) its 1929 level, and GDP per head falling by 16 percent between 1929 and 1932; but unemployment remained consistently lower than in the two Anglo-Saxon powers – 0.6 percent between in 1930–1933, as against 14.7 percent in Britain and 18.2 in the United States; and 3.5 percent in 1934–1938, as against 8.7 percent in Britain and 18.3 percent in the United States (Bairoch 1997).

The perception of the nature of the financial crises and the causes of the depression was also different. Far more emphasis was put on the stock market crisis in the United States than in either Britain or France. This is not surprising given the role played by the Wall Street crash of 1929 in the narrative of the financial crisis and the Great Depression – the excesses of the 1920s, characterized by the euphoria of growth, speculative fever, inexperience, and also, at times, fraud, and the inevitable, and spectacular, bubble burst (Galbraith 1954). No such crash took place in either the London Stock Exchange or the Paris Bourse, which were not really part of the drama. But it also true that the stock market fell much deeper in the United States: in New York, the market was down 89 percent from its 1929 peak when it bottomed out in July 1932, while in Paris, stocks had fallen by 57 percent during the same period; they didn't reach their lowest point before August 1936, soon after the signature of the Matignon agreements, having by then lost three-quarters of their value since 1929. In Britain, shares continued to fall until June 1932, but never lost more than a third of their value (Sauvy 1984). Moreover, with the democratization of share ownership in the United States, losses were borne by a much wider investing public than in France or even Britain. There might have been as many as 15 million holders of securities in America in 1925 (the highest estimate), though no more than 600,000, or barely 4 percent, were believed to own 75 percent in value. Nevertheless the figure of the small investor held a central place in the collective imagination of the day, and his or her participation in subscriptions was sought after, as shown by the low value of various shares.

The financial scandals that shocked American public opinion in the 1930s and shaped its vision of the crisis tended to revolve around stock market operations and the deceit of small investors by unscrupulous

Regulatory Responses to the Great Depression 359

bankers.[8] This was clearly the case during the Senate investigation into banking and stock exchange practices, which greatly contributed to the pillorying of Wall Street's leading bankers and financiers, with its findings echoed in the popular press and resonating with the public. The hearings went on for two years, from April 1932 to May 1934, and took a particularly energetic turn with the appointment, in January 1933, of Ferdinand Pecora as counsel of the Senate Committee on Banking and Currency. Witnesses included Charles Mitchell, chairman and former president of National City Bank and its investment bank affiliate, National City Company; Albert Wiggin, his counterpart at Chase National Bank and Chase Securities Corporation; Jack Morgan, senior partner of J.P. Morgan & Co., and Thomas Lamont, the firm's foremost partner; Albert Kahn, a senior partner in Kuhn Loeb & Co., Wall Street's second largest investment bank; Clarence Dillon, senior partner of Dillon Read & Co., another top investment bank, which rose to prominence in the 1920s; Richard Whitney, chairman of the New York Stock Exchange; and others (Carosso 1970; Fraser 2005; Chernow 1990; van B. Cleveland and Huertas 1985).

These Wall Street grandees were not held responsible for bringing their banks to the brink of collapse: the major banks had indeed survived four banking crises and, however weakened, still stood firm in the depth of the depression. They were mainly accused of malpractices judged harmful to investors and condemned by a public opinion increasingly hostile to the financial world. Abuses, breach of trust, and self-enrichment, in various guises, were recorded. National City Company, for example, advised investors to buy South American stock (especially two loans for Minas Geraes, a state in the Brazilian Republic, and three Peruvian government loans) that it had issued, and knew to be low grade, with no information about risks or market conditions. Kuhn Loeb issued investors in a holding company (the Pennroad Corporation) not with actual shares but with certificates carrying no voting rights. Dillon Read structured the capital of two of its investment trusts (United States and Foreign Securities Corporation and United States and International Securities Corporation) in such a way as to exert complete control with less than 10 percent of their capital. J.P. Morgan allocated the stock of one its holding companies (the Alleghany Corporation) to a "preferred list" of

---

[8] The importance of identifiable victims and villains to form symbols in narrative framing is discussed elsewhere in this volume in more detail. See, e.g., Mayer's Chapter 6, this volume; Birkland and Warnement's Chapter 5, this volume.

influential friends. National City Bank, through its affiliates, took part in stock pools, especially in copper; and National City Company was used to trade in the stock of its parent company in order to prop it up. Albert Wiggin sold short the shares of his own bank, Chase National, in the midst of the stock market crash, netting four million dollars. Charles Mitchell, whose earnings exceeded one million dollars in 1929 (his basic salary was 25,000) did not pay any income tax in that year, nor did Jack Morgan and his partners in 1931 and 1932 (Carosso 1970).

Financial scandals took a more political turn in France. One of the most resounding was the failure of the Banque Oustric in November 1930 and the revelations of corruption of several politicians. First in line was Raoul Péret. As finance minister in 1926, he authorized the listing on the Paris Bourse of Snia Viscosa, an Italian synthetic fibers company, days before leaving his post. And he became legal adviser to the Banque Oustric the following year. Obtaining such a listing was extremely difficult, due to the government's efforts to check capital flight. Péret's decision, taken against the recommendation of his advisers, marked the beginning of the big time for Albert Oustric, the son of a café proprietor in Carcassonne. His standing rose considerably, he gained control over a number of companies, mainly through the purchase of founders' shares, and was admitted to the board of some of France's leading companies, including the Automobiles Peugeot, which were to suffer heavy losses as a result of their involvement in his affairs. Oustric was becoming increasingly reckless, paying inflated prices to gain control of an ever-growing number of businesses. The bubble burst with the fall of the stock market, and on November 7, 1930, the Banque Oustric went into liquidation, dragging other banks down with it, whose resources had been heavily drawn upon by Oustric.

The "Stavisky affair" broke out a few years later, on January 8, 1934, when the body of Alexandre Stavisky, a Ukrainian-born businessman with connections in political and media circles, was discovered in a chalet in Chamonix. A few days earlier, on December 24, 1933, the manager of the Crédit municipal de Bayonne, a semi-public local bank, had been arrested. He was involved with Stavisky in the issue, through his bank, of false bonds amounting to a total of 200 million francs, all with the tacit agreement of the deputy and mayor of Bayonne, Dominique-Joseph Garat; another bank linked to Stavisky, the Crédit municipal d'Orléans, had also issued 70 million francs of such bonds. Thanks to his political connections, Stavisky had been able to place these bonds, which did not enjoy any state guarantee, with insurance companies.

Moreover, despite being charged for fraud and forgery, Stavisky was not being tried, thanks to support in high places involving the left of center Radical Party in power. On February 6, 1934, a demonstration against "the thieves," organized by the extreme right parties, ended in 16 deaths after the guard fired at the crowd, and led to the resignation of Radical Prime Minister Edouard Daladier, intensifying political tensions in France.

In Britain, the revelations of frauds and embezzlement, such as those following the collapse of the company promoter Clarence Hatry in 1929, remained primarily financial scandals, despite heavy losses suffered by shareholders.

In the end, Britain was more concerned with the fate of the pound than its fairly stable banking system. In France, politico-financial scandals tended to overshadow a lingering though never really acute banking crisis. In the United States, the banking system remained, as in 1907, the main culprit, despite the creation, in the interval, of the Federal Reserve, which was supposed to have stabilized it. More than elsewhere, the blame was put on the speculative excesses of the 1920s and the financial crisis, and more generally the depression, linked to the Wall Street crash of October 1929 (Fraser 2005).

III

Regulatory measures were adopted in different conditions in the United States, France, and Britain. In the United States, they almost immediately followed the high point of the banking crisis in March 1933: less than four months after the bank holiday declared by Roosevelt, the Securities Act, on the capital market, and the Banking Act, on banks, had become effective. They were completed by further legislation in the following years, all in a climate of intense and widespread hostility towards Wall Street, including in the White House. In France, legislation was passed several years later, in 1941, after the country had been defeated and was occupied by Germany, transferred its capital to Vichy, and promoted a "national revolution." Nevertheless, similar measures had been envisaged by the Popular Front government in 193–1938, reflecting a broad consensus towards reform. And the banking law was renewed, with minor amendments, after Liberation in 1944.

In the United States, a series of radical reforms were introduced within the framework of the New Deal. The Securities Act, passed in 1933, contained various provisions aimed at improving the quality of information about the securities offered and traded on the stock exchange. The Banking Act, better known under the name of its two promoters as the Glass-Steagall Act, decreed the complete separation of commercial banking activities (taking deposits and making loans) from investment banking activities (issuing, distributing, and trading securities), including if these activities were shared between parent companies and subsidiaries or through either the cross-holding of shares or overlapping directorships; it also introduced federal deposit insurance.

Other laws completed this New Deal legislation, in particular the Securities Exchange Act of 1934, which created the Securities and Exchange Commission (SEC); the Banking Act of 1935, which reformed the Federal Reserve System, centralizing the conduct of monetary policy with the seven members of the Board of Governors in Washington; and the Investment Companies Act of 1940, which codified the rules governing investment companies.

While the Securities Act had little incidence on the activities of investment banks, the Banking Act transformed the banking profession in the United States: the commercial banks parted from their subsidiaries involved in securities transactions, whereas the vast majority of private banks opted for investment banking. The major exception was J.P. Morgan & Co, Wall Street's most famous bank, which chose to become a commercial bank, a decision that led several partners to resign and to found an investment bank, Morgan, Stanley & Co (Chernow 1990). Federal deposit insurance was compulsory for banks that were members of the Federal Reserve System, but optional and conditional for the others. The insured institutions paid a premium based on a percentage of their total assets, as a contribution to a guarantee fund intended to pay the depositors of a bankrupt bank. Six months after voting on the law, 14,000 banks had already decided to insure their clients, for a maximum sum of 5,000 dollars per deposit. Another federal regulation – Regulation Q – set a maximum interest rate that the banks could pay on savings deposits. Despite its impact on American banking, the separation between commercial and investment banking did not really address the main cause of the banking crises: most of the small banks that failed between 1930 and 1933 were only commercial banks, and they failed because of the depression, their fragility, and the failure of the Federal Reserve to come to their aid. Conversely, the large New York and major city banks, which had securities affiliates, survived the crisis. And while there might have been conflicts of interests between commercial and investment banking, there is no evidence that the banks engaged in the two activities took more risks in the sort of securities they underwrote and marketed than specialized investment banks (White 1986; Kroszner and Rajan 1994). On the other end, the introduction of deposit insurance, in other words the government commitment to make banks safer to depositors, could justify a measure limiting risk taking, though as a federal measure, it had long met with strong opposition. How then to account for these regulatory responses? The decision was the result of a number of interrelated factors, including the understanding of the banking crises, the impact of public opinion, the political climate, and the role of key individuals.

The severity of the banking crises left most commentators, regulators and practitioners at first incredulous and then mainly helpless in the face of the disaster. Proposals for reshaping the financial system were influenced by previous experiences of banking crises and banking reform and prevailing opinions in matters of banking practices. In that respect, the orthodox view regarding a sound banking system was still very much influenced by the "real bills doctrine" and based on the functioning of the English banking system, where commercial banks confined themselves to short-term loans for commercial purposes and didn't engage in any type of investment banking activities. This was, in particular, the view held by H. Parker Willis, a professor at Columbia University and one of country's leading experts on financial matters. Pushed somewhat to the background by the move of American banks towards a "universal" banking model and the boom of the 1920s, the old banking orthodoxy strongly reasserted itself in the aftermath of the stock market crash of 1929, blaming speculative excesses on deviations from this line. Despite the changes that had taken place in American banking for over a quarter of a century, a new banking theory had yet to emerge (Perkins 1971). The alternative was to more tightly regulate the existing system, in particular the investment banking affiliates of the commercial banks – as advocated by a number of banks' representatives.[9] It didn't prevail in large part because of the severity of the crisis, which strongly fueled anti-banker feelings and called for more radical solutions.

The sequels to the previous financial crisis, the panic of 1907, also played a role in the framing of the new regulations. In 1912, the Pujo Committee – named after its chairman Arsène Pujo – was appointed to investigate the heavy involvement of bankers on the boards of manufacturing companies

---

[9] This was for example the position of L.E. Wakefield, the chief executive of the First Bank of Minneapolis, a strong advocate of government regulation in the investment banking field (Perkin 1971, 514).

and the concentration of issues in the hands of a few investment banks. The inquiry findings might have been inconclusive, yet by then a sizeable part of the population appeared to believe in the existence of a "money trust" controlling the country's business (Fraser 2005). The conviction that financial power was excessively concentrated was reinforced by the Wall Street crash, the banking crises and the depression, providing further arguments in favour of splitting banking activities.

The press amplified these analyses, though only to a certain extent.[10] Of the "Big Five" newspapers (*Chicago Tribune, Los Angeles Times, New York Times, Wall Street Journal,* and *Washington Post*), only the *New York Times* tended to publish articles on proposed legislation, painting government actors in positive light and hardly delving into Wall Street's side of the story.[11] Conversely, the *Wall Street Journal* published more articles favoring the status quo and denying the need for regulation from Washington.[12] These were, not surprisingly, the two newspapers most interested in covering financial matters, especially the regulation debate. *The Chicago Tribune, Los Angeles Times* and *Washington Post*, were more centrist, publishing articles focusing primarily on facts rather than policy; and when policy issues were discussed, articles tended to cover both sides of the story roughly equally. On the other hand, the press coverage of the Pecora inquiry undoubtedly had a strong impact on the outcome of the banking legislation, exacerbating anti-banker sentiments and forcing politicians to take greater account of public opinion (Calomiris and White 2000).

Interest groups were thus overshadowed by public opinion and in any case, many commercial and investment bankers were increasingly favourable to the new legislation. The former, with little in the way of business in investment banking, saw an opportunity to reduce their costs and offload their investment banking activities; while the latter were increasingly fearful of the competition from commercial banks' affiliates (Bentson 1990). Even the leaders of the large integrated banks, not least National

---

[10] I am grateful to Joe Oehmke of Duke University for conducting the research in the Historical Newspaper Archives.

[11] See, e.g., any of the articles about Fletcher; Norbeck Blames the Stock Market, http://search.proquest.com/docview/99710822/13EBF97AFBA5FF3573A/27?accountid=10598; Urges Real Value of Stock Be Listed, supra; and similar.

[12] See, e.g., Irving Fisher Back Mitchell, *WSJ*, Apr. 9, 1929, http://search.proquest.com/docview/130641603/13EC00C3CAE74D3DFA3/3?accountid=10598; Hamilton Talks on Wall Street, *WSJ*, Oct. 19, 1929, http://search.proquest.com/docview/130756410/13EC00C3CAE74D3DFA3/10?accountid=10598; Whitney's Views on Trading Bill, supra.

Citibank and Chase National Bank, anticipated the trend and announced the demerger of their security affiliates before the passing of the Act.

On the political stage, the narrative of the crisis emphasizing bankers' speculative excesses was strongly supported by the Democratic Senator Carter Glass – the main architect of the Federal Reserve in 1913, a former secretary of the Treasury, and widely seen as the preeminent expert on banking and financial matters in the Senate. His long-term adviser was the economist H. Parker Wills. For Carter Glass, security affiliates led to unsound practices on the part of commercial banks and should be separated from them. His proposal was part of the Democratic campaign in 1932 and was personally endorsed by Roosevelt. The November election thus marked a decisive shift in favor of the new regulation (Perkins 1971).

The other narrative of the crisis related to the losses suffered by depositors, which mounted to unprecedented levels between 1929 and 1933 and spread across the entire country (Calomiris and White 2000). This considerably strengthened the position of the supporters of deposit insurance, for whom it appeared profoundly unjust that depositors should bear the losses caused by bankers' malpractices. Such feelings reached their paroxysm in early 1933, with the deepening of the banking crisis and the hearings of the Pecora inquiry. Yet there were still fierce opponents to deposit insurance, including President Roosevelt, his Secretary of the Treasury William Woodin, Senator Carter Glass, and the American Bankers' Association, mainly on the ground that it would encourage moral hazard.[13] The main advocate of deposit insurance was the Democrat Representative Henry Steagall, chairman of the Committee on Banking and Currency, and a staunch supporter of small single unit banks, seen as the prime beneficiaries of deposit insurance.

The Glass-Steagall Act was the result of a political compromise, as the separation between commercial and investment banking would only be voted in the House of Representatives if the bill included deposit insurance, and vice versa. Both measures had been contemplated for a long time. Both measures, especially deposit insurance, appeared as a solution to prevent the recurrence of banking crises. Yet only exceptional circumstances could lead to their adoption.

Financial regulation did not go so far in the major European economies, including Germany, where universal banking managed to survive. In France, the law introduced by the Vichy government, upheld and

---

[13] Bankers also considered that deposit insurance meant that well-managed banks would subsidize poorly run ones (Calomiris and White 2000).

completed in 1944, controlled and regulated banking activities that until then had been open to any newcomers. Henceforth, banks had to be registered according to their type of activity. The law made a clear distinction between an investment bank and a deposit bank. However, this distinction had a long tradition in France and was often referred to as the "doctrine Henri Germain," from the name of the founder and chairman of the Crédit Lyonnais, who shaped it in the wake of the financial crisis of 1882. The new law also defined a number of specialized institutions, according to their operations or clientele, including finance companies and discount houses. One of the consequences of the new banking law, and of the anti-Semitism of the regime, was the disappearance of specialized private banks, mostly Jewish. Henri Ardant, the leading banker under Vichy and guiding spirit behind the reform, attempted to rationalize the banking system by getting rid of what he called its "most sordid" elements – no less than 63 percent of French banks. However, small banks still enjoyed protection in Vichy France and the number of local banks only fell by 22 percent (from 262 to 205) between 1940 and 1944 (Andrieu 1990).

In France, the regulatory response was thus not a response to the banking a crisis. It was a belated acceptance of the need to regulate the banking industry after some seventy years of inconclusive and highly politicized debates – since the beginning of the Third Republic in 1870 (Andrieu 1990). Until the 1941 legislation, banking remained totally unregulated in France. Even the accession to power of the Popular Front in June 1936, at the height of the depression, did not change the situation, despite the left's declared aim of "controlling the economy." The Popular Front acted decisively regarding the central bank, swiftly reforming the status of the Banque de France – though coming short of nationalizing it – in order to reduce the influence of the regents, its directors who mostly belonged to the *haute banque* and symbolized the enduring power of the "200 families." On the other hand, the Popular Front dithered about the commercial banks, which were fiercely opposed to any form of legislation. Significantly, they contended that the relatively mild character of the banking crisis in France vindicated the smooth working of the country's banking system. A committee charged with investigating the "organization of credit" was appointed by the government and convened between November 1936 and June 1937, but saw little cooperation from the banks, was violently attacked by the right-wing press, and never published a report (Andrieu 1990). When financial regulation was eventually introduced in 1941, it was as a result of the trauma of defeat and occupation rather than as a response to the financial crisis of the 1930s.

Britain, for its part, steered clear of the trend towards greater regulation of the banks, probably because there had not been any bank bankruptcies during the 1930s, the financial system was more specialized than elsewhere, and the Bank of England effectively monitored it to ensure that it was working properly.

IV

With the exception of the United States, the measure of financial regulation taken in the wake of the Great Depression appears relatively mild and little intrusive. As far as European countries were concerned, state intervention in banking affairs was as much a result of the depression as a consequence of the economic and political context of the 1930s and, even more, World War II.

Britain is a case in point. From an informal regulatory framework, mainly based on the personal suasion of the Bank of England's Governor, it emerged from the war with a nationalized central bank, though its governing structure at first remained virtually unchanged; and clearing banks still in the private sector but under the Treasury's and the Bank of England's strict control – as Keynes put it, in no need to be nationalized as in actual fact they already had been. The London Stock Exchange was not only tightly regulated by the authorities, but its dealings were regarded with suspicion – options, considered highly speculative, were only reintroduced in May 1958 after an interruption of 19 years. In effect, the London Stock Exchange set itself up as the regulator of the securities market, with all the caution and conservatism that that implies (Michie 1999).

In France, most of the financial sector came under state control after Liberation. The Bank of France was nationalized, together with the four big deposit banks (Crédit lyonnais, Société générale, Comptoir national d'escompte de Paris, and Banque nationale pour le commerce et l'industrie) and all major insurance companies – though the *banques d'affaires*, led by the Banque de Paris et des Pays-Bas, remained in private hands. The state's grip ended up stifling the Parisian capital market, not only when it came to foreign issues, but also issues by French companies. The Paris Bourse became pretty sluggish. Having lost two-thirds of the nominal value of its securities through nationalizations, it went through a "long depression" that lasted until the 1980s (Straus 1992; Feiertag 2005).

The "lessons" from the Great Depression must thus be put in their proper context. The regulations that characterized the third quarter of the twentieth century were the result of an exceptional historical period marked by two world wars, the redrawing of international boundaries, a devastating economic crisis, massive political upheavals, and shifts in ideological outlooks – the "Thirty Years War" of the twentieth century. This led to an ideological shift, which, combined with a generational change, favored state intervention and a more organized form of capitalism. Even in the United States, where the regulatory framework was essentially set up in the wake of the crisis, the Glass-Steagall Act appears to have been more ideological than pragmatic and was rooted in Americans' instinctive distrust of financial concentration and power, already denounced by the Pujo Commission of Enquiry at the beginning of the century. Reactions to the crisis, in particular the necessity to do something, were thus motivated by a combination of economic and political considerations and, despite national differences, had their roots in a common Zeitgeist – a sense that solutions should be found in state intervention rather than market mechanisms.

References

Accominotti, O. 2012. "London Merchant Banks, the Central European Panic and the Sterling Crisis of 1931," *Journal of Economic History*. 72(1): 1–43.

Andrieu, C. 1990. *Les banques sous l'occupation. Paradoxes de l'histoire d'une profession*. (Paris: Les Presses de Sciences).

Bairoch, P. 1997. *Victoires et déboires. Histoire économique et sociale du mode du XVIe siècle à nos jours*, vol. III. (Paris: Gallimard).

Balderston, T. 1994. "The Banks and the Gold Standard in the German Financial Crisis of 1931," *Financial History Review*. 1(1): 43–68.

Bentson, G. 1990. *The Separation of Commercial and Investment Banking: The Glass-Steagall Act Revisited and Reconsidered*. (Oxford: Oxford University Press).

Bernanke, B. 2000. *Essays on the Great Depression*. (Princeton: Princeton University Press).

Bonin, H. 2002a. *La Banque nationale de crédit. Histoire de la quatrième banque de dépôts française en 1913–1932*. (Paris: PLAGE).

Bonin, H. 2002b. *La Banque de l'union parisienne (1874/1904–1974). Histoire de la deuxième banque d'affairs française*. (Paris: PLAGE).

Bordo, M. D. and H. James (2009). The Great Depression Analogy. NBER Working Paper 15584. www.nber.org/papers/w15584.

Born, K. E. 1967. *Die Deutsche Bankenkrise 1931*. (Munich: R. Piper and Co.).

Cairncross, A. and B. Eichengreen. 1983. *Sterling in Decline: The Devaluations of 1931, 1949, and 1967*. (Oxford: Oxford University Press).

Calomiris, W. and E. N. White. 2000. "The Origins of Federal Deposit Insurance," in C. W. Calomiris ed., *U.S. Bank Deregulation in Historical Perspective*. (Cambridge: Cambridge University Press): 193–200.

Carosso, V. 1970. *Investment Banking in America: A History*. (Cambridge, MA: Harvard University Press).

Cassis, Y. 2011. *Crises and Opportunities. The Shaping of Modern Finance*. (Oxford: Oxford University Press).

Chernow, R, 1990. *The House of Morgan: An American Banking Dynasty and the Rise of Modern Finance*. (London: Grove Press).

Eichengreen, B. 1992. *Golden Fetters. The Gold Standard and the Great Depression 1919–1939*. (Oxford: Oxford University Press).

Feinstein, C., P. Temin, and G. Toniolo. 1997. *The European Economy between the Wars*. (Oxford: Oxford University Press).

Feiertag, O. 2005. "The International Opening Up of the Paris Bourse: Overdraft-Economy Curbs and Market Dynamics," in Y. Cassis and E. Bussière (eds.), *London and Paris as International Financial Centres in the Twentieth Century*. (Oxford: Oxford University Press): 229–46.

Fraser, S. 2005. *Wall Street: A Cultural History*. (London: Faber and Faber).

Friedman, M. and A. J. Schwartz. 1963. *The Great Contraction, 1929–1933*. (Princeton, NJ: Princeton University Press).

Galbraith, J. K. 1954. *The Great Crash. 1929*. (Boston: Mariner).

James, H. 1984. "The Causes of the German Banking Crisis of 1931." *Economic History Review*, 37(1): 68–87.

Kennedy, S. E. 1973. *The Banking Crisis of 1933*. (Lexington: University Press of Kentucky).

Kroszner, R. S. and R. G. Rajan. 1994. "Is the Glass-Steagall Act Justified? A Study of the U.S. Experience with Universal Banking before 1933," *American Economic Review*, 84(4): 810–32.

Kunz, D. B. 1987. *The Battle for Britain's Gold Standard in 1931*. (London: Routledge).

Laufenburger, H. 1940. *Les banques françaises*. (Paris: Librairie du Recueil Sirey).

Michie, R. 1999. *The London Stock Exchange. A History*. (Oxford: Oxford University Press).

Perkins, E. J. 1971. "The Divorce of Commercial and Investment Banking: A History," *The Banking Law Journal.* 88(6): 501–26.

Rappoport P. and E. White. 1994. "Was the Crash of 1929 Expected?" *American Economic Review.* 84(1): 271–81.

Roberts, R. 1992. *Schroders: Merchants and Bankers.* (London: Palgrave Macmillan).

Sauvy, A. 1984. *Histoire économique de la France entre les deux guerres,* vol 3. (Paris: Economica).

Sayers, R. S. 1976. *The Bank of England, 1891–1944.* (Cambridge: Cambridge University Press).

Schnabel, I. 2004. "The German Twin Crisis of 1931," *Journal of Economic History.* 64(3): 822–71.

Schubert, A. 1991. *The Credit-Anstalt Crisis of 1931.* (Cambridge: Cambridge University Press).

Straus, A. (1992). "Structures financiéres et performances des entreprises industrielles en France dans la seconde moitié du XXe siècle," *Entreprises et Historie,* 2: 19–33.

Van B. Cleveland, H. and T. F. Huertas. 1985. *Citibank 1812–1970.* (Cambridge, MA: Harvard University Press).

White, E. 1984. "A Reinterpretation of the Banking Crisis of 1930," *Journal of Economic History.* 44(1): 119–38.

White, E. 1986. "Before the Glass-Steagall Act: An Analysis of the Investment Banking Activities of the National Banks," *Expolorations in Economic History.* 23: 33–54.

White, E. 1990. "The Stock Market Boom and Crash of 1929 Revisited" *Journal of Economic Perspectives.* 4(2): 67–83.

Wicker, E. 1996. *The Banking Panics of the Great Depression.* (Cambridge: Cambridge University Press).

Williamson, P. 1992. *National Crisis and national Government: British Politics, the Economy and Empire, 1926–1932.* (Cambridge: Cambridge University Press).

14

Financial Decommodification

Risk and the Politics of Valuation in US Banks

Bruce G. Carruthers

During the Global Financial Crisis of 2008, the American Bankers' Association (ABA) publicly pressured the Financial Accounting Standards Board (FASB) and the Securities and Exchange Commission (SEC) to suspend and/or modify "fair value" accounting rules. The gist of the ABA's argument was that continuing to mark the accounting value of bank assets to current market values had introduced unhealthy volatility into the financial system and had severely undermined the already precarious stability of US banks. Given the extraordinary economic conditions in 2007–2008, financial markets were said to be "illiquid," "thin" and "distressed." Under these circumstances, the bankers claimed, market prices had diverged far from "fundamental" or "intrinsic" asset values, even assuming there were enough transactions to generate a market price. Suspending or weakening "fair value" rules would enable bank valuations to reflect the true worth of assets rather than the "fire sale" prices that then prevailed in markets. Modification would also, the banks claimed, forestall downward spirals where banks would have to raise money in order to comply with bank capital standards, and since this required asset dumping in distressed markets (at "fire sale" prices), asset prices would continue to drop and the problem of non-compliance would worsen rather than improve. These valuation difficulties were particularly acute for over-the-counter (OTC) financial derivatives and swaps (bespoke bilateral contracts between a dealer-banker and a particular client), which seldom had a benchmark market price and were generally enshrouded in opacity. Supposedly, OTC participants could privately manage risk through the appropriate use of measures like credit ratings, or the posting of collateral, but the events of 2008 showed that such measures were inadequate.

The ABA's intervention was remarkable in that a fundamental mode of capitalist valuation was so publicly questioned by a group of prominent and powerful capitalists. Among other things, fair value accounting rules were seen as truly objective, realistic, and less prone to manipulation by interested parties (Power 2010).

In response to this political pressure, and in the context of the global financial crisis, regulators loosened bank accounting rules, and banks gained a greater measure of flexibility in how they valued their own assets. Researchers responded as well, and scholars have debated whether fair value accounting standards indeed worsened the crisis, as the banks claimed at the time (Badertscher, Burks, and Easton 2012; Bougen and Young 2012; Heaton, Lucas and McDonald 2010; Ryan 2008; Wallison 2008). But the more general issues raised by this episode stem from the interface between valuation and risk assessment, and how it was expressed through emergent tensions over the application of "fair value" or "mark-to-market" accounting rules to financial institutions. The banks seemingly proposed a simple but radical solution to the problem of their exposure to excessive market risks: disengage the banks from the market. They suggested that there was a valuational alternative to market price, and that some kinds of valuation threatened to undermine financial stability. However, in an era in which neoliberal ideas were ascendant, why wouldn't market prices be regarded as the most accurate measure of economic value? Surely "underpriced" assets simply offer profitable arbitrage opportunities that, when exploited, will reduce the mispricing until the price differential disappears, and a similar dynamic occurs for assets that were "overpriced." If such pricing mistakes are self-correcting, why the need to change accounting rules? Furthermore, do not accurate valuations enable market actors to make rational decisions and so increase overall market efficiency? And don't market prices offer the best and truest indicators of asset value (or, as the efficient-markets approach puts it, don't they fully reflect all available information).

A number of conflicting imperatives operated at the interface between asset valuation and risk management. Accounting is primarily about valuation (what are a firm's assets worth? Is a bank solvent or insolvent?), and the disclosure requirements mandated during the New Deal were intended to ensure that investors could obtain credible, accurate, and timely information about the financial status of publicly traded firms. With accurate and complete information, investors can make rational decisions and a firm's counterparties can take whatever measures they deem to be in their self-interest. In the case of banks, bank examiners assess a financial institution's assets and liabilities to determine its solvency, and to take appropriate action if it is insolvent. The value of bank assets is sometimes hard to measure, especially for long-term illiquid items like loans. A traditional "originate and hold" bank will keep loans on its own balance sheet until maturity and may never seek a market valuation by trying to sell the loan. If the loan maintains its "performing" status, however, the bank can calculate the net present value of the payment stream and so determine the value of a particular asset. Those asset values can then be summed to calculate the total value of assets. Banks that securitize their assets (the "originate and distribute" model) regularly obtain market valuations because they don't keep loans on their balance sheets – rather, they sell them off. But what happens when "accurate" valuations produce instability by allowing market risks to affect bank solvency? This could occur if accounting rules transmitted unstable market values into highly variable bank asset values (Plantin, Sapra, and Shin 2008, 88).

In making sense of why the banks sought to modify accounting rules in 2008, I use the concept of "decommodification," borrowed from Karl Polanyi (1944). Polanyi recognized that pure *laissez-faire* capitalism was a historical rarity that arose only under particular conditions. And even when those conditions were met, the situation wasn't a stable one. According to Polanyi, completely free and competitive markets threaten to undermine their own social and political foundations: "To allow the market mechanism to be sole director of the fate of human beings and their natural environment, indeed, even of the amount and use of purchasing power, would result in the demolition of society" (Polanyi 1944, 73). As a result, nineteenth-century industrial societies witnessed a "double movement" in which the market was simultaneously extended and restricted. In order to expand markets, in other words, certain of their effects had to be tempered. The basic elements of Polanyi's argument have been very influential in the literature on the welfare state (e.g. Esping-Anderson 1990; Brady, Seeleib-Kaiser, and Beckfield 2005; Bolzendahl 2010; Martin 2004), and the concept specifically refers to the necessity of mitigating the effects of free labor markets if they are to remain politically sustainable. In particular, capitalist democracies (where wage-earners have the right to vote) have developed a number of policies to ensure that labor markets are not entirely free or perfectly competitive. These include minimum wage laws, unemployment insurance, workplace safety rules, employment protections, old-age pensions, and so forth. In some measure, therefore, labor is "decommodified," and household earnings are stabilized in the face of labor market instabilities with policy that puts a "floor" underneath. Overall, Polanyi explained, "Society protected itself against the perils inherent in a self-regulating market system (Polanyi 1944, 76).

The concept of decommodification is more general than its current application in welfare state research suggests, and indeed Polanyi himself emphasized its relevance to the markets for the three general factors of production: land, labor, and capital. In the latter case, he claimed that: "... the market administration of purchasing power would periodically liquidate business enterprise, for shortages and surfeits of money would prove as disastrous to business as floods and droughts in primitive society" (Polanyi 1944, 73). In other words, Polanyi's claims about the necessity of decommodification to stabilize financial markets directly contradict arguments made by the advocates of vigorous financial deregulation. Policy-makers like Alan Greenspan, who were so influential during the 1990s and early 2000s, clearly thought that the imperatives of rational self-interest and profit maximization ensured that self-regulating financial markets would efficiently and stably allocate society's savings. Consequently, many New Deal regulations, established in the wake of the financial collapse of the early 1930s, were repealed or weakened at the end of the twentieth century. Here, I will use Polanyi's concept to make sense of what banks were seeking during 2008: a way to insulate their assets from the vagaries of the market. In effect, they wanted to "decommodify" bank assets so as to put a "floor" under their valuation.

In extending the usage of the concept of "decommodification," I point to some telling parallels between the events of 2008 and an earlier historical episode involving a similar political imperative. The push to decommodify was not just a crazy impulse of 2008 but rather reflected a deeper pattern. During the Great Depression, and more recently, the political incentive to stabilize financial and housing markets necessitated the suspension of strict market valuation.

Specifically, during the 1930s federal bank examiners and policy-makers began to differentiate between the market value of assets and their "intrinsic" or "fundamental" value, and deliberately used the latter measure (which was higher in dollar value) when assessing the worth of bank assets and home mortgages (Stuart 2003, 45–46). They did so in order to help stabilize the banking system. In similar fashion, the Home Owners' Loan Corporation (HOLC) was established to bolster housing and mortgage markets by refinancing homeowners whose homes has collapsed in value. HOLC's key intervention assumed that market prices did not reflect the fundamental value of a home, and in making a new mortgage it would use the latter rather than the former in determining the size of the loan it would make.

Among other things, the market is a valuational device, and through "price discovery," it gauges the worth of objects. In detaching asset valuation from market price, advocates of decommodification were forced to turn to valuational alternatives: other credible (or at least plausible) ways to determine what something was worth. Credit ratings, quantitative risk models, valuation techniques, and collateral all functioned as cognitive and operational devices to help measure value and manage risk. Edwin Hutchins' (1995) study of navigation underscored that human cognition does not happen within the confines of a single human brain. Rather, cognition is distributed across and between people and a set of cognitive devices (like maps, charts, timepieces, compasses, rangefinders, GPS systems, or a rendering of the night sky). Similarly, risk assessment did not occur solely within the brains of bankers, risk officers, CEOs or CFOs, but rather in a distributed fashion that involved key models, assumptions, and pieces of information. By the early 2000s, the machinery of risk assessment was highly elaborate, quantitatively sophisticated, and enjoyed the endorsement of figures like Alan Greenspan and other top policy-makers. But in 2008 it became apparent that risks had in fact been misrecognized, and that the consequences were not only born privately, but publicly. Extending Hutchins, I refer to this as "distributed misrecognition." Asset valuation formed one simple but foundational piece of this risk assessment machinery: what was an asset worth? How can one manage risk if one doesn't know what is at risk?

FAIR VALUE ACCOUNTING TODAY

Purchased assets can be valued at their historic cost (what a firm paid to acquire them in the first place) or at their current market price. Only in rare circumstances will these two numbers be exactly equal. Using current market price means applying a "fair value" or "mark to market" standard. In 2006, FASB issued FAS No. 157 to clarify the definition and application of "fair value" measurement, specifying that "fair value" meant the price at which the reporting entity could sell an asset (i.e. an exit price) in an orderly market (see www.fasb.org/summary/stsum157.shtml).[1] When such exit

---

[1] The exact wording is: "Fair value is the price that would be received to sell an asset or paid to transfer a liability in an orderly transaction between market participants at the measurement date" (FASB 2006: FAS157-8).

prices are unavailable, FAS 157 stipulated a "fair value hierarchy" that gave reporting entities a menu of alternative inputs for valuation, and the order in which they were to proceed through the alternatives (FASB 2006: FAS157-12).[2] The issuance of FAS 157 culminated a shift toward fair-value accounting that had occurred over several decades (Tweedie and Whittington 1984: Chapter 7; Kemp 2005; Heaton, Lucas and McDonald 2010; Georgiou and Jack 2011). Other statements of the fair value standard included FAS No. 115, issued in 1993 (which accounted for investments in debt and equity securities), FAS No. 133, issued in 1998 (on derivatives and hedging activities), and FAS No.159, issued in 2007 (concerning financial assets and liabilities).

One proximate motivation to adopt "fair value" standards for financial reporting in the US stemmed from the experience of the Savings-and-Loan crisis of the 1980s. Reliance on historical cost accounting allowed S&L's and their regulators to overlook the extent of the problem with S&L balance sheets and postpone the moment of reckoning (Georgiou and Jack 2011, 317; Plantin, Sapra, and Shin 2008, 86). A "fair value" standard would have revealed the financial losses earlier, prompted more timely remedial action, and allowed for a less costly resolution. But the reliance on historical cost was itself the product of an even earlier set of problems. Shortly after its establishment in the 1930s, the SEC cracked down on the widespread practice among public utility companies to revalue their fixed assets upward in order to manipulate their reported profits (Walker 1992; Georgiou and Jack 2011, 316). From the 1940s until the 1970s (when inflation became a real problem), the SEC strongly encouraged firms to use historical cost as the basis for asset valuation (Heaton, Lucas and McDonald 2010, 66–67).

To be sure, even today fair value standards are not applied to all assets. Financial institutions begin by classifying their assets into three overall

---

[2] FASB states (FASB 2006: FAS157-12, FAS157-13) that level 1 inputs involve: "quoted prices (unadjusted) in active markets for identical assets or liabilities that the reporting entity has the ability to access at the measurement date." Absent level 1 inputs, reporting entities can then look to level 2 inputs, which are: " ... inputs other than quoted prices included within Level 1 that are observable for the asset or liability, either directly or indirectly." This would include quoted prices for similar (but not identical) assets. Without Level 1 or Level 2 inputs, the entity may turn to Level 3 inputs, which are: "... unobservable inputs for the asset or liability." These unobservable will reflect: "... the reporting entity's own assumptions about the assumptions that market participants would use in pricing the asset or liability (including assumptions about risk)" (FASB 2006: FAS157-15). Some have termed Level 3 valuation as "mark to model." It does grant the reporting entity considerable discretion in valuation (Laux and Leuz 2010, 97).

categories: those that are held-to-maturity, those that are traded, and those that are available for sale. The first group consists of assets that the bank intends to keep on its balance sheet, and they are valued at their historical cost. This value is adjusted only if the assets become permanently impaired (as, for example, when a debtor defaults on a loan or bond issue). The other two categories are marked to market, and so their value is adjusted as market prices change. But, depending on the availability of market prices, firms may resort to some other level of the fair value hierarchy.

FAS 157, which became effective in November of 2007, arrived just in time for the financial crisis and was soon put to the test. In the fall of 2008, US banks expressed their unhappiness about fair value accounting rules, claiming that their application was endangering the stability of banks. In a letter to the SEC, the ABA flatly stated that: "One key factor that is recognized as having exacerbated these [financial] problems is fair value accounting" (ABA 2008c, 1). The ABA pointed out how difficult it was to find market prices when markets were illiquid: "As financial markets thin out or even seize up, as trades become fewer and more volatile, and in general as trading values become increasingly unreliable, it is daily more apparent that for many assets, especially under current conditions, there is not a true 'fair value'" (ABA 2008c, 2). This kind of mispricing wasn't a minor problem that solved itself, for in fact, the ABA claimed, it worsened the already bad conditions in financial markets: "Mark to market based on exit-price in an illiquid market results in an unrealistic downward bias, which reduces transparency and can have serious public policy implications" (ABA 2008c, 2). In various communications, the ABA acknowledged the worth of fair value standards during normal market operations, but asserted that financial markets in 2008 were "dislocated," "seized," "illiquid," "frozen," and "distressed." Under these problematic circumstances, market prices would not reflect an asset's "true value" and as a "pro-cyclical" policy, fair value standards were counterproductive (see, e.g., ABA 2008a; 2008b; 2008d).

Banks weren't the only ones to worry about the effects of fair value accounting. Numerous scholars, policy analysts and commentators weighed in on the issue (IMF 2008; Plantin, Sapra, and Shin 2008; Ryan 2008; Haldane 2009; Pozen 2009; Wallison 2009; FCIC 2011, 226–7; Bougen and Young 2012). A number of politicians called for the outright suspension of fair value accounting (Cheng 2009, 7). Although the bankers pushed hard to weaken fair value accounting, others pushed back. And it was some time before detailed academic studies of the role of fair value accounting in the crisis were completed (e.g. Bhat, Frankel, and Martin 2011; Heaton, Lucas, and McDonald 2010; Laux and Leuz 2010). In the meantime, however, the crisis dictated that something be done quickly. To begin, the Emergency Economic Stabilization Act of 2008 (the so-called TARP bailout) included a provision requiring the SEC to study mark-to-market accounting standards and make recommendations.

The study was quickly completed and issued December 30, 2008 (SEC 2008a). It concluded that fair value standards did not have a significant impact on bank failures, that they enabled investors to obtain accurate financial information, and that there were no really good alternatives to fair value. The report recommended that FAS No. 157 be "improved" rather than "suspended," with a particular focus on the valuation of assets for which markets are illiquid or inactive (SEC 2008a, 7–9).

At the same time that the SEC was conducting its study, the FASB proposed a FASB Staff Position (FSP) to amend FAS No. 157 by clarifying how it would apply to assets in inactive markets (FSP FAS 157-d). The FASB proposal received many comments, including from the ABA. The ABA underscored again what it believed to be a: "... bias toward observable market data when the quality or usefulness of that data is questionable" (ABA 2008b, 1). One telling comment from a financial firm, forwarded by then Congressman Mark Kirk (R-IL), criticized the proposed FSP on the grounds that it really didn't offer any relief to troubled financial institutions (Wehmer and Dykstra 2008). The letter went on to make specific reference to the valuation practices of HOLC during the financial crisis of the 1930s, commending HOLC for recognizing that: "... a more realistic concept of fair value was necessary and more meaningful than the highly distressed prices at a point in time during a market crisis (Wehmer and Dykstra 2008, 1). The letter suggested that FASB follow HOLC's example and develop its own "intrinsic value concept." In April of 2009 the FASB issued three final staff positions on the matter of fair value accounting, and although it did not reject fair value methods, in the circumstances of distressed sales or inactive markets, it did offer some flexibility to reporting entities (Heaton, Lucas and McDonald 2010, 68).

Subsequent research offered mixed results on the actual effects of fair value standards during the financial crisis. Both Laux and Leuz (2010) and Badertscher, Burks, and Easton (2012) concluded that fair value rules did not force banks to dump assets and so did not play a substantial role in worsening the crisis. However, Bhat, Frankel, and Martin (2011) studied bank holdings of mortgage-backed securities (MBS) and found that price drops prompted asset sales by banks, and furthermore that the effect was weaker once mark-to-market rules were eased in April of 2009. Merrill et al. (2012) show that capital-constrained insurance companies were more likely to sell off residential mortgage-backed securities (RMBS), and at lower prices, than their less capital-constrained counterparts. They conclude that fair value accounting rules led to asset fire sales.

Separate from their actual effects, fair value accounting rules were claimed by some powerful interest groups to be destabilizing. Transmitting problematic market prices instantly and directly into bank balance sheets would hurt banks, the claim went, particularly when those prices emerged out from "distressed," "illiquid," or "frozen" markets. And although fair value wasn't abolished, it was modified in ways that gave banks some welcome flexibility. But the larger significance of this episode stems from the way in which political resources were mobilized to alter regulatory and accounting rules so as to insulate financial institutions from market instability. Even though the logic of market valuation had deliberately been inscribed within financial accounting rules (via FAS No. 157 and other measures) the effects were troublesome enough to motivate corresponding mitigation. And these compensatory actions were justified on the grounds that when markets were behaving in a disorderly and dysfunctional manner, banks had to be protected from them. Using the terminology of Polanyi, financial assets were partly "decommodified."

NEW DEAL BANK STANDARDS

The crisis of 2008 is not the first episode to provoke attempts to decommodify the financial system. As Wehmer and Dykstra (2008) suggested in their letter to FASB, similar problems arose during the 1930s. During the Great Depression, corporate bonds constituted a substantial proportion of the total assets of US banks. As the economy crashed, the solvency of banks depended on the value of the assets on their balance sheets, and as the value of these assets decreased, greater and greater numbers of banks failed. Following the emergency bank holiday of March 1933, 3,460 banks were suspended in the third quarter of 1933. Consequently, the total number of banks within the Federal Reserve System declined from 8,929 in 1928 to 5,606 in 1933. Over the same period, the number of non-member banks declined from 16,869 to 8,601.

The collapse of the worth of bank assets, and the subsequent collapse of many banks, put tremendous pressure both on the US financial system and on the regulators and politicians who oversaw it.

One response to these problems involved a change in bank examination rules. This comports with the observation by Birkland and Warnement that policy changes in response to crisis events need not emanate from the legislative branch. Instead, policy-makers at multiple levels of government have a strong incentive to be seen as proactive problem solvers and do not necessarily passively await direction from superordinate government branches (Birkland and Warnement, Chapter 5 this volume). Traditionally (and before the era of federal deposit insurance), it was the task of federal or state bank examiners to assess the actual financial status of a bank, to appraise bank assets and liabilities, and to close down banks that were insolvent (Jones 1940, 183). But by relaxing the financial standards they applied, bank regulators could introduce a measure of forbearance that would make it easier for marginal banks to survive. Before 1930, the Comptroller of the Currency valued bank-owned securities (like bonds) at their market value when calculating the overall worth of bank assets during a bank examination. As market prices plummeted, the value of total bank assets declined and more banks had negative worth (liabilities greater than assets). And with the collapse of the stock market, it was clear that bank balance sheets could not be repaired by having banks simply raise more capital from investors. But starting in 1931, as the bond market collapsed, the Comptroller initiated a new policy that made use of the ratings produced by agencies like Moody's, Poor's Publishing, and Standard Statistics (which later formed Standard and Poor's). This change built on a new valuation method devised by the Federal Reserve Bank of New York to use bond ratings as a way to estimate the overall quality of a bank's bond portfolio (Harold 1938, 160–161).[3] Gustav Osterhus, the bank official who invented the new method, styled it as a "mechanical method designed to disclose true quality of a bank's investment account" (Osterhus 1931, 67).[4] After the

---

[3] Flandreau et al. (2011) examine the poor performance of bond rating agencies during the 1930s and wonder how ratings could have been given regulatory standing in precisely the same decade. But this move was not unprecedented for the New York Federal Reserve Bank, perhaps because of its close ties to Wall Street and familiarity with current financial practices. Issued soon after the founding of the Federal Reserve System, a New York Fed circular on the topic of eligible paper (i.e. what member bank assets could be brought to the Fed discount window for rediscounting) shows that mercantile agency credit ratings factored into the determination of eligibility (Federal Reserve Bank of New York 1915). This was done without strong evidence of the predictive accuracy of credit ratings. Furthermore, rating agencies were increasingly a point of reference in US courts (Flandreau and Stawatyniec 2013).

[4] He also criticized market prices as a way to value bonds on the grounds that prices simply reflect the overall judgments of investors, and most investors know little. In fact, they: "... do not investigate carefully" (Osterhus 1931, 110).

Comptroller's new policy, bonds rated BBB and higher could be valued at their *historical cost* (i.e. what the bank originally paid for them) rather than at their current market value (Morton 1939, 277; Harold 1938, 26; Jones 1940, 187). The Federal Reserve Board granted additional official salience to bond ratings when it started in 1933 to publish tables that showed how current bond yields varied across different rating categories, using Moody's system of Aaa, Aa, A, and Baa for "investment grade" securities (see Federal Reserve Board 1933, 483). Many state bank regulators subsequently followed the Comptroller's new policy. Indeed, in 1938 the National Association of Supervisors of State Banks affirmed the use of bond ratings to group bank bond holdings into four different classes, with the highest rated bonds (i.e. given AAA, Aa, A, or Baa ratings) being valued at their historic cost (Morton 1939, 280; Jones 1940, 194). Highly rated bonds would no longer be marked to market.[5]

New Deal bank regulations also looked to the other end of the bond-rating spectrum. The Banking Act of 1935 gave to the Comptroller of the Currency a new power to impose restrictions on the kinds of securities a national bank could purchase on its own account. In early 1936, the Comptroller applied this power when issuing a new rule that prohibited the purchase of securities that were "distinctly or predominantly speculative," or of an even lower standard (Atkins 1938; Palyi 1938, 70; Comptroller of the Currency 1937, 21–22). The real problem with US banks, the Comptroller asserted in a speech to the California Bankers Association made in May 1936, lay chiefly on the asset side of the balance sheet, not on the liability side. During the 1920s, too many banks had invested in too many "speculative" bonds, and it was necessary to prevent them from doing so again (Federal Reserve Board 1936, 422–423). The rule also explicitly referred to the agency rating manuals when defining the meaning of terms like "speculative" or "predominantly speculative" (which, in the Moody's rating scale, meant ratings below the Baa level). Since bank examiners had little experience with the detailed assessment of financial securities, the application of ratings to bond portfolios was done quite mechanically (Atkins 1938, 14). Most state bank regulators followed the federal lead and adopted similar prohibitions (Palyi 1938, 73). Although bond ratings had been published and used by private investors since 1909, this direct incorporation of ratings into bank regulatory machinery increased their significance many-fold. Low-rated

---

[5] In addition to protecting bank assets from the collapse of the bond market, this measure also put a premium on high ratings.

bonds were sanctioned as much as high-rated bonds were encouraged. Highly rated bonds could be protected from low market valuations.

An important additional step was taken in 1938, when the Uniform Agreement on Bank Supervisory Procedures tried to standardize bank examination procedures across multiple regulatory agencies, including the Federal Reserve, Comptroller of the Currency, and Federal Deposit Insurance Corporation (FDIC). The new common standard put bank assets into overall categories, depending on the bank examiner's judgment of the relevant level of risk or impairment. Assets in the top category were to be shown at book value, without adjustment for market depreciation (Simonson and Hempel 1993, 255). In appraising assets (both loans and securities), examiners were to consider them: "... in the light of inherent soundness rather than on a basis of day to day fluctuations" (Board of Governors 1938a, 564). More generally, regulators affirmed that the overall solvency of the banking system: "... should not be measured by the precarious yardstick of current market quotations which often reflect speculative and not true appraisals of intrinsic worth" (Board of Governors 1938a, 564). Instead of its normally dry bureaucratic language, here the Federal Reserve Bulletin utilized distinctive terminology to separate market prices from "intrinsic worth" or "inherent soundness," and to downplay the significance of market price. When these different measures didn't align, the agreement clearly signaled to regulators and examiners that they should overlook low asset prices and emphasize instead their "intrinsic worth."[6] However, the Federal Reserve Board was emphatic that ignoring market price did not mean a "relaxation of standards" (Board of Governors of the Federal Reserve System 1938b, 39). Many US state-banking authorities followed the lead of federal regulators and henceforth the distinction between "investment grade" and "below investment grade" became enormously consequential.

Bond market prices dropped at the outset of the Depression, and so these new valuation policies in effect allowed federal and state bank regulators to overlook the substantial decline in the value of bank bond assets, and to treat a particular bank as if it were in better financial condition than it actually was. Market valuation of bank bonds was partly suspended, contingent on the private evaluations produced by the bond rating agencies. In addition, new prudential standards for banks

---

[6] Bach (1949, 275–276) also notes the regulatory difference between "intrinsic soundness" and "current market value," and the importance of recognizing the former during a recessionary period.

constrained their own bond portfolios to exclude assets that were, again relying on the judgment of the rating agencies, "below investment grade" (Fons 2004). For the first time, bond ratings were given regulatory standing in a manner that deliberately protected banks from the vagaries of the bond market and forestalled insolvency. Such regulatory forbearance kept a bank open until a future time when, it was hoped, its assets would have recovered their "true" value. Yet this change was not simply an ad hoc lifeline thrown to banks, for as Jones (1940, 186) explained, "... bank assets have a particularly indeterminate value in periods of crisis. Current market prices cannot be accepted because markets are demoralized, and many classes of bank assets, e.g. most bank loans and real estate, cannot be sold at any price." In other words, during a crisis, previously trustworthy market signals became inaccurate and unreliable as measures of value, reflecting a "demoralized" collective sentiment rather than economic fundamentals. Since market price is the default measure of value, departure from this standard would presumably be justified only under special circumstances. When the Federal Reserve Board reconsidered the Uniform Agreement in 1949, during the post-war boom, the insulation of highly rated bonds from valuation by market price was simply continued (Clayton 1949).[7]

NEW DEAL MORTGAGE STANDARDS

The prospect of a total collapse of the banking system during the Great Depression prompted "decommodification" of bank bond portfolios. Other markets closely linked to the financial system also teetered on the edge, and were similarly insulated from pure market forces. Some financial institutions invested in home mortgages, and so were adversely affected by the decline in housing prices. Real estate market prices dropped, homeowners failed to make their mortgage payments and even defaulted, loans shifted from "performing" to "non-performing" status, and so the financial institutions which made home mortgage loans moved toward insolvency. In fact, the entire housing market collapsed as residential property construction fell 95 percent between 1928 and 1933 (Jackson 1980, 424).

---

[7] There were some minor modifications of the labels of different groups of bonds, and group 2 ("substandard" quality) securities were to be valued at market prices rather than at an 18-month moving average of market prices (Federal Reserve Board 1949, 776–777). Most importantly, however, highly rated bonds were not to be valued at their market prices.

The Home Owners' Loan Corporation (HOLC) was established as a short-term measure in 1933 to bolster the housing and home mortgage markets by helping to refinance homeowners on easier terms (longer-term loans worth up to 80 percent of the value of the home). HOLC was explicit in not taking current market value as the measure of what a house was worth (and therefore, how large of a new loan could be made to the homeowner): "The Corporation will not depend upon the technical market value. On the other hand, it will give equal weight to (1) present market value; (2) cost of reproduction of the property, less depreciation; and (3) a capitalization of the reasonable monthly rental value of the property over a period of the past 10 years" (HOLC 1933, 2). In effect, this more complicated method of appraisal allowed HOLC to derive estimates that were significantly higher than current market values (Wickens 1937, 76–77; Fishback, Rose, and Snowden 2013, 74–75). But this higher value was considered the "normal" value of a home, in contrast to its current market value (Fishback, Rose, and Snowden 2013, 59). With a higher appraisal, HOLC could extend larger loans secured by real estate and provide more support to the mortgage and housing markets. Similarly, the Emergency Farm Mortgage Act of 1933 supported additional loans to aid farmers. As with any type of mortgage, the size of the loan was conditional on the value of the underlying property that secured the loan, so the appraised value of a farm dictated the magnitude of financial assistance. As Gaddis (1935, 469) notes, the Act of 1933 changed appraisal methods, shifting away from current values to "normal values," where the latter did not reflect the collapse in prices that afflicted rural real estate markets during the early 1930s.

The Federal Housing Administration (FHA) was established in 1934 and it also was intended to support housing and mortgage markets (Weimer 1937). One of its primary activities involved the provision of mortgage insurance for home mortgages that conformed to new and more affordable mortgage standards (longer-term, higher loan-to-value ratios, and self-amortizing). The new standards significantly reduced the required monthly payments and allowed buyers to finance a greater proportion of the value of their new home. In short, homeownership became easier. Despite reducing a borrower's monthly financial burden, however, the appraised value of a home remained a key variable. Here the FHA made a telling accommodation. Its 1936 Underwriting Manual explicitly laid out some "axioms of valuation": "Valuation endeavors to estimate prices which are fair and warranted, that is, prices which represent the worth at the time of appraisal of the future benefits which will arise from ownership, rather than prices which can be obtained in the market" (quoted in Stuart 2003, 47). Clearly, official policy at the FHA was that market price need not be the best estimate of value, and appraisers would retain a meaningful level of discretion in making their estimates.[8] In a manner that paralleled the bank regulators, the FHA also included credit agencies in the mortgage-application evaluation procedure (Stuart 2003, 91). When evaluating an applicant, a lender could turn to one of three credit-reporting agencies (which evaluated individual borrowers rather than corporate bond issues).

Home mortgages, and the residences that secured them, represented another important economic asset where regulators and government agencies, pressed by the realities of economic collapse, distinguished between market prices and "intrinsic values," and chose to favor the latter when market prices were too low. Economic value was no longer strictly dictated by the market because the latter was generating current market prices that were politically intolerable. Faced with catastrophe, policy-makers and regulators instead embraced a variety of alternative valuation methods that in some manner captured what they deemed to be the "true" value of an asset. The beneficiaries of this decommodification included banks with non-performing mortgages on their books, but also home-owners, mortgagees, and others with an interest in the vitality of the real estate market.

FINANCIAL DECOMMODIFICATION

It is no coincidence that during the Great Depression and the Great Recession (the worst downturn since the Great Depression), regulators, examiners, agencies, and policy-makers worked to disconnect the valuational machinery of the financial sector from a market that was variously characterized as "insolvent," "illiquid," "demoralized," and "distressed." Markets expand and contract, and severe contractions produce outcomes that are politically unpopular: widespread insolvencies, failures, foreclosures, and defaults. Market instability per se was less important than instability that drove down prices and begat economic decline. Rather than simply submit to the harsh strictures of the market (as Treasury Secretary Andrew Mellon once

---

[8] FHA guidelines also played an important role in encouraging residential segregation, by having underwriting standards that favored homogeneous neighborhoods over those that were mixed either by race or class (Jackson 1980, 430–431).

recommended),[9] however, various parties deployed alternative valuational rules that enabled them to act with some authority as if assets were worth significantly more than what current market prices indicated. The rejection of the market was only partial, however, because bond ratings themselves were "market based" in the sense that they were produced in and for private investors and financial markets. This shift valuation did not represent a wholesale rejection of the market, but rather a re-articulation of the relationship between financial institutions and markets.

In some instances, the new rules for the first time imported into public regulation the bond ratings produced by private rating agencies. This had the effect of dispersing valuation so that public oversight partnered with private evaluation to determine the effective worth of assets and the solvency of financial institutions. Such valuational fiat stemmed from attempts to stabilize the financial system and the housing market. And unlike many of the policies that "decommodified" labor (e.g. minimum wage laws, unemployment pensions) the measures that decommodified the financial system were implemented bureaucratically via regulatory modification, mostly out of the spotlight of overt political contention.

The decision to incorporate credit ratings directly into financial regulations ultimately proved to be a fateful one. Indeed, according to Partnoy (2009) the decision to give regulatory standing to credit ratings in the 1930s inadvertently worsened the crisis of 2008. Distributing the valuation of assets between public and private organizations increased the significance of the judgments made by bond rating agencies, and magnified the effect of "mistaken" ratings.

Following the lead of federal bank regulators, many other regulatory agencies in and outside of the US adopted ratings to set prudential standards, protect investors, and govern market access (FCIC 2011, 119; Langohr and Langohr 2008, 430–439). During the 1990s, ratings were also transplanted into the private contractual machinery at the core of the burgeoning OTC derivatives markets: under an ISDA master agreement, swap transactions typically required each side to post collateral, contingent on their ratings, as a type of private risk management (Cunningham and Werlen 1996; Gregory 2010, 65–66; ISDA 1996, 11, 20, 49–50). To receive a ratings downgrade from Moody's or S&P signaled greater

---

[9] Mellon famously recommended to "liquidate labor, liquidate stocks, liquidate farmers, liquidate real estate. It will purge the rottenness out of the system." See https://fee.org/articles/andrew-mellon-the-best-treasury-secretary-in-us-history/.

risk and meant having to post more collateral (which, in fact, was the proximate cause of the collapse of AIG in 2008). By the end of the twentieth century, many parties in financial markets depended on ratings to calibrate and manage risk.

Coming from both public regulatory and private contractual sources, the imperative to achieve "AAA" rating helped to fuel the frenzy of financial engineering that investment banks undertook to turn pools of subprime mortgages into high-grade securities. Credit ratings played a central role in the securitization process in which assets were pooled, tranched, rated, and sold to investors around the world in form of ABSs, RMBSs, CDOs, CDO2s, and other exotic instruments (MacKenzie 2011). Rating agencies were paid directly by issuers to rate securities (the "issuer pays" model), and so it was clear that agency's real customers were the issuers of securities, not the buyers (FCIC 2011, 118–119). Nevertheless, buyers relied on ratings to determine the riskiness of an investment, with AAA signaling the very safest investments. And often buyers were constrained by prudential regulations to rely on ratings. In cooperation with rating agencies, securitizers worked to create new instruments that would have the highest possible ratings, and so would be most attractive to investors (SEC 2008b).

Whether or not it makes use of credit ratings in an alternative valuational regime, decommodification need not be the only way to stabilize markets. Bank capital standards create financial "cushions" that can absorb losses and prevent a financial institution from collapsing. Deposit insurance protects banks from bank runs caused by panicky depositors. Other prudential regulations attempt to impose some measure of asset diversification among banks. Likewise, housing markets can be boosted in a variety of ways (e.g. via tax expenditures like the home mortgage interest deduction for federal income taxes). But in these two historical episodes, during the 1930s and the 2000s, US regulators and policy-makers were prompted to "decommodify" assets whose production, distribution, and evaluation had previously been largely governed by market forces. Both times, the goal was not simply to smooth or stabilize asset values whether they went up or down, but to ensure that they didn't go down too far. Decommodification was, in other words, asymmetric.

This chapter does not aspire to offer an exhaustive list of instances where people have attempted to decommodify financial institutions (I imagine a careful survey of previous financial crises would likely reveal other cases). Nor is it the only situation where people have challenged the economic value bestowed by the operations of the market. Consider, for example, the moral economy standard of a "fair price" invoked in early modern bread riots and by subsistence peasant farmers (Thompson 1971; Scott 1976). In this case, there was little appeal to a popular notion of "fair price" for financial assets or homes. Rather, the current market price was rejected as an appropriate or accurate measure of "true value" because of the unusually disturbed status of the market, not because of any moral implications or qualms about fairness. Presumably, the return to market "normality" would restore market price to its favored position as the standard for valuation.

CONCLUSION

In an era where markets were ascendant, and where some version of *laissez-faire* economic doctrine (known as "neoliberalism" in the 2000s) was dominant, important economic interests worked to suspend market valuation. This intervention wasn't motivated by a deep critique of capitalism, nor by a sense that the rules of some moral economy had been transgressed, nor even by a narrow conviction that the efficient-markets hypothesis was wrong. Rather, market valuation was suspended or weakened because markets were deemed to be (temporarily) dysfunctional, and therefore market prices no longer measured the actual, underlying worth of assets. Instability was the consequence of market prices unhinged from true value, and during a financial crisis, such instability became politically intolerable, not just to the bankers but to all others with an interest in functioning financial systems. Over the twentieth century, as more and more American households became engaged with the financial system, either by using financial services or possessing financial assets, those with an interest in functioning finance grew in numbers. By applying a different valuational standard, *illiquid but solvent* banks could "ride out" the crisis. Of course, *illiquid and insolvent* banks might postpone the inevitable, but their day of reckoning would eventually come. Unfortunately, however, by allowing insolvent banks to continue to operate, the eventual cost of failure (to creditors, taxpayers, insurers) often increases.[10]

Advocates argued that default valuation rules ("fair value accounting" or its equivalent) had to suspended, and they recognized that the onus was on them to make the case for change. Using distinctive and resonant terminology ("illiquid," "distressed," "dislocated," "seized up"), they

---

[10] Witness regulatory forbearance and the costs of the S&L crisis in the 1980s.

discounted current market prices and proposed alternatives. But, as with Justice Potter Stewart's recognition of pornography ("I know it when I see it"), there was no precise *ex ante* definition of market distress. Market distress, the *sine qua non* of this valuational intervention, was in the eye of the beholder, and not everyone saw eye to eye on this issue. Hence political action to frame the issue and mobilize support had to precede valuational modification, if the latter were to succeed (see Mayer's Chapter 6, this volume for more detail on framing). And as a practical matter, valuational alternatives relied on available cognitive devices like bond ratings and quantitative financial models. Unfortunately, these had their own limitations, which contributed to a collective "misrecognition" of value and risk during the 2008 crisis.

To view this particular intervention as a form of "decommodification" links it to public policies that normally are not associated with financial regulation, particularly those that concern wages, pensions, and labor protections. Such a linkage holds out the possibility of a comparative analysis of decommodification, across issue areas. Decommodification provides a way for market societies to tolerate the risks and instabilities that markets generate. Some of these risks are more problematic than others, and so many countries have developed programs to decommodify labor markets. Many countries have also developed ways to stabilize their financial systems, although many of the policy measures taken during the 1930s were repealed during deregulation at the end of the twentieth century.

As conceived by Polanyi, decommodification was part of a "double movement" with respect to the expansion and contraction of markets; clearly the politics vary dramatically. Popular demonstrations by workers in support of higher wages, at one end of the spectrum, contrast sharply as a mode of collective action from highly technical memos on accounting methods submitted to regulatory agencies, at the other end. Furthermore, low-income wage- earners are a much bigger voting bloc than bank employees, and so are more likely to have political voice in a capitalist democracy. So both the means and the outcome of decommodification will vary depending on what (or who) is being decommodified. Nevertheless, both types of decommodification aimed at tempering the instabilities of labor and financial markets by focusing on the bottom of the wage or price distribution. Policies tried to raise up the bottom rather than pull down the top. And although the proximate beneficiaries varied between traditional and financial decommodification (workers vs. banks), the financial crises of 1933 and 2008 showed that many have an interest in a stable financial system, not just the banks.

Financial decommodification via the weakening of "fair value" accounting standards offered a quick way to protect financial institutions from "distressed" markets. It can help impede the propagation of bank failures and slow down cascading insolvency, which benefits the financial sector but also those who depend on credit (e.g. homeowners). Modification of accounting rules possesses little political salience so such measures can be taken without much political debate or democratic consideration. Furthermore, it doesn't provoke the same kind of political opposition as traditional decommodification. Employers often oppose minimum wage laws or higher minimum wages because they will face higher labor costs. They oppose workplace protections that impede the free operation of "employment at will" (e.g. the unconstrained right to terminate employees). By contrast, new valuation standards don't impose burdens that are obvious and immediate enough to mobilize widespread political opposition. Social movements do not spring up in defense of "technicalities" like mark-to-market accounting rules. This is not to say that financial decommodification is costless, of course, merely that the discussion of the relative costs and benefits of such a policy is likely to be a technocratic one dominated by experts. Some issues, including accounting standards, may be "policies without publics" in which reform is largely left to technocrats with limited engagement by wide constituencies (Birkland and Warnement, Chapter 5, this volume). Other issues galvanize the general public, such that policy-makers may face pressure to respond to their changed risk perceptions, even when experts disagree with the public's risk assessment (Weber, Chapter 4, this volume).

References

American Bankers Association (ABA). 2008a. Letter to Mr. Robert Herz, Chairman, Financial Accounting Standards Board. August 7, 2008. Washington, DC: ABA.

American Bankers Association (ABA). 2008b. Letter to Mr. Russell G. Golden, FASB Technical Director, Financial Accounting Standards Board. October 9, 2008. Washington, DC: ABA.

American Bankers Association (ABA). 2008c. Letter to Mr. Jim Kroeker, Deputy Chief Accountant, US Securities and Exchange Commission. November 13, 2008. Washington, DC: ABA.

American Bankers Association (ABA). 2008d. Letter to The Honorable Henry M. Paulson, Jr., Secretary of the Treasury. November 25, 2008. Washington, DC: ABA.

Atkins, Paul. 1938. "The Official Supervision of Bank Security Portfolios," *Bankers' Magazine* July: 13–19.
Bach, G. L. 1949. "Bank Supervision, Monetary Policy, and Governmental Reorganization," *Journal of Finance* 4(4): 269–85.
Badertscher, Brad A., Jeffrey J. Burks, and Peter D. Easton. 2012. "A Convenient Scapegoat: Fair Value Accounting by Commercial Banks during the Financial Crisis," *Accounting Review* 87(1): 59–90.
Bhat, Gauri, Richard Frankel, and Xiumin Martin. 2011. "Panacea, Pandora's Box, or Placebo: Feedback in Bank Mortgage-Backed Security Holdings and Fair Value Accounting," *Journal of Accounting and Economics* 52: 153–73.
Board of Governors of the Federal Reserve System. 1938a. *Federal Reserve Bulletin*. Washington, DC: Federal Reserve.
Board of Governors of the Federal Reserve System. 1938b. *Twenty-Fifth Annual Report of the Federal Reserve System*. Washington, DC: Federal Reserve.
Bolzendahl, Catherine. 2010. "Directions of Decommodification: Gender and Generosity in 12 OECD Nations, 1980–2000," *European Sociological Review* 26(2): 125–41.
Bougen, Philip D., and Joni J. Young. 2012. "Fair Value Accounting: Simulacra and Simulation," *Critical Perspectives on Accounting* 23: 390–402.
Brady, David, Martin Seeleib-Kaiser, and Jason Beckfield. 2005. "Economic Globalization and the Welfare State in Affluent Democracies 1975–2001," *American Sociological Review* 95(4): 921–48.
Cheng, Kang. 2009. "Mark to Market or Mark to Expectation?" *Commercial Lending Review* January-February: 3–7.
Clayton, Lawrence. 1949. Memorandum to Board of Governors, from The Marriner S. Eccles document collection, from FRASER, http://fraser.stlouisfed.org/docs/historical/eccles/046_20_0002.pdf.
Comptroller of the Currency 1937. *The Seventy-Fifth Annual Report of the Comptroller of the Currency*. Washington, DC: Government Printing Office. https://fraser.stlouisfed.org/files/docs/publications/comp/1930s/compcurr_1937.pdf
Cunningham, Daniel, and Thomas Werlen. 1996. "Derivatives and the Reduction of Credit Risk," *International Financial Law Review* 15: 35–36.
Esping-Anderson, Gøsta. 1990. *The Three Worlds of Welfare Capitalism*. Princeton: Princeton University Press.
Federal Reserve Bank of New York. 1915. "Eligible Paper," Circular No. 25, June 19, 1915. New York: Federal Reserve Bank of New York.

Federal Reserve Board. 1936. *Federal Reserve Bulletin June 1936.* Washington, DC: US Government Printing Office.

Federal Reserve Board. 1949. *Federal Reserve Bulletin July 1949.* Washington, DC: US Government Printing Office.

Financial Accounting Standards Board (FASB). 2006. "FAS 157: Fair Value Measurements," Norwalk CN: FASB.

Financial Crisis Inquiry Commission (FCIC). 2011. *The Financial Crisis Inquiry Report.* New York: PublicAffairs.

Fishback, Price, Jonathan Rose, and Kenneth Snowden. 2013. *How the New Deal Safeguarded Home Ownership.* Chicago: University of Chicago Press and NBER.

Flandreau, Marc, Norbert Gaillard, and Frank Packer. 2011. "To Err Is Human: US Rating Agencies and the Interwar Foreign Government Debt Crisis," *European Review of Economic History* 15: 495–538.

Flandreau, Marc, and Joanna Kinga Sławatyniec. 2013. "Understanding Rating Addition: US Courts and the Origins of Rating Agencies' Regulatory Licence (1900–1940)," Working Paper No.11/2013. Geneva: Graduate Institute of International and Development Studies.

Fons, Jerome S. 2004. "Tracing the Origins of 'Investment Grade'," *Special Comment,* New York: Moody's Investors Services.

Gaddis, P. L. 1935. "Appraisal Methods of Federal Land Banks," *Journal of Farm Economics* 17(3): 469–80.

Georgiou, Omiros, and Lisa Jack. 2011. "In Pursuit of Legitimacy: A History Behind Fair Value Accounting," *British Accounting Review* 42: 311–23.

Gregory, Jon. 2010. *Counterparty Credit Risk: The New Challenge for Global Financial Markets.* New York: Wiley.

Haldane, Andrew G. 2009. "Fair Value in Foul Weather," Speech to the Royal Institute of Chartered Surveyors, London, November 10 2009, London: Bank of England.

Harold, Gilbert. 1938. *Bond Ratings as an Investment Guide: An Appraisal of Their Effectiveness.* New York: Ronald Press.

Heaton, John C., Deborah Lucas, and Robert L. McDonald. 2010. "Is Mark-to-Market Accounting Destabilizing? Analysis and Implications for Policy," *Journal of Monetary Economics* 57: 64–75.

Home Owners' Loan Corporation (HOLC). 1933. *Federal Relief for Home Owners.* Washington, DC: Government Printing Office.

Hutchins, Edwin. 1995. *Cognition in the Wild.* Cambridge: MIT Press.

International Monetary Fund (IMF). 2008. *Global Financial Stability Report.* Washington, DC: IMF.

International Swaps and Derivatives Association, Inc. (ISDA) 1996. *Guidelines for Collateral Practitioners.* New York: ISDA.

Jackson, Kenneth T. 1980. "Federal Subsidy and the Suburban Dream: The First Quarter- Century of Government Intervention in the Housing Market," *Records of the Columbia Historical Society* 50: 421–51.

Jones, Homer. 1940. "An Appraisal of the Rules and Procedures of Bank Supervision, 1929–39," *Journal of Political Economy* 48(2): 183–98.

Kemp, M. H. D. 2005. "Risk Management in a Fair Valuation World," *British Actuarial Journal* 11(4): 595–712.

Langohr, Herwig M. and Patricia T. Langohr. 2008. *The Rating Agencies and Their Credit Ratings*. New York: Wiley.

Laux, Christian and Christian Leuz. 2010. "Did Fair-Value Accounting Contribute to the Financial Crisis?" *Journal of Economic Perspectives* 24(1): 93–118.

MacKenzie, Donald. 2011. "The Credit Crisis as a Problem in the Sociology of Knowledge," *American Journal of Sociology* 116(6): 1778–1841.

Martin, Cathie Jo. 2004. "Reinventing Welfare Regimes: Employers and the Implementation of Active Social Policy," *World Politics* 57(1): 39–69.

Merrill, Craig B., Taylor D. Nadault, Rene M. Stulz, and Shane Sherlund. 2012. "Did Capital Requirements and Fair Value Accounting Spark Fire Sales in Distressed Mortgage-Backed Securities?" National Bureau of Economic Research Working Paper 18270, Cambridge MA: NBER.

Morton, Walter A. 1939. "Liquidity and Solvency," *American Economic Review* 29(2): 272–85.

Osterhus, Gustav. 1931. "Flaw-Tester for Bond Lists," *American Bankers Association Journal* August: 67–110.

Palyi, Melchior. 1938. "Bank Portfolios and the Control of the Capital Market," *Journal of Business* 11(1): 70–111.

Partnoy, Frank. 2009. "Historical Perspectives on the Financial Crisis: Ivar Kreuger, the Credit- Rating Agencies, and Two Theories about the Function, and Dysfunction, of Markets," *Yale Journal on Regulation* 26(2): 431–43.

Plantin, Guillaume, Haresh Sapra, and Hyun Song Shin. 2008. "Fair Value Accounting and Financial Stability," *Financial Stability Review* 12: 85–94.

Polanyi, Karl. 1944. *The Great Transformation*. Boston: Beacon Press.

Power, Michael. 2010. "Fair Value Accounting, Financial Economics and the Transformation of Reliability," *Accounting and Business Journal* 40(3): 197–210.

Pozen, Robert C. 2009. "Is It Fair to Blame Fair Value Accounting for the Financial Crisis?" *Harvard Business Review* November: 84–92.

Ryan, Stephen G. 2008. "Accounting in and for the Subprime Crisis," *Accounting Review* 83(6): 1605–38.
Scott, James C. 1976. *The Moral Economy of the Peasant: Rebellion and Subsistence in Southeast Asia.* New Haven: Yale University Press.
Securities and Exchange Commission (SEC). 2008a. *Report and Recommendations Pursuant to Section 133 of the Emergency Economic Stabilization Act of 2008: Study on Mark-to-Market Accounting.* Washington, DC: SEC.
Securities and Exchange Commission (SEC). 2008b. *Summary Report of Issues Identified in the Commission Staff's Examinations of Select Credit Rating Agencies.* Washington, DC: SEC.
Simonson, Donald G. and George H. Hempel. 1993. "Banking Lessons from the Past: The 1938 Regulatory Agreement Interpreted," *Journal of Financial Services Research* 1993: 249–67.
Stuart, Guy. 2003. *Discriminating Risk: The US Mortgage Lending Industry in the Twentieth Century.* Ithaca: Cornell University Press.
Thompson, E. P. 1971. "The Moral Economy of the English Crowd in the Eighteenth Century," *Past and Present* 50: 76–136.
Tweedie, D. P. and G. Whittington. 1984. *The Debate on Inflation Accounting.* Cambridge, UK: Cambridge University Press.
Walker, R. G. 1992. "The SEC's Ban on Upward Asset Revaluations and the Disclosure of Current Values," *Abacus* 28(1): 3–35.
Wallison, Peter J. 2008. "Fair Value Accounting: A Critique," *Financial Services Outlook* July: 1–8.
Wehmer, Edward J. and David A. Dykstra. 2008. Letter to Mr. Russell G. Golden, FASB Technical Director, Financial Accounting Standards Board. October 9, 2008. Lake Forest IL: Wintrust Financial Corporation.
Weimer, Arthur M. 1937. "The Work of the Federal Housing Authority," *Journal of Political Economy* 45(4): 466–83.
Wickens, David L. 1937. "Developments in Home Financing," *Annals of the American Academy of Political and Social Science* 190: 75–82.

15

Euro Area Risk (Mis)management

Barry Eichengreen

INTRODUCTION

Other chapters in this volume consider manmade disasters related to the production of energy – the Santa Barbara and Gulf oil spills and the possibility of nuclear power plant accidents in Japan, Germany, and France – focusing on how risks were managed, why dangers were overlooked, and how risk-management practice was revised following incidents. Additional chapters consider financial disasters – the depression and financial crisis of the 1930s and the global credit crisis of 2007–2008 – and similarly ask how risks were managed, why dangers were overlooked, and how risk-management practice changed subsequently. This chapter considers another manmade disaster, the European sovereign debt and banking crisis. It too asks how economic and financial regulators perceived risks and sought to regulate them before the fact. It discusses why the framework they implemented failed to avert the worst. Next it tries to understand the response to the disaster in terms of regulatory, institutional, and policy reform and concludes with some speculations about whether the new framework will be more successful than its predecessor in achieving its goals.

Before proceeding, a few clarifications are in order. First, while the preceding paragraph, like much of the literature, refers to Europe, the focus in this paper is on the Euro Area, the 17 countries that (at the time of writing) have adopted the euro. But it acknowledges that the Euro Area is embedded in a larger institutional entity, the European Union, something that presents both constraints and opportunities for risk management and reform. Second, the crisis that is its subject is better referred to in the plural, as crises or at least as a crisis with multiple dimensions: sovereign debt crisis, banking crisis, growth crisis, and political crisis. These dimensions, while distinct, interact and reinforce one another in both negative and positive ways.

The next section starts with a timeline of the crisis, framed so as to highlight the emphases in subsequent sections.[1] I then describe how risks to economic and financial stability were perceived prior to the onset of the crisis and the institutions and policies put in place to contain them. This if followed by a section that asks why those institutions and policies failed to avert the worst. The next section describes the institutional and policy reforms undertaken in response to the crisis. The final section concludes.

The argument runs as follows. In thinking about the nature and immediacy of the risks confronting their monetary union, the views of the epistemic community of European policy-makers were heavily shaped by history, by faith in the operation of financial markets, and by expert opinion from likes of the rating agencies. These three factors caused them to focus on a narrow and specific set of risks, namely that excessive budget deficits and their monetization would give rise to high inflation, at the expense of other sources of risk, including private debt and high bank leverage. When Europe was then sideswiped by Lehman Bros. and the Global Financial Crisis, these private-debt and high-bank-leverage problems came to the fore, and the monetary union descended into serious crisis. While that crisis spawned much discussion of institutional reform, policy-makers continue to focus, to a remarkable extent, on excessive deficits at the expense of other risks. This too reflects the powerful role of historical perception as a framing device.

TIMELINE OF THE CRISIS

Initially it was possible for European officials and commentators to regard what was still quaintly referred to as "the Subprime Crisis" as an exclusively American affair. This impression began to change in the summer of 2008 with the revelation that IKB (formally, the Dusseldorf-based *Bank fur deutsche Industrieobligationen*, or Bank for German Industry

---

[1] Any chronology of the crisis that is less than book-length is necessarily partial and selective. In selecting events to emphasize, I am consciously attempting to set the stage for the remainder of the paper, as noted, rather than to be comprehensive. I am also aware that this account fails to capture the high drama of the crisis, but then I am not trying to write a thriller.

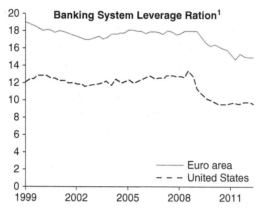

¹Ratio of adjusted assets (total assets minus intangible asses) to Tier 1 capital (capital plus reserves minus intangible assets).

FIGURE 15.1 Bank leverage ratios in the Euro Area and the United States

Obligations) was heavily invested in US subprime securities and had to be rescued by the German government.[2] This was followed in August with the news that BNP Paribas was suspending three in-house hedge funds due to problems with their holdings of US mortgage securities. Suddenly it was clear that not only did the US have a shadow banking system, but also that the US had a shadow banking system heavily invested in subprime-related claims. The infection had not yet reached Europe's large universal banks, but the news was a reminder that leading banks in a number of European countries, starting with Spain and Ireland, were heavily exposed to the domestic real estate sector.

Further questions then arose in September with the failure of Lehman Bros., which caused securitization and wholesale money markets in the United States and globally to seize up. The fact that Euro Area banks were even more highly leveraged than their US competitors and relied even more heavily on wholesale funding now came into focus as a risk to stability (see Figure 15.1).[3]

Yet despite these warning signs there was no sense of impending financial crisis. There was, to be sure, an appreciation of the potential

---

[2] In fact, IKB had been one of the principal "customers" on the other side of John Paulson's Abacus trade that was notoriously intermediated by Goldman Sachs (see Zuckerman 2009).

[3] Data for the figures in this paper, other than Figure 15.6, is from Barkbu, Eichengreen, and Mody (2013).

for liquidity problems, given that much of the wholesale funding on which Euro Area banks depended was denominated not in euros but in dollars.[4] But this problem was addressed by arranging dollar swap lines with the Federal Reserve System. There was awareness that a deep decline in trade and output might result in more nonperforming loans, but this danger could be contained by avoiding mutually destructive resort to trade protection and arranging an internationally coordinated fiscal initiative, as the G20 did under Gordon Brown's leadership in November 2008. With hindsight, a more differentiated response would have been better, with more stimulus by countries with relatively strong banking systems and less from countries with large contingent banking and financial liabilities. It would have been better had there been more attention to banking-system vulnerabilities from the start. But European leaders, having been socialized to expect trouble in the fiscal accounts, not in bank balance sheets, failed to anticipate the risks.

More consonant with policy-makers' presumptions was the revelation in 2009 that the Greek government had fudged its budget statistics. Following its October victory in snap elections, the new government of George Papandreou acknowledged that its predecessor had vastly understated the budget deficit. In response to the admission that the Greek deficit for 2009 was in fact 12.7 percent of GDP, not 3.7 percent, the rating agencies sharply downgraded Greek bank and sovereign debt. The European Commission also issued a report condemning "severe irregularities" in the Greek government's budgetary procedures.[5]

With hindsight, it is peculiar that neither the rating agencies nor the Commission had a glimmering that something was amiss. But having received this wake-up call, they now began to worry about the finances of other European countries such as Portugal, Spain, and Ireland, which it was thought might be harboring similar financial problems.[6] Contagion had never been high on the watch list of risks to sovereign credit worthiness in the Euro Area. In retrospect, this too seems peculiar, given how interest rate compression had spread contagiously across member states in the good times before 2008 (see Figure 15.2). But from this point, with

---

[4] On this previously underappreciated feature of European banking, see Shin (2012).
[5] In addition, in November 2009 Dubai World, a conglomerate owned by the government of Dubai, requested a six-month debt moratorium of its creditors. The news unsettled financial markets, leading to an increase in risk aversion.
[6] The idea of a wake-up call effect originates in the literature on the Asian financial crisis and has been formalized by Ahnert and Bertsch (2012).

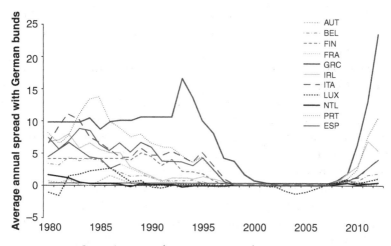

FIGURE 15.2 Sovereign spread compression in the Euro Area (1990–2011)

contagion seemingly operating in reverse, the risk was foremost in officials' minds.

There then followed, starting in early 2010, a complicated charade in which Greece sought to restore investor confidence by unilaterally raising taxes and cutting public spending while denying that it would turn to the Commission, the European Central Bank, and the International Monetary Fund for assistance. The Commission and the ECB, for their part, denied that emergency financial assistance would be extended by themselves or, worse, in conjunction with an extra-European entity like the Fund. All the while, negotiations proceed behind the scene. Already in February 2010, in response to a public-sector wage freeze, mass demonstrations, and a 24-hour strike closed schools and halted the air-transport system. This highlighted the political dimension of the crisis; it was a warning that governments pursuing policies of austerity and demanding painful sacrifices of their constituents might not retain political support for long. And if their constituents turned against them, the consequences for the stability and, indeed, the existence of the Euro Area were, at a minimum, uncertain.

On April 11, 2010 EU leaders announced the long-denied bailout for Greece, with a one-third contribution from the IMF. Greece requested activation and disbursement of funds two weeks later, prompting Standard & Poor's to downgrade Greek debt to junk. For good measure it also downgraded Portuguese and Spanish debt. In May, the 27 EU member states then agreed to create a special-purpose financial vehicle, the European Financial Stability Facility (EFSF), authorized to borrow up to €440 billion to fund this and other loans[7]. In July, in response to more bad news about the Greek economy, the package was revised, its size being increased by 40 percent. In August, with more signs that the confidence crisis was spreading to Italy and Spain, the ECB announced its intention of buying their sovereign bonds in order to put a ceiling on spreads. Italy and other countries passed austerity budgets. The extension of EFSF and ECB support occurred against the backdrop of objections, generally by the political opposition, in other Euro Area countries such as Germany. Those objections again flagged the political dimension of the crisis; they raised unsettling questions about how far and under what conditions stronger European countries would be prepared to aid their weaker brethren.

The answer to this last question came into focus at the end of October, when German Chancellor Angela Merkel's governing coalition, looking forward to state elections in 2011, endorsed the idea that bondholders should be forced to take losses in any future rescue of a European sovereign. The position may have been expedient politically and admirable economically (many economists would endorse the idea of haircuts for junior creditors on moral hazard grounds), but coming at this point in the crisis it was destabilizing. Bondholders fearing that they would be first to be sacrificed in the event of additional difficulties rushed to dump the bonds of other potential crisis countries.

Ireland, already in the throes of a property market collapse and incipient banking crisis, was affected most immediately, with Irish spreads rising to a 600-point premium over German bunds. No longer able to tap the markets, Dublin was forced into talks with the EU, the ECB, and the IMF (the so-called Troika). The Irish program controversially required the government, which had assumed custodianship of its insolvent banks, to service bank debt to senior unsecured bondholders in full, saddling the Irish sovereign with a crushing burden. The bondholders in question were disproportionately other European banks; in addition, banks in other Euro Area countries relied for their own funding on the same kind of senior unsecured debt. Thus, whatever else one might say about it, this

---

[7] That this was necessarily a facility of the 27 EU member states and not just the 17 Euro Area members was a source of differences of opinion when disbursement decisions were taken. This is an example of the additional complications arising from the fact that the Euro Area was embedded in the European Union.

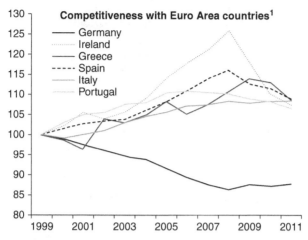

FIGURE 15.3 Divergence and adjustment of unit labor costs

decision reflected a new awareness on the part of policy-makers of the vulnerability of the European banking system and the danger of contagion.

What investors, regulators and journalists had previously perceived as a crisis limited to Greece was now more than that: the sovereign creditworthiness of Ireland and, prospectively, Portugal and Spain was cast into doubt. Still there was little awareness at this point of the development of a systemic crisis that would ultimately infect the entire Euro Area. Officials could point to the fact that the Euro Area was continuing to grow, by 0.6 percent in the first quarter of 2011 and 0.2 percent in 2011Q2. The problem was that growth was on a declining trend; it would go negative in 2011Q4. Moreover, the Euro Area average was held up by strong growth in Germany, which benefited from earlier reforms and strong demand for its exports of capital goods from China. The crisis countries, for their part, were already on the verge of or in outright recession. The adjustment of relative unit labor costs, while underway, was painfully slow, given social resistance to nominal wage cuts (see Figure 15.3).

Meanwhile, policies of fiscal consolidation designed to reduce debt burdens sucked spending out of the economy. This growth crisis reinforced the debt, banking, and political crises. Slow growth depressed the real estate market and weakened bank balance sheets; the weak bank

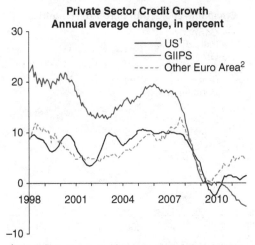

FIGURE 15.4 Change in private sector credit in the Euro Area and the United States

lending that resulted then further weakened growth (see Figure 15.4). Slow growth shrank the denominator of the debt/GDP ratio, frustrating efforts to stabilize and reduce debt ratios; failure to progress on the debt front meant high spreads that in turn made borrowing and investing costly, further depressing growth. Finally, slow growth undermined support for reform and heightened political uncertainty; political uncertainty clouded the policy future, which further weakened demand and growth.

Still, it was Greece that dominated the headlines through the first half of 2011. The country repeatedly missed its budget targets. Public opposition to spending cuts and tax increases continued to mount, fueling speculation that the country might be forced to restructure its debt or leave the Euro Area. In April, Portugal acknowledged that it too had lost access to financial markets and applied for assistance. The EU, ECB and IMF extended it a €78 billion rescue package in May, adopting the 2/3–1/3 formula adopted earlier for Greece. In July, in the face of even more violent demonstrations, the Greek parliament voted to implement an additional round of spending cuts, and further augmentation of the Greek program was agreed.

But if additional money had been committed, nothing fundamental had changed. Institutional investors, aware of the point, now turned their attention to Italy and Spain. Like Greece, Ireland, and Portugal, these countries had heavy debts, bleak growth prospects, and banking-sector and political weaknesses. But unlike Greece, Ireland, and Portugal, they were too big to save by existing rescue mechanisms, given political resistance in Germany to debt mutualization and ongoing transfers. Yields on Spanish and Italian government bonds rose sharply starting in August. Higher debt service costs implied even larger deficits, creating the danger that investor fears might prove self-fulfilling. Rising spreads and falling bond prices also posed problems for banks holding reserves in the form of government bonds – reserves that they could now only turn into liquid form at the cost of major losses.

The autumn saw no let-up in the drumbeat of bad news. S&P downgraded seven Italian banks and the Italian government in September. It cut Spain's credit rating by one notch and put the country on negative outlook in October. In response to protests and demonstrations against the budget cuts required to keep official assistance flowing, on November 1, 2011, Greek Prime Minister Papandreou announced a referendum on the agreement between the Greek Government and its creditors. He revoked the proposal two days later in response to pressure from French and German governments, which feared that a negative result might create expectations of similar outcomes elsewhere. Political noise then grew louder when Italian Prime Minister Silvio Berlusconi announced his resignation on November 8, heightening uncertainty about what kind of Italian government would follow. The cacophony diminished only slightly when the economist Mario Monti was asked to form a technocratic government a week later.

Much of the strain fell on the Euro Area's wholesale-funding-dependent banks. With the development of doubts about the staying power of the Euro Area, the interbank market dried up, as Northern European and US banks with available liquidity increasingly hesitated to place it with their illiquid Southern European counterparts. To prevent the run by wholesale depositors from becoming self-fulfilling, the European Central Bank stepped into the breach. It offered to provide liquidity to financial markets through a program of Long-Term Refinancing Operations, extending funding to the banks at concessional interest rates of 1 percent for up to three years. The banks used their excess liquidity to purchase government bonds, which lowered spreads for the moment. The central bank's action was controversial. It was criticized in some circles for recapitalizing the banks through the back door and surreptitiously taking the pressure off governments. In other circles it was applauded as a first step toward transforming the ECB into a true lender of last resort.

Despite these actions, economic and political crises only worsened in 2012. Growth was worse, with GDP growth Euro Area–wide turning sharply negative after the first quarter. Unemployment was worse, averaging more than 10.9 percent Euro Area–wide in the first half of the year and 11.5 percent in its second. The credit situation was worse, as the rating agencies downgraded additional members of the core (France and Austria) and cut Portuguese government debt to junk-bond status. S&P twisted the knife by downgrading the European Financial Stability Facility, raising questions about whether it would be able to finance additional rescues.

Political uncertainty was also worse. In early May a majority of Greeks voted for parties that rejected the country's rescue agreement with the Troika; inability of the parties to form a coalition forced a second election in June. In the second election, New Democracy, which supported the rescue package and Euro Area membership, received the most votes of any party, walking the country back from the brink. But there was no let-up in political uncertainty with the election in France of the Socialist candidate Francois Hollande, whose rhetoric was pro-growth but whose intentions remained obscure.

Nor was there a decline in economic and financial uncertainty. In March Greece took a modest step forward by restructuring the debt of its government to private creditors. In doing so it took advantage of the fact that most of that debt had been issued under Greek law, permitting the bonds to be retrofitted with restructuring-friendly collective action clauses by an act of Parliament.[8] Officials insisted that the Greek case was sui generis; there would be no more restructurings either in Greece or in other countries. This assertion, however, belied the fact that a significant fraction of the government's debt was in official hands, not least as a result of the ECB's Long-Term Refinancing Operations, making the actual reduction in the debt burden much less than the headline figure of 74 percent and suggesting that further restructurings might ensue. The pronouncement also ignored that the finances of other countries bore an increasing resemblance to Greece.

Spain was a prime case in point. In May the Spanish government was forced to nationalize the country's largest mortgage lender, Bankia. Spanish government debt was then downgraded by the US rating agency

---

[8] A small share of the debt had been issued abroad under English law, but following convention in the London market, this already included collective action clauses. For details see Zettelmeyer, Trebesch, and Gulati (2012).

Egan-Jones, which pointed to the so-called diabolic loop running from banking-sector problems to sovereign debt problems and back. On June 9, the Spanish government requested €100 billion of emergency assistance from the European Union to recapitalize its banks. Spanish Prime Minister Mariano Rajoy attempted to differentiate his country by arguing that its problems were limited to the banking sector. He sought political cover for his government by suggesting that the full conditionality of a Troika program was not justified. His first argument was undermined not just by the diabolic loop between the banks and the sovereign, but also by the large ongoing deficits of the country's provincial governments. His second argument foundered on the unwillingness of the creditor countries, for political reasons, to offer support without full IMF and EU conditionality.

The second half of 2012 was then dominated by two events that diminished the risks of sovereign default and Euro Area break-up. The first was agreement at an EU summit on June 29 to establish a banking union, complete with single supervisor, common deposit insurance scheme and bad-bank resolution mechanism. This agreement was seen as permitting the direct injection of EU funds into the banks from the European Financial Stability Facility and subsequently the European Stability Mechanism. Since the banks would now be supervised and regulated by an agency of the EU, it would be possible to relax existing practice whereby all funds supplied for bank recapitalization, as for other purposes, were obligations of the sovereign. In turn this promised to break the diabolic loop between sovereign and bank credit worthiness.

But as the year progressed it became clear that a vision of banking union constituted at best a long-term prospect. The member states with the least pressing problems were reluctant to cede supervisory authority to the EU. It was not clear that an EU entity – ultimately the ECB was designated – could quickly develop the capacity to supervise scores of systemically significant banks. It was not even clear that EU leaders fully understood what they had signed up for, under duress, at their emergency summit in late June. For all these reasons, backtracking on the June 29 commitment proceeded apace.

The second key event was ECB President Mario Draghi's statement to an investor conference on July 26 that the central bank was ready to "do whatever it takes to preserve the euro." What it might take wasn't specified, but the implication was that the central bank was prepared to purchase sovereign debt in unlimited quantities on behalf of countries requesting assistance. The details emerged in September with the announcement of the bond-buying program, known as Outright

Monetary Transactions, or OMT. The ECB Governing Council explained that it would purchase bonds of one to three years' maturity, but on the secondary market only, thereby paying obeisance to the provision in the central bank's statute prohibiting direct money financing of governments. The rationale, again designed to conform to the ECB's mandate, was that unreasonably high bond spreads were disrupting the transmission of monetary policy. Purchases would be conditional, however, on a country requesting assistance, negotiating a precautionary arrangement or adjustment program with the European rescue fund, and adhering to the terms of that agreement.

This commitment settled the markets even in the absence of an application from Spain or another embattled member state. By the end of November, Spanish bond spreads had fallen to an eight-month low. European leaders asserted that the crisis was over. Yet none of the underlying conditions had been cured. The recession continued to deepen: new data in February showed that Euro Area GDP fell by 0.3 percent in 2012Q4, far exceeding official forecasts. With disappointing growth came disappointing progress on restoring debt sustainability. Banking problems then surfaced in Italy's oldest bank, the Banca Monte dei Paschi di Sienna, and in the Dutch financial services group SNS Reaal. And political noise rose in advance of the Italian elections scheduled for February.

Neither the debt crisis, nor the banking crisis, nor the growth crisis, nor the political crisis had been solved. It was far from clear that the patient was cured. Indeed, at this point it was no longer clear that the patient was even in remission.

PERCEIVED RISKS TO ECONOMIC AND FINANCIAL STABILITY PRIOR TO THE CRISIS

In November 2007, looking forward to the tenth anniversary of the euro, the European Commission convened a conference to commemorate the first decade of the single currency. The subsequent volume, presumably intended to appear in 2009 on the anniversary in question, was published as Buti et al. (2010). The timing of the convocation was notable: the Subprime Crisis had erupted in the United States, as falling real estate prices had begun creating problems in securitization markets, leading to the collapse of prominent investment funds with positions in mortgage-related securities. But, at this point, financial instability was still seen as limited to the US. It had not yet migrated to Europe.

The Buti et al. volume thus provides a window into contemporaneous perceptions of the risks to Euro Area stability. At one level, it points to the perception that serious risks were absent. There was no discussion of a crisis of sovereign debt sustainability of the sort that could lead a Euro Area government to restructure its debt – no discussion of "modalities" such as whether official as well as private debts should be restructured.[9] There was no discussion of the special difficulties of containing and resolving a banking crisis in a monetary union, or of whether the absence of a banking union with a single supervisor, single deposit insurance scheme and single resolution mechanism was a key shortcoming.

Nor does the volume in question consider the role that might be played by the European Commission, the European Central Bank and the International Monetary Fund in organizing an emergency response to a European financial crisis. There was no discussion of the role of the ECB as lender of last resort. There was no analysis of "convertibility risk," of the possibility that a participant might choose or be forced to leave the monetary union, or what this might imply for the stability of the other members' finances. There was no discussion of TARGET 2 imbalances, the assets and liabilities accumulated by the members of the European System of Central Banks through the operation of their real-time gross-settlement system, imbalances that grew explosively once private capital flows shut down.

The absence of discussion of these questions in the Buti et al. volume appears remarkable, with benefit of the hindsight that subsequent experience provides. It is indicative of the failure of the epistemic community of Euro Area experts (the transnational network of knowledge-based experts helping decision-makers managing the monetary union) to define the problems they faced and devise solutions to them.[10] Instead, contributors built on the theory of optimum currency areas (OCAs) to describe how unemployment and inflation differentials across Euro Area member states might be higher than analogous differentials across the 50 US states, reflecting lower levels of labor mobility in Europe and the absence of federal taxes and transfers to compensate for the loss of monetary

---

[9] An issue that proved especially controversial in the context of the Greek sovereign debt restructuring of 2012; see above.

[10] This definition of epistemic communities is from Haas (1992). Some readers may think that I make too much of one conference volume. But not only is Buti et al. a publication of the European Commission, but this nearly one-thousand-page tome arguably assembles the collective wisdom of virtually all the leading thinkers on the single currency at the time the conference in question was held. I was there; I plead guilty.

independence at the member-state level.[11] They focused on the ECB's monetary policy operating strategy comprised of two pillars —a de facto inflation target and, in Bundesbank tradition, a monetary target— asking how the two guideposts might be reconciled, and on the credibility and transparency of the central bank's communications strategy. The focus was on inflation risk, that is, on whether the ECB had established a credible commitment to low inflation and what sacrifices might be required to maintain it. Chapters on enlargement and governance considered whether the conduct of monetary policy would become more complex and potentially unmanageable as additional member states joined the currency area, given the representation of every national central bank on the ECB board and its system of unweighted voting. The chapters on the euro and financial markets were triumphal; they described how the advent of the single currency worked to facilitate the construction of uniform benchmarks and well-defined yield curves and to foster the emergence of deeper and more liquid bond markets. They described the progress of the euro in attaining an international- and reserve-currency role. The chapters on structural reform, if not exactly triumphal, left a positive impression. The euro, they concluded, had been a force for reform. Members of the euro area had been impelled to redouble their efforts to reform product and labor markets, given the absence of the exchange rate as an instrument of adjustment at the national level, although it was acknowledged that more remained to be done.

The closest the authors came to anticipating serious risks was in the chapters on fiscal policy and on growth and volatility. The concern with fiscal policy hardly comes as a surprise, since this source of risk had been a preoccupation of the euro's founders. Fiscal preconditions for membership (the "convergence criteria") had been included in the Maastricht Treaty, sealing the agreement to move to monetary union in 1993. They required aspiring national participants to bring their sovereign-debt-to-GDP ratios down to 60 percent (that being the EU average when the treaty was negotiated) and their consolidated-deficit-to-GDP ratios down to 3 percent (that being the level needed to keep debt ratios stable given conventional assumptions about nominal interest rates and income growth).[12] But there was wiggle room in the application of the criteria. That they were then waived to permit the participation of Italy, Belgium, and other high-debt countries as founding members of the Euro

---

[11] The locus classicus of the theory of optimum currency areas is Mundell (1961).
[12] See Buiter, Corsetti, and Roubini (1993).

Area only reinforced the concerns of policy-makers and experts over fiscal risks.

The result was the Stability Pact (subsequently the Stability and Growth Pact), negotiated in 1997 and entering into force in 1998, empowering the European Commission and Council of Ministers to monitor the fiscal performance of members after joining the monetary union and to issue recommendations for policy action to keep debts and deficits within their respective limits. But when France and Germany violated the 60 and 3 percent reference values in 2004 and no action was taken, and similarly when punitive proceedings were started against Portugal in 2002 and Greece in 2005 but no financial penalties were levied, academics, investors, and other commentators raised questions about the effectiveness of the process. In response, the enforcement criteria were adjusted. Specifically, they were adjusted to account for the business cycle, acknowledging that excessive debts and deficits might reflect unfavorable external conditions rather than the decisions of national governments.[13] In practice, the additional discretion only made the pact more difficult to enforce.

This preoccupation with the risks of excessive deficits was rooted in the belief that they might create pressure for the European Central Bank to monetize public debts. The consequence would be high inflation, the worst fear of the German, and in particular, the Bundesbanker in the street. With hindsight we can say that this fear was exaggerated. Although the euro crisis has had many adverse consequences, high inflation is not one.[14] The possibility that excessive debts might require restructuring that can damage bank balance sheets and expose the absence of an adequate mechanism for recapitalization was not meaningfully discussed. The fact that large deficits ultimately posing a threat to debt sustainability might result not from the profligacy of governments but from a banking crisis and associated recapitalization costs was not part of the conversation.

The key chapters of Buti et al., with hindsight, were those by Barrell and coauthors on convergence and growth, and by Gerlach and Hoffman on implications for international stability.[15] Barrell et al. acknowledged

---

[13] Failure to do so having been one of two factors that had led officials to hesitate in applying penalties to previous national violators, the self-interested political power of the large member states France and Germany being the other.

[14] It can always be objected that the inflationary explosion is just around the corner, but with the crisis in its fifth year at the time of writing, the case is becoming progressively harder to make.

[15] See Barrell et al. (2010) and Gerlach and Hoffman (2010).

that the growth performance of the Euro Area was mixed. They then went on to consider channels through which the single currency influenced growth outcomes. They suggested that, past disappointments notwithstanding, the single currency was likely to exercise a positive impact on growth in the future, especially in the larger countries at the core of the monetary union.

What was absent was an awareness that the "cohesion process" (the convergence of levels of per capita income and economic structure through fast growth in the poorer members of the Euro Area periphery) was at this very time grinding to a halt. The second generation of work on the theory of optimum currency areas had pointed to the endogeneity of the OCA criteria.[16] As incomes and economic structures converged, a common monetary policy would become more appropriate for the member states, and the monetary union would operate more smoothly, this new literature suggested. A leading indicator of mounting strains would have been evidence that convergence had halted and, indeed, shifted into reverse (see the bottom panel of Figure 15.5). But this evidence, while accumulating, was not marshaled.

Gerlach and Hoffman's conclusion was that "macroeconomic stability ha[d] increased since the inception of EMU" (2010, 648). They drew this conclusion both for nominal stability – for the volatility of interest rates and inflation – and for real stability, notably the stability of consumption. The greater stability of consumption was in turn attributed to improved macroeconomic policies and to a monetary policy that was synchronized across countries. Euro-related increases in financial integration, moreover, were credited with enhancing opportunities for international risk sharing. That is to say, as a result of the euro's role in fostering an integrated capital market, households were better able to borrow and lend so as to buffer the impact on consumption of idiosyncratic income and output shocks. This analysis, encapsulating the conventional wisdom circa 2007, did not anticipate that capital flows within the Euro Area might amplify imbalances. It may be too much to ask, as Queen Elizabeth II did, why economists failed to anticipate the risks to economic and financial stability. But the absence of any discussion of those risks in a capstone chapter on stability and volatility in the Euro Area is striking.

Two summary impressions flow from this account of the views of the epistemic community of expert analysts and policy advisors on the eve of the crisis. First, opinions were heavily informed by past risks, notably risks

[16] See Frankel and Rose (1996).

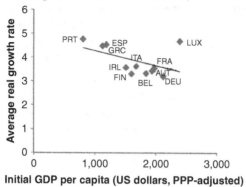

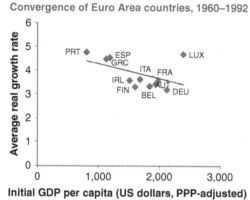

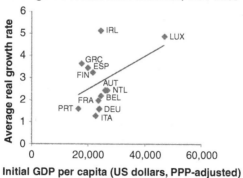

FIGURE 15.5 Convergence of Euro Area countries (1960–92)

of inflation and excessive deficits, that had preoccupied policy debates during the formative years when key officials had received their educations and cut their teeth as policy-makers. Inflation had been a problem prior to the advent of the single currency, in the Euro Area periphery in the 1970s and 1980s and in Germany in earlier decades. The inflation in question was ascribed to pressure on central banks to monetize government budget deficits, in turn focusing attention on the risks of excessive debts. This backward-looking analysis caused the epistemic community of analysts and policy-makers to inadequately appreciate more serious risks emanating from other sources. As noted in the earlier chapter contributed by Thomas Birkland and Megan Warnement, policy change is "filtered by pre-existing ideological or epistemological commitments" and "events are unlikely to overcome these forces unless they are sufficiently severe to overcome these constraints" (Birkland and Warnement, Chapter 5, this volume).

Second, the relatively smooth performance of the Euro Area in its first 10 years led analysts to focus on minor dangers rather than catastrophic tail risks. That the theory of optimum currency areas directed attention to "run of the mill" business cycle disturbances, as opposed to full-blown financial crises, further encouraged this perspective. Analysts focused on the ups and downs of business cycles and the convergence or divergence of inflation and interest rates, not on the risk of a catastrophic financial crisis.

BASES FOR RISK ASSESSMENT

One cannot understand the halting, slow motion character of Europeans responses to their sovereign debt crises without coming to terms with historical perceptions. History played an important role in shaping how the epistemic community of analysts and policy-makers perceived risks to the stability of the currency area. Specifically, historical experience with inflation rooted in chronic public-sector imbalances pointed to debts and deficits as the principal source of risks. There was nothing unusual about the influence of history in this context. Experience with high inflation in the 1920s, a phenomenon also linked to chronic budget deficits, similarly informed and constrained policy responses to the Great Depression of the 1930s (Eichengreen 1992). The fact that the second half of the twentieth century was a period of relative stability for banking systems in Europe – though there were bank failures, none was serious enough to pose risks to systemic stability – encouraged policy-makers and analysts to minimize this source of stability risk.

Financial market outcomes further shaped views of potential risks. So long as market prices – interest rates on government bonds for example – indicated that investors were sanguine about the prospects, policymakers concluded that those prospects must be good. In part, their faith in market discipline reflected intellectual trends, for example the spread of the efficient-markets hypothesis in academia.[17] In part, it reflected the rapid growth of the financial sector, its profitability and its political influence in the years leading up to the crisis. And, in part it reflected fading memories of earlier periods like the late 1920s and early 1930s when financial markets had malfunctioned with disastrous consequences.

To the extent that information on economic and financial conditions was costly to assemble and process, analysts relied on "delegated monitors" that had sunk the costs of investing in a monitoring technology and were in the business of selling assessments based on this expertise. These were the credit rating agencies.[18] Ratings were seen as providing a leading indicator of future problems. They were incorporated into regulatory practice: they were the basis for attaching risk weights to complex securities on the balance sheets of Euro Area banks. The longer European banks and government bonds were rated AAA, the more remote the risk of serious problems and losses appeared.

With hindsight it is clear that these were thin reeds on which to hang assessments of stability risk. A substantial literature, motivated initially by experience during the Asian financial crisis, questions the value added by credit ratings. It suggests that ratings respond to events in financial markets and the economy rather than anticipating them. They have little predictive content relevant to the future evolution of asset prices and economic and financial conditions above and beyond that contained in current asset prices. Ratings tend to be lagging rather than leading indicators and as such can have a self-reinforcing impact on crisis risk.[19] Partnoy (1999) provides an early, post-Asian-crisis critique of the rating agencies, while De Haan and Amtenbrink (2011) review their operation from a specifically European perspective.

Similarly, the efficient-markets hypothesis should have been thought of as no more than a point of departure for thinking about financial markets

---

[17] See inter alia Fama (1970).
[18] The seminal work on delegated monitoring (e.g. Diamond 1984) focuses on banks, but much the same argument can and has been made about rating agencies.
[19] See Ferri, Liu, and Stiglitz (1999).

and their information content. Subsequent research has not been kind to the most basic tenet of the hypothesis: namely, that future asset prices should not be predictable on the basis of current asset prices but rather should follow a random walk.[20] Detailed empirical work has turned up a substantial number of "anomalies" in the behavior of asset markets and prices hard to square with the presumption of market efficiency. Those anomalies in turn have spawned a subfield of behavioral finance in which scholars study how investors use rules of thumb, past experience and intuition as guides to decision-making, leading to overconfidence, projection bias and related behaviors and to market outcomes at variance with the efficient-markets view.[21] The bottom line is that current market conditions are not always a reliable guide to what comes next.

The abrupt and violent reaction of sovereign spreads across Southern Europe to news about the size of the Greek budget deficit is hard to square with the efficient-markets view.[22] It is not as if there was simultaneous news of larger than expected budget deficits in other countries. The more plausible interpretation is that news that the published Greek figures were less reliable than previously thought led investors to question the reliability of the figures published by other Southern European governments. This reaction can be related to theories of herding and "informational contagion" in financial markets (see, e.g., Dungey et al. 2005), where market participants with incomplete information or facing costs of processing it reassess the situation of a country on the basis of new information about other countries, or where market participants reassess the condition of a financial institution on the basis of new information about other financial institutions.

The "wake-up call hypothesis" – the idea that investors devote costly time and attention to assessing risks to stability only when the immediacy of those risks is brought to their attention – finds support in the data on sovereign spreads in the Euro Area. These have been analyzed previously by a number of investigators such as Alessandrini et al. (2012). Their analysis is updated and extended in Tables 15.1 and 15.2 below. There I

---

[20] A compendium of the recent evidence is Lo and MacKinlay (2001).
[21] For an introduction, see Shleifer (1999).
[22] But not with prior historical experience: in an early contribution to the literature Sy (2001) similarly found that Asian spreads were excessively low relative to fundamentals before the Asian crisis and then excessively high immediately following its outbreak. Mishkin (2003) similarly points to the role of asymmetric information that led investors to magnify relatively small risks in all Asian countries during the crisis.

TABLE 15.1 *Determinants of 10-year government bond spreads relative to German bunds 2005Q1–2011Q4*

Regressor	Coefficient	Standard Error	T-statistic
Global risk aversion	-0.29	4	-0.074
Primary balance	-8.73	1.445	-6.04***
Public debt	0.80	0.24	3.29**
Growth	-21.34	2.58	-8.27**
Bid-ask	7.93	0.27	28.9***
r-squared=0.8526			
Adjusted r-squared=0.8499			
F-statistic: F(5,274)=316.9			
p-value<2.2e-16***			

Significance codes: 0 '***' 0.001 '**' 0.01 '*' 0.05 '.' 0.1
Constant term included but not reported.
Source: See text and Appendix A.

TABLE 15.2 *Determinants of 10-year government bond spreads relative to German bunds, 2005Q1-2011Q4 with post-2008Q2 interaction effects*

Regressor	Coefficient	Standard Error	T-statistic
Global risk aversion	6.97	16.57	0.421
Primary balance	-0.70	2.78	-0.253
Public debt	0.22	0.34	0.635
Growth	0.13	4	0.032
Bid-ask	9.18	18.59	0.494
D	17.5	15.59	1.124
Global risk aversion: D	-4.55	16.96	-0.268
Primary balance: D	-10.13	3.09	-3.28**
Public debt: D	2.02	0.43	474***
Growth: D	-27.99	4.76	-5 87***
Bid-ask: D	-2.07	18.59	-0.111
r-squared=0.9135			
Adjusted r-squared=0.9099			
F-statistic: F (11,268)=257.3			
p-value<2.2e-16***			

Significance codes: 0 '***' 0.001 '**' 0.01 '*' 0.05 '.' 0.1
Constant term included but not reported.
Source: See text and Appendix A.

estimate a standard model of the determinants of sovereign spreads (10-year government bond spreads for Euro Area members relative to German bunds on quarterly data since 2005).[23] Table 15.1, with estimates for the entire period 2005Q1 through 2011Q4, seemingly supports the hypothesis of market discipline: spreads fall significantly with the primary budget balance (larger budget surpluses make for lower spreads); they rise with the public debt ratio (heavier debts make for wider spreads); they decline with economic growth (faster growth makes servicing debt easier); and spreads rise with the bid-ask spread (shallower markets with larger transactions costs make for higher spreads).

But Table 15.2, where these same variables are in addition interacted with a dummy variable for 2088Q3 and the remainder of the sample, shows that the explanatory power is entirely concentrated in the period following the failure of the two BNP Paribas funds in the summer of 2008. The primary balance, debt ratio, and growth rate are all significant and matter in the expected way after 2008 but not before. It is as if investors suddenly awoke to the relevance of such factors to default risk, where they had been sleeping soundly through their signals before.[24]

The same is true of informed press commentary. Prior to 2009, the major financial news organs in the United States and Britain contained no mentions of the possibility of a Euro Area debt crisis; Table 15.3 lists the first mentions of this and related phrases in the *Financial Times, New York Times* and *Wall Street Journal*. Similarly, there was little mention of this possibility prior to 2009 on the Internet, but an explosion of references thereafter (see Figure 15.6).

Finally, while historical experience can provide a useful frame through which to assess current circumstances and risks, history does not speak for itself. There exist a plethora of potential historical analogies with which current circumstances may be compared. Cognitive scientists and others who have studied analogical reasoning have sought to understand why specific historical analogies or precedents are chosen by policy-makers to the neglect of others. They point to the disproportionate influence of "searing" or "molding" events of "transcendent importance" (Zelikow

---

[23] Sources and definitions of variables are in Appendix A.

[24] In related work, Georgoutus and Migiakis (2010) consider the behavior of sovereign spreads (relative to bunds) from 1992 through 2009 – that is, both before and after monetary unification – using regime-switching regression. They allow the data to tell them when the switches occurred – they do not simply assume that the answer is 1999.

TABLE 15.3 *First mention of crisis phrases*

Greece Debt Crisis

"Internet Giants," New York Times, 1/22/2009

But there is a catch: Germany's membership in the Euro zone. This presents complications because of the perilous fiscal and economic positions of several other members. Portugal, Greece, and Spain have had their debt downgraded recently by Standard & Poor's and have payments deficits around 10 percent of GDP, while Italy has enormous public debt and poor budgetary control. www.lexisnexis.com/lnacui2api/api/version1/getDocCui?lni=4VF4-XKN0-TW8F-G16F&csi=6742&hl=t&hv=t&hnsd=f&hns=t&hgn=t&oc=00240&perma=true

"Once a Boon, the Euro is Now a Burden for Some," New York Times, 1/24/2009

Last week, Standard & Poor's downgraded Greek debt to A−, and the gap between the interest rate it pays on its bonds, versus what richer countries like Germany pay, is nearly 3 percentage points, the widest in the euro zone. www.lexisnexis.com/lnacui2api/api/version1/getDocCui?lni=4VFK-27F0-TW8F-G11K&csi=6742&hl=t&hv=t&hnsd=f&hns=t&hgn=t&oc=00240&perma=true

Credit Risk Diverges Across Eurozone," Financial Times, 7/21/2008

CDS prices have risen since June 5, when Jean-Claude Trichet, European Central Bank president, stepped up warnings on inflation. Since then, the cost to insure German debt against default has risen by EUR1,000 to EUR6,000 for EUR10m of debt. In contrast, the cost to insure Greek debt has risen EUR16,000 to EUR51,000. It has risen EUR15,000 for Italy, EUR14,000 for Portugal, EUR13,000 for Spain and EUR10,000 for Ireland. www.lexisnexis.com/lnacui2api/api/version1/getDocCui?lni=4T1N-8FY0-TW84-P00H&csi=293847&hl=t&hv=t&hnsd=f&hns=t&hgn=t&oc=00240&perma=true

"Italian and Greek Bonds Underperform," Financial Times, 11/21/2008

Italian and Greek bond yields widened sharply in relation to German yields this week as investors grew wary of buying securities of governments with high levels of debt. www.lexisnexis.com/lnacui2api/api/version1/getDocCui?lni=4TYW-VMB0-TW84-P15W&csi=293847&hl=t&hv=t&hnsd=f&hns=t&hgn=t&oc=00240&perma=true

"Emerging Market Assault Comes to Euro Zone," Wall Street Journal, 10/24/2008

Greek government bonds are the latest target. Ten-year Greek debt yielded 1.18 percentage points more than similar German debt on Friday. That spread was in double digits as recently as Monday. http://blogs.wsj.com/economics/2008/10/24/emerging-market-assault-comes-to-euro-zone/?KEYWORDS=Greek+debt

(continued)

TABLE 15.3 *(continued)*

Greece Debt Crisis

Portugal Debt Crisis

"Internet Giants," *New York Times,* 1/22/2009

But there is a catch: Germany's membership in the Euro zone. This presents complications because of the perilous fiscal and economic positions of several other members. Portugal, Greece, and Spain have had their debt downgraded recently by Standard & Poor's and have payments deficits around 10 percent of GDP, while Italy has enormous public debt and poor budgetary control. www.lexisnexis.com/lnacui2api/api/version1/getDocCui?lni=4VF4-XKN0-TW8F-G16F&csi=6742&hl=t&hv=t&hnsd=f&hns=t&hgn=t&oc=00240&perma=true

"Europe Watches as Portugal, Its Deficit Rising, Struggles to Restore the Economy," *New York Times,* 2/10/2010

Portugal's debt is expected to rise to 85 percent of gross domestic product this year, from 76.6 percent in 2009, because of rising unemployment and government spending on infrastructure projects like dams, hydroelectric power systems and a high-speed rail line to Madrid. www.lexisnexis.com/lnacui2api/api/version1/getDocCui?lni=7XS2-NGB0-Y8TC-S343&csi=6742&hl=t&hv=t&hnsd=f&hns=t&hgn=t&oc=00240&perma=true

"Capital Rules Undermine Banks," *Financial Times,* 1/16/2010

The stock fell nearly 5 percent on Friday on concerns about Portugal's debt burden and budget deficit. www.lexisnexis.com/lnacui2api/api/version1/getDocCui?lni=7XJP-B7P1-2PB1-83VT&csi=293847&hl=t&hv=t&hnsd=f&hns=t&hgn=t&oc=00240&perma=true

"Portugal Debt Chief: We're Not Greece," *Wall Street Journal,* 3/1/2010

The European Commission has given Portugal until 2013 to bring its deficit below the 3 %-of-GDP threshold required by European Union rules. In recent months, rating agencies have warned of possible downgrades to Portugal's ratings, citing rapid deterioration of public accounts and historically low economic growth. http://online.wsj.com/article/SB10001424052748703943504575099562100208294.html?KEYWORDS=%22Portugal+debt%22

Spain Debt Crisis

"Internet Giants," *New York Times,* 1/22/2009

But there is a catch: Germany's membership in the Euro zone. This presents complications because of the perilous fiscal and economic positions of several other members. Portugal, Greece, and Spain have had their debt downgraded recently by Standard & Poor's and have payments deficits around 10 percent of GDP, while Italy has enormous public debt and poor budgetary control. www.lexisnexis.com/lnacui2api/api/version1/getDocCui?lni=4VF4-XKN0-TW8F-G16F&csi=6742&hl=t&hv=t&hnsd=f&hns=t&hgn=t&oc=00240&perma=true

(continued)

"Wall Street Pushes Higher After 2 Days of Losses," New York Times, 12/10/2009

The advance ended a day of back-and-forth trading Wednesday as investors grew cautious about rising government debt levels in Spain, Greece and other countries. Investors have been looking for safety after the credit rating agency Standard & Poor's reduced the outlook on Spain's debt rating Wednesday. www.lexisnexis.com/lnacui2api/api/version1/getDocCui?lni=7X8V-D8G0-Y8TC-S24N&csi=6742&hl=t&hv=t&hnsd=f&hns=t&hgn=t&oc=00240&perma=true

"Eurozone Exports Plunge 4.7% in month," Financial Times, 1/17/2009

Standard & Poor's, the rating agency, has said it might downgrade Spain's debt ratings because of deteriorating public finances. www.lexisnexis.com/lnacui2api/api/version1/getDocCui?lni=4VD2-62H0-TW84-P06F&csi=293847&hl=t&hv=t&hnsd=f&hns=t&hgn=t&oc=00240&perma=true

"Euro-zone Countries Feel the Debt Crunch," Wall Street Journal, 9/30/2008

FRANKFURT – European governments are finding it harder and more expensive to issue debt as investors become choosier about which governments they fund. The Belgian and Italian governments both barely got through their latest bond issues Monday, with Belgium selling only the minimum it had intended and Italy attracting weak demand. Spain and France may face similar struggles at their government bond auctions Thursday. http://online.wsj.com/article/SB122271427415867 31.html?KEYWORDS=Spain±debt

Ireland Debt Crisis

"S&P Lowers Ireland's Debt Rating a Second Time," New York Times, 6/9/2009

The debt rating of financially stricken Ireland was downgraded Monday by Standard & Poor's for the second time this year as concerns grew about the cost of the government's bailout of banks. www.lexisnexis.com/lnacui2api/api/version1/getDocCui?lni=7VWK-MMF0-Y8TC-S0B4&csi=6742&hl=t&hv=t&hnsd=f&hns=t&hgn=t&oc=00240&perma=true

"Dublin's Dilemma," Financial Times, 1/16/2009

Last week, Standard & Poor's downgraded Ireland's debt outlook from stable to negative and warned it might cut its sovereign debt rating. www.lexisnexis.com/lnacui2api/api/version1/getDocCui?lni=4VCV-6Y00-TW84-P16G&csi=293847&hl=t&hv=t&hnsd=f&hns=t&hgn=t&oc=00240&perma=true

"Ireland Pins Hopes on Global Recovery," Wall Street Journal, 4/14/2009

Has Ireland done enough? Last week's emergency budget and new bank-bailout plan has created some room to maneuver its economy and banking system to safer ground. But it is unlikely to be enough to spare Ireland from a debt crisis if the world economy doesn't start to pull out of its slump early next year. http://online.wsj.com/article/SB123966815674115381.html?KEYWORDS=%22Ireland±debt%22

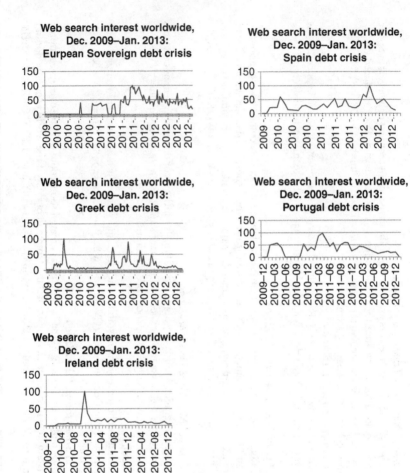

FIGURE 15.6 Internet Incidence of Euro crisis-related terms (Google Trends)

1999). As noted by the political scientist John Kingdon, such experiences and events may play a particularly important role in agenda setting (Birkland and Warnement, Chapter 5, this volume). From this point of view it is unsurprising that the German hyperinflation of 1922–1923 continued to shape the outlook of policy-makers and analysts – especially in Germany – fully eight decades later. Analysts similarly point to the failure of policy advisors, who are in the business of providing simple messages, to provide decision-makers with a portfolio of potential historical analogies pointing to a range of potential risks.

THE INSTITUTIONAL AND POLICY RESPONSE

The most controversial aspect of the response to the materialization of crisis risk has been policies of austerity and structural reform. Impassioned debate continues, in the journals and at the polls, about whether fiscal consolidation, achieved through a combination of tax increases and spending cuts or spending cuts alone, can be expansionary, or whether the fixation on balancing budgets has consigned the Euro Area to a self-inflicted Depression.[25] Analysts similarly disagree on whether measures deregulating labor markets and increasing competition in product markets increase productivity and growth in the long run but cause dislocations and the separation of employees in the short run.[26] The literature on these questions is vast, and I do not attempt to add to it here. Rather, I focus on institutional and policy initiatives at the level of the Euro Area and ask whether they are likely to significantly diminish crisis risk going forward.

A first set of institutional reforms seeks, predictably, to strengthen oversight of national fiscal policies. To facilitate surveillance, member states have agreed to a "European semester" with common timelines for the formulation of fiscal policies. This has led to agreement on a so-called six-pack of legislative measures adopted by the Commission in December 2012. The six-pack provides for more automatic application of the EU's fiscal rules – stronger enforcement of the Stability and Growth Pact. It adds a rule for the growth of government expenditure to existing regulations pertaining to deficits and supplements the deficit criteria with rules for the level and rate of change of debt. It unpacks the procedure for the application of fines, with the goal of making financial penalties more realistic. It encourages the adoption of national fiscal frameworks and procedures that emphasize transparency and medium-term planning. Finally it creates a Macroeconomic Imbalances Procedure that widens

---

[25] On the expansionary fiscal consolidation view see Alesina and Ardagna (1998). Their hypothesis has been prominently questioned by Blanchard and Leigh (2013). My own critique is in Eichengreen (1998).

[26] OECD (2012) is probably the most comprehensive compendium of evidence on the short-run effects of labor market reform. A more technical approach is Veracierto (2007).

the remit of surveillance to encompass external as well as internal imbalances.

The six-pack was then supplemented by a "two-pack" of regulations designed to further strengthen budgetary surveillance. Its first element requires member states to publish medium-term fiscal plans and for those plans to be assessed and conformance to be monitored by the Commission. The Commission is then empowered to request, where necessary, a revision of the national budgetary plan. The second regulation envisages new provisions allowing the Commission and the Council to engage in even closer surveillance of countries in economic or financial difficulty.

Finally, all EU members other than the UK and the Czech Republic have signed a Treaty on Stability, Coordination and Governance that provides for a "fiscal compact" obliging the signatories to adopt balanced budget rules at the national level. The incentive to comply is that only countries that have ratified the treaty and adhere to its terms will have access to the European Stability Mechanism.

This is not the first time member states have sought to strengthen the Stability and Growth Pact or to increase its automaticity. Experience shows that governments are reluctant to sanction their neighbors, given the solidarity and consensus decision-making that characterizes EU affairs. In addition they fear the precedent; they worry that they might be next. Delegating these decisions to an independent entity is the textbook solution to this time-consistency problem. Unfortunately, the obvious delegate, the Commission, is not independent of politics. Commissioners are appointed by agreement with their governments, following a long-established formula designed to ensure balanced national representation. There are proposals for Europe-wide election of the president of the Commission that envisage giving her the power to appoint a team of commissioners.[27] Other proposals imagine delegating responsibility for enforcing the Stability and Growth Pact to an independent committee of experts with dedicated funding, serving long terms in office, and enjoying the autonomy necessary to enforce its provisions. But national governments remain jealous of their fiscal prerogatives. And there is the worry that a proliferation of independent economic agencies – the ECB is one, the committee of fiscal experts would be another – will only widen

---

[27] See, e.g., Berglof et al. (2003).

Europe's democratic deficit. As a result, there has been a reluctance to go down this road.

Commentators despairing of the capacity of the European Union and Euro Area to conduct meaningful surveillance and impose sanctions on its members sometimes recommend relying on market discipline.[28] Currently, investors are acutely aware of the fiscal shortcomings of member states. Lacking a national central bank to backstop national debt markets, members of the Euro Area feel market discipline intensely. The only problem, in this view, is that the EU has not stood behind the no-bailout clause of the Maastricht Treaty. It has repeatedly bailed out member states in difficulty, which weakens the response of financial markets when problems appear on the horizon. The resulting recommendation is that members should recommit to the no-bailout rule. Governments that have run irresponsible fiscal policies should be allowed, indeed forced, to default. Know that this is the practice, market discipline will be strong and suffice to head off problems in all but the most extreme cases.

This approach involves a heavy dose of wishful thinking, for two reasons. First, fear of contagion and systemic repercussions, justified or not, render Euro Area members reluctant to contemplate default by one of their number. In some circumstances, default (nee "restructuring") may be unavoidable. More often, however, it is ruled out first as impossible and then as exceptional. The consequences of default are necessarily uncertain. The non-zero probability of an undesirable outcome encourages officials to put off the day of reckoning in the hope that favorable news averting the need to restructure may turn up. It is unrealistic, in other words, to imagine that recourse to multilateral rescues can be ruled out.

In addition, the idea that market discipline is always and everywhere effective is fanciful. As we saw in earlier in the chapter, market discipline is sporadic. Investors may have been awake to the risks in 2009–2012, but they were blissfully unaware of them for much of the preceding decade.

A further approach to fiscal discipline in the Euro Area has emphasized strengthening the national institutions and procedures through which budgetary decisions are made. Academics and officials have recommended that governments should be required to publish consistent

---

[28] An example of the market discipline approach is Mangenelli and Wolswijk (2007).

medium-term plans; budgeting procedures should be transparent; and more agenda-setting power should be delegated to the prime minister or finance minister in order to address the common-pool problem in which every spending ministry seeks more resources without adequate regard to the implications for the overall balance. Some observers have even suggested that in the extreme, key fiscal decisions, such as the overall size of the deficit, could be outsourced to an independent fiscal commission following practice in countries like Chile.[29] Some analysts have recommended this approach for years.[30] The six-pack alludes to the desirability of these reforms. The Treaty on Stability, Coordination and Governance potentially takes a step in this direction by linking access to emergency assistance to institutional and procedural reform at the national level, although it focuses too much on balanced budget rules and too little on more flexible but equally effective approaches like the establishment of fiscal councils.

This focus on fiscal issues in post-crisis reform efforts suggests that the Euro Area's disproportionate concern with budget deficits is not yet a thing of the past. As indicated by Mayer (Chapter 6, this volume) narrative framing also helps explain the focus on budget deficits and continues to incline European officials toward addressing past problems as opposed to likely future risks.

More promising from this point of view, in principle if not in practice, is the effort to create a new supervisory framework for financial institutions and markets, the financial sector having been deeply implicated in the crisis. Before 2009, financial surveillance was the province of national authorities with relatively weak incentives to share information on potential risks in their financial sectors. February 2009 saw the initiation of a new European System of Financial Supervision with two pillars: a micro-prudential pillar comprised of three new supervisory authorities, the European Banking Authority, the European Insurance and Occupational Pensions Authority, and the European Securities and Markets Authority; and a macro-prudential pillar concerned with systemic issues and placed in the European Systemic Risk Board. Time will tell whether these new structures succeed in exercising sufficiently vigorous surveillance of financial institutions and markets and in requiring prompt corrective action. But by design they are

---

[29] A good summary of the Chilean approach is Frankel (2012).
[30] See von Hagen and Harden (1994) and Wyplosz (2005).

squarely directed at the key risks that were unappreciated in the run-up to the recent crisis.

Problems of excessive leverage and undercapitalization in the banking sector are to be further addressed by the EU's new Capital Requirements Regulation and Capital Requirements Directive IV, which will hold banks to stricter capital standards by transposing the Basel III Capital Accord into EU law.[31] Under these regulations, banks will be required to hold higher-quality capital (more common equity and fewer "hybrid instruments" that are not truly loss absorbing in a financial crisis). This is a useful increase in capital standards from previous levels, at least in comparison with independent proposals.[32]

But adherence to the new capital standard would have done little to effectively insulate troubled banks in the recent financial crisis. The new Capital Requirements Directive continues to rely on credit ratings and risk weightings. Insofar as credit ratings are procyclical, so is the new capital-requirements regime. The problems created by the existence of the shadow banking system, highlighted by the failure of the two BNP Paribas hedge funds in 2008, remain unaddressed. In any case, the new requirement will be phased in over a ten-year period.

Regulators were conscious of the trade-off between strengthening the banks' capital position now to render them more resilient to crises, on the one hand, and the fact that forcing banks to raise more capital now would limit their lending and hinder recovery from the crisis. Observers disagree about whether they got the balance right.[33] Critics of the Basel III approach continue to press for more radical reforms, ranging from narrow banking and measures designed to break up too-big-to-fail banks to the adoption of a US-style Volcker rule limiting proprietary trading by deposit-taking financial institutions. For reasons that are not self-evident, these radical proposals have gained more purchase outside the Euro Area in countries like the UK and Switzerland.[34]

---

[31] The basic reference on Basel III and its progress is Basel Committee on Banking Supervision (2012).

[32] In addition, the directives provide for a liquidity coverage ratio, under which banks are required to maintain liquid balances ("liquidity reserves") to protect against net liquidity outflows and difficulty-causing maturity mismatches. The liquidity standard is a positive response to the previously underappreciated risk posed by reliance on short-term wholesale funding.

[33] For a review see Casselmann (2013).

[34] One might argue that the reasons referred to in this sentence include policy-makers clinging to the dominant conceptual frameworks of the earlier period, the influence of big banks on policy, or a desire to protect domestic banks against foreign competition and

For the moment, responsibility for implementing the new capital regulations and directives continues to reside with national authorities. Thus, the risk remains that national regulators will fail to take into account the Euro Area–wide implications of their decisions. They may delay imposing more stringent requirements on their banks in an attempt to give them a competitive advantage relative to their foreign competitors. They may hesitate to share information that paints domestic banks in an unfavorable light, the existence of the European Banking Authority notwithstanding. The only reliable solution to these problems is a single supervisor with authority over the banking system of the entire area. And, as noted previously, resistance to the creation of such a supervisor is considerable. In addition, there is the further complication of whether banks inside the single market but outside the monetary union, UK banks in particular, should be subject to common supervision.

A final institutional reform, noted previously, was the creation of the European Financial Stability Facility, now the European Stability Mechanism in its permanent incarnation. This was a response to the recognition that, for reasons of history and design, the European Central Bank is limited in its ability to act as a lender of last resort to sovereigns and banks. Yet the crisis, by pointing up the risk of self-fulfilling debt and bank runs, underscored the critical need in a monetary union for a lender-of-last-resort facility. The ESM is now available to carry out this role. It can provide loans and precautionary assistance (the promise of loans). It can purchase bonds on both the primary and secondary markets and, in principle, inject capital directly into Euro Area banks. Loans are contingent on the negotiation of an adjustment program with the Commission, the ECB and the IMF. The ESM has a paid-in capital of €80 billion and €620 billion of callable capital, both from Euro Area countries, giving it a lending capacity of some €500 billion. To some eyes these might seem like impressively large numbers. But, in fact, they are only 4 to 5 percent of Euro Area GDP – less, in other words, than an adequately funded lender of last resort may require.

The Commission and ECB acceded to the involvement of the IMF in the Greek and Irish rescues only reluctantly, but this has turned out to be a happy arrangement from the European point of view. The Europeans have

---

therefore a reluctance to subject them to more demanding capital standards. But the point is that none of these arguments plausibly applies less powerfully to the UK and Switzerland than the members of the Euro Area.

been able to delegate much of the high-visibility day-to-day monitoring of program countries to the Fund, deflecting some of the associated political controversy, without having to give up much in terms of program design. The arrangement has been less happy from the Fund's standpoint. It has contributed money and taken political flack without having much capacity to shape program design. Some critics have suggested that the IMF should, in the future, avoid putting itself in the position where it is a minority participant in a program arranged jointly with a regional entity like the EU.[35] The Fund has become more insistent in raising objections to the European approach to structural adjustment over time. This raises questions about whether the existing division of labor between Euro Area and multilateral institutions will prove stable.

CONCLUSION

Not all risks that materialize can be anticipated, and not all risks that are anticipated can be avoided. The Euro Crisis reflects the materialization of a set of risks that was neither anticipated nor avoided. In thinking about the nature and immediacy of the risks confronting their monetary union, the views of the epistemic community of European thought leaders and policy-makers were heavily shaped by history, by faith in the operation of financial markets, and by expert opinion from inter alia the rating agencies. The first factor directed their attention to a narrow and specific set of risks, namely that excessive budget deficits and their monetization would give rise to high inflation. The response was a set of mechanisms and procedures intended to provide surveillance and force corrections when deficits were excessive – mechanisms and procedures that, in the event, fell short. The second and third factors reinforced their neglect of other sources of risk and led them to underestimate their immediacy. The result was that when Europe was sideswiped by the subprime meltdown and the failure of Lehman Bros., it descended into a very serious crisis.

The argument in this paper, like any argument in political economy, has limits. For example, it paints the views of European policy-makers and market participants with a broad brush. In reality, there always was and always will be some diversity in their views of the nature of the risks to Euro zone stability. The fact that assessments differ in turn points to the limitation of the thesis that those views are importantly shaped by

---

[35] This is the recommendation of Goldstein (2011).

perceptions of history, faith in market forces, and confidence in expert opinion. It is a reminder that history is not a given; rather, the "lessons of history" are drawn by those who interpret it, and different historians interpret it differently, leading to different lessons. Faith in the markets was never absolute; some European officials and, for that matter, investors had less such faith than others. The same is true of trust placed in the rating agencies; the criticism to which the agencies were subjected following the Asian crisis of 1997–1998 is a reminder that skepticism about their assessments was, at least in some circles, of long standing. At the same time, the dominance of a particular point of view, which emphasized inflation and fiscal risks to the neglect of other potentially destabilizing factors, suggests that there was a dominant historical narrative, a prevailing faith in the markets, and a general willingness to trust the rating agencies. Why dissenting voices were not heard more clearly is an important question in and of itself.[36]

That crisis spawned much talk and some progress in the direction of institutional and policy reform. Policy-makers continue to focus on excessive deficits. They have further strengthened procedures for identifying such deficits and correcting them. We will see how effective the traditional approach, so strengthened, is the next time it is tested. In addition, they have strengthened oversight of European banks and financial markets and created an unprecedented rescue mechanism, the European Stability Mechanism, for banks and governments that lose access to the markets. These innovations go at least some way in addressing sources and forms of crisis risk that were neglected before.

But earlier problems remain. Rules for capital adequacy continue to rely on commercial credit ratings, which are less than reliable. Policy-makers continue to rely on the disciplining role of the market, which is sporadic at best. The focus on excessive deficits remains excessive. In all these respects, old habits die hard. It is not surprising, in this light, that Europe has still not succeeded at drawing a line under its crisis.

---

[36] This question deserves a separate paper of its own. One might point to the failure of economists and other policy analysts to undertake a serious study of the relevant history; in the absence of such serious study, "stylized [historical] facts" continued to dominate. One might point to lobbying and argumentation of those who profited from the operation of lightly regulated markets in proselytizing in favor of lightly regulated markets. One might point to lack of competition in the rating agency industry and the tendency for the authorities to hard wire (utilize) commercial credit ratings in their regulation. At this point these are mere speculations.

APPENDIX A: VARIABLE DESCRIPTIONS, SOURCES, AND SUMMARY STATISTICS

			Mean	
Variable	Definition	Source	2000q1–2011q4 (480 obs.)	2005q1–2011q4 (280 obs.)
Spread	Difference in yields to maturity of 10-year government bonds of the Euro member countries relative to Germany's; (Country x's bond yield–German bond yield)*100; Quarterly averages from daily data	Global Insight	67.52	102.17
Global risk aversion	Negative of 1st Component of the PCA of 4 measures of risk: OEX volatility index, BofA Merrill Lynch US Corporate AAA effective yields, BofA Merrill Lynch US Corporate BBB Effective Yields and Euro-Yen Implied 3-month exchange volatility; Quarterly averages from daily data	OEX volatility index and Euro-Yen exchange volatility from Bloomberg; Corporate bond yields from FRED	0.0766	−0.168
Bid-ask	Bid-ask spread in 10-year government bond market relative to German values; Ireland uses 9-year government bond; Bid-ask spread=(bid-ask)*100; Country x's value–German value; Quarterly averages from daily data	Bloomberg	2.078	3.561

(continued)

(continued)

			Mean	
Variable	Definition	Source	2000q1-2011q4 (480 obs.)	2005q1-2011q4 (280 obs.)
Primary balance*	Quarterly primary surplus as a % of GDP; Country x's value-German value; the variable is a (2-1-2) moving average; Not seasonally adjusted	Eurostat-Quarterly non-financial accounts for general government-Net lending/Net borrowing;	-0.82	-2.39
Public debt	Quarterly gross government debt as a % of GDP; Country x's value-German value	Eurostat	3.69	2.71
Growth	Quarterly GDP at market prices; % change compared to corresponding period of the previous year; Country x's value-German value; not seasonally adjusted	Eurostat	0.36	-0.61
Inflation	Monthly HICP Index, 2005=100; Quarterly averages from monthly data; country x's index-German index	Eurostat	-0.48	1
Labor productivity	Quarterly data; Real labor productivity per employee; % change on previous period; seasonally adjusted and adjusted by working days; Country x's value-German value	Eurostat	0.03	-0.03

(continued)

(*continued*)

Variable	Definition	Source	Mean	
			2000q1-2011q4 (480 obs.)	2005q1-2011q4 (280 obs.)
Current account balance	Quarterly data; Net current account balance, as a % of GDP; Partner=all countries of the world; Country x's value −German value	Eurostat	−5.69	−8.25
Liabilities to German banks	Quarterly data; Country x's financial liabilities vis-à-vis German banks over German GDP	BIS-International bank claims, consolidated; Ultimate risk basis; Type of reporting banks: Domestically owned banks; Reporting country: Germany; Foreign claims		14.16
PRR	Political Risk Rating (0–100) Quarterly averages from monthly data: Country x's value−German value	The PRS Group	−1.49	−2.44

* Primary Balance data for 2012Q1, Q2 is available from the German statistical office in their press releases.

431

References

Ahnert, Toni, and Christoph Bertsch (2012), "A Wake-Up Call: Information Contagion and Speculative Currency Attacks," unpublished manuscript, London School of Economics and University College, London (November).

Alesina, Alberto, and Silvio Ardagna (1998), "Tales of Fiscal Adjustment," *Economic Policy* 13(27): 488–545.

Alessandrini, Pietro, Michele Fratianni, Andrew Hughes Hallett and Andrea Filippo Fresbitero (2012) "External Imbalances and Financial Fragility in the Euro Area," unpublished manuscript, Indiana University and George Mason University (May).

Barkbu, Bergljot, Barry Eichengreen, and Ashoka Mody (2013), "The Euro's Twin Challenges: Experience and Lessons," unpublished manuscript, International Monetary Fund, University of California at Berkeley, and Princeton University (January).

Barrell, Ray, Dawn Holland, Iana Liadze, and Olga Pomerantz (2010), "The Impact of EMU on Growth and Stability in Europe," in Marco Buti, Servaas Deroose, Vitor Gaspar, and Joao Nogueira Martins (eds.), *The Euro: The First Decade*, Cambridge: Cambridge University Press: pp. 607–37.

Basel Committee on Banking Supervision (2012), "Progress Report on Basel III Implementation," Basel: Basel Committee (October), www.bis.org/publ/bcbs232.pdf.

Berglof, Erik, Barry Eichengreen, Guido Tabellini, and Charles Wyplosz (2003), *Built to Last: A Political Architecture for Europe*, London: Centre for Economic Policy Research.

Blanchard, Olivier, and Daniel Leigh (2013), "Growth Forecast Errors and Fiscal Multipliers," IMF Working Paper 13/1 (January).

Buiter, Willem, Giancarlo Corsetti, and Nouriel Roubini (1993), "Excessive Deficits: Sense and Nonsense in the Treaty of Maastricht," *Economic Policy* 8: 57–100.

Buti, Marco, Servaas Deroose, Vitor Gaspar, and Joao Nogueira Martins (eds.) (2010), *The Euro: The First Decade*, Cambridge: Cambridge University Press.

Casselmann, Farina (2013), "Financial Services Regulation in the Wake of the Crisis: The Capital Requirements Directive IV and the Capital Requirements Regulation," Working Paper no.18, Institute for International Political Economy, Berlin.

De Haan, Jakob, and Fabian Amtenbrink (2011), "Credit Rating Agencies," DNB Working Paper no. 278, Amsterdam: De Nederlandsche Bank (January).

Diamond, Douglas W. (1984), "Financial Intermediation and Delegated Monitoring," *Review of Economic Studies* 51: 393–414.
Dungey, Mardi, R. Fry, B. González-Hermosillo, and V. Martin (2005), "Empirical Modeling of Contagion: A Review of Methodologies," *Quantitative Finance* 5: 9–24.
Eichengreen, Barry (1992), *Golden Fetters: The Gold Standard and the Great Depression 1919–1939*, New York: Oxford University Press.
Eichengreen, Barry (1998), "Comment on Alesina, Perrotti and Tavares," *Brookings Papers on Economic Activity* 1: 255–60.
Fama, Eugene (1970), "Efficient Capital Markets: A Review of Theory and Empirical Work," *Journal of Finance* 25: 383–417.
Ferri, G., L. Liu, and J. Stiglitz (1999), "The Procyclical Role of Rating Agencies: Evidence from the East Asian Crisis," *Economic Notes* 28: 335–55.
Frankel, Jeffrey (2012), "Chile's Countercyclical Triumph," *Foreign Policy Online* (June 27), www.foreignpolicy.com/articles/2012/06/27/chile_s_countercyclical_triumph.
Frankel, Jeffrey, and Andrew Rose (1996), "The Endogeneity of the Optimum Currency Area Criteria," NBER Working Paper no.5700 (August).
Georgoutsus, Dimitris, and Petros Migiakis (2010), "European Sovereign Bond Spreads: Monetary Unification, Market Conditions and Financial Integration," Bank of Greece Working Paper no. 115 (June).
Gerlach, Stefan, and Matthias Hoffmann (2010), "The Impact of the Euro on International Stability and Volatility," in Marco Buti, Servaas Deroose, Vitor Gaspar, and Joao Nogueira Martins (eds.), *The Euro: The First Decade*, Cambridge: Cambridge University Press: 648–69.
Goldstein, Morris (2011), "The Role of the IMF in a Reformed International Monetary System," Paper prepared for the Bank of Korea Research Conference, Seoul (June).
Haas, Peter (1992), "Introduction: Epistemic Communities and International Policy Coordination," *International Organization* 46: 1–35.
Lo, Andrew, and Craig MacKinlay (2001), *A Non-Random Walk Down Wall Street*, Princeton: Princeton University Press.
Mangnelli, Simone, and Guido Wolswijk (2007), "Market Discipline, Financial Integration and Fiscal Rules: What Drives Spreads in the Euro Area Government Bond Market?" Working Paper no.745, Frankfurt: European Central Bank (April).
Mishkin, Frederic. (2003), "Financial Policies and the Prevention of Financial Crises in Emerging Market Countries," in Martin Feldstein (ed.), *Economic and Financial Crises in Emerging Market Countries*, Chicago, University of Chicago Press: 93–154.

Mundell, Robert (1961), "The Theory of Optimum Currency Areas: An Eclectic View," *American Economic Review* 53: 717–25.

OECD (2012), "The Short-Term Effects of Structural Reforms," OECD Economics Department Working Paper no. 949 (March).

Partnoy, Frank (1999), "The Siskel and Ebert of Financial Markets: Two Thumbs Down for the Credit Rating Agencies," *Washington University Law Quarterly* 77: 619–712.

Shin, Hyun (2012), "Global Banking Glut and Loan Risk Premium," revision of the Mundell-Fleming Lecture presented at the 2011 IMF Annual Research Conference, Princeton University (January).

Shleifer, Andrei (1999), *Inefficient Markets: An Introduction to Behavioral Finance*, New York: Oxford University Press.

Sy, Amadou (2001), "Emerging Market Bond Spreads and Sovereign Credit Ratings: Reconciling Market Views with Economic Fundamentals," IMF Working Paper no.01/165 October.

Veracierto, Marcelo (2007), "On the Short-Run Effects of Labor Market Reforms," *Journal of Monetary Economics* 54: 1213–29.

Von Hagen, Juergen, and Ian Harden (1994), "National Budget Processes and Fiscal Performance," *European Economy Reports and Studies* 3: 311–408.

Wyplosz, Charles (2005), "Fiscal Policy: Institutions versus Rules," *National Institute Economic Review* 191: 70–84.

Zelikow, Philip (1999), "Thinking About Policy History," *Miller Center Report* 14, pp. 5–7.

Zettelmeyer, Jeromin, Christoph Trebesch, and Mitu Gulati (2012), "The Greek Debt Exchange An Autopsy," unpublished manuscript, European Bank for Reconstruction and Development, University of Munich and Duke University (September).

Zuckerman, Gregory (2009), *The Greatest Trade Ever The Behind-the-Scenes Story of How John Paulson Defied Wall Street and Made Financial History*, New York: Broadway Books.

16

The Regulatory Responses to the Global Financial Crisis

Some Uncomfortable Questions

Stijn Claessens and Laura Kodres

INTRODUCTION

This chapter identifies some of the current key reform challenges for creating stable, yet efficient financial systems.[1] It does so in light of lessons from the recent and past financial crises and using insights from analytical and empirical studies. The general objective of possible reforms is clear: to reduce the chance and costs of future systemic financial crises in the most efficient manner, that is, at the lowest costs to economic growth and welfare more generally. The most important conceptual and practical challenge identified in the chapter is that policy-makers (and market participants) need to think more about the system as a whole when engaging in their risk monitoring efforts and financial system reforms. Although some policy-makers have adopted this mindset, many are still questioning its usefulness. However, the crisis has made clear that, in spite of what appeared to be individually sound and well-supervised financial institutions, well-functioning financial markets, well-diversified risks, and robust institutional infrastructures, systemic risks emerged, yet went undetected or not addressed for some time and then created great havoc.

Here, despite some decent progress in a few areas, the sad news is that the general approach to reforms is largely still been based on an outmoded

---

[1] Paper prepared for the project: "Recalibrating Risk: Crises, Perceptions, and Regulatory Change" and presented on September 19–20, 2013, at the authors' meeting at Duke University. We like to thank the conference participants, and especially our discussant Lawrence Baxter, and Ed Balleisen, and Kim Krawiec, as well as colleagues in the IMF for their comments. This paper is based on information only up to April 2014. Views expressed do not necessarily represent those of the IMF, its Board of Directors, or its management, IMF policy, or the views of the Bank for International Settlements.

and by now largely repudiated conceptual framework of regulations, which does not start from the "system-wide" characteristics of risks and often misses key risks. Systemic risk in modern financial systems arises endogenously and cannot just be captured by individual institutions' balance sheets, or specific market or asset price-based measures alone, especially when these metrics are static or backward looking. A system approach is all the more necessary as modern financial intermediation processes add newer elements that do not always fit into the traditional, silo-based ways of formulating micro-prudential, bank- or market-based regulations and conducting institution-based or market-specific supervision. Reform approaches need to be more holistic – examining the interactions between and across institutions, markets, participants, and jurisdictions, and across types of risks (e.g. market, credit, liquidity, and operational). Moreover, approaches need to actively anticipate the side effects of one regulation or action on others, both within and across jurisdictions.

In addition to lacking a focus on systemic risks, many reform areas have lagged for other reasons: a lack of a specific enough analytical framework and appropriate data with which to evaluate the possible costs and benefits of various regulations and their interactions, making reform steps consequently unclear; and a lack of practical methods of implementation or enforcement of conceivable reforms. We realize that there will always remain such and other constraints on knowledge and data, but we argue that these constraints should be more explicitly acknowledged. The outcome should be that policy-making takes a more "Bayesian" approach where reforms are implemented in areas where knowledge is greater, while in other areas both a more "experimental" approach is taken and more resources – data, analyses – are invested to clarify the best approach. And, in the end, there needs to be adequate recognition that institutional, political, and other constraints will affect the final reform choices and the degree to which regulations are actually enforced. As such, and despite various efforts, financial crises will likely recur. There is thus a need to enhance crisis management (including resolution and transparent burden sharing), again both within and across jurisdictions.

The chapter begins by reviewing the most common explanations for the recent financial crisis, which tend to stress both causes common to many other, past financial crises and a set of new causes. The exact weight to put on each of these causes is not clear, however. It also briefly reviews the main financial reforms, highlighting the areas where progress has been greatest.

The paper then frames the overall challenges in developing policies that will prevent future crises, considering three perspectives: taking a system-wide view and addressing market failures and externalities; improving incentives at all levels (i.e. including market participants, other monitors, and supervisory agencies); and improving data and analyses to reduce the unknowns. It recognizes that these general principles do not suffice to determine specific reforms. It thus ends by suggesting specific further steps that can be undertaken, within all the constraints, to improve financial policy-making. That said, we do not flesh out specific reforms, but simply intend to provide ways in which the principles could be met.

The next section assesses progress in three areas corresponding to the perspectives identified in the previous section: first, pursuing a system-wide view – adopting macro-prudential policies, reducing pro-cyclicality, and addressing the shadow banking and OTC derivatives markets; second, encouraging more prudent banking, reducing the too-big-to-fail problem, improving regulatory governance, and achieving better international financial integration; and third, getting more data and conducting better analyses. Unfortunately, rigorous theoretical analysis of recent and historical experiences remains in short supply, as does relevant evidence about the impact that various new regulations and requirements have on the risks of new financial crises. As a result, we caution that in designing reforms, policy-makers have to be more explicit about the analytical, practical and data constraints, and the many remaining known unknowns and unknown unknowns. The section ends therefore with a (sober) message: given the likely inability to prevent all future financial crises, there is a need to enhance crisis management and resolution as part of the ongoing reform agenda.[2] The last section concludes with some general lessons.

WHAT CAUSED THE GLOBAL FINANCIAL CRISIS? AND WHAT IS THE STATE OF AFFAIRS OF REFORM?

Analyzing the policy responses needed to prevent future financial crises has to start with an analysis of the causes of crises, most notably, but not solely the most recent one, the global financial crisis. While its exact causes will be continue to be debated, it is clear that this crisis, like others, had multiple and interlinked causes, some common to other financial crises

---

[2] This realization also implies that "policy regret" may be anticipated. See Lori Bennear, Chapter 2, in this volume.

and others unique. We can group them into four common causes and four unique causes (see further Reinhart and Rogoff 2009; Calomiris 2009; Claessens and Kose 2014; Eichengreen 2002; Almunia et al. 2010, and Claessens et al. 2010, the latter on which this section draws, for reviews of the causes of financial crises in general and the most recent specifically). We next review briefly the main regulatory responses to date.

Common Causes

The first common cause stressed in most accounts of the recent crisis is the occurrence of a credit boom or, more generally, rapid financial expansion. Credit booms are often associated with deterioration in lending standards – as observed in the subprime lending in the United States. While booms do not always cause crises, they do make them more likely (Dell'Ariccia et al. 2012) and most financial crises are in some way related to credit extension to borrowers that become non-performing. Moreover, credit booms are typically associated with high leverage, which is why they can be so dangerous.

A second, and often related, "common" cause is rapid asset price appreciation, with housing the most common asset. House prices in the United States rose more than 30 percent from 2003 to the onset of the crisis. In many other markets, such as Ireland and Spain, prices rose even more. Because houses are used as collateral underpinning mortgage credit, their rising values facilitate accelerating credit extension, and hence are often associated with a rapid growth in household credit and increased leverage, all of which further heightens the risks and adverse consequences of a subsequent bust.[3]

The creation of new instruments whose returns rely on continued favorable economic conditions stands out as a third frequently invoked cause of crises. In this instance, the rapid growth of structured credit products – such as collateralized debt obligations (CDOs) and the like – depended in complex ways on the payoffs to other assets (see IMF 2009; Fostel and Geanakoplos 2012). Often the risks associated with the new products are not fully comprehended or appreciated, or are simply explained away by key institutional players such as rating agencies, adding to instability.

---

[3] See Kiyotaki and Moore (1997) the seminal work demonstrating this property; and Fostel and Geanakoplos (2013) for a more recent review of leverage cycles.

Financial liberalization and deregulation constitute a fourth commonly identified contributor to crisis conditions. Observers have emphasized such moves as the removal of barriers between commercial and investment banking in the United States and the greater reliance of banks on internal risk management models, all of which occurred without a commensurate buildup in supervisory capacity. Conversely, regulation and supervision were slow again to catch up with new developments, in part due to political processes and capture, and failed to restrict excessive risk-taking. Risks, notably in the 'shadow banking system' but also at large, internationally active banks, were permitted to grow without much oversight, leading eventually to both bank and nonbank financial instability (see Wellink 2009).

New Causes

Of the new causes, the first and most significant was the widespread and sharp rise of households' leverage and subsequent defaults on (housing) loans. While other crises have been associated with real estate booms and busts, most of those centered on excessive commercial real estate lending and rarely on households. The collapse of the subprime market and the vicious cycle of falling house prices was a catalyst for the crisis in the United States. It triggered similar declines in housing markets in many advanced countries (Ireland, Spain) as well as some emerging markets that had seen booms.[4] By directly involving so many homeowners, this crisis became far more complicated. There are no established best practices for how to deal with large-scale households' defaults and associated potential future moral hazard problems, and equity and distributional issues. What is clear is that restoring households' balance sheets will take a long time, extending the economy recovery period.

A second new aspect was how increased leverage manifested itself across a wide range of agents – financial institutions, households – and markets. While a buildup in leverage was not new, the extent of many classes of borrowers' dependence on finely priced, illiquid collateral limited the system's ability to absorb even small shocks. This led to a rapid decline in collateral values (notably of houses and their related structured credit products), which shook confidence. Fear of counterparty defaults in

---

[4] A few countries, notably Korea and Iceland, have seen household leverage-induced financial difficulties, but advanced economies seldom witnessed such widespread household distress outside of the Great Depression.

major financial institutions – that were highly leveraged, thinly capitalized, short of funding liquidity and had extensive off-balance sheet exposures – rose dramatically early on in the crisis, freezing market transactions and making valuations of underlying assets even more problematic. The emergence of systemically important non-bank financial institutions (MMMFs, finance companies, insurance companies (e.g. AIG), and investment banks) added to overall risks, and in some cases required public backstops for the first time. The systemic vulnerabilities that were building up eventually helped turn a liquidity crisis into a solvency crisis.

A third new element has been increased complexity and opacity, resulting largely from the US private label securitization of weak credits, the explosive growth in derivatives globally, and the murky operations of the shadow banking system. While the originate-and-distribute model of securitized mortgages held the promise of better risk allocation, it turned out that risks were less widely distributed than envisaged and incentives to properly assess risks, including by rating agencies, were undermined. The complexity of the securitized products made it much more difficult to know their true value and who incurred the various risks. Hence, the solvency of financial institutions that were thought to own them quickly became questioned. The complex use of asset-backed commercial paper (ABCP) backed by CDOs and other Mortgage-Backed Securities (MBS) – with their differential maturities of liabilities and assets, added the risk of rollovers to a loss of confidence in the values of the underlying assets. In the run-up to 2007–2008, huge sums from US and Euro Area money market funds flowed into bank commercial paper and short-term debt, while extensive use of repurchase agreements and rehypothecation strategies[5] generated long chains of borrowings for the support of other trading book assets in large, interconnected securities dealers and banks. These developments fostered excessive use of short-term wholesale funding in various forms that was not well understood, setting the stage for a confidence crisis.

Fourth, international financial integration had increased dramatically over the decade before the crisis. Global finance no longer involves just a few players, but many from various markets and different countries. Many mortgage-backed securities and other US-originated instruments were held in other advanced economies and by the official sector in several

---

[5] Rehypothecation refers to the re-use of collateral in other repurchase or securities lending agreements.

emerging markets – and funded by dollar-based liabilities in other, non-dollar-based countries. Cross-border banking and other capital flows had increased sharply, notably for and among advanced European countries. While these developments undoubtedly had benefits during "normal" times, they quickly translated the turmoil in the United States into a global crisis. Subsequently, turmoil in countries in the euro area led to multiple rounds of cross-border spillovers and further crises. The various intense links meant not only that disturbances quickly spread, but also made co-coordinated solutions much more difficult to implement. More generally, there may have been "too much finance," in that finance had grown big and complex, and provided many products that offered little real added value but generated many risks.

The exact weights of each of these and other causes remain unclear, generating many questions as to why this crisis has been so bad and so long. Other contributing factors suggested by scholars include too loose monetary policy and weaknesses in fiscal policy, such as generous tax deduction of interest, but since these factors have been present in previous cycles it is difficult to conclude that they are much to blame. Nevertheless, it is generally agreed that the causes were many and the "solutions" to prevent future crises will equally have to be found in a combination of important changes to national and international regulatory frameworks, the conduct of monetary policy and fiscal policies, and legal and institutional environments (see Viñals et al. 2010 for an overview of the overall policy agenda). Below we focus on the financial regulatory agenda, acknowledging policy changes are also needed in other areas.

Regulatory Responses to Date: Where Are We Now?

Policy-makers have sought to rectify the damage done to financial systems and economies by enacting a large set of financial reforms, both at the international and domestic level. The informal group of regulators and central bank experts that had been meeting in Basel prior to the crisis became more formal in April 2009 through the establishment of the Financial Stability Board (FSB). The FSB now coordinates the work of national financial authorities and standard-setting bodies at an international level. It brings together national authorities responsible for financial stability, albeit mostly from G-20 countries. Some of the key reforms that have been finalized under FSB guidance and are being implemented can be summarized as follows (for more details, see the latest FSB progress report to the G-20, September 5, 2013, on which this section draws, and for example Atlantic Council et al. 2013).

- Adoption of Basel III capital requirements, including a countercyclical capital buffer and a surcharge for globally systemically important financial institutions (G-SIFIs), both of which represent a first international attempt to institute a macro-prudential tool.[6]
- Agreement reached on one of two envisioned liquidity standards – the Liquidity Coverage Ratio (LCR).[7]
- Some progress on reducing too-big-to-fail, with the identification of G-SIFIs, domestically systemically important banks (D-SIBs), higher capital adequacy requirements and more intense supervision, and some reforms of national resolution schemes (including bail-in instruments) so that failing institutions can be resolved without wider disruptions.[8]
- Enhancements to the "securitization model."[9]
- Adoption of principles for sound compensation practices, to avoid perverse incentives for risk-taking.[10]
- Agreement in principle on similar treatment of some types of financial transactions under US Generally Accepted Accounting Principles and International Financial Reporting Standards.

---

[6] The rules (see Basel Committee on Banking Supervision [BCBS] 2011) are: a 4.5 percent basic and a 2.5 percent conservation buffer requirement for all banks; a 2.5 percent countercyclical buffer in the boom phase of the financial cycle; and for some banks (designated as systemic), an up to 2.5 percent systemic surcharge. Altogether, the highest minimum requirement in the form of common equity (Tier 1) would be 12 percent. In addition to this would be 1.5 percent alternative Tier 1 equity and 2 percent Tier 2 (hybrid) forms of capital. These ratios all apply to risk-weighted assets. Additionally, a simple leverage requirement, ratio of (common) equity to total assets, has been adopted. Besides raising the level of requirements, at least as important, Basel III requires better forms of capital, especially more core equity, rather than the hybrid forms of equity that were much used before the crisis.

[7] The LCR, announced early 2013 by the BCBS, requires banks to have enough liquidity, defined as having on balance sheet certain assets (High Quality Liquid Assets) and access to some facilities (including some forms of central bank liquidity), to cover 30 days of outflows. The Net Stable Funding Ratio (NSFR), still under discussion, aims for better structural asset and liability maturity matches.

[8] See FSB (2013a).

[9] Credit rating agencies are asked to disclose more; formal rules requiring the retention of underlying assets have been instituted in various jurisdictions; and accounting information on off-balance sheet vehicles, such as Special Investment Vehicles (SIVs) and conduits, must be consolidated.

[10] See FSF (2009).

- Some closure of data gaps, for example, the beginning of harmonized collection of improved consolidated data on bilateral counterparty and credit risks of major systemic banks (for the major 18 G-SIBs and 6 other non-G-SIBs from 10 jurisdictions).[11]
- Some OTC derivatives reforms.[12]

WHY HAVEN'T WE MADE MORE PROGRESS? AN ANALYTICAL FRAMING

The reforms to date, in light of the diagnosis of the crisis, provide some insights into what more might be needed. To identify, evaluate, and prioritize further specific reforms is challenging, however, as the "right" tools can be hard to identify and conceptual and practical issues raise many difficult trade-offs. There clearly is much "path-dependency" in that reforms undertaken to date can constrain choices going forward and a radical rethinking might not be feasible technically or politically. Furthermore, countries differ in many dimensions, suggesting reform choices will vary, possibly greatly.

Determining approaches and constraints to reform is nevertheless best done with a clear framework in mind. The general analytical approach this paper uses can be summarized under three themes: think system-wide and try to explicitly address market failures and externalities; improve incentives, individually and collectively, of all those involved in finance; and, collect more, higher-quality data and conduct better analyses of that information. At the same time, the paper stresses the importance of acknowledging that many risks may remain, in part due to unknowns, so one also needs to proceed cautiously and plan (better) for future crises.

Think System-Wide and Address Market Failures and Externalities

The crisis has made clear that in spite of what appeared to be individually sound and well-supervised financial institutions, risks that were thought to be well diversified, and institutional infrastructures that appeared to be robust, systemic risks nonetheless emerged, went undetected for some

---

[11] See further Heath (2013) and FSB-IMF (2013).
[12] For requirements of the reporting and centralized clearing of some types of OTC derivatives in some jurisdictions, as well as guidelines and minimum standards for centralized counterparties (CCPs) by the Committee on Payments and Settlement Systems and the Technical International Organization of Securities Commissions (CPSS-IOSCO 2012). For a review of recent OTC derivatives reforms see the FSB (2013b).

time, and then created great havoc. Since then, through better analytical modeling, information gathering, identification, and monitoring as well as a focus on macro-prudential policies, systemic risk has received a greater focus. Yet, these efforts do not suffice. A perspective that acknowledges much more explicitly the interactions, market failures, and externalities is still needed. This system view should include but not just be limited to regular (public) financial stability reviews, large-scale stress tests, and other such analyses.[13] Such reviews and analyses should be an integral part of a broader process by which all supervisory agencies consider their roles primarily to oversee (a segment) of the financial system in its entirety, and only secondarily the individual institutions or agents within certain markets. Any micro-prudential supervisor, for example, should consider, and be equipped to address if necessary, the systemic consequences of the institution she reviews.

The system-wide view is not just needed for supervision, but also for the design of regulations. Conceptually, it is now well recognized that even fully effective regulation (and supervision) at the individual level (alone) does not assure a safe financial system (see Brunnermeier et al. 2009; Osiński et al. 2013 for a general discussion; and De Nicolò et al. 2012 for an analytical review of a macro-prudential versus a micro-prudential perspective on financial stability and regulation). One obvious reason is the various fallacies of composition. Bank A can have liquidity insurance from bank B, and bank B from bank A, allowing both to satisfy a micro-prudential liquidity requirement, yet in aggregate, liquidity risk obviously still remains. More generally, the high degree of interconnectedness of financial systems and the large scope for market failures and externalities make a system-wide perspective necessary for financial stability, both at national and international levels.

Currently, regulations and other requirements are, however, largely designed from a micro-prudential perspective. It can even be the case that such micro-prudential requirements, even when well designed, make the system as a whole more, instead of less, risky.[14] Some regulations can lead

---

[13] As with all efforts, publication of a financial stability report and other such analyses does not necessarily contribute to financial stability as Čihák et al. (2012) find. When of higher quality (clear, consistent, etc.), however, publication is more likely positively associated with financial stability.

[14] For instance, the zero-risk weight on sovereign debt in bank capital requirements arguably encouraged larger holdings among European banks than would otherwise have been the case in the crisis deleveraging process and hence large, simultaneous losses when sovereign downgrades took place.

to more procyclicality, as for example has been argued in case of Basel II.[15] And without a system-wide view on both private and public provision of liquidity, a micro-prudential liquidity rule can act perversely – as when all banks have to meet a requirement at the same time. Reducing the risks of a crisis requires therefore a system perspective combined with a (macro-prudential) toolkit, some of which has to be global given the close connections these days among financial systems and through international markets (and this may have to involve as well capital flows management tools). And it requires proper institutions to assure, besides system-wide risk monitoring, the necessary remedial actions.

Improve Incentives

Improving incentives, rather than prescribing specific behavior, is obviously, to an economist at least, the best way to enhance financial sector performance and ensure greater financial stability. This "incentives view" applies to direct market participants, to what can be called "auxiliary monitors," and to regulators and supervisors. Direct market participants include owners, creditors (including deposit insurance agencies), managers as well as staff (e.g. the "traders") of financial institutions; the many, often atomistic participants in financial markets; and the numerous final users of financial services – households, corporations, sovereigns, others. Incentives – the possible gains and losses they face, including the chance of sanctions for (criminal) wrongdoing – drive these agents' actions, including how they manage risks and serve (or not) as mechanisms to absorb shocks. Because of the diversity of modern finance alone, no single economic or financial "model'" can capture the motivations and incentives of each of these agents. And clearly there are many behavioral and other non-economic aspects that drive the decision-making of agents for which economic "models" do not exactly apply (and knowledge is otherwise as of yet limited). Nevertheless, altering these incentives through the "right" regulations and policies is likely to bring about

---

[15] One notable reason is that Basel II encourages the use of VaR models, which are often used with similar inputs, including a short time frame, which induces more risk-taking, as volatility of asset prices fall in an upswing, and a common withdrawal from risks, as volatility rises in a downturn. Some other (capital and liquidity) regulations can also, by inducing more common behavior, increase overall risks (e.g. by focusing on risk weights, rules can induce too much investment in some asset classes). More generally, even when rules encourage diversification at the individual firm level, they may reduce useful diversity at the system level.

a better (that is, both more stable and efficient, and fairer) financial system. As such, we devote considerable attention to reviewing existing knowledge on incentives of the direct participants (see further World Bank 2013 on the role of incentives for a sound and efficient financial system).

Generally, less attention has been given to the incentives of auxiliary monitors. These agents include rating agencies, accounting and auditing firms, various elements of the institutional infrastructure for financial markets (e.g. clearing houses, CCPs) as well as the financial press and other "whistleblowers." They can all play useful roles in creating a safer financial system by exercising market discipline, identifying problems and risks at both the micro and system level. Whether these agents can identify important risks, will voluntarily reveal them and act on them will depend on their incentives. An examination of the incentives facing rating agencies revealed their contributing role in perpetrating the financial crisis (e.g. Partnoy 2010) and part of the reform agenda underway is consequently aimed to remedy this. There are many other auxiliary monitors, though, which arguably also failed in their roles or had a conflicted set of incentives, for which reforms still have to start.[16] Yet, others have been surprisingly strong in their roles as monitors, even when not charged with monitoring formally. Many cases of malfeasance have, for example, been discovered by employees and the press (see Dyck et al. 2010).

If incentives of direct market participants and auxiliary monitors fail to detect and act on risks that can become systemic, regulatory and supervisory agencies become the last, but important, defense. Weaknesses in supervision and capture of agencies, nationally and internationally, however, have at times also adversely affected financial stability (and possibly as well as the efficiency of provision of and the access to financial services by many groups in society). Capture of regulatory, supervisory, and other public oversight agencies occurs in many ways and can undermine financial stability and efficiency.[17] Regulators, supervisors, and many other

---

[16] This relates to the role of self-regulatory organizations (SROs) in finance. In many other industries, such SROs can help exercise discipline, in part as the collective reputation of the industry depends on the behavior of individual members (see other contributions in this volume). This reputational channel appears to depend, however, on the industry facing potential competitive threats, which may not work equally well in financial services industries that are often essential and therefore protected to some degree, including by a public safety net. The international nature of the financial services industries makes the model also harder to implement. See Omarova (2010) for a proposal for a new paradigm for SROs to account for systemic risk.

[17] Some forms of capture are subtle: insiders – both people within the financial industries and important users – set the rules, standards, and institutional designs, mostly to benefit

officials who failed in their public policy roles, have suffered little *ex-post* cost (in terms of loss of jobs, for example). At the same time, few if any officials receive any reward for discovering risks early or attempting to flag imminent problems. As such, enhancing national and international regulatory governance and accountability must anchor any incentive approach to reduce the chances of financial crises.

Realize Risks, Known and Unknown, Will Remain

As in other industries (e.g. nuclear, health, food) and with other types of man-made and other risks (e.g. climate, spread of diseases, weather), there has to be the realization that, even with better incentives and a more system-wide view, many risks will remain. Some will constitute risks that explicitly or implicitly will be deemed to be "acceptable" – since a fully "fail-proof" financial system may not be the most efficient in delivering economic growth or other desirable outcomes. Optimizing welfare in the presence of full information about risks can after all mean a genuine trade-off between efficiency and stability. And insuring explicitly against some risks may not be efficient or actually create more moral hazard.[18]

More resources, analyses and data can help reduce to some degree the set of currently unknown risks that are deemed unacceptable. Some old and new risks, including perhaps those (deliberately) hidden, can be discovered using a more eclectic way of doing "prudent" oversight, that is, not (just) relying on formal risk indicators or rules, but using more "market intelligence." For instance, usually asking "why"-type questions of intermediaries or market participants is helpful, for example, why are some users willing to manufacture or buy some new product?[19]

---

themselves. As rents arise, the costs of financial services increase and access declines for some groups. In some cases, capture occurs in very blatant ways, such as corruption, which includes not only "stealing" (as when state-owned banks lend to cronies who subsequently default) but also the misallocation of resources. Gains from capture often occurs *ex post* – through, for example, bailouts induced by the moral hazard of too-big-to-fail financial institutions or more relaxed monetary policy and fiscal policies to deal with (the risks of) a systemic financial crisis. In addition to capture, there can be group-thinking (see further Barth et al. 2012; and the Section "Improving Incentives" below).

[18] See, e.g., Carolyn Kousky, "Revised Risk Assessments and the Insurance Industry" (Chapter 3), in this volume for more on the challenges of insuring catastrophic risks.

[19] This relates to work by Ayres and Braithwaite (1995) on the balance between formal rules and informal regulatory governance (regulators "kicking the tires" to keep abreast of what is going on), and to the empowerment of both private and public interest groups in the regulatory process, including by encouraging effective industry self-regulation.

Some (perhaps many) risks though will remain undiscovered, not just because of a lack of attention by markets, supervisory agencies and others, but because they are not easily recognizable. Indeed, sometimes these (system) risks of a (new) product are not even known by the purveyor. Other risks will come from new sources or arise from existing sources anew, such as unforeseen interactions between markets and agents, or side effects of new regulations. Some systemic events will not be anticipated in any way ("Black Swans").[20] Because many risks remain, contingency planning and the ability to respond to (the onset of) financial crises with flexibility will remain needed. And effectively and efficiently mitigating the impact of crises when they occur will have to remain an important policy area too.

Adapt Approaches and Avoid Fallacies

While useful starting points, these general considerations do not suffice to determine specific reforms. That still requires much more analysis and work, including notably adaptation of approaches to country circumstances. Here constraints are numerous, as examples in a number of areas show (see Box 16.1 below). Further progress to overcome these constraints is needed to avoid at times mistaken approaches and fallacies. Even though general prescriptions are not possible (or useful), recommendations in terms of process can still be made, including: adopting a framework for regular consultation and coordination across regulators, possibly even cross-border, and with financial services providers and users; and conducting from time to time a review of financial regulations from both development and stability points of view.

SOME REFORMS TO ACHIEVE GLOBAL FINANCIAL STABILITY

With this general framing and keeping in mind the many constraints and trade-offs, we next discuss general remedies to the three main areas identified above on which forward progress is still needed in many jurisdictions: reforms mitigating systemic (as opposed to idiosyncratic) risk; alterations in incentive structures; and better data and information to

---

[20] Because of their severity and relative rarity, limited data is a frequent feature of such risks. See Kousky, Chapter 3 in this volume (discussing cases in which the risk distribution is unknown); Lori Bennear, Chapter 2 this volume (discussing methods by which experts estimate risks when limited data exists).

BOX 16.1 *Overall approaches to determine specific reforms*

- *Overall consistency across reforms.* This consistency, especially when reforms proceed on many fronts as they have recently, is often not assured. As one example, there is a tension between liquidity regulations (the LCR) and components of the resolution regime (bail-in requirements). Another, related problem is that regulators in one area do not necessarily talk to those in another area – for example, resolution authorities and banking supervisors (even if they are in the same building) or accountants (requiring consolidated treatment of SIVs and conduits), banking supervisors (assigning risk weights to securitized products), and securities regulators (insisting on more "skin in the game" for securitization). These and many other such examples show the complexities of designing and implementing financial reforms at the same time as well as balancing various trade-offs, such as between assuring financial stability and having efficient financial services provision to support economic growth. Unfortunately, policy-makers do not always discuss, assess, or even recognize many of these complexities and trade-offs.
- *Timing of reforms and implementation.* Consistency is also necessary with respect to timing. Some "fixes" are hindering the current economic recovery (such as higher capital ratios that are leading to deleveraging through asset sales or less credit creation). Other reforms aimed to support a recovery, such as credit enhancing policies (such as (temporarily) lower risk weights on SME loans), may lead to excessive risk-taking, since they purposely underprice risk relative to its true price. More generally, if reforms are too slow, risks will build up again; if reforms are too fast, the real economy fails to recover: a "just right" approach requires a lot of judgment and flexibility.
- *Migration and global consistency.* Despite, or perhaps because of, a "global" (e.g. at least G-20 or G-25) representation within the FSB for regulations to be "cleared," there is (still) a tendency to adopt the lowest common denominator or to negotiate specific one-off exceptions. An example is that for some concentrated market activities (e.g. OTC derivatives)

BOX 16.1 (cont.)

migration and fragmentation are constantly issues for the private sector, with pressures on their regulators to favor their own jurisdiction. Protecting the financial system can then conflict with making markets (especially market infrastructure, such as CCPs) more competitive. Competition among countries and more generally can lead to lower standards and higher risks. Minimums (of capital, risk-management standards, leverage, remuneration, and so on) are meant to help avoid a race to the bottom, but require all jurisdictions to actively enforce the minimums. At the same time, some countries are aiming to be "super safe" and hence are going much beyond the agreed-upon standards. For some, this includes segmenting (parts of) their system, which raises many questions, including at what point the benefits of an open, global financial system with free movement of resources and ample risk diversification begin to be outweighed by protectionism.

- *Cost-benefit analysis.* Regulators/supervisors are thinking about cost-benefit analysis, but usually in a very narrow way, with many worried about raising the costs of intermediation while the recoveries in many crisis-hit countries remain weak. Policy-makers should be thinking long-term – through the business cycle – adjusting implementation time frames, but not the final goal. And they should explicitly acknowledge that ever more complexity in rules has not just direct costs, but can even increase financial stability risks. Related is the need to avoid a *narrow view of the crisis "causes."* The notion of choosing "winners and losers" among the various activities depending on whether they were viewed as "causing" the financial crisis is a concern.[21] For example, new rules have made most private label securitization more expensive (and possibly uneconomic) for securitizers and potential purchasers are not participating due to the negative perception such activities engenders – even though restarting securitization could

---

[21] For a more detailed discussion of such narratives, see discussion by Mayer in Chapter 6 of this volume.

BOX 16.1 (cont.)

help the economic recovery. Moreover, introducing too much rigidity in rules hinders future crisis management. For instance, the Dodd-Frank Act disallows the Federal Reserve System from providing liquidity to certain entities, even in an emergency, without the Treasury Secretary's "permission." To "tie the hands" of some authorities in such a way to prevent moral hazard issues from arising may at the end of the day cause more panic than it prevents when financial stress arises. Restricting business activities (the Volcker rule in the United States, the Vickers Commission in the United Kingdom, and Liikanen report in the European Union) all similarly have the problem that they attempt to isolate the "risky" activities from the banking system, but this only moves the risks (and only if effective) and doesn't necessarily lessen them for the system as a whole.

reduce unknowns. Many of these areas are interdependent – without advancements in all three areas, any one set of reforms may only marginally improve global financial stability.

Adopting a System-Wide View

Macro-prudential policies. Consistent with the greater appreciation of *externalities and market failures*, a new area of "macro-prudential" policy-making has emerged (IMF 2013a and 2013b review; see Claessens et al. 2011 for a collection of papers). There are many dimensions to having a macro-prudential approach, varying from better identifying risks, to building more robust institutional infrastructures (like more use of CCPs), to adopting new, system-oriented policies aimed at reducing excessive procyclicality and risks, and designing the institutional framework for operating them. The starting point and most complex issue, as already noted, is to better understand the dimensions of systemic risks and have associated warning signals.[22]

---

[22] See further IMF (2012a) and Blancher et al. (2013) for various types of systemic risk monitoring tools and Arsov et al. (2013) for comparisons across a set of indicators regarding their effective prediction of financial distress.

Despite much discussion and some tentative steps forward, as of yet approaches remain largely micro-prudential. For the most part, Basel III is micro-prudentially oriented. It, appropriately, targets the quantity and quality of bank capital as these institutions' lack of good capital made them vulnerable during the crisis. However, more capital only helps cushion an individual institution's losses and hence the systemic nature of multiple and simultaneous bank distress is only partially addressed. As for liquidity risk, the determinants of the Net Stable Funding Ratio (one of the two liquidity risk components of Basel III) are not yet finalized and various parts look watered down already. Again neither element – the Liquidity Coverage Ratio nor the Net Stable Funding Ratio – firmly counters banks' potential to generate systemic liquidity risk *ex ante*, although with high enough ratios the chance of a systemic liquidity event is lessened.

As regard to possible macro-prudential *policies*, a broad distinction can be made between those that aim to reduce risks arising from procyclicality (the time series dimension) and those arising from interconnections (the cross-sectional dimension). So far, only a few macro-prudential tools have been adopted and mostly only for banks. Notably, Basel III contains the countercyclical capital buffer to account for the procyclicality of credit extension and the systemically important capital surcharge that tries to address the over-weight importance of too-big-to-fail institutions. The calibration and effectiveness of these surcharges, and macro-prudential policies tools more generally, is, however, yet to be fully determined, with the calibration mostly based on rough estimates so far. While countercyclical buffers have been used, notably in Spain, where the evidence suggests some effectiveness (e.g. Saurina 2009; Jiménez et al. 2012), they did not stop a banking crisis from occurring.

Many other tools, ranging from adjustments in loan-to-value ratios (to limit real estate lending during booms to avoid busts) to levies or taxes (to reduce the incentives for whole-sale funding or to offset the TBTF subsidy), have been mentioned as potential macro-prudential tools. Some of these have been studied (see, e.g. Lim et al. 2011, for the effectiveness of various macro-prudential tools in a cross-country context; Crowe et al. 2011, on the use of macro-prudential policies for mitigating real estate booms and busts; and Claessens et al. 2013, for cross-country work on how macro-prudential policies affect banks' riskiness). These tools are in the correct direction as they attempt to put in place incentives that will lower systemic risks. Nonetheless, much still remains to be determined before their effective use can be assured, including their calibration to country characteristics and circumstances (see IMF 2013a; IMF 2011).

Other important elements of a macro-prudential framework include issues of the regulatory governance (who is in charge, including as regards to cross-border aspects; see further Nier et al. 2011), and their relationships and interactions with other polices (notably micro-prudential, monetary, and fiscal policies). So, while the greater emphasis on macro-prudential policies is promising, and some emerging market countries seem to have used such policies effectively, it is still too early to rely on them heavily, also as their costs – including (indirect) adverse effects on resource allocation – are not well known. This is notably so in advanced economies: with their more sophisticated financial systems, where arbitrage and avoidance are serious problems. It will hence remain important to not rely on macro-prudential policies too much and complement them with tools such as banking system stress tests (which can also be viewed as a macro-prudential tool).

Procyclicality. Another element of the system-wide approach that could reduce the frequency and depth of crises is to reduce the procyclicality of financial markets by structural means. Some forms of procyclicality are embedded in market practices – including compensation practices, risk management tools, such as traditional Value-at-Risk (VaR) modeling and credit risk modeling, and margining and collateral practices applied in a number of markets, notably derivatives markets. Procyclicality can also be induced by regulations such as accounting and valuation practices, capital and liquidity requirements, risk weights, provisioning requirements, and deposit insurance schemes (that lower premiums in boom times and raise them in bust times). And still other forms seem more behavioral in nature – for example, the tendency for investors to buy as asset prices are rising.

Typically (or at least in the pre-Global Financial Crisis era) compensation packages had a bonus component solely based on "returns" without considering risks. A bonus pool was built up during the year, based on the trading or other profits that a business unit accumulated and then it was dispersed at the end of the year or the beginning of the next. Little attention was paid to the risks involved in gaining those profits or whether the risks would later materialize from transactions taken during a previous time period. Since profits normally expand during an economic upswing, the procyclicality of compensation schemes is built into the system.

A first step to remove this procyclical element is to allocate compensation on a risk-adjusted profits basis. A second and even better step is to do so "through the cycle" and pay only a portion of the profits in any given year of the cycle with the remaining amounts used to absorb losses occurring later. Some of these notions have been instituted – some institutions now pay only a portion of the bonus pool out in a given year (usually with a three-year horizon), some tie it to options on their stock price, and some have a "high water" mark that hold some of the bonus pool back in case losses later materialize.

Although total pay packages have become less bonus-oriented, payouts from bonus pools are still largely short-term and large relative to base pay. Most firms are reluctant to risk-adjust bonuses because they are unsure whether their risk models are accurate enough for compensation purposes. They are also concerned that other firms will continue to pay on a non-risk-adjusted, return-only basis and hence they might lose their best talent to better-paying firms. There may be a need for a mandatory, coordinated compensation scheme with risk-adjusted bonus payments to overcome this incentive. Even then, limited liability, for the institution and clearly for the employee, makes risk-based packages in general less than perfectly incentive compatible.

Risk management systems themselves can be procyclical. For instance, low volatility of and low correlations across asset prices during economic upswings mean that risks are underestimated in VaR models that use only a limited historical period to calibrate potential losses. Regulations that strongly encourage firms to use only short periods (one year) of historical data for regulatory capital purposes also encourage procyclical trading behavior. To the extent that many financial institutions use similar models and hold similar positions, overall procyclicality increases.[23] Credit models also typically measure the probability of default at a point in time rather than "through the cycle." Even credit rating agencies that claim to rate through the cycle are not, in fact, doing so.[24] Such problems can be ameliorated by encouraging longer-term horizons for risk modeling, but for newer products it is harder without historical price data.

Accounting standards and fair-value accounting (FVA) also contribute to procyclicality. With asset values increasing in upswings and decreasing in recessions, there is a natural procyclical tendency built into the asset side of balance sheets when assets are mark-to-market. Although this

---

[23] The use of a "stressed" VaR in Basel 2.5 that requires additional capital (calculated on a continuous basis) to be added to other market-based capital requirements may help mitigate some of the procyclicality embedded in VaR-based capital requirements.

[24] See IMF (2010) and Kiff et al. (2013) for empirical evidence.

could be offset in part if liabilities were also mark-to-market, few businesses extend this practice to both sides of their ledgers. Balance sheets therefore tend to expand in upswings and contract in downturns (Adrian and Shin 2010). Book equity values (the residual of assets less liabilities) tend also to be procyclical (IMF 2008).

A corollary to FVA is that many other practices, such as using margins and haircuts on collateral, are dependent on mark-to-market values. So when collateral looks highly valuable, the margin or haircut required for a borrower to post declines (Geanakoplos 2010). This occurs in repo transactions, securities lending, collateral posted at central banks, centralized counterparties (CCPs), and stock and derivatives exchanges. Hence a number of practices reinforce procyclicality.

On the regulatory side, earlier regulation was known for its procyclicality. Basel II capital requirements were highly criticized, even at their inception, for being procyclical – the amount of capital needed during an upswing became less and less as the value of risk-weighted assets rose. The notion that buffers should be built up during the good times for use in the bad times was viewed as a preferred outcome, but the regulation was not constructed in way to codify this notion. Similarly, loan loss provisioning practices have this characteristic – as loans look safer during an upswing, less specific (and general) loan loss provisions are made since the borrower is viewed as more likely to be able to pay interest and principal on the loan. Only when bad times hit, does it become clear that not enough had been put aside for the larger share of non-performing loans. In both cases, a more "through-the-cycle" notion needs to be instilled. And in many countries, the accounting and tax systems do not allow or discourage through-the-cycle loan loss provisioning.

Basel III has gone some way to ameliorate this problem with a countercyclical capital buffer added onto the usual minimum regulatory capital standard. The ability to provision against future loans is encouraged, but some accounting practices have not made commensurate progress. The IASB is only now discussing the use of an "expected loss" concept that would allow financial institutions to align the supervisory definition of "all possible default events for the life of the financial instrument" with accounting definitions.

Shadow banking. The crisis revealed many systemic problems arising from so-called shadow banking activities, notably in the United States (see Claessens et al. 2014 for a collection of papers on shadow banking). The FSB (2012) defined shadow banking as "credit intermediation involving entities and activities (fully or partially) outside the regular banking system." This definition is meant to include any nonbanks that are active in one or more of the following activities: maturity or liquidity transformation, leverage, and credit risk transfer. As such, it is a broad definition that captures many forms of financial intermediation that are important for economic growth, but not necessarily of systemic size or importance. So which shadow banking activities that are systemic enough to be regulated needs to be determined separately. Indeed, some activities can be systemically stabilizing if they provide alternative financing arrangements when one set of institutions or activities become unable to perform their normal functions. And strictly speaking the FSB definition ignores shadow banking activities that occur *within* banks and that rely on the (implicit) safety net provided to banks (see also Claessens and Ratnovski 2013 on how to define shadow banking).

While many of the previously identified risky shadow banking activities are lower today, the cyclical conditions associated with their lull are dissipating and some of the activities are picking up again. As such, going forward, shadow banking can again become a source of systemic risk. Furthermore, some countries (like China) are experiencing increases in forms of financial intermediation labeled as shadow banking, which could prove to be of systemic concern. How to monitor and regulate these new forms is thus a policy issue of much debate, in part because shadow banking is so broadly defined.

In generally, what specific aspects of shadow banking can lead to systemic risks is not clear. Neither has it been established whether shadow banking is best regulated indirectly, that is, by putting limits on the banking system which most often supports it or whether the "system" should be regulated directly, by for example, curbing certain activities via various means. Without addressing these issues, but on the basis of the crisis evidence, the FSB has identified five work streams that would require the initial attention of policy-makers, with reforms started in some of the more obvious activities and progress in some others since.

One work stream is to examine the connections between regulated banks and shadow banks. So far, alongside the Basel Committee, the FSB has proposed restrictions on regulated banks' large exposures and equity investments to shadow banks (two of the main connections identified), with work underway on defining exposure limits to funds and securitized vehicles. This indirect method can reduce risk to the banks from shadow banking activities – as it removes a source of contagion to banks' balance sheets, but may push systemic risks elsewhere.

A second work stream involves money market mutual funds (MMMFs). FSB has tasked the International Organization of Securities Commissions (IOSCO) to develop guidelines for MMMFs. The United States (the largest market for MMMFs) has shortened allowable asset maturities, thereby limiting maturity transformation, but has made no decision about whether constant Net Asset Values (NAVs) should remain a mainstay attribute of the sector nor whether, if a constant NAV remains, liquidity buffers or other capital-type regulations should be put into place.[25] As such, this remains a critical point of weakness to address – as noted, the run on MMMFs in fall of 2008 aggravated the financial turmoil.

The identification of other nonbank financial institutions that act like shadow banks and potentially pose systemic risks is a third work stream, but this one is far from being finalized. Monitoring exercises by the FSB are more detailed each year, but mostly measure shadow banking by examining assets under management from national flow-of-funds data, hence it presents a very limited view of the risks they may pose. Agreement on a framework for other financial institutions, with more emphasis on functions rather than legal form, has been reached in principle, but it has proven hard to put to work without the appropriate data. Case studies and specific subsectors are currently under investigation.

Securitization makes up a fourth work stream. While some of the more obvious risks have been addressed in some jurisdictions, there is now a patchwork of retention rules that either provide avenues for regulatory arbitrage or make securitization uneconomical. The rules are not related, necessarily, to the risks that the originators face, and thus only partially create incentives to originate or monitor loans that are placed into securitization products.[26] Increased disclosure and capital-based risk weights applied to the products has made them more costly to issue – so much so that some previous originators find it uneconomic to do so. That said, it is difficult to nail down whether the moribund securitization market (especially in the United States) is a result of its tarnished reputation and weak demand or over-regulation.

---

[25] A constant NAV means that one share is priced to equal one dollar and "breaking the buck" refers to a situation in which the assets can no longer maintain the one share equals one dollar convention.

[26] See Kiff and Kisser (2010) for an analysis that shows retention rules can alter the incentives of those involved in securitization.

A final work stream involves repo and securities lending markets. Some risks in tri-party repo markets have been subdued given less intra-day counterparty risk taken by the tri-party repo agents in the United States, but no agreement exists on whether minimum haircuts should (or could) be established.[27] There has been only limited discussion of countercyclical margin requirements even though this was identified as a contributing cause of the crisis.[28] In this area, a quantitative impact study is underway to evaluate the effects on minimum haircut standards, including numerical haircut floors, but the study does not include haircuts on government securities (the largest repo instrument).

The New York Federal Reserve has collected information about the types of collateral underpinning repos, but it is still too coarse to identify the risks. Hence, the FSB's latest set of recommendations include three focal points: (1) more granular exposure data from the largest international financial institutions, (2) trade-level (flow) data of outstanding balances in repo markets, and (3) an initiative to aggregate and compare trends in securities financing markets at the global level by the FSB. Other recommendations focus on disclosure to end-investors and others, as well as implementation of regulatory regimes that meet minimum standards for cash collateral reinvestment. Since repo activity (and its associated risks) is on the upswing again, this work has a high priority.

OTC derivatives markets. The OTC derivatives markets represent one of the most globally active markets, allowing the transmission of risks across markets, institutions, and jurisdictions. In good times, the lack of transparency about the location and extent of risk taking is acceptable to its participants, but during volatile times this feature can be debilitating. While jurisdictions are making progress in OTC derivatives reform, progress is very uneven across jurisdictions and not all G-20 countries have implemented earlier commitments. Without a coordinated response, there

---

[27] The tri-party repo market is one in which a custodian bank or international clearing organization (the tri-party agent) acts as an intermediary between the two parties to a repo transaction. The tri-party agent is responsible for the administration of the transaction, including collateral allocation, marking to market, and substitution of collateral.

[28] While minimum haircuts in securities markets have not yet been adopted, the Basel III LCR does define stressed haircuts on particular instruments and against counterparties (e.g. 100 percent for banks/banks or banks/other financial institutions, or 100 percent on lower rated corporate debt or gold, or 25 percent on higher rated corporate securities, 0 percent for government securities), which have to be held through the cycle and also applied to securitized products including repos. The same haircuts will likely be applied within the NSFR framework to banks' assets and liabilities.

is increasing concern about the inconsistency of rules and the migration of trading to less regulated jurisdictions as costs are rising in important centers – this dichotomy is particularly noticeable across the Atlantic where the United States and Europe are moving in different directions.[29] A recent study examining how the costs and benefits to the currently formulated reforms are likely to affect economic output concluded that the benefits outweigh the costs, but by a relatively small amount – about 0.12 percentage points more GDP growth per year over the long run when the reforms have been fully implemented and their full economic effects realized.[30] That said, many end users claim that they will stop using derivatives altogether due to higher costs of trading (a factor not accounted for in the study), which presumably would reduce their ability to lay off risks.

Of particular difficulty has proven to be the process to calibrate regulations for bilateral collateral requirements, capital charges for non-collateralized trades, and collateral to be held in CCPs so as to engineer incentives to move standardized OTC derivatives to CCPs where multilateral netting can lower bilateral exposures. Progress on getting trades into trade repositories (reporting venues) is much better, but getting the information out of trade repositories to those responsible for examining risks has not appreciably improved. Currently, restrictions on data usage mean only a few regulators get information, and only about their own institutions, so third-party interconnectivity is not visible to anyone. Efforts to loosen these restrictions are ongoing.

Improving Incentives

Incentives associated with banking system reforms. Albeit with some important differences across countries, banks generally still undertake the largest share of financial intermediation. To assure safer and more efficient banking systems, higher capital and stricter liquidity requirements are very much at the center of current regulatory efforts, with Basel III at its core. Furthermore, a minimum leverage ratio has been agreed upon. Such an approach may help remove some of the gaming of the more complicated risk-based capital system, though it is currently being viewed as an ancillary "cap" on overall leverage. The analytical and empirical support for these formal requirements from an incentive point of view are, however, less clear

---

[29] See, e.g., Atlantic Council et al. (2013).
[30] See the MAG on Derivatives (2013).

than perhaps often thought. For example, as is, the analytical literature has offered relatively limited guidance on the exact incentive effects of capital requirements. Some theoretical analyses have even found that higher requirements can have perverse effects (for an early version of this argument, see Genotte and Pyle 1991). Furthermore, risk-taking incentives are affected by both a bank's current capital adequacy and its franchise value of future profit and growth opportunities, with possibly opposing effects.[31]

While the incentive effects, in terms of reduced risk-taking, are perhaps less clear, most research does acknowledge the obviously beneficial effects of higher capital in terms of buffers as well as easing the need for public intervention in weak banks. By having more capital, it is easier for a bank to absorb losses when hit by an adverse shock, thereby helping it lower the risk of default and permitting it to "go on." This can be privately beneficial for all stakeholders combined, as it preserves specialized knowledge and franchise value and avoids the direct and indirect costs of a potential bankruptcy. Yet, owners and managers may not fully internalize these deadweight costs, in part due to the presence of a public safety net, and hence choose to hold too little capital. This can thus justify government-mandated and enforced capital adequacy requirements. Capital also facilitates and eases interventions when a bank is a "gone concern." It helps to protect debtholders, including (small) depositors and their "agent," the deposit insurance scheme, from the consequences of distress. And it can help regulators define measured shortfalls, indicating when to intervene from a legal perspective and discipline the regulators to intervene in a timely fashion. For these reasons, most research supports some government-mandated and enforced capital requirements, as private costs seem low and social benefits considerable.

While the case for good liquidity management at the micro-prudential, individual bank level is obvious, the analytical case for liquidity requirements from a system point of view is less clear. This is not surprising as the concept of liquidity at a system level is very complex and not well defined, making liquidity requirements not easy to design from an incentive or a buffer point of view. And current academic thinking on liquidity seems both less well advanced and less reflected in regulations being adopted or

---

[31] One other indication of the difficulty with capital adequacy requirements is that the risk models that banks use show very large differences, i.e. banks apply very different risk weights for the same asset. The BCBS (2013) study asked 15 large banks in nine countries to calculate the total capital required to support the same hypothetical trading portfolio. The results ranged from €13m to €35m, and the variation within individual asset classes – such as credit risk or interest rate portfolios – was in several cases more than eight times.

underway.³² Overall, research suggests no clear form for liquidity requirements. It does acknowledge that the current design of liquidity rules, given the interactions across financial institutions and with retail customers, may be (even) less likely effective than capital regulation. A liquidity problem develops into a systemic problem much faster than a solvency problem. A rethinking of how best to "tax" an institution or market for contributing to a lack of liquidity, given the contingent nature of the problem, may end up looking more like an insurance-type charge or levy than a consistent "buffer"-type surcharge.

Almost regardless of the exact design, banks and others in the financial services industries object to the new capital and liquidity requirements. Their arguments mostly rest on the increased costs of financial intermediation and the resulting adverse impact on the real economy. The arguments are not particularly strong. For one, it is worth recognizing that many banks already hold capital and liquidity buffers above the requirements currently considered and would not need to change their operations. And even for those banks affected, most analyses finds small costs of reasonably higher requirements (see Santos and Elliott 2012; and BCBS 2010). And, even then, banks could adjust along several margins, some of which may further improve stability, say if they curtail activities "underpriced" before the new regulations (e.g. lending to marginally productive sectors).³³ Moreover, since there are cases where countries have raised capital well beyond the minimums (e.g. Switzerland for their two systemically important banks) – a race to the bottom is not always evident.

---

[32] Regulations are largely for example aimed at banks, i.e. they tend to try to address funding liquidity, but are less able to affect market liquidity, which is likely a big deficiency given the increasing importance of capital markets. There are also many cross-border issues in liquidity, important as well during the recent financial crisis – such as the shortage of dollar funding, but these are even less well understood. The behavioral components leading to the start of a run – a tipping point – is seldom discussed and in the past it was assumed "savers" are depositors in banks, whereas in today's markets the risks is more of large, wholesale providers of "liquidity" that may run. Measures of systemic risk are still relatively untested and their ability to signal difficulties much in advance of a period of liquidity distress is limited.

[33] BCBS analyzes the costs to the real economy of higher requirements. A one percentage point increase in the capital ratio is estimated to translate into a median 0.09 percent decline in the level of output at the end of an eight-year period relative to the baseline. The impact of meeting the liquidity requirement is estimated to be of a similar order of magnitude, at 0.08 percent. As a stronger banking system should be expected to reduce the occurrence and severity of crises, albeit these gains are hard to quantify, there are likely net positive gains.

A bigger long-term issue may be the "dis-intermediation" triggered by higher requirements, where activities migrate to less regulated parts of the industry. Only to the extent this raises new forms of systemic risk, however, should this be a source of concern. A more important worry is that in the transition to moving to higher requirements, adverse effects may be large, undermining the economic recovery. This is hard to judge, in part because how banks will adjust remains unclear – raising capital or deleveraging – and because the costs of raising capital or liquidity quickly are not well known. MAG (2010) nevertheless estimates that a 1 percentage point increase in the target ratio of tangible common equity to risk-weighted assets leads to a reduction in the annual GDP growth rate of 0.04 percentage points over a four and a half years period. These transition costs seem reasonably low.

Incentives to limit "too-big-to-fail" institutions. Many large financial institutions (especially G-SIFIs) benefitted in the past from government support – indeed, the majority of recapitalization and guarantees support in the financial crisis went to them. And today many still benefit from an implicit safety net subsidy. This subsidy has been estimated to amount to be up to 100 basis points, or up to $10 billion per banking group (as the average balance sheets for a SIFI is about $1 trillion) (Ueda and Weder di Mauro 2012). The (continued) large size of this "subsidy" indicates how distorted the provision of financial services is and how much taxpayers continue to be at risk. Hence the creation of a safer system requires reducing the incentives for institutions to become too-big-to-fail (TBTF). Clearly, this goal implies a broad agenda with multi-pronged solutions. Some elements have been set in motion, but many other important reforms are still needed (see further IMF 2014).

As a start, the new capital and liquidity regulations are more likely to be binding on those institutions that have implicitly (or explicitly) benefitted from their size. Furthermore, the global systemically important banks (G-SIBs) on which extra capital will be imposed have been defined. Also guidance about how to identify domestic systemically important banks (D-SIBs) has been promulgated. Most jurisdictions, though, have yet to implement final rules and higher capital requirements for D-SIBs. And rules for defining systemically important non-banks have only just started. Systemically important insurance companies, for example, are only just now being identified (the difficulty has been to define and identify "non-traditional insurance business," the root cause of AIG's problems prior to its bailout). Also, there is still no global agreement on how to deal with CCPs which can be systemically important financial market infrastructures.

While preventative tools, such as higher capital and liquidity buffers and more intrusive supervision, could help *ex ante*, reducing the too-big-to-fail problem will also need to include the assurance that an individual institution's failure can occur without damaging the rest of the financial system. Today, it is far from clear whether existing regulations will provide institutions with enough incentives to avoid failure and whether supervisors will be willing to stand aside, and use their resolution powers as prescribed. A smooth process of unwinding *ex post* requires enhanced resolution frameworks and enough loss-absorption capacity, including a minimum amount of bail-in debt to encourage better risk-taking, and a loss-sharing arrangement so unsecured creditors bear the risks that they legally agreed to assume. Requiring *ex ante* contractual new capital raising arrangements, such as those embedded in contingent capital (CoCo) type instruments, and improving the design of the public safety net to make transparent which deposit holders receive preferential treatment (as done in some areas of the Dodd-Frank Act) would also be helpful. Other tools, such as living wills or rapid resolution plans, may help *ex ante* to encourage simpler, more resolvable, institutions.

The global nature of large financial institutions, however, continues to raise thorny problems about resolution policies. Many coordination failures can arise for troubled institutions operating across borders, as governments have political incentives to protect their own constituencies. While an agreement on a comprehensive framework has been reached in 2011, in the form of the "Key Attributes of Effective Resolution Regimes" (FSB 2011; IMF 2012b), the agreement largely calls for harmonized resolution regimes, rather than addressing the issue of cross-border resolution, and there are as of yet few details about this issue specifically. For instance, questions on which agreement has yet to be reached include: how to deal with bail-in debt (should there be a minimum amount and how big should it be?), asset encumbrance (should there be a constraint and if so how much?) or depositor preference (who should be covered and what will be its effect on unsecured debt holders?). Importantly, cross-border burden sharing issues have yet to be addressed even in Europe, which is otherwise moving ahead with a banking union, including a single supervisory mechanism.

Incentives for regulating the regulators. Improving regulatory governance is clearly necessary, given the many supervisory failures before and during the recent crisis. While this is a complex and multi-faceted problem (with many political economy aspects) and should be considered in a broader context of government vs. market failures, some steps to enhance regulatory governance seem feasible. For one, in many countries, agencies lack sufficient legal, financial, and operational independence from the financial services industry and legislative bodies, and operate under political economy pressures more generally. Funding independence can be an important element to secure intellectual and operational independence and hence improve the process for formulating and implementing good regulation (Fullenkamp and Sharma 2012).

At the same time, formal public oversight of regulators and supervisors as to their performance is often minimal with essentially no consequences for poor performance (few supervisors have gotten "fired"). Through objective assessments and regular checks, weaknesses in their independence, accountability, integrity, and transparency of operations could be brought out and corrected through new laws or self-imposed new internal practices. Some of this is already done in the IMF's and World Bank's *Review of Standards and Codes* and in some peer reviews, but more emphasis could be put on assessing the effectiveness of "governance" of regulators and the transparency of "processes" (and the link to outcomes).

With better regulatory governance in place, one could have less emphasis on formal rules and give more discretion to supervisory agencies. This could perhaps avoid the proliferation of rules that may add more costs than they provide benefits and may even increase overall risks (see Haldane and Madouros 2012). Of course, such greater discretion may have to come with limits in other ways (for example, it could be balanced with some formal triggers, as has been done in the United States through the FDICIA which codified prompt corrective action, or PCA). It could also be combined with greater use of market signals, such as declines in stock prices or increases in interest rates on repriced subordinated (or other classes) of junior debt. Either way, such thresholds can be useful disciplining devices for supervisors, even (or especially) in cases where large, systemic banks run into some difficulties.

Better governance should also involve more transparency in the design of rules, with more views (allowed) to be expressed and greater participation by the public. Better and maybe new institutions are needed. Despite the inefficiencies, distortions and costs, the general public is little involved in financial sector matters, both because it is poorly informed about some of the problems – financial systems and regulations are complex – and because it is not easily mobilized. At the heart of the issue is that the incentives (benefits) for correcting problems are too diffuse so that any single individual has very little to gain by themselves. The new Consumer Financial Protection Bureau in the United States can be seen as an attempt to create a counterforce to insiders designing and applying the rules for the financial sector. Although few other such bureaus exist so far, and the one in the United States remains very incipient, it could be a sensible model, as it replicates what often exists for other products (for example, consumer product safety bureaus). There may be a role for public financial support (perhaps through grants) for (new and existing) non-partisan, non-profit groups to represent the interest of the general public in financial reform, including in the area of macro-prudential policy.[34]

There could also be additional forms of formal oversight, both before and after financial crises or events. For example, some academics have proposed a "sentinel" – an informed, expertly staffed and independent institution evaluating financial regulations and regulatory actions from the public's point of view (Barth et al. 2012). Although hard to design in such a way as to avoid "group-think," it is worth considering. Perhaps requiring formal, *ex ante* Food and Drug Administration–style approvals of new financial instruments could be a more modest, yet still useful, concept to ensure that financial services are not only "safe" for the general public, but also socially valuable.[35] Or – and maybe more realistically, as each new financial service would be hard to approve *ex ante* – an agency could be set up to systemically investigate and report on financial "failures." Such a National Transportation Safety Board–like agency would be better than financial crises commissions, which are too ad hoc and often have too little standing (see Fielding et al. 2011). Complementary countries could engage in an "incentive audit," as proposed by Čihák et al. (2013). This would entail reviewing regulations so to have at their core the objective of addressing incentives on an ongoing basis.

Regulatory governance issues also arise, albeit with even more complexity, in an international context. Overseers often fail in their

---

[34] The US Federal Trade Commission's Consumer Protection Agency is a case in point. Formed in the 1970s to represent the consumer interest in the regulatory proceedings of other agencies, it remains a focal point for consumer complaints for a number of industries. The newly created Americans for Financial Reform, a coalition of more than 250 national, state, and local groups advocating for reforms in the financial sector could be a model for such finance-oriented groups.

[35] The Commodity Futures Trading Commission used to require futures exchanges to justify new futures contracts by demonstrating a public benefit.

(macroeconomic and financial stability) surveillance roles. More attention has been placed on international governance and legitimacy in recent years, and some progress is being made to broaden the set of stakeholders (as reflected in the greater role of the G-20). Peer reviews on countries' reform progress are underway in some areas, but their effectiveness is not yet clear. Still, formal governance has proven hard to change (witness, for example, the tediousness of the ongoing governance and quota debate for the IMF).

One clear means of improving international decision-making would be to open up further the standards-setting processes, especially by broadening membership of some groups and soliciting public inputs from the end users of finance more explicitly (although many small users will need support given the technical nature of the discussions). Although transparency has improved, more is still needed at the international level on how decisions are reached and the information on which they are based. The countries requesting deviations or exceptions from the established guidelines should be required to identify themselves and publicly provide their economic (as opposed to political) rationales.

Incentives for better international financial integration. Coordination improved during the early days of the crisis, but has largely lapsed since. Financial regulation and supervision remain largely national. Still there has been some progress. Supervisory colleges have been set up for the G-SIBs; there is some (but not enough) sharing of information on bank exposures across jurisdictions. The most important issue that needs to be tackled convincingly is cross-border burden sharing – both governmental costs of supporting cross-border banks and the allocation of remaining assets in the situation of a resolution or liquidation. The banking union in the euro area is a step in the right direction, but much of its details still need to be worked out. More generally, supervision of parent and cross-border subsidiaries relationships for financial institutions is still murky (e.g. which entity should or can hold capital, pay dividends, or fund assets under various circumstances) and the thinking about these issues is often domestically oriented – again resulting from the lack of a governmental burden sharing arrangements.

International coordination of the activities of global markets, including debt issuance, trading, OTC derivatives reforms, and a host of reporting and disclosure issues is even less well developed. Some guidelines have been issued by IOSCO and the Committee on Payments and Settlement Systems (CPSS), but they are not focused on systemic risk regulation, but on level-playing field considerations and establishing minimum requirements of various types (often devolving to the lowest common set to which all can agree). The thinking is just beginning about how interconnections across jurisdictions may alter how crucial institutional infrastructures might be affected under stress.

The FSB was set up, at least in part, to ensure better international coordination across financial regulatory regimes – and it has had some success in doing so. However, as with most international bodies (e.g. the Basel Committee, IOSCO, and the CPPS) decision-making has to occur by consensus and the FSB has no mechanisms for enforcement of its guidance beyond peer pressure. Consideration of a body with global jurisdiction and authority has been bandied about, but the crisis did not provide enough impetus to develop such an agency.

Better Data and Information to Reduce the Unknowns

The goals of preventing financial crises and accurately assessing the efficiency of financial services require much better data (not necessarily more) in multiple dimensions, for both the private sector and supervisory agencies. Those tasked with evaluating individual financial institutions require better financial statements, since some aspects of financial services provision remain obscure ("buried in footnotes"). And forward-looking risk analysis is generally lacking (although it has improved over time). Even better disclosures, moreover, will not improve decision-making unless this more salient information becomes embedded in the decision-making processes of financial firms and their customers. Some research has documented that only when this "embedding" is complete do disclosures begin to accomplish their stated policy goals.[36]

At the system level, regulators should especially seek better and more disaggregated information on the costs of financial services, data on (aggregate and bilateral) exposures, including in shadow banking and OTC derivatives markets, and the extent of use of new instruments. Another area crying out for more data collection concerns more granular international capital flows in their various forms and cross-border exposures (see Cerutti et al. 2011).[37]

---

[36] See Weil et al. (2006) for a fuller discussion of this concept using eight case studies ranging from corporate finance to health and safety. They demonstrate the importance in tailoring the regulatory disclosure regime to how users do (or should) make decisions using the information so that the public policy goals are met.
[37] For what information is needed for financial stability analysis see Kodres (2013) and for current progress with the G-20 Data Gaps Initiative see Heath (2013).

Raw data can be useful, but more and better analyses are at least as important. Besides trying to predict individual defaults and systemic risks using balance sheet related measures, including bilateral exposures, there have been a host of new systemic risk measures, such as the MES, CoVar, and SLR, which appear promising.[38] There remains, nevertheless, a large need for the development of better indicators and tools that can signal risks in a more timely fashion and evaluate their usefulness in various circumstances. Much of this work will likely remain confidential (to supervisory agencies), but information that does not compromise individual privacy concerns or unduly damages competition can be usefully made public to enhance market discipline. Analyses can for example be included in (global) financial stability reviews, which should also be conducted more frequently and in general be better resourced.

Financial system stress tests offer another means of trying to discern the impact of a systemic or tail event. Regulators should make the further development of techniques and data for financial system stress tests a priority; they should also conduct such tests more regularly. And finally, while the development and use of these and many other, formal analyses will be useful, it will remain important to combine them with "market intelligence" to gain a deeper understanding of why some risks are undertaken as well as to spot newly emerging risks. Observing new trends and talking to a variety of market participants and end users often produces useful "soft information" that lead to more formal data requests and analysis.

Generally though, modesty will remain important. As noted, we do not (yet) know many of the reasons why systemic risks build up, how these varied risks interact, or, more generally, how to avoid crises. This lack of knowledge even applies to the effects of what we think are the right incentives (e.g. does higher capital really lower incentives for risk taking overall?). And it applies to how to exactly design, calibrate, and use tools

---

[38] These are market-based risk measures that develop use market (as opposed to accounting) data to measure systemic risk. The Marginal Expected Shortfall (MES) measure has been developed by Acharya et al. (2010). The MES of a financial institution is its short-run expected equity loss conditional on the market taking a loss greater than its value-at-risk at a specific (predefined, tail) percentile. The Conditional Value at Risk (CoVaR) has been developed by Adrian and Brunnermeier (2011) and represents the value-at-risk (the loss) in the financial system conditional on one institution being under distress. The Systemic Liquidity Risk (SLR) is a global indicator of systemic liquidity stress developed by Severo (2012). It measures the breakdown of arbitrage conditions in major markets using the first principal component of a number of arbitrage violations in international financial markets.

such as macro-prudential policies without inducing unintended consequences, including, for example, the migration of risks to less regulated areas or to institutions or individuals unable to manage them. The lack of knowledge also extends to the drivers and the buildup of risks. How do buildups of risks in insurance and pension systems exactly come about? How, for example, do endogenous tail risks develop, say, through shadow banking?

Even as we obtain higher-quality data and refine our analyses of these significant issues, formal knowledge will always remain limited, especially with rapidly evolving financial systems. This does not mean that we must resign ourselves to periodic financial crises. There may be scope to better use existing (collective) information and analyses that is already out there. Market participants, for example, have tried to develop "model-free" indicators to warn of the next "Black Swan" that will adversely affect markets (e.g. the number of Google hits of the words "crisis" and the like). Needless to say, not all of these efforts have been successful – many have analytical failings and most suffer from "in-sample" biases, making them less useful for predictions. Nevertheless, there are perhaps more ways to extract information from public sources than what has been done to date.

It may also be feasible to better use other sources of information or develop some markets that can reveal unknown or show unexpected risks. For example, the Iowa Electronic Markets allow traders to buy and sell, among other things, political elections results or economic indicators.[39] These aggregators can usefully provide additional information to that from financial markets or other sources. It may also be possible to develop new markets that can both serve to indicate the presence of systemic risks as well as to lay off some risks (e.g. see Brunnermeier et al. 2013, for ideas on a systemic liquidity risk measure of this type). Such indicators and markets could indirectly help to reduce the risk of financial crises though, careful attention to the types of participants and structure is important to avoid manipulation.

Assume Crises Will Recur, and Improve Crisis Management

Unfortunately, even with improvements in all these areas, crises will likely recur. How one responds to crises will thus remain important. Here the recent record is better (say compared to what happened in the Great

---

[39] See www.tippie.uiowa.edu/iem

Depression) but still relatively poor. Interventions are often too late, too timid, and not well coordinated. This leads to a larger final taxpayer bill and higher economic costs – in the form of lost output. There is thus a need to do better.

There are relatively well-known lessons here at both the national and international level that could be applied (better). The main one is the need to absorb any losses resulting from the crisis – whether in the financial, corporate, or household sector or at the sovereign level – as quickly as possible. In practice this means quickly recapitalizing banks when needed; having strong, efficient, less creditor-biased resolution and restructuring mechanisms to resolve overly indebted corporations and households; and to quickly restructure sovereign debt if necessary, including through the use of concerted mechanisms (such as collective action clauses and the like). Another general lesson is the need for the capacity to efficiently and flexibly respond to a crisis. The large, but unplanned role of central banks during the (ongoing) crisis in advanced countries demonstrates the need to have this spare capacity. While there are trade-offs here – too much spare fiscal capacity may introduce moral hazard – some ability of the central bank to manage unanticipated contingencies is nevertheless important.

CONCLUSIONS: WHAT DO WE HAVE TO DO IN ORDER TO DO BETTER?

Given the many similarities in their run-ups, one would hope it should be possible to prevent financial crises. Yet, to date, that seems to have been an impossible task (of course there is a counterfactual – that many crises have been avoided – but it is hard to proof). Indeed one of the main conclusions of any review of the abundant literature on financial crises (e.g. Claessens and Kose 2014; Reinhart and Rogoff 2009; Allen and Gale 2007; Kindleberger 1978) is that it has been hard to beat the "this-time-is-different" syndrome. This, as aptly described by Reinhart and Rogoff (2009: 2), is the belief that "financial crises are things that happen to other people in other countries at other times; crises do not happen to us, here and now. We are doing things better, we are smarter, we have learned from past mistakes."[40] Although often preceded by similar patterns, policy-makers tend to ignore the warnings and argue that: "the current boom,

---

[40] See also Reinhart and Rogoff (2013), which has the apt title "Banking Crises: An Equal Opportunity Menace."

unlike the many booms that preceded catastrophic collapses in the past (even in our country) is built on sound fundamentals ... " (Reinhart and Rogoff 2009: 15). Leading up to a crisis, it is often claimed that the reasons for apparent vulnerabilities are different from those of the earlier episodes. Before the latest episode, the notions that risks were well-diversified across agents and advances in risk management techniques and institutional frameworks were used to justify the belief that "this time is different."

After crises, however, reforms remain often incomplete. One of the difficulties in making overall progress is that crises tend to instill forward momentum on obvious failings, but often ignore the underlying, deeper causes. Moreover reform processes (especially in advanced economies) take significant time for construction, debate, refinement, and implementation during which the public cries for reform diminish and financial sector lobbyists regroup to water down the reforms they perceive as lowering their profitability. The energy for reforms wanes and the perception of the benefits become distant memories. Rationales for enhancing crisis management and resolution also appear less urgent as the immediate crisis fades into the background.

This pattern suggests that one should be quite modest about the depth and impact of many financial reforms in beating the "this-time-is-different" syndrome. Indeed, many of the incentives for risk buildup are still present – despite regulatory reforms. Hence, to be more successful, the starting point must be a better understanding of people's mindset and behavior. Moreover, a deeper understanding about why the previous set of rules has been unsuccessful in preventing crises is needed. And it is important to look carefully for all signs of risks and allow different views to be heard. It appears that prior to many crises, a small minority of onlookers *do* observe that a crisis is coming but they either do not have the incentive to try to prevent it (perhaps because they benefit from the buildup of risks or the crisis itself) or do not have the means of convincing others of their insights.

The most recent crisis has convinced many policy-makers and academics that the financial sector paradigm that emerged over the past quarter of a century is due for changes, both to mitigate the frequency and severity of financial crises and to reorient the financial sector toward activities that benefit society at large. To achieve these objectives, reforms must shift toward how benefits are allocated and risks occur, which in turn means rethinking both governance strategies and the fashioning of incentives. Changing governance will be complex and require altering both the set of stakeholders involved (including governments) as well as the processes that set the rules of the game. Although many stakeholders are involved in financial services, not all are well represented.[41] Improving governance and processes thus requires greater representation of some groups, especially those that are currently not present in the discussions (such as households and other end users).

Representation and governance is largely an (international) political economy question, on which economists traditionally have had little to say, but they can nevertheless raise questions. How can relevant parties, including the general public, be better mobilized to demand a bigger say in discussions? How can one better harness the power of nongovernmental organizations, such as 99%–type movements and other such groups, so that they advocate for a better balance of benefits and risks in finance? Of course, it is also relevant to better understand existing stakeholders' objectives and views. How uniform or diverse are they actually? Does the lack of an effective voice from emerging markets in global regulatory reform for example derive from their diversity, as groups like the BRICs are not necessarily unified in their views? Would it be helpful if their views coalesced better?

At this juncture, several years beyond the height of the crisis, the financial reform agenda is still only half-baked at best. As noted above, some reforms gesture in the right direction, but don't go far enough or have not been implemented fully. Others are either in conflict with one another or appear to have unintended consequences. Policy-makers continue to face severe constraints including complicated governance frameworks, unfavorable structures of political economy, limited knowledge, and stiff political opposition to implementation of reforms. Aside from the greater understanding of incentives of agents and the political economy of reform, how can we make progress on the designs of reform from analytical and empirical points of view? One way of visualizing the efforts so far and what to do next is to consider two dimensions of regulatory reform. One dimension could be the degree of knowledge about what needs to be done – as a gradient from "knowns" to "unknowns." A second dimension would be the practical ability to formulate the regulations – as a gradient from "actionable" to "unactionable." Of course these two dimensions are

---

[41] In most countries, while providers of financial services are well represented, users, notably households, but also many institutional investors, are much less so. Much regulation is also determined through groups, such as the Basel Committee on Banking Supervision, where advanced countries dominate, with emerging markets much less represented than their current economic sizes warrants, and low-income countries hardly represented. With the ongoing shifts in income and financial assets toward emerging markets and developing countries, these discrepancies are likely to increase.

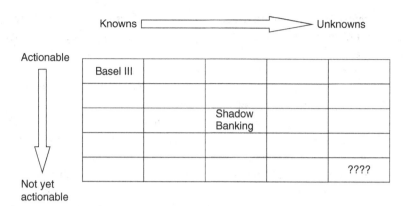

FIGURE 16.1 Dimensions of regulatory reform

not separable (a point driven home by Figure 16.1 below) since how actionable a policy is will depend on the state of knowledge. When knowledge is based on "soft" information as opposed to "hard" data, political constraints become more difficult to overcome and policies appear less actionable.[42]

So far the upper left box of the "knowns" and "actionable" figure has been the focal point, as it has been relatively easy. Regulation on bank capital is a good case in point. Larger capital buffers are known to help mitigate losses and there was already a large set of regulations dealing with bank capital. So tweaks to this area of regulation are relatively easy to define, explain to the relevant agents, including lawmakers, and accomplish. Moving down the diagonal of this "matrix" might be the topic of shadow banking. We know something about how financial institutions operated in this area in this crisis and how bank-like products emerged, but not everything is known, in part because of limited models and insufficient data. The basis on which regulation can be formulated may thus be only partially actionable, as the slower progress on shadow banking reforms shows.[43]

---

[42] Agur and Sharma (2013) make these points.
[43] For instance, the repo market in the United States is known to have procyclical haircut (margin) practices and there has been discussion about how to ensure that these do not become too low in the upswing of the credit cycle. But, as yet, no one is quite sure whether such a rule will not distort the market in a way that is more perverse. Neither is it clear how to impose a minimum floor because a repo is not necessarily initiated from one side (borrower) or other side (lender) of the market. Moreover, since these transactions are not on any organized exchange or location, enforcement is problematic.

There are many areas that need attention, unfortunately, in the lower quadrant of the matrix: areas where a deep understanding of the problem is still nascent and actionable policies are lacking. For instance, many may feel uncomfortable about the speed and degree of automation of transactions in stock and foreign exchange markets or in the ETF markets with its broad retail participation. We do not know, though, if it would be useful to put "sand in the gears" of the trade execution system (e.g. put so-called latency limits on High-Frequency Trading) or whether that would cause more harm than good (e.g. not just higher spreads/costs or lower liquidity, but even more volatility). Also, although we have a vague unease that there may be a "tipping point" in these fast-moving markets, we do not know how to identify it or what would happen if the market suddenly passes such a point.

The real issue, then, is how to gain enough understanding and practical knowledge to move further down the diagonal of this matrix. One way forward is to design ways of connecting the increasing number of measures of systemic risk directly to mitigation tools. So far this is done in a relatively simplistic way (e.g. size, interconnectedness, and substitutability are the sole criteria for G-SIBs/G-SIFIs) without really linking the "systemicness" to the tool (except for the rather coarse way in which Basel III assigns a systemic surcharge). If overseers could directly see the marginal contribution of each individual institution or agent to systemic risk, then one can devise a 'cost' ("levy") that will provide an incentive to lessen that contribution, and thereby internalize the externality.

Even though the point is obvious, it bears repeating that all this requires the right information and ability to analyze it – without these basic building blocks the development and implementation of better policies will be inhibited. Here again incentives will play a role. Confidentiality agreements, the power from holding on to data and information, and the incentive to keep embarrassing information about the (missed) risks of individual institutions hidden, all stymie better understanding of the evolving financial systems. Independent and accountable institutions, whether national, regional or global, must receive the legal and administrative wherewithal to gather sufficient data and identify emerging risks.

In closing, to move forward to reduce systemic risks requires attention to three basic lessons.

- While much progress has been made since the crisis, policy-makers (and market participants) need to think even more system-wide in their risk monitoring efforts and reforms. This systemic view should include not only many (new) forms of analysis, but also become

a process in which supervision is primarily geared to oversee the financial system in its entirety. And a systemic view has to include the adoption of macro-prudential and other policies that explicitly address market failures and externalities.
- Incentives matter, yet they are not nearly well-enough incorporated into current regulations. Many problems will not be solved until one better understands the incentives of all those involved and regulations better align incentives with goals. Here, the ability to fine-tune regulations is likely to be low – given information constraints, the lack of appropriate data and information (including "soft," qualitative information). Hence regulators would do well to take a "do not harm" oath in setting policies – using basic principles and simple measures when information on effectiveness is lacking.
- Risks and uncertainty will remain, in part as a conscious risk-return trade-off and in part as there will always be unknown unknowns – be they tipping points, fault lines, or spillovers – and more data and information are clearly needed. It will thus pay (probably literally) to have a "plan B" – good crisis management plans for when preventive measures fail and risks occur. These plans need to be integral part of the design of the financial system as a whole, not improvisations after the fact.

With these basic components, we believe faster forward progress could be made than is currently the case.

References

Acharya, Viral, Christian Brownlees, Robert Engle, Farhang Farazmand, and Mathew Richardson, 2010. "Measuring Systemic Risk" in Viral Acharya, Thomas Cooley, and Mathew Richardson (eds.), *Regulating Wall Street: The Dodd-Frank Act and the New Architecture of Global Finance.* Hoboken, NJ: John Wiley and Sons: 87–120.

Adrian, Tobias, and Markus K. Brunnermeier, 2011, "CoVaR," NBER Working Paper 17454.

Adrian, Tobias, and Hyun S. Shin, 2010, "Liquidity and Leverage," *Journal of Financial Intermediation* 19(3), pp. 418–37.

Agur, Itai, and Sunil Sharma, 2013, "Rules, Discretion, and Macroprudential Policy," IMF Working Paper, 13/65. Washington, DC: International Monetary Fund.

Allen, F., and D. Gale, 2007, *Understanding Financial Crises,* Clarendon Lectures in Finance. Oxford, UK: Oxford University Press.

Almunia, Miguel, Agustín Bénétrix, Barry Eichengreen, Kevin H. O'Rourke, and Gisela Rua, 2010, "From Great Depression to Great Credit Crisis: similarities, differences and lessons." *Economic Policy* 25 (62): 219–65.

Arsov, Ivailo, Elie Canetti, Laura Kodres, and Srobona Mitra, 2013, "'Near Coincident' Indicators of Systemic Stress," IMF Working Paper, 13/115. Washington, DC: International Monetary Fund.

Atlantic Council, Thompson Reuters, and The City UK, 2013. "The Danger of Divergence: Transatlantic Financial Reform and the G-20 Agenda," The Atlantic Council of the United States. Washington, DC.

Ayres, Ian, and John Braithwaite, 1995, *Responsive Regulation: Transcending the Deregulation Debate.* Oxford, UK: Oxford Socio-Legal Studies.

Barth, James, Gerard Caprio, and Ross Levine, 2012, *Guardians of Finance.* Cambridge, MA: MIT Press.

BCBS (Basel Committee on Banking Supervision), 2010. An Assessment of the Long-Term Economic Impact of Stronger Capital and Liquidity Requirements at www.bis.org/publ/bcbs173.pdf.

BCBS (Basel Committee on Banking Supervision), 2011. Basel III: A Global Regulatory Framework for more Resilient Banking Systems (Dec 2010, revised June 2011) at www.bis.org/publ/bcbs189.pdf.

BCBS (Basel Committee on Banking Supervision), 2013. Regulatory Consistency Assessment Programme (RCAP) – Analysis of Risk-Weighted Assets for Market Risk at www.bis.org/publ/bcbs240.htm.

Blancher, Nicolas, and others, 2013, "SysMo – A Practical Approach to Systemic Risk Monitoring," IMF Working Paper 13/168. Washington, DC: International Monetary Fund.

Brunnermeier, Marcus, Andrew Crockett, Charles Goodhart, Avinash D. Persaud, and Hyun Shin, 2009, *The Fundamental Principles of Financial Regulation.* Geneva, Switzerland: The International Center for Money and Banking Studies.

Brunnermeier, Marcus, Arvind Krishnamurthy, and Gary Gorton, 2013, "Liquidity Mismatch Measurement," in *Risk Topography: Systemic Risk and Macro Modeling,* eds. by M.K. Brunnermeier and A. Krishnamurthy. Chicago, IL: NBER/University of Chicago Press: 99–112.

Calomiris, Charles W., 2009, "The Subprime Turmoil: What's Old, What's New, and What's Next," *Journal of Structured Finance,* 15(1): 6–52.

Cerutti, Eugenio, Stijn Claessens, and Patrick McGuire, 2011, "Systemic Risks in Global Banking: What Available Data Can Tell Us and What

More Data Are Needed?" in *Risk Topography: Systemic Risk and Macro Modeling*, ed. by M.K. Brunnermeier and A. Krishnamurthy. Chicago, IL: NBER/University of Chicago Press: 235–60.

Čihák, Martin, Aslı Demirgüç-Kunt, and R. Barry Johnston, 2013, "Incentive Audits: A New Approach to Financial Regulation," World Bank Policy Research Working Paper 6308. Washington, DC: World Bank).

Čihák, Martin, Sònia Muñoz, Shakira Teh Sharifuddin, and Kalin Tintchev, 2012, "Financial Stability Reports: What Are They Good For?" IMF Working Paper 12/1. Washington, DC: International Monetary Fund.

Claessens, Stijn, Douglas D. Evanoff, George G. Kaufman, and Laura Kodres (eds.), 2011, *Macro-Prudential Regulatory Policies: The New Road to Financial Stability*. New Jersey: World Scientific Studies in International Economics, Pte. Ltd.

Claessens, Stijn, Douglas D. Evanoff, George G. Kaufman, and Luc Laeven (eds.), 2014, *Shadow Banking Within and Across National Borders*. New Jersey: World Scientific Studies in International Economics, Pte. Ltd.

Claessens, Stijn, Swati Ghosh and Roxana Mihet, 2013, "Macro-Prudential Policies to Mitigate Financial System Vulnerabilities," *Journal of International Money and Finance*, 39: 153–85.

Claessens Stijn, and M. Ayhan Kose 2014, "Financial Crises: Explanations, Types, and Implications," in Stijn Claessens, M. Ayhan Kose, Luc Laeven, and Fabián Valencia (eds.), *Financial Crises: Causes, Consequences, and Policy Responses*, Washington, DC: IMF: 3–60.

Claessens Stijn, M. Ayhan Kose, and Marco Terrones 2010, "The Global Financial Crisis: How Similar? How Different? How Costly?," *Journal of Asian Economics* 21: 247–64.

Claessens, Stijn and Lev Ratnovski, 2013, "What Is Shadow Banking?" http://voxeu.org/article/what-shadow-banking.

Committee of Payments and Settlement Systems and the Technical Committee of the International Organization of Securities Commissions (CPSS-IOSCO), 2012, "Principles for Financial Market Infrastructures," April. Basel: Bank of International Settlements.

Crowe, Chris W., Giovanni Dell'Ariccia, Deniz Igan, and Pau Rabanal, 2011, "How to Deal with Real Estate Booms: Lessons from Country Experiences" IMF Working Paper, No. 11/91. Washington, DC: International Monetary Fund.

De Nicolò, Gianni, Giovanni Favara, and Lev Ratnovski, 2012, "Externalities and Macroprudential Policy," IMF Staff Discussion Note 12/05. Washington, DC: International Monetary Fund.

Dell'Ariccia, Giovanni, Deniz Igan, Luc Laeven, Hui Tong (with Bas Bakker and Jerome Vandenbussche), 2012, Policies for Macrofinancial Stability: How to Deal with Credit Booms, IMF Staff Discussion Note 12/05. Washington, DC: International Monetary Fund.

Dyck, Alexander, Adair Morse, and Luigi Zingales, 2010, "Who Blows the Whistle on Corporate Fraud," *Journal of Finance*, 65: 2213–53.

Eichengreen, Barry, 2002, *Financial Crises: And What to Do about Them*. Oxford, UK: Oxford University Press.

Fielding, Eric, Andrew W. Lo, and Jian Helen Yang, 2011, The National Transportation Safety Board: A Model For Systemic Risk Management, *Journal of Investment Management*, 9(1): 17–49.

Fostel, Ana and John Geanakoplos, 2012, "Tranching, CDS and Asset Prices: How Financial Innovation Can Cause Bubbles and Crashes," *American Economic Journal: Macroeconomics*, 4(1): 190–225.

Fostel, Ana and John Geanakoplos, 2013, "Reviewing the Leverage Cycle," Working paper, George Washington University, September, http://home.gwu.edu/~afostel/forms/wpfostel5.pdf.

FSB (Financial Stability Board), 2011, "Key Attributes of Effective Resolution Regimes for Financial Institutions," October, www.financialstabilityboard.org/publications/r_111104cc.pdf.

FSB (Financial Stability Board), 2012, "Strengthening Oversight and Regulation of Shadow Banking." Consultative Document.

FSB (Financial Stability Board), 2013a, Progress and Steps Toward Ending "Too-Big-To-Fail" (TBTF), Report to the G20, September 2, 2013, www.financialstabilityboard.org/publications/r_130902.pdf.

FSB (Financial Stability Board), 2013b, Sixth Progress report on OTC Derivatives Reform Implementation, September 2, www.financialstabilityboard.org/publications/r_130902b.pdf.

FSB-IMF, September 2013, Fourth Progress report on Data Gap Initiative, Washington, DC. www.imf.org/external/np/g20/pdf/093013.pdf.

FSF (Financial Stability Forum), 2009, Principles for Sound Compensation Practices, April 2, 2009, www.financialstabilityboard.org/publications/r_0904b.pdf.

Fullenkamp, Connel and Sunil Sharma, 2012, "Good Financial Regulation: Changing the Process is Crucial," International Centre for Financial Regulation/Financial Times Essay, February 7.

Geanakoplos, John, 2010, "The Leverage Cycle," in D. Acemoglu, K. Rogoff and M. Woodford (eds.), *NBER Macroeconomic Annual 2009*, vol. 24, Chicago, IL: University of Chicago Press: 1–65.

Genotte, Gerard, and David H. Pyle, 1991, Capital Controls and Bank Risk, *Journal of Banking and Finance*, 15: 805–24.

Haldane, Andrew G. and Vasileios Madouros. 2012. "The Dog and the Frisbee." Speech at the Federal Reserve Bank of Kansas City's 36th economic policy symposium, "The Changing Policy Landscape," Jackson Hole, Wyoming, August 31.

Heath, R., 2013, "Why Are the G-20 Data Gaps Initiative and the SDDS Plus Relevant for Financial Stability Analysis?" IMF Working Paper 13/6. Washington, DC: International Monetary Fund.

International Monetary Fund, 2008, "Fair Value Accounting and Procyclicality," Chapter 3 in the Global Financial Stability Report. Washington, DC: International Monetary Fund.

International Monetary Fund, 2009, "Restarting Securitization Markets: Policy Proposals and Pitfalls," Chapter 2 in the Global Financial Stability Report. Washington, DC: International Monetary Fund.

International Monetary Fund, 2010, "The Uses and Abuses of Sovereign Credit Ratings," Chapter 2 in the Global Financial Stability Report. Washington, DC: International Monetary Fund.

International Monetary Fund, 2011, "Towards Operationalizing Macroprudential Policies: When to Act?" Chapter 3 in Global Financial Stability Report. Washington, DC: International Monetary Fund.

International Monetary Fund, 2012a, "Macrofinancial Stress Testing— Principles and Practices." Washington, DC: International Monetary Fund.

International Monetary Fund, 2012b, "The Key Attributes of Effective Resolution Regimes for Financial Institutions— Progress to Date and Next Steps." Washington, DC: International Monetary Fund.

International Monetary Fund, 2013a, "Key Aspects of Macroprudential Policy," Board paper. Washington, DC: International Monetary Fund.

International Monetary Fund, 2013b, "Key Aspects of Macroprudential Policy – Background paper." Washington, DC: International Monetary Fund.

International Monetary Fund, 2014, "Global Financial Stability Report, Chapter 3 – How big is the implicit subsidy given to too-important-to-fail banks?," forthcoming April. Washington, DC: International Monetary Fund.

Jiménez, Gabriel, Steven Ongena, José Luis Peydró, and Jesús Saurina, 2012, "Macroprudential Policy, Countercyclical Bank Capital Buffers and Credit Supply: Evidence from the Spanish Dynamic Provisioning Experiments," Barcelona GSE Working Paper No. 628.

Kiff, John, and Michael Kisser, 2010, "Asset Securitization and Optimal Retention," IMF Working Paper 10/74. Washington, DC: International Monetary Fund.

Kiff, John, Michael Kisser, and Liliana Schumacher, 2013, "The Effects of Through-the-cycle Rating Methodology," IMF Working Paper 13/64. Washington, DC: International Monetary Fund.

Kindleberger, Charles, 1978, *Manias, Panics, and Crashes: A History of Financial Crises*, New York: Basic Books, revised and enlarged, 1989, 3rd ed. 1996.

Kiyotaki, Nobuhiro and John Moore, 1997, "Credit Cycles," *Journal of Political Economy*, 105: 211–48.

Kodres, Laura, 2013, "Data Needed for Macroprudential Policymaking," in Margarita S. Brose, Mark D. Flood, Dilip Krishna, and Bill Nicholls (eds.), *The Handbook of Financial Data and Risk Information* (Cambridge: Cambridge University Press): 566–92.

Lim, Cheng-Hoon, F. Columba, A. Costa, P. Kongsamut, A. Otani, M. Saiyid, T. Wezel, and X. Wu, 2011, "Macroprudential Policy: What Instruments and How to Use Them, Lessons from Country Experiences," IMF Working Paper 11/238. Washington, DC: International Monetary Fund.

Macroeconomic Assessment Group (MAG, formed by Basel Committee on Banking Supervision and the Financial Stability Board), 2010. "Assessing the macroeconomic impact of the transition to stronger capital and liquidity requirements – Interim Report," at www.bis.org/publ/othp10.htm

MAG on Derivatives, 2013 "Macroeconomic Impact Assessment of OTC Derivatives Regulatory Reforms," August. Basel: Bank of International Settlements. www.bis.org/publ/othp20.htm.

Nier, Erlend W., Jacek Osiński, Luis I. Jácome, and Pamela Madrid, 2011, "Institutional Models for Macroprudential Policy," IMF Staff Discussion Note 11/18 and Working Paper 11/250. Washington, DC: International Monetary Fund.

Omarova, Saule T., 2010, "Rethinking the Future of Self-Regulation in the Financial Industry," *Brooklyn Journal of International Law*, 35 (3): 665–706,

Osiński, Jacek, Katharine Seal, and Lex Hoogduin, 2013, "Macroprudential and Microprudential Policies: Towards Cohabitation," IMF Staff Discussion Note 13/05. Washington, DC: International Monetary Fund.

Partnoy, Frank, 2010, "Overdependence on Credit Ratings Was a Primary Cause of the Crisis," in Lawrence Mitchell and Arthur Wilmarth (eds.), *The Panic Of 2008: Causes, Consequences, and Implications for Reform*. Northhampton, MA: Edward Elgar Press: 116–31.

Reinhart, Carmen, and Kenneth Rogoff, 2009, *This Time is Different: Eight Centuries of Financial Folly*. Princeton, NJ: Princeton University Press.

Reinhart, Carmen, and Kenneth Rogoff, 2013, "Banking Crises: An Equal Opportunity Menace," *Journal of Banking and Finance*, 37(11): 4557–73.

Santos, André Oliveira, and Douglas Elliott, 2012, "Estimating the Costs of Financial Regulation," Staff Discussion Note12/11, IMF. September. Washington, DC.

Saurina, Jesús, 2009, Loan Loss Provisions in Spain. A Working Macro-Prudential Tool. *Revista de Estabilidad Financiera*, 17: 11–26.

Severo, Tiago, 2012, "Measuring Systemic Liquidity Risk and the Cost of Liquidity Insurance," IMF Working Paper 12/194. Washington, DC: International Monetary Fund.

Ueda, Kenichi, and Beatrice Weder di Mauro, 2012, "Quantifying Structural Subsidy Values for Systemically Important Financial Institutions," IMF Working Paper 12/128. Washington, DC.

Viñals, José, Jonathan Fiechter, Ceyla Pazarbasioglu, Laura Kodres, Aditya Narain, and Marina Moretti, 2010, "Shaping the New Financial System," Staff Position Note, 10/15. Washington, DC: International Monetary Fund.

Weil, David, Archon Fung, Mary Graham, and Elena Fagotta, 2006, "The Effectiveness of Regulatory Disclosure Policies," *Journal of Policy Analysis and Management*, 25(1): 155–81.

Wellink, A.H.E.M, 2009, "The Future of Supervision," Speech given at a FSI High Level Seminar, Cape Town, South Africa, January 29, www.dnb.nl/en/news/news-and-archive/speeches-2009/dnb212415.jsp.

World Bank, 2013, *Global Financial Development Report, Rethinking the Role of the State in Finance*, Washington, DC: World Bank.

PART V

CONCLUSIONS

17

Institutional Mechanisms for Investigating the Regulatory Implications of a Major Crisis

The Commission of Inquiry and the Safety Board

Edward J. Balleisen, Lori S. Bennear, David Cheang, Jonathon Free, Megan Hayes, Emily Pechar, and A. Catherine Preston

As we have seen, events like the 2008 financial crash, the BP-Deepwater Horizon Gulf oil spill, or the Fukushima nuclear accident trigger massive public attention and often significant regulatory reactions. Once media coverage and the crystallization of public opinion anoint such events as crises that require priority consideration, policy-makers have to discern a way forward. This process typically involves some effort to investigate the recent events and identify their causes. The bodies responsible for such retrospective analysis usually look ahead as well as back. They make judgments about whether policy-makers should revise their risk assessments in light of events, recalibrate their views of trade-offs among competing policy goals, and reconstruct strategies of risk management.

There are many ways to structure investigation into the causes and policy implications of crisis events. In some cases, governments rely on the standard institutions of policy-making. Investigatory bodies at every level of government pride themselves in their ability to perform probing, incisive studies that reveal pivotal evidence and offer relevant analysis for the formulation of policy recommendations. The applicable legislative committees ask staff to undertake extensive background studies and hold a series of hearings in the usual course of business. They then publish extensive reports to guide and justify legislature reforms, or legislative inaction. Executive agencies responsible for mitigating or preventing relevant risks may pursue similar inquiries, whether based on staff research or the work of outside experts, and either alone or through the auspices of cross-agency task forces.

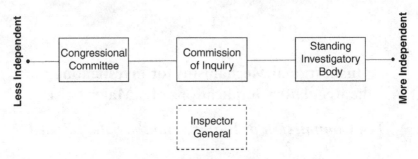

FIGURE 17.1 Spectrum of independence possessed by policy investigatory bodies in the United States

The "normal" channels of policy assessment, however, have limitations. They sometimes lack expertise with regard to the issues at hand, and inevitably take place in a context of partisan politics. They also place the responsibility for policy analysis in the hands of the very institutions whose prior choices failed to prevent the crisis event. Officials within those institutions have strong incentives to shape explanatory narratives so as to deflect blame for the events that have brought such significant social and economic costs.

These shortcomings have frequently led governments to shy away from the typical institutional channels of democratic governance as the most appropriate policy coroners to undertake a crisis event "autopsy." On many occasions, governments have instead turned to ostensibly more independent mechanisms of investigation and policy analysis. This chapter examines two of these alternative investigatory modes – the one-off Commission of Inquiry (COI) and the standing independent safety board. The special commission has been the more common form of investigating major crises.

Indeed, the United States employed this investigative institution after both the Deepwater Horizon blowout and spill, and the 2008 financial meltdown. The safety board has nonetheless become an entrenched institution for assessing the causes of various kinds of transportation and industrial accidents, and deserves close attention as well, since it offers intriguing advantages for retrospective inquiries into the causes and policy implications of major crises.

Figure 17.1 above provides an overview of the different types of investigatory bodies used by the United States government to evaluate policy issues raised by crisis events, placed along a spectrum of independence from governmental influence. We see congressional committees as the least independent, often characterized by partisan political influences and conflicts among committee members. COIs typically possess a good bit more independence, though a number of social science studies have chronicled their vulnerability to political manipulation (Ashforth, 1990; Hanlon, 1991; Centa and Macklem 2001; McGarity 2005; Tama 2011). Standing accident or crisis investigatory bodies, by contrast, usually retain considerable autonomy. Examples of these bodies include the United States National Transportation Safety Board, the United States Chemical Safety and Hazard Investigation Board, the New Zealand Transport Accident Investigation Committee, and the Dutch Safety Board, among others. As permanent bodies, these organizations run independent investigations on incidents in specific sectors (most frequently transportation). They combine extensive methodological experience, technocratic expertise, and an ethos of dispassionate, factual evaluation. These organizations have, on the whole, developed a strong reputation for technical competence, professionalism, and commitment to safety improvement. But the extension of this approach beyond discrete accidents or incidents to larger crisis events such as financial crises may pose additional challenges.

Some authors have also identified Inspector Generals as a potential alternative to Commissions of Inquiry. Inspector Generals (IGs) are embedded within a government agency and serve as internal watchdogs. Although technically part of the agency, they usually have secure budgets, receive long-term appointments, and do not report to agency heads. While these institutional features provide a great deal of independence, the purpose of these institutions is not usually to investigate crisis issues or policy problems. Instead, Inspector Generals generally focus on preventing fraud and abuse of power in government departments (Centa and Macklem 2001). The success of these groups in effectively investigating government matters, however, could be used as a benchmark to develop independent crisis investigation bodies. Due to the fundamentally different missions of IGs, the remainder of this chapter focuses on COIs and safety boards.

The central goal of this chapter is to prompt more careful thinking about the best way to conduct official policy inquiries into major crisis events, drawing on the now considerable historical record of both crisis inquiry commissions and permanent accident investigation boards. These two institutional mechanisms of fact-finding and policy analysis each have important advantages and notable drawbacks; neither unambiguously offers the most sensible mode of policy analysis in every context of crisis. We aim to furnish guides to their relative strengths and weaknesses, while also identifying best practices for institutional design and day-to-day operations. When policy-makers confront the question of how to learn from a major crisis, there is no simple algorithm that can replace political and policy judgment about the most appropriate mode of inquiry. But the specific frameworks for any COI or permanent accident/crisis investigatory board clearly matter, and we aim to suggest how.

SPECIAL COMMISSIONS OF INQUIRY

In the aftermath of a crisis event that places significant political pressure on policy-makers, political elites often respond by appointing ad hoc investigatory bodies commonly known as Commissions of Inquiry (COI), Royal Commissions, or Blue Ribbon Panels. Governments establish these ad hoc commissions to uncover facts and identify relevant broader contexts that help to make sense of a crisis event, to identify what went wrong, and to provide policy recommendations to address failings in the government or other organizations. Governments tend to create COIs when normal investigatory procedures or judicial inquiries seem to be inadequate to the task at hand, either because of the need for technical expertise or because of a desire for greater independence from governmental bodies (Cooray 1985). Scholarly assessments of COIs show that they can be a meaningful addition to policy-making, especially by providing the valuable insight of experts outside of the government (Tama 2014).

The existing literature on COIs mostly focuses on reviewing specific commissions, evaluating their use in specific domains like national security (Farson and Phythian 2010; Kitts 2006) or critiquing them as a political tool for blame avoidance in the government (Sulitzeanu-Kenan 2010). Thus few authors have attempted to provide a broad description of these institutions, and even fewer have evaluated how effective these commissions are at addressing crisis events specifically. Drawing especially on English-language social science scholarship, this section furnishes a descriptive analysis of Commissions of Inquiry in several countries as a policy mechanism for responding to crisis events. After a brief history of COIs, we compare different formats in several English-speaking countries and identify their strengths and weaknesses in developing workable policy recommendations after crisis situations.

COIs, we argue, have proved especially influential when they are transparent and open through public hearings and broad stakeholder involvement, when recommendations are unanimous and favor reorganization over legislation, when they move relatively quickly to take advantage of policy windows, and when they possess sufficient flexibility and autonomy from the government to maintain independence.

COIs have been an integral part of governance in many countries for centuries, but took on their modern form in Britain during the nineteenth century, chiefly as a means to investigate and address a variety of socioeconomic challenges posed by industrialization (Ashforth 1990). The tradition of Commissions of Inquiry quickly spread outside of the UK, first to other Anglo countries in the nineteenth century, and soon to other governments and even the United Nations thereafter. In several countries, legislatures have formalized the role of commissions as independent investigatory bodies that probe matters of public and political concern through a series of governmental statutes that stipulated basic procedural frameworks: Canada in 1846, Australia in 1902, New Zealand in 1908, the UK in 1921, and the US more recently in 1972 (Kitts 2006; Sulitzeanu-Kenan 2010). The contemporary prevalence of this mode of government investigation varies from country to country, becoming more common recently in Canada and Australia, and less common in the United Kingdom. Nonetheless, COIs as a policy tool continue to be important institutional bodies to aid in governance and public policy-making worldwide (Rowe and McAllister 2006).

Every country that uses COIs as an investigatory policy tool structures them somewhat differently. We focus here on COIs found in the English-speaking countries of the United States, the United Kingdom, Australia, Canada and New Zealand, where they tend to be most prevalent. These countries typically use COIs in roughly similar forms. Figure 17.2 displays the variety of COIs found in each of the five countries, including the statute giving the inquiries legitimacy, and the appointing body (signifying whether the commission is appointed by the executive or legislative branch).

In the United States, COIs at the national level have two forms – an executive form (Presidential Advisory Commissions) and a legislative form (Congressional Commissions) (Hanlon 1991). These commissions receive a mandate to investigate enduring policy issues or crisis events and prepare reports, through which they educate the president, lawmakers, and the public about an issue, and typically propose a set of policy recommendations. Individual states also sometimes appoint commissions to investigate crisis events, as Alaska did after the Exxon Valdez oil spill in

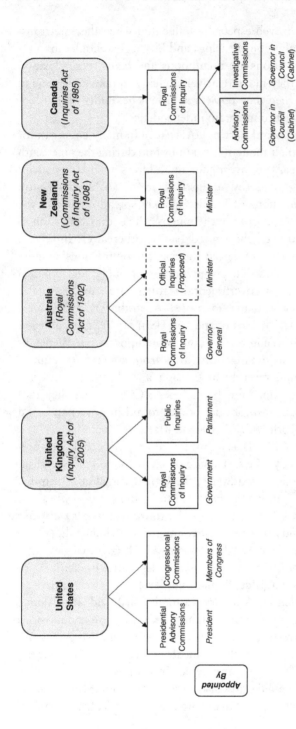

FIGURE 17.2 Forms of national commissions of inquiry in the five case countries

1989 and as Colorado did after the Columbine school shooting in 1999. Whether appointed by the executive or the legislative branches, American COIs, like their counterparts elsewhere, seek to provide a highly visible, time-limited forum for examining important issues, on the basis of greater expertise than may be available in the legislature or executive agencies (Glassman and Straus 2013).

Throughout the British Commonwealth, COIs mostly take the form of either Royal Commissions, appointed by the government of the day and often led by a single person, or Public Inquiries, which receive a mandate from the legislature or civil bureaucracy (Gilligan 2002; Salter 2003; Prasser 2005). In the United Kingdom and New Zealand, the use of Royal Commissions has been supplanted in the past 50 years by a growth of Public Inquiries. In Canada and Australia, Royal Commissions of Inquiry have remained a popular investigatory tool throughout the twentieth century, with the former appointing 74 Royal Commissions since 1970, and the latter appointing 89.

One of the most attractive features of a COI is its flexibility in form and structure, allowing the commission to adapt to the needs of distinct crises. As the UK Council on Tribunals concluded after an inquiry into potentially setting standard procedures for COIs:

It is clear that the infinite variety of circumstances that may give rise to the need for a major public inquiry make it wholly impracticable to devise a single set of model rules or guidelines that will provide for the constitution, powers and procedure of every such inquiry. (Bradley 2003, 36)

Although COIs can vary enormously in structure, size and mandate, commissions appointed to respond to crisis events usually share several features. Figure 17.3 outlines the basic process that a COI undertakes when investigating a crisis event. First, the appointing body selects the commissioners and sets the mandate. This charge varies depending on the relevant statute and crisis circumstances, but generally identifies the focus of the investigation, sets a reporting deadline, and provides funding. The investigation itself is undertaken by the commissioners, supported by a staff of varying size. Hearings, if a COI holds them, can be public, private, or a mix of both. Finally, the commission members submit a report of their findings and policy recommendations to the appointing body. Although most COIs make great efforts to reach unanimity, sometimes COIs publish reports with dissenting opinions (as occurred with the 2011 American Financial Crisis Inquiry Commission). In some cases, commission members also attempt, on their own initiative, to monitor the implementation of their

FIGURE 17.3 Summary of the commission of inquiry process

policy recommendations. The 9/11 Commission, for example, recently published a Tenth Anniversary Report Card, noting that the federal government had adopted 32 of 41 recommendations (National Security Preparedness Group 2011).

To evaluate the characteristics common among COIs, we assessed the workings of selected crisis-driven appointed COIs over the past 50 years in the United States, Australia, New Zealand, Canada, and the United Kingdom. Table 17.1 presents data about several key characteristics of these commissions, including mode of appointment, length, mandate, size, subpoena power, and unanimity. We see the following patterns emerging from this survey:[1]

Appointing Body: More than half of the American crisis-response COIs in the sample were presidentially rather than congressionally appointed. This trend seems to be shifting, however, with a majority of Congressional Commissions appointed since 1995. In other English-speaking countries, there has been a growing tendency to move away from single person-led Royal Commissions, appointed by the full government, to Public Inquiries launched by a single parliamentary minister.

Duration: The COIs in the sample averaged approximately 15 months in duration, with the shortest lasting only three months (the US Rogers Commission investigating the space shuttle Challenger crash), and the longest

---

[1] Intriguingly, although Australia has been especially inclined to appoint Royal Commissions, it has infrequently used them as a policy tool to address crisis situations. Of the 42 Australian commissions appointed since 1970, only four investigated crisis events, none of which had widespread national impacts: the Royal Commission into Aboriginal Deaths in Custody of 1991; the Commission of Inquiry into the Relations Between the CAA and Seaview Air of 1994; the HIH Insurance Group Failure Royal Commission of 2001; and the Equine Influenza Inquiry of 2008.

TABLE 17.1 *Major commissions appointed in response to crisis events (1960–2010)*

Commission	Country	Date	Duration	Appointment	Size	Subpoena	Unanimous
Warren Commission (*The President's Commission on the Assassination of President Kennedy*)	US	November 29, 1963	10 months	Presidential	7 Commissioners; 12 staff	Yes	Yes
Kerner Commission (*National Advisory Commission on Civil Disorders*)	US	July 28, 1967	7 months	Presidential	11 Commissioners; 115 staff	Yes	Yes
Scranton Commission (*President's Commission on Campus Unrest*)	US	June 13, 1970	3 months	Presidential	9 Commissioners; 172 staff	Yes	Yes
Kemeny Commission (*The President's Commission on the Accident at Three Mile Island*)	US	April 1, 1979	5 months	Presidential	12 Commissioners; 16 staff	Yes	Yes
The Mahon Royal Commission (*Royal Commission to Inquire into and Report upon the Crash on Mount Erebus, Antarctica, Of A DC10 Aircraft Operated by Air New Zealand Limited*)	NZ	April 21, 1980	11 months	Government/ Prime Minister	1 Commissioner; 3 staff	Yes	Yes
Rogers Commission (*Presidential Commission on the Space Shuttle Challenger Accident*)	US	February 6, 1986	3 months	Presidential	13 Commissioners	No	Yes
The Piper Alpha Cullen Inquiry (*Public Inquiry into the Piper Alpha Disaster*)	UK	January 19, 1988	1 year, 1 month	Government/ Prime Minister	1 Commissioner	Yes	Yes

(continued)

TABLE 17.1 (continued)

Commission	Country	Date	Duration	Appointment	Size	Subpoena	Unanimous
Alaska Oil Commission (In response to the Exxon Valdez oil spill)	US	May 1, 1989	8 months	Gubernatorial	7 Commissioners; 11 staff	Yes	Yes
Seaview Commission (Commission of Inquiry into the Relations between Civil Aviation Authority and Seaview Air)	AUS	October 25, 1994	2 years	Government/ Prime Minister	1 Commissioner	Yes	Yes
Aspin-Brown Commission (Commission on the Roles and Capabilities of the US Intelligence Community)	US	March 1, 1995	1 year	Joint (Presidential/ Congressional)	17 Commissioners*; 23 staff *Loss of one commissioner during commission process. Final report produced by 17		Yes
Ladbroke Grove/ Southall Rail Cullen Inquiry (Ladbroke Grove and Southall Rail Joint Inquiry)	UK	October 8, 1999	1 year, 2 months	Government/ Prime Minister	1 Commissioner; 24 staff	Yes	Yes
9/11 Commission (National Commission on Terrorist Attacks Upon the United States)	US	November 27, 2002	1 year, 8 months	Congressional	10 Commissioners; 80 staff	Yes	Yes
Major Commission (Commission of Inquiry into the Investigation of the Bombing of Air India Flight 182)	CA	June 21, 2006	4 years	Government/ Prime Minister	14 Commissioners; 35 staff	Yes	Yes

(continued)

Financial Crisis Inquiry Commission (*The National Commission on the Causes of the Financial and Economic Crisis in the US*)	US	May 9, 2009	1 year, 9 months	Congressional	10 Commissioners; 87 staff	Yes	No
Deepwater Horizon Commission (*The National Commission on the Deepwater Horizon Oil Spill and Offshore Drilling*)	US	May 21, 2010	7 months 22 days	Presidential	7 Commissioners; 80 staff	No	Yes
Royal Commission on the Pike River Coal Mine Tragedy	NZ	November 29, 2010	1 year, 11 months	Government/ Prime Minister	3 Commissioners; 4 staff	Yes	Yes

lasting four years (the Canadian commission to investigate the bombing of Air India flight 182). Commissions responding to crisis events tend to take less time to produce final reports than those examining more enduring policy issues, perhaps reflecting post-crisis public demands for rapid policy response.

Mandate: Most COIs receive the same basic instructions: investigate what happened, identify causes, and develop policy recommendations to prevent the specific crisis event from occurring again. In essentially every COI in the sample, commission work remained strictly independent from any judicial proceedings, and so could not legally implicate any individual for wrongdoing. However, COI findings can and do lead prosecutorial or judicial bodies to take-up criminal investigations.

Size: The UK and New Zealand tend to have smaller commissions, with inquiries in the UK being led primarily by one individual supported by a large staff.[2] In the US, the average number of COI commissioners is 10.3, with most made up of between seven and thirteen members. Support staff varies widely as well in the sample, anywhere from a contingent of only four staffers in the New Zealand Pike River Coal Mine Royal Commission to 172 staff and contributors on the Scranton Commission following the 1970 Kent State shootings.

Methods: Investigative tools include intensive reviews of official records, informal interviews, testimony under oath, and expert reports. A key issue concerns public access to any hearings. Although some COIs in the sample, such as the Warren Commission investigating the assassination of President John F. Kennedy, chose to keep their hearings closed, most embraced the principle of transparency and kept at least some of their hearings and depositions open to the public. Since 1990, most COIs have opted for a mix of both public and private hearings in order to maintain some level of openness, while also protecting the testimonies of their witnesses on sensitive matters.

Subpoena Power: Most COIs in the sample possessed the authority to compel testimony from governmental officials and industry representatives. But some, like the Deepwater Horizon Commission, appointed to investigate the oil spill in the Gulf of Mexico in 2010, never received subpoena power despite several congressional attempts to provide it (Capps 2010). Internationally, commissions and inquiries established by statute generally can compel cooperation from corporations, government bodies, and individuals (Gilligan 2002).

Unanimity: Most COIs in the sample generated unanimous final reports, but there have been exceptions. Four members of the 2009 United States Financial Crisis Inquiry Commission, for instance, formally dissented from the majority findings. Other COIs, such as the Aspin-Brown Commission investigating the US Intelligence Agency in 1995, have also documented dissent among commissioners (Johnson 1995). Such dissent, even if not formally captured in the commission's final report, can lower the ambition and reduce the credibility of the commission's recommendations.

---

[2] Several scholars have criticized the appointment of sole commissioners to conduct major inquiries, citing concerns about the potential death or incapacitation of a sole commissioner during the investigation, and arguing that "the agreement of competent minds is to be preferred to the view of a single person" (Keith 2003, 169).

The record concerning the translation of COI recommendations into policy-making is decidedly mixed. In the sample of commissions responding to crisis events presented in Table 17.1, every commission prompted some kind of new legislation or bureaucratic reorganization based on its recommendations. At least sometimes, as with the American 9/11 Commission, recommendations can spur dramatic reforms, in this case landmark intelligence reorganization, a doubling of the intelligence budget, new airline security mandates and, the formation of the Department of Homeland Security (National Security Preparedness Group 2011).

Several characteristics render COIs more influential at generating policy change. COIs appointed by executives appear to generate significant policy change more frequently than those created by legislatures alone. Usually, executive-appointed commissions complete their work more quickly, allowing their recommendations to feed into post-crisis policy deliberations before attention shifts elsewhere. By contrast, legislative commissions (especially in the US) have often had to cope with drawn out appointment processes and stronger partisan divides in the composition of membership. These partisan divisions generate more contention within the commission and less agreement with regard to policy recommendations (Johnson 1995).

COIs that recommend a reorganization of the government instead of legislative changes seem especially likely to have their recommendations adopted. Indeed, among sample COIs, the recommendations most frequently embraced by the government involve institutional reorganization. Thus the highly successful Rogers Commission primarily called for organizational changes within NASA. Alternatively, the Deepwater Horizon Commission made a series of substantive proposals for legislative overhaul of offshore oil drilling regulation, with almost no resulting congressional action (NASA 1987; Graham and Reilly 2014). Reorganization has several benefits, including placing attention on questions of institutional structure, rather than assessment of mistakes by individuals, and overcoming partisan gridlock over more substantive issues (Tama 2014).

Lastly, COIs have had greater success when their investigation gains a reputation for inclusiveness and transparency. Adherence to these values enhances public confidence in an investigation and can greatly bolster its policy recommendations (Terry and Nantel 2014). COIs with public hearings that involve the interests of all sides of an issue are more likely to be taken seriously by the public, industry, NGOs, and the government, and are better able to recommend effective policy changes. The Major Commission's investigation into the bombing of Air India Flight 182, for example, prided itself on holding no private or secret hearings. When its report was published, the government took immediate policy action based on its recommendations, with quick legislative progress on each of the six broad areas of national security. The Warren Commission of 1963, on the other hand, held mostly private hearings; its report was met with controversy and failed to silence conspiracy theories surrounding the event. Although private hearings are sometimes necessary in order to protect the testimonies of witnesses and families, commissions with public hearings more frequently achieve public standings as truly deliberative proceedings that incorporate all relevant stakeholders (Resodihardjo 2006).

Two late twentieth-century British COIs, both led by the Scottish judge William Cullen, suggest how investigations into major disasters can frame eventual policy reforms. The first of these investigated the 1988 Piper Alpha oil rig explosion in the North Sea, which destroyed the rig and killed 167 workers. The second examined the 1999 Ladbroke Grove rail accident, which resulted in 31 deaths and more than 500 injuries. In each case Cullen managed thorough inquiries by comparatively large staffs, identified shortcomings in maintenance, safety management, and regulatory oversight. Cullen's recommendations for offshore drilling prompted tougher regulations the transfer of North Sea safety responsibility out of the Department of Energy so as to remove the conflict between production and safety goals. His report also resulted in the alignment the philosophical approach to offshore oil regulation with the approach that had already been developed and used onshore, namely a shift of responsibility for safety away from regulators writing proscriptive rules toward industry developing site-specific risk plans (Bennear 2015). His later report on rail safety led to new safety rules and eventually to the formal separation of regular safety inspection and accident investigation, through Parliament's creation of the Rail Accident Investigation Branch in 2005.

The flexible nature of COIs means that their form and power will vary from crisis to crisis. Appointing a COI can also be risky for a government; by giving control of suggesting policy improvements to an independent commission, a government puts itself at risk of having its own policies sharply criticized. Despite this political risk, there are a number of advantages of COIs as a policy tool that drive governments to appoint them in response to crisis events, including a desire to transcend narrow electoral politics, a need to incorporate neutral experts, and a need to quickly address issues that the public has deemed of great moment.

Independent commissions have the considerable advantage over traditional government investigations of sidestepping partisan politics, allowing for the evaluation of policy issues through more dispassionate eyes. Politics can sometimes generate controversy in the appointment of a commission, as occurred between the executive and legislative branches over the appointment of the Aspin-Brown commission. But COIs generally carry out independent deliberations that allow their members and staffs to investigate and deliberate outside of the partisan pressures that so often confront government bureaucracies or legislative committees. Independent COIs can also avoid legislative stalemates on divisive issues that demand immediate attention, while avoiding tensions between bureaucracy and legislative policy-making. As Virginia Congressman Frank Wolf observed in 2011, "Overall, Congress is dysfunctional, partisan, and polarized, and it isn't getting anything done. We need commissions to break out of divisive partisanship" (Tama 2011, 32).

A second major advantage of COIs concerns their inclusion of ostensibly neutral experts to provide critical analysis of facts and context related to a given crisis event. Governments often face limitations of resources when confronting important political issues, especially when relevant expertise is not present in the bureaucracy or is spread across several agencies. The mandate of commissions allows them to expand their knowledge base by drawing on expertise from multiple agencies within and outside of the government for short-term appointments. Commissions can also incorporate a variety of voices from business, labor unions, NGOs, and other interested parties in a neutral atmosphere frequently not present in the legislative realm (Inwood and Johns 2014). In this manner, COIs are able to bring a fresh perspective to policy and expand the realm of potential solutions beyond the prevailing intellectual approaches within the government bureaucracy. This feature is especially valuable in many crisis situations when the government is charged with investigating highly technical issues with multiple stakeholders, such as in the Three Mile Island nuclear disaster or a major plane crash such as the Air India bombing of a Canadian airliner over Irish waters in 1985 (Ashforth 1990).

The flexibility of COIs allows their members to respond quickly to mounting public and media attention in the wake of a crisis, while still undertaking a thorough investigation that incorporates comparatively long time horizons. Unlike legislative or bureaucratic bodies that have competing priorities, commission members are more singularly focused on investigating and addressing one particular issue or related set of issues, and are better able to devote specific time to a full and complete investigation (Centa and Macklem 2001). By having a single mandate that allows the commission to focus on one topic, COIs are able to focus their attention and craft the most effective policy recommendations (Tama 2014).

Despite these advantages, scholars and policy practitioners have raised a number of criticisms of COIs as policy tools. Some commissions gravitate toward less ambitious policy recommendations because their commissioners want their recommendations to be adopted, both to further the public interest and their own personal reputations. Commission members accordingly have significant incentives to avoid making controversial policy recommendations in order to improve their chances of being seen as successful at policy implementation. This orientation often leads to negotiations and compromise, a process through which "objective inquiry" can suffer (Ashforth 1990, 14). We find evidence of this in our study as COIs with unanimous recommendations are more likely to see their recommendations adopted. But consensus requires compromise and may involve silencing advocates for more significant changes.

An additional criticism focuses on the potential introduction of political bias, whether through the appointment mechanism or the framing of mandates and budgets. Indeed, governments sometimes limit the scope of a COI's investigation or cut it short (Gilligan 2002). New York Governor Andrew Cuomo, for instance, compelled the shutdown of the Moreland Commission to Investigate Public Corruption in April 2014, a year before its originally scheduled completion, leaving several critics to insist that the commission's work remained unfinished (Lovett 2014).

Critics of COIs also worry that elected officials often also deploy them as a political tool of blame avoidance, using them to shift public criticism and deflect pressures for immediate action. In at least some contexts, COIs can function mostly as an exercise in venue alteration, allowing elected officials to replace volatile, critical audiences, such as the media and a newly attentive public, with a slower-moving and more predictable policy audience – the COI (Sulitzeanu-Kenan 2010). Some authors have suggested that the Krever Commission, a Canadian commission set up to investigate the crisis of a tainted blood supply, was set up to deflect criticism in the midst of a general election (Centa and Macklem 2001). As an Eisenhower White House aide reflected in the early 1970s, "[Commissions] could be used to put a problem on a shelf...If you were rid of something for a year or two it might go away or administrations might change and it would be somebody else's problem" (Wolanin 1975, 23).

One might conclude from this review that democratic governments should continue to deploy COIs in the aftermath of major crises, but take pains to set them up in accordance with best practices. That is, governments should move quickly to get COIs up and running, ensure their independence and relevant expertise, give them sufficient budgetary resources and investigative authority, require appropriate levels of transparency, and establish some means for COI members to monitor and report on governmental responses to their recommendations. Certainly, if governments turn to a COI to perform post-crisis policy autopsies, we would strongly encourage them to adhere to these elements of institutional design.

At the same time, the most well-constructed COIs still have important limitations. Many, and possibly most commission members and staff will be engaging in this sort of inquiry for the first time, and so will take time to find their investigative bearings. Almost by definition, the staff will not be able to draw on institutional memories of analogous investigations. Nor will they be able to take advantage of long-established relationships with relevant parties, such as government agencies, leading firms in the relevant industry, unions, and NGOs. If the institutional independence of a COI translates into isolation, it can compromise its capacity to create a full factual record and place that evidence in appropriate context. Any COI also faces the challenge of building institutional legitimacy from scratch. Again, by definition, it cannot draw on a previous track record or established reputation. And compared to a standing bureaucracy, it will face significant constraints in both keeping track of how governmental bodies deal with its recommendations and liaising with the legislative committees or administrative agencies that wish to adopt those recommendations.

Interviews with senior officials of the NTSB, conducted in June 2014, echoed all of these points. Joseph Sedor, the National Transportation Safety Board (NTSB)'s Chief of Major Aviation Investigations, noted that in his encounters with commissions used in foreign accidents, "the strengths are just not there." These commissions lack "relationships" with other key governmental players; they cannot draw on a clear "investigation structure"; they are "not as familiar with the process" and so struggle "to deal with the chaos" surrounding an accident investigation. As a result, commission-based inquiries "take longer" and rarely develop a sensible strategy of public outreach (Sedor and Schulze 2014). Dr. Joseph Kolly, Director of the NTSB's Office of Research and Engineering, further stressed that COIs

do not have the staying power to do the necessary follow-up that would need to occur for years and years beyond. You can identify the problem, you can make the recommendations, but it is not going to get solved at the end of that, so there needs to be a lot of follow up, and you need to have that knowledge base still hanging together to take it all the way through and even beyond (Kolly 2014).

These concerns help to explain why the British government repeatedly turned to a highly respected judge, Lord Cullen, to oversee a variety of inquiries into high profile industrial accidents. After the first inquiry, Cullen could bring managerial expertise, reputational capital, and at least some seasoned staff members to later investigations. An alternative institutional possibility, however, would be to rely on a standing public body to fulfill the responsibilities of post-crisis policy coroner. Indeed, most industrialized democracies have already created this sort of policy coroner, especially with regard to major transportation accidents, such as airline crashes, major train derailments, and pipeline explosions. The workings of these institutions, such as the NTSB, also deserve careful evaluation.

THE PERMANENT INVESTIGATORY SAFETY BOARD

Relatively autonomous accident investigation bodies have a long tradition in the transportation sector, especially with regard to aviation, where any single accident can shape public perception of the whole industry's risk. As early as 1944, with the establishment of the International Civil Aviation Organization (ICAO) to promote commercial aviation, the international community fashioned an agreement among member countries on rules and standards for crash inquiries. A central feature of the agreement was the strict separation of technical investigations from judicial inquiries. For the subsequent seventy years, this clear delineation between learning from accidents to prevent future recurrence and the attribution of blame in specific cases has been conducive to the rapid identification and adoption of safety improvement measures. With the rise of nationalized airlines, governments and regulatory agencies became interested parties in the investigations. Hence, ICAO further stipulated that the investigations be conducted independently, to avoid conflicts of interest and foster dispassionate analysis.

Many countries have since institutionalized independent aviation accident investigation bodies. Stoop and Kahan (2005) note that airline safety performance has improved by a factor of 200 since 1945, making aviation the safest mode of travel. In the hopes of replicating aviation's success, industrialized democracies have set up numerous independent accident investigation bodies for other modes of transportation. The jurisdictional scope and institutional design of safety boards, however, has varied widely, allowing for comparative analysis, assessment of best practices, and consideration of whether this institutional approach makes sense for a wider array of crises. We begin our discussion with the issue of jurisdictional scope for safety boards. We then engage in two instructive institutional comparisons. The first considers the US National Transportation Safety Board, which has developed an especially strong international reputation for competence, with the more recently established US Chemical Safety and Hazard Investigation Board (CSB), which, although modeled on the NTSB, has struggled to match its record. The second contrasts the Dutch and Swedish Safety Boards, which each possess unusually broad legal jurisdiction, but have interpreted that authority very differently.

In large European countries like the UK and France, authorities have opted for single-modal safety boards, with separate investigatory agencies for each type of transport. Thus the United Kingdom has created Accident Investigation branches (AIB) for each of mode of transportation, in each case after an especially grave accidents led to public outcry.[3] In France, civil aviation incidents are investigated by the Bureau of Enquiry and Analysis for Civil Aviation Safety (*Bureau d'Enquêtes et d'Analyses pour la Sécurité de l'Aviation Civile*, BEA), an agency established in 1946. Analogous agencies investigate maritime incidents (the Maritime Bureau of Enquiry and Analysis, or *Bureau d'Enquêtes sur les Événements de Mer*, created in 1997) and land transport accidents (the Land Transport Accident Investigation Bureau, or *Bureau d'Enquêtes sur les Accidents de Transport Terrestre*, established in 2004 as a response to a pair of major accidents).

The single-modal approach has worked well for these countries. The UK's Air Accident Investigation Branch (AAIB) has an international reputation for excellence, as does France's BEA (Jupp J. 2005; Sedor and Schulze 2014). Indeed, this record of strong institutional performance has

---

[3] The Air AIB (AAIB) was first established within the British Army during World War I. Following World War II, the AAIB was transferred to the Ministry of Civil Aviation. Following the Herald of Free Enterprise ferry disaster in 1987, the government built on this example, establishing the Marine AIB (MAIB) two years later. Similarly, the Rail AIB (RAIB) was set up in 2005 after the 1999 Ladbroke Grove rail accident, and a subsequent COI led by Lord Cullen. One of Cullen's key recommendations was the establishment of a Rail AIB separate from the government's regulatory functions, modeled along that of the aviation industry.

led single-modal agencies to oppose consolidation into multi-modal boards on the grounds that such a move threatened to dilute hard won expertise. Still, it is possible that these countries have missed out on synergies of combined AIBs covering multiple modes.

Such synergies have long been pursued by the United States through the NTSB, which pioneered a multi-modal institutional approach from its inception in 1967. Congress established the NTSB as an investigation agency with oversight of accidents in all forms of transportation that reach across state lines; this move was part of a wider effort to create a national Department of Transportation that could better coordinate the sometimes conflicting regulatory and promotional efforts of the transportation industry. The NTSB has since cultivated an enviable record of influence, with the federal government and other stakeholders consistently adopting more than 80 percent of its recommendations. NTSB's basic structure has been emulated by such countries as Norway, Belgium, Denmark, Hungary and Latvia, which have all created multi-modal investigation boards. In some cases, as with Norway's Accident Investigation Board (AIBN), the emergence of a multi-modal structure emerged through a long period of deliberation and incremental accretion of modes.

The United States has experimented with pushing the basic model of independent safety investigation boards even further, with the establishment of the Chemical Safety and Hazard Investigation Board in 1998, which has the authority to examine a wide array of industrial accidents. Three European countries, Sweden, the Netherlands, and Finland, have taken even bolder steps, giving their safety investigation boards jurisdiction over accidents in other sectors such as healthcare, defense, manufacturing and construction. We have pursued especially detailed research into the Swedish and Dutch multi-sectoral boards, to which we shall return below. First, we give extended attention to the US experience with independent accident boards, which suggests crucial prerequisites for their effective operation.

The US Experience with Permanent Investigation Boards

Established by Congress in 1967, the NTSB soon forged an enviable reputation for timely, objective, and constructive analysis, which influenced the creation of similar transportation investigation boards in Europe and Asia. Impressed by this record, Congress sought to create a similar agency for the US chemical industry following a series of major chemical plant explosions during the 1980s. Lawmakers clearly saw the independence and institutional focus of the NTSB as a key explanation for its operational successes. As first-year Democratic Senator Joe Lieberman argued on the floor of Congress in 1989, not only was a chemical safety agency necessary, but it was also imperative that Congress establish an agency whose independence could enable it to "inspire the same public confidence when it investigates industrial accidents as [the NTSB] does when it investigates transportation accidents" (Lieberman 1989: S14538).

As a result, the two agencies have practically identical organizational structures. As Figure 17.4 indicates, each board is comprised of four members and a Chair, appointed by the President. Both boards make recommendations and issue reports based on research conducted through the office of a Managing Director, who oversees day-to-day operations of the multiple divisions within each agency. The NTSB and CSB also share similar congressional mandates. As investigative boards, neither has the authority to establish or enforce rules and standards, and thus only generates analytical investigative reports and related policy suggestions to the appropriate regulatory agencies, stakeholders, and/or Congress. A defining feature of the NTSB since 1974, this complete independence from regulatory authority permits the Board to investigate and critique both public and private entities.

Despite the high hopes that accompanied the creation of the CSB as an independent agency, the Board has not yet fulfilled lawmakers' expectations that it would achieve the respect and influence of the NTSB. As Figures 17.5 and 17.6 demonstrate, the NTSB has been extremely productive since its inception, averaging more than 1,000 completed investigations each year. Additionally, the annual percentage of recommendations issued by the NTSB and later adopted by relevant governmental actors has remained high, never dropping below 70 percent (NTSB Safety Recommendations Database). Although the CSB has achieved a similar percentage of favorable conclusions to its recommendations, its output has been considerably lower. Between 2009 and 2012, the CSB completed only 15 of its 37 planned investigations (EPA, 2013). Even prior to 2009, however, the CSB rarely drew praise for its productivity. On two separate occasions, in 2000 and 2008, the Government Accountability Office (GAO) has chastised the CSB for underproduction, poor management, and (perhaps most damningly) a lack of rules governing conflicts of interest. In both instances, the GAO found the performance and management of the CSB so alarming that it recommended oversight by an external Inspector General (GAO 2000; 2008). Even more recently, a House Committee on Government Oversight and Reform

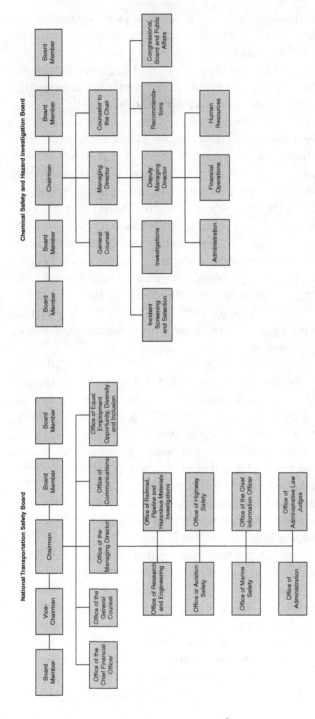

FIGURE 17.4 Organizational charts, National Transportation Safety Board (2014) and Chemical Safety and Hazard Investigation Board (2013)

The Regulatory Implications of a Major Crisis

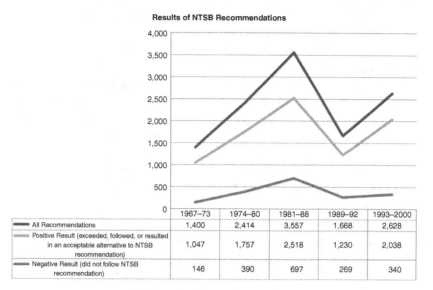

FIGURE 17.5 Results of NTSB recommendations, 1967–2000

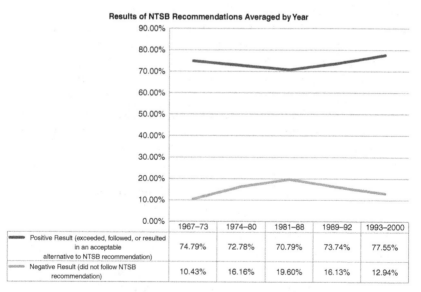

FIGURE 17.6 Results of NTSB recommendations averaged by year, 1967–2000

report, published in conjunction with a hearing that explored charges of whistleblower reprisals, found "serious management deficiencies" within the CSB (Committee on Government Oversight and Reform 2014, 5).

A comparison of the origins and evolution of the NTSB and CSB offers important clues as to why the former achieved a prominence so far unmatched by its successor, despite the intentions of legislators like Senator Lieberman. The global influence of the NTSB makes this question all the more pressing. An understanding of how agencies with similar structures and mandates developed such disparate levels of respect and influence can provide policy-makers with a better sense of the institutional and administrative factors that determine the effectiveness of independent investigative boards.

Three defining features differentiate the NTSB and CSB. First, the NTSB developed a set of investigative procedures that industry structure and institutional capacity made unavailable to the CSB. Secondly, we suggest that both boards were the product of specific historical and political moments. Whereas the NTSB benefited from forty years of policy experimentation and emerged in a climate of regulatory expansion, Congress established the CSB during an era of deregulation and highly politicized budgetary debates. This very different context of governance ultimately resulted in a nine-year gap between its authorization and activation, as well as comparatively parsimonious budgets. These constraints hampered CSB's capacity to fulfill its legislative mandate. Finally, we also find that organizational impact depends heavily on the quality and policy entrepreneurship of its leadership. The history of both agencies demonstrates that interpersonal conflict and weak leadership can paralyze or at least compromise the capacity of organizations to attain their objectives.

Commercial Aviation Safety, Industry Promotion, and the "Party System"

The NTSB's origins in commercial aviation safety continue to shape its institutional structure and priorities. Although the Board has the authority to determine which accidents it investigates in other modes of transport, it is required by law to issue reports on every plane crash either occurring within US territory or involving a US company. Furthermore, as one NTSB official notes, the airline industry has "grown up" with safety regulations (Sedor and Schulze 2014). This concurrence between the development of regulation and the maturation of the industry as

The Regulatory Implications of a Major Crisis 509

a whole has had disadvantages and advantages. Because government intervention into commercial aviation began during the industry's inception, initial regulatory efforts were often complicated by the dual goals of promoting the development of the airline industry and establishing an effective safety regime. Ultimately, however, the synergy, rather than tension, between these two goals resulted in collaborative strategies that are essential to the success of the NTSB.

As early as the 1910s, the US national government adopted numerous policies to promote the emerging air industry. Through various Acts of Congress, officials sought to foster "the science and art of aeronautics," advance air navigation, and improve safety, especially by creating mechanisms to study and learn from air accidents. These moves drew strong support from industry leaders like members of the American Aeronautical Chamber of Commerce, which began to call for "some sort of bureau of civil aeronautics" in 1922. Four years later, Congress established a Commercial Aviation Bureau in the Commerce Department, which had authority to investigate air accidents and publicize their causes. Nonetheless, the early history of the industry was also marked by considerable strain between industry promotion and safety regulation (*New York Times* 1922; US Senate 1926; Institute for Government Research 1930–1931).

As commercial aviation matured, policy-makers increasingly recognized the need for an independent agency focused solely on safety issues. A 1936 Senate investigation, for example, found that "personal, promotional, and political" biases within the Commerce Department consistently hampered the effectiveness of the Bureau of Air Commerce, especially with regard to safety policy (US Senate 1932). This problem recurred for several decades, despite periodic legislative and executive attempts to separate the administrative offices responsible for industry promotion and accident investigations. In 1938, for example, Congress passed legislation that created a separate Air Safety Board. Yet other regulatory officials tended to view the Board members as unwelcome monitors and tended to ignore their recommendations. Two years later, President Roosevelt attempted to remedy this tension through institutional reorganization. Roosevelt discontinued the Air Safety Board and divided the remainder of the Aeronautics division into two new agencies, the Civil Aeronautics Administration (CAA) and Civil Aeronautics Board (CAB). Both agencies remained within the Commerce Department for administrative purposes, though the CAB, which was charged with accident investigation, safety rule-making, and rate adjudication, received the status of an independent agency (US Senate 1940; *Washington Post* 1941).

The burden that the new structure placed on the CAB regularly occasioned complaints from inside and outside the industry. Critics, including airline pilots and legislators, argued that the tasks of promoting air travel, setting rates, and investigating accidents not only ran the risk of being contradictory, but also were too diverse and complicated for one agency to manage simultaneously (*The Sun* 1940). Not surprisingly, these critiques became especially salient in the wake of airline crashes, such as a spate that occurred in 1947 (Air Policy Commission, 1948).

Nonetheless, the promotional mandate of the Commerce Department complicated attempts to enact more stringent safety regulations. Even though members of the CAA and CAB were technically not subordinate to the Secretary of Commerce, the Secretary often held enough power and influence to disrupt their activity. Following revelations that officials in the Commerce Department had repeatedly interfered with CAA and CAB operations, Congress established the Federal Aviation Agency – later the Federal Aviation Administration – in 1956 (US Senate 1956) and gave it the regulatory authority formerly held by the CAA.

The difficulty of maintaining the independence of a safety board housed in a larger department devoted to promotional goals continued to be an issue even after the creation of the NTSB. In 1966, Congress established the Department of Transportation (DOT) in response to President Lyndon Johnson's declaration of the need for a new department to improve policy coordination between the federal government's various transportation agencies. The act also created the NTSB as an independent agency within DOT and transferred the aviation accident investigation duties from the CAB to the new Board. Soon, however, the NTSB's position in DOT proved to be just as problematic as the CAB's relationship to the Department of Commerce. By 1972, NTSB officials were already publicly calling on Congress to remove the Board from DOT entirely. According to NTSB officials, the Board's connection to the DOT created an "appearance of a lack of independence," which was "nearly as detrimental as ... actual infringement" (*Los Angeles Times* 1972). Those concerns together with accusations of executive meddling (which, notably, came in the midst of the Watergate Scandal) prompted Congress to pass the Independent Safety Board Act of 1974, which removed the NTSB from DOT and finally established a completed independent transportation safety board.

Although the dual goals of aviation safety and promotion of the airline industry often created conflict, they were not necessarily diametrically opposed. Early on, industry leaders recognized that commercial aviation would benefit from government-funded research and industry-wide collaborations that improved both the quality and safety of air travel (Vietor 1990; *New York Times* 1922; Institute for Government Research 1930). As early as the 1920s, leaders in commercial aviation recognized that "a sincere effort at careful flying and good equipment" was essential to instilling trust in potential customers (Post 1927). Accordingly, during the 1920s, they established trade organizations like the Aeronautics Chamber of Commerce of America, through which plane manufacturers, airlines, pilots and air traffic controller unions, university researchers, and private institutes created a framework of collaborative research, and actively engaged with government agencies like the CAB (Klemin 1927).

This cooperative spirit created the groundwork for the investigative strategy that has become the NTSB's trademark. After receiving notification of an accident, the NTSB dispatches a "Go Team," which may vary in size and expertise depending on the particular nature of the incident, to the accident site. The team is headed by a chief investigator, who, once on location, coordinates efforts between NTSB staff and stakeholders such as the airline, manufacturer, labor unions, state and local officials, regulatory institutions, and media outlets.

The NTSB regularly mobilizes the technical expertise and human resources of these other stakeholders, through a strategy it refers to as the "Party System." Because it realizes that no 400-person agency could fully understand the technical details of every possible transportation accident, the NTSB invites parties it believes can provide a significant contribution to participate in the first phase of its investigations, which involves data and evidence collection. In order to preserve the Board's objectivity, however, party members may only participate in this initial segment of the investigation. Once the NTSB moves on to its analysis phase, parties no longer have any role in its work.

Observers often cite independence from both industry and regulators as the NTSB's defining characteristic, as well as the feature that lawmakers should replicate most diligently when designing new investigatory agencies. But interviews with NTSB officials corroborate the arguments put forward by several scholars that the Party System is a vital component of the Board's success (Fielding et al. 2011; Hart 2014; Terry and Nantel 2014). Though the origins of the Party System strategy are unclear in both

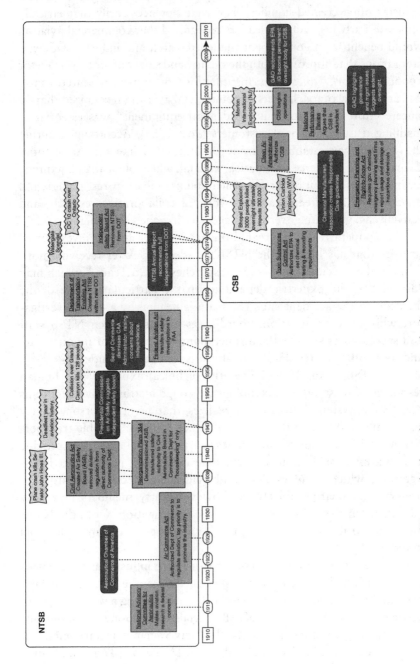

FIGURE 17.7 History of independent investigatory bodies in American transportation and chemical safety

the literature and institutional memory of the NTSB, it seems to have emerged from the collaborative relationship between federal agencies and industry leaders during the early years of aviation safety regulation (Baxter 1995; Fielding et al. 2011; Bowling 2005; Hart 2014; Kolly 2014; Schulze and Sedor 2014). A clear understanding of the long history of aviation safety, including the enduring difficulties of balancing the dual goals of safety and promotion, is essential to make sense of the NTSB's distinctive organizational culture and its collaborative relationship with the airline industry. Attempts to improve aviation safety while also encouraging the young industry's development often created tension between federal officials over whether one goal should be given precedence over the other. But once the aviation industry recognized that increased safety was essential to growth, it developed a collaborative research structure that agencies like the CAB could draw upon when conducting their own investigations.

Chemical Safety and the Budgetary Concerns of the 1990s

While the NTSB benefitted from forty years of regulatory experimentation alongside a young and growing industry, the CSB faced the difficult task of adapting its practices to a well-established sector in a relatively short period of time. Further complicating matters, the early CSB also suffered from the pressures for budgetary austerity that dominated politics in the 1990s, as well as a policy context that increasingly viewed many regulatory frameworks with suspicion. Bipartisan concern over the growing federal debt initially stalled and ultimately crippled the CSB with a lack of resources, an unclear mandate, and uneasy relationships with both industry and other federal agencies. Already on an unstable foundation, early management conflicts and interpersonal issues created an antagonistic culture that the agency has continuously struggled to overcome.

Unlike aviation, industrial chemical production existed for centuries before lawmakers sought to constrain it through the auspices of technocratic regulatory institutions. Although the industry originated in the early industrial revolution and took its modern shape during the Great Merger Movement of the early twentieth century, it was not until a massive explosion at a Union Carbide plant in Bhopal, India in 1984 killed 2,000 people and injured hundreds of thousands that lawmakers began calling for a chemical safety board akin to the NTSB. Within 18 months of the Bhopal explosion, another major accident occurred at another Union Carbide plant in West Virginia that injured 135 people (Belke and

Dietrich 2005). While the Bhopal disaster raised concerns because of its magnitude, the proximity of the West Virginia explosion increased the public pressure on both industry leaders and US lawmakers to reform chemical safety standards. Cognizant of the damage these accidents were causing to the industry's image, the Chemical Manufacturers Association drafted a set of "Responsible Care" guidelines to promote safety and positive community relations within its membership.

In spite of this move toward self-regulation, lawmakers also began pushing for federal oversight. In 1986, Congress passed the Emergency Planning and Community Right-to-Know Act (EPCRA) in 1986 to address the lack of public information regarding the chemicals and potential hazards housed at industrial plants, and to improve coordination of emergency response efforts. Legislators shaped these provisions to address policy shortcomings that had been exposed by the incidents at the Union Carbide factories in India and West Virginia. A series of minor chemical accidents in the late 1980s continued to fuel lawmakers' desire for tighter regulations and an independent investigatory agency. Accordingly, the Clean Air Act Amendments of 1990 required that OSHA and the EPA increase chemical safety and environmental standards and authorized the creation of the CSB.

Although Congress authorized the CSB in 1990, philosophical and budgetary concerns delayed its activation until 1998. Dubious constitutional concerns originally delayed appointments to the CSB under the first Bush administration (Lieberman 1991; *Orlando Sentinel* 1991). Then, the Clinton administration's focus on budget reduction presented an equally substantial obstacle to the Board's launch. Driven by the desire to make government "more effective, economical, and efficient," Clinton initiated a National Performance Review in early 1993 to target unnecessary and wasteful federal programs. Following phase two of the review in 1995, the Performance Review task force proposed a number of institutional changes, including the termination of the CSB due to perceived redundancies between its investigative function and those of OSHA and the EPA (Reylea 1993).

Budgetary stringency continued to hamstring the CSB after it was finally launched in 1998. The CSB's initial funding and staff levels remained tiny in comparison to those originally afforded the NTSB. In real terms, the CSB's 1998 budget was only one-fifth of that of the NTSB's in 1967, which allowed the CSB to hire only one-tenth the number of staff. Later congressional appropriations kept these ratios fairly constant (NTSB 1968; CSB 1998).

The CSB's lack of adequate resources has been exacerbated by regular congressional pressure to investigate accidents not necessarily within its purview. For instance, while the Board originally deemed the 2010 Deepwater Horizon explosion to be outside its mandate, lawmakers eventually compelled a CSB investigation because no other independent agency existed to carry out an assessment of its causes. Similar investigations, such as the 2008 combustible dust explosion at the Imperial Sugar plant in Georgia, led one former investigator to suggest that the CSB is misnamed, and should instead be called the "Industrial Safety Board" (Vorderbrueggen Interview 2014).

Early ambivalence toward the CSB not only hampered its budget and staffing, but also robbed the agency of the chance to take advantage of the public concern that spurred its creation in the early 1990s. Consequently, by the time the Board began operation in 1998, it entered an increasingly antagonistic climate. Instead of capitalizing on the wave of safety reforms fueled by the disasters of the mid-1980s, the Board spent the decade languishing in an administrative netherworld. During the passage of the Clean Air Act Amendments, political will and industry support were aligned in a way that resembled early efforts to regulate civil aviation safety. If the CSB had become active at the point of its authorization by Congress, it may have more easily integrated itself into the emerging regulatory framework, replicating the trajectory of the NTSB. But its delayed activation forced the new CSB leadership to develop organizational strategy amid a highly contentious policy environment, in which established networks of industry and regulatory officials had created an increasingly antagonistic atmosphere. In fact, the activation of the CSB came in the wake of renewed calls for an independent safety board following a joint report by OSHA and the EPA in 1997 that charged the industry with poor safety practices. The harshness of these critiques prompted industry leaders to question the objectivity of the two agencies and to urge lawmakers to fund the independent safety board, so that they would not have to deal with OSHA or EPA investigators (*Washington Post* 1997). So while some industry voices viewed the CSB as a potential partner in safety analysis, key regulatory agencies tended to see it as unnecessary. These officials simultaneously worried that it might prove too close to industry.

After appointments and appropriations brought the Board to life in 1998, tension among Board members inhibited the agency's ability to fulfill its mandate, and created long-standing institutional challenges. In late 1999, Board members Gerald Poje, Andrea Kidd Taylor, and Irv

Rosenthal objected when chair Paul Hill Jr. submitted a request to Congress to double the agency's budget without their approval (Crow 2000). Escalating disagreements over the responsibilities of the Board caused both sides to seek clarification from the Department of Justice Office of Legal Counsel (OLC) on the statutory relationship between the Chair and the rest of the Board. In 2000, OLC released what became known as the "Moss Opinion," which determined that while the "day-to-day administration of ... Board matters and execution of Board policies are the responsibilities of the chairperson ... substantive policy-making and regulatory authority is vested in the Board as a whole" (Moss 2000). This diffusion of authority at the CSB distinguished it from the NTSB, whose chairperson retained a wider degree of unilateral power over policy and administration (Hart 2014). In addition, disputes over the precise boundaries of the CSB chairperson's authority have continuously generated tensions between CSB Board Members (Cheang 2015). Most recently, Board Member Beth Rosenberg resigned over a number of complaints with the Board's management, including Chairman Rafael Moure-Eraso's noncompliance with the Moss Opinion (COGR 2014).

Outside of a short period in the mid-2000s, when the CSB achieved a relatively productive and positive working environment under the leadership of Carolyn Merritt, it has suffered from tight budgets and staffing vacancies, non-collaborative relationships with industry and regulatory stakeholders, and conflicts regarding CSB leadership and professional mandates. The lack of steady leadership not only constrained operations at the Board level, but also affected performance and morale throughout the organization. Whereas experts cite employee satisfaction among the NTSB's greatest strengths, poor management and low staff retention continue to plague the CSB (Fielding et al. 2011; GAO 2008; EPA OIG 2013; COGR 2014).

The broader historical contexts in which the NTSB and CSB were founded also help to account for the differences in their effectiveness and reputational capital. The NTSB benefitted from a period of regulatory growth during the Johnson and Nixon Administrations, which permitted the organization to receive the fundamental political and industry support vital to its early existence. The late 1960s and early 1970s was also a period characterized by policy innovations in numerous agencies through the federal and state governments, a state of affairs that encouraged policy entrepreneurship. This support and experimental climate was particularly important during the early

1970s, when the NTSB weathered a conflict between high-ranking staff and management that hampered investigations and delayed reports (NYT 1974; Chicago Tribune 1977). When the CSB experienced similar issues in the late 1990s and early 2000s, it had not only a more inchoate organizational culture, but also a less robust political buffer to insulate it. Formed in an era of deregulation, the CSB was set-up without the resources, regulatory authority, or creative leadership necessary to replicate the successes of the NTSB. The vastly different historical contexts in which the two boards were created profoundly shaped their institutional trajectories.

Differences in Strategic Orientation and Collaboration with Industry

Some of the CSB's shortcomings arguably relate to policy choices unrelated to broader historical contexts. Despite the decision to model the CSB on the NTSB, CSB's early leaders opted for quite different interpretations of investigative independence, which in turn encouraged distinctive investigative practices. For the NTSB, independence never implied isolation, and the Board has always worked closely with stakeholders throughout the process to address gaps in institutional capacity. The party system, widely perceived as one of the NTSB's strongest attributes (Baxter 1995; Fielding et al. 2011), has no counterpart within the CSB, which has chosen to rely entirely on its own inspections. Indeed, in a 2009 response to GAO recommendations, the CSB vigorously defended its refusal to emulate the NTSB in this regard, asserting that the Board "has correctly interpreted its Congressional mandate by independently investigating major accidents and hazards in depth, rather than attempting to serve as a clearinghouse for numerous, disparate, and often superficial reports from other organizations" (GAO 2008: 68).

Figure 17.8 depicts the interactions that the two Boards have with a wide variety of stakeholders from Congress and the Executive Branch, to local governments and private industry firms. The NTSB has developed a much more interconnected web, with firms both supporting NTSB investigations and receiving Board recommendations. The CSB on the other hand, is much more one directional. It issues recommendations but receives limited contributions from invested parties. Part of the CSB's isolation stems from the organization's already complicated relationship with the EPA. In 2004, the EPA Office of Inspector General (EPA OIG) received temporary congressional appropriations to serve as the primary oversight body for the CSB. Citing continued organizational

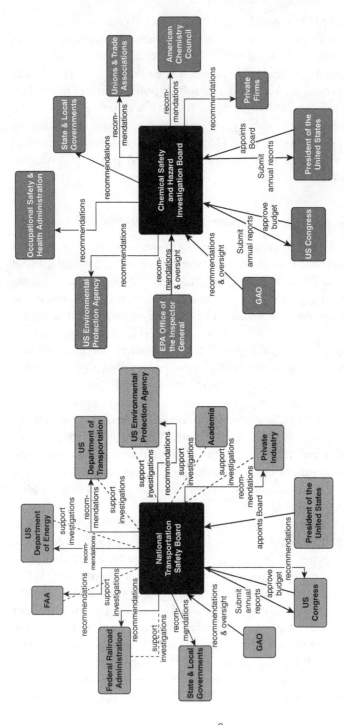

FIGURE 17.8 Relationship maps between interested parties, NTSB and CSB

issues and slow development, the GAO suggested that the EPA OIG become the CSB's permanent ombudsman. The Board rejected the GAO's suggestion, questioning the independence of the EPA OIG since the CSB regularly issues recommendations to its parent organization. While the EPA OIG has yet to have been formally granted permanent oversight status, this relationship between the two organizations remains a constant source of tension, resulting in a 2014 Congressional hearing stemming from the Board's failure to respond to EPA OIG requests (GAO 2008; COGR 2014). In addition to oversight from the EPA OIG, the CSB is also subject to much stronger institutional oversight from the Government Accountability Office (GAO), Congress and the Executive branch. Comparatively, the NTSB is reviewed only by Congress and occasionally the GAO, upon request by a member of Congress.

These alternate interpretations of independent accident investigation at least partly reflect structural differences in the transportation and chemical industries. First, the number of chemical firms greatly outnumbers the number of airlines, major bus lines, railroad companies, or pipeline companies. Since the early twentieth century, only a handful of airlines have controlled the commercial passenger aviation market. The small number of prominent firms encouraged industry-wide collaboration on safety research, development, and regulation. Similar consolidation occurred in other transportation sectors. Conversely, the chemical industry is composed of more than 10,000 firms, both large and small (data from the Commerce Department, "Select USA"). The diversity within the industry makes it less equipped than commercial aviation to coordinate safety standards. Because of their lack of resources and expertise, moreover, smaller firms often view regulatory constraints as impositions that compromise their ability to operate profitably.

Secondly, because fatal transportation accidents are highly visible and directly impact travelers, they can dramatically affect public perception of the entire industry. Chemical accidents, while sometimes more disastrous, generally affect employees or nearby residents, but not consumers of the chemicals themselves. News coverage of a disastrous plane crash, psychologists have found, significantly increases the perceived risk of air travel. By contrast, a fatal explosion at a chemical plant producing compounds used in pesticides would not necessarily lead to a decrease in sales of that particular pesticide. As a result, executives of transportation firms perceive themselves as sharing a community of fate with regard to safety, and so have a much higher incentive to invest in research and development on improving systemic safety than those in the chemical industry. The priorities of the NTSB and transportation industry leaders, therefore, are more aligned than those of the CSB and the chemical industry. This allows the NTSB to draw upon a broader network of safety experts who not only offer knowledge and skills critical to the investigation process, but who also are generally more receptive to safety recommendations.

Furthermore, safety developments and self-regulation in the chemical industry developed in response to concerns about growing pressures for strong public safety regulation rather than from anxieties about nervous consumers. Following the massive chemical explosion in Bhopal, the chemical companies worried that their reputation for "ruthless, uncaring ambition," would increase community resistance to future plant construction and hinder industry development (Hook 1996: 1138). To re-establish public trust and credibility, the Chemical Manufacturers' Association developed the Responsible Care program in 1985, an industry-wide system of self-regulation. But unlike the early air safety efforts that eventually evolved into the NTSB, the chemical industry did not view the CSB's investigations of chemical process safety as an essential component of industrial development. Throughout the 1990s, the industry lobbied for the activation of the CSB not out of economic self-interest, but to remove existing investigatory authority from adversarial regulatory bodies such as OSHA and the EPA (Andrews 1993).

The close working relationship between industry officials and NTSB staff also stems from shared professional backgrounds. At the time of our interviews, all NTSB board members and department directors had an extensive background in transportation and engineering. Trained as engineers, NTSB leadership understands transportation accidents from a perspective similar to the industry it regulates. With a shared set of priorities, values, and training, NTSB directors are able to communicate effectively with industry leaders and create cohesive action plans. By contrast, the majority of CSB leadership comes from backgrounds in public health or environmental toxicology, disciplines seemingly at odds with the goals and actions of the chemical industry. CSB officials and staff also tend not to share regulatory priorities with the industry they investigate and often have had ties to labor organizations that are perceived to impact their objectivity (Cheang 2015). The differences in background, training, and perspective between the CSB leadership and executives within the industry it has fostered adversarial rather than collaborative relationships, a very different dynamic than exists between the NTSB and transportation firms.

The Regulatory Implications of a Major Crisis

PERMANENT SAFETY BOARDS WITH BROAD JURISDICTION: THE SWEDISH AND DUTCH EXPERIENCES

As with the US NTSB and CSB, the Dutch and Swedish Safety Boards have similar mandates and formal organizational structures, but have exercised their authority quite differently. A comparative history of these two European investigative bodies reinforces our argument that organizational culture has profound implications for the effectiveness of safety investigation boards. This comparison further underscores the salience of policy entrepreneurship and leadership to autonomous investigative bodies.

In Sweden, as in many countries, a more expansive institutional approach to accident investigations emerged gradually, with the state establishing a series of independent bodies for specific contexts, and then pulling them together into a single agency. The Swedish Parliament established an initial safety board in 1957 with the sole focus on military aviation. By the 1970s, a combination of greater technological sophistication and several accidents involving passenger airlines led the Accident Board to expand its charge to include civilian incidents to "prevent reappearance of identical accidents" and to reinforce public confidence in the air sector (Jakobsson 2011: 2010).

Discussions over the creation of a Swedish accident board for maritime incidents began as early as 1958, with Parliament creating the formal structure for such a board in 1963. The government, however, did not actually appoint permanent members to this entity before the mid-1970s, perhaps because of resistance by industry. Even after the initial organization of the maritime board, its permanency remained in some doubt.

The Swedish politics of safety investigation shifted as a result of several high profile accidents that that occurred in 1977 and 1978, including a large landslide and several restaurant and hotel fires. These events heightened public concerns over risks, generating a broad consensus around the advisability of an investigatory board with a broader mandate. The Swedish Parliament accordingly drew up plans for a State Catastrophe Board, which would ideally facilitate a "transfer of experience" from investigations of transportation-related accidents to investigations of industrial or construction-related accidents. The idea was to achieve a better understanding of the ways in which materials, construction and human interactions relate to one another in an incident to minimize mortal and financial damages.

Initially, the government intended the Swedish Accident Investigation Board, the Swedish Maritime Accident Investigation Board, and the State

Catastrophe Board, to work side by side, each focused on a specific niche. In 1986, however, fiscal pressures led Parliament to consider the possibility of combining these agencies to achieve cost savings. The resultant report found that the Swedish Maritime Accident Investigation was far too slow in producing its reports and would benefit from a merger with the agency concerned with aviation. The report further found that the procedures of the State Catastrophe Board were almost identical to those of the Swedish Accident Investigation Board. Despite the opportunities for savings and operational improvements through consolidation, the various accident investigation boards strongly preferred to remain separate.

Several accidents in 1986, however, accelerated the pace towards merger despite the boards' preference for autonomy. These included the Prime Minister's assassination, a radioactive pollution spill, and a school bus brake failure that resulted in several deaths (Jakobsson 2011). Parliament's decision to impose consolidation was predicated partly on the assumption that "uniformity in accident investigation methods" would lead to an improved safety culture and that the new entity could overcome any lingering institutional differences.

The resulting merger occurred in 1990, with a new *Statens haverikommission*, or Swedish Accident Investigation Board (SHK) taking over all assignments. Drawing its organizational structure from the Multi-Modal model championed by the NTSB, SHK encompasses more than just transportation accidents, as it has a charge to respond to accidents of all sorts, on the basis of their seriousness, measured either by loss of life or resulting damages (Pollack 1998). In light of this broad mandate, the SHK relies heavily on external experts when it treads outside its traditional areas of competence. These specialists work alongside SHK investigators to assist in building the technical expertise for an effective investigation (Stoop and Roed-Larsen 2009). This allows the SHK to comprehensively examine a wide range of incidents despite a relatively small number of permanent employees (Pollack 1998). Currently the SHK employs close to 40 permanent employees, of which only 24 are investigators.

Limited English-language sources complicate assessment of the SHK's effectiveness in terms of the proportion of its recommendations that have been implemented by the Swedish government or pivotal industries. Nonetheless, the SHK's own documentation does permit analysis of the extent to which the agency has branched out beyond transportation. Of the 206 incident reports that the SHK published from 1996 to 2012, 79 percent were aviation accidents, 14 percent took place in the maritime arena, only 1 percent concerned rail incidents, and just 6 percent involved "other" events. This catchall category included a discotheque fire, but also a collision between train wagons and a tanker truck, a bus fire, and an accident involving a paraglider, all incidents also involving transportation in some way.

Thus despite the holistic charge of the SHK, investigation into non-transportation incidents remains exceedingly rare. This lack of emphasis on "other" incidents is reflected in the organization's allocation of resources. Only one staff member has the responsibility for "fire, road, rescue and other accidents." In reviewing the SHK's budgets for the years 2011–2013, the funds set aside for "other" incidents also paled in comparison to those allocated to aviation. It is possible that there have been significantly fewer incidents to investigate in the "other" category. But since 1990, the Swedes have continued to use Commissions of Inquiry to investigate non-transportation accidents. The SHK, then, has not delivered on the Parliament's charge to extend investigations beyond transportation.

The Dutch have had an even longer history than Sweden of investigation agencies for transportation accidents (e.g. the Inland Waterways Disaster Committee in 1931, the Civil Aviation Board in 1937, and the Railway Accidents Inquiry Board in 1956). Following two serious accidents (aviation and rail) in 1992, the government chose to merge these agencies in 1999, with the formation of the Transport Safety Board. Beyond the modal integration, Parliament also increased the investigation agency's independence and sought to focus its work on safety improvement. Parliament simultaneously clarified the separation of accident investigation from the judicial process.

In the aftermath of a fireworks factory explosion and a deadly café fire in 2000, the government established the Dutch Safety Board in 2005, vesting it with a mandate to carry out independent investigations of disasters, accidents and near accidents in all policy sectors, including the fields of transportation, defense, industry, health care, the environment, and crisis management. In considering proposals to create a more all-encompassing DSB, the Dutch Parliament focused especially on mechanisms of institutional design that would foster independence. Thus the legislature took care to make the DSB an autonomous organization, answerable only to Parliament and so not subordinate to any minister. The DSB also received wide discretion in the selection of events to investigate (apart from the aviation incidents subject to international obligations), as well as clear authorization to enter places, issue subpoenas, and call on other Dutch ministries for the deployment of experts where necessary. Underscoring the importance of keeping its activities separate from the judicial process, Parliament directed DBS to withhold information from the judicial authorities unless it uncovered evidence of egregious crimes such as murder or terrorism.[4] Once launched, the DSB employed about seventy-five full-time staff, including approximately forty investigators. Like the NTSB, the DBS intentionally lacked any overt power to impose adoption of its recommendations.

Although more than 80 percent of DSB's investigations deal with aviation incidents, there have been more than 40 investigations of accidents in non-transport sectors, including healthcare (e.g. salmonella in smoked salmon, October 2012), crisis management (e.g. safety of asylum seekers, April 2014), construction (e.g. collapse of stadium roof, July 2011), and industry (e.g. fatal accident in manure silo, February 2014). A Belgian study (Navamar 2011) on the DSB assessed its reports as consistently attaining a high quality. The study also concluded that the Dutch Board did its work efficiently, since dispersed investigations by multiple bodies would have required more resources. The DBS also appears to have achieved considerable domestic legitimacy, since the Dutch government has implemented the vast majority of its recommendations, at least in part.

Media accounts of the DSB, moreover, usually treat it as an authoritative source. One can see this implicit respect for DSB judgments in contexts as various as IT infrastructure security (Essers 2012), healthcare (Radio Netherlands 2012) and food safety (Dutch Parliament 2013). DSB investigation reports regularly received respectful citations in trade publications, especially for those in the aviation sector, as well as academic papers. For example, the trade journal *Flight International* (2009) reported on DSB's investigation into an accident of the Royal Netherlands Air Force Apache helicopter, and that in response to the findings, the air force amended some of its procedures and was working to identify other improvements. In *Safety Science*, DSB's report on a crane accident was largely reproduced in an article by Swuste (2013). DSB's investigations in non-traditional areas such as food safety also appear to be highly regarded by scholars (van den Akker and Lange 2014).

Although the DSB is the newest of the multi-sectoral investigation agencies, it has been held up as a model by countries considering the approach. When the Polish government tasked the Gdansk University of Technology (GUT) to consider the model of integrated transport safety

---

[4] For example, in the investigation into the 2009 crash of a Turkish airliner, DBS refused the Attorney General access to the black box data, on the grounds that its investigation would be hampered by the threat of legal action.

management for Poland, GUT researchers turned to the experiences of DSB, along with NSTB, in identifying key requirements of a proposed transport safety board (Zukowska, Michalski, and Krystek 2010). Similarly, in a study into the establishment of an independent investigation board for Belgium commissioned by the Ministry of the Interior, the Leuven Institute of Criminology held up the DSB as a model institution.

The potential synergies of having an agency span several categories of accidents, however, do not arise automatically. As the Swedish case indicates, simply housing the separate sectors into the same office does not create an effective multi-modal agency. Institutions have to intentionally foster cross-team collaboration, exchange of knowledge, and the cultivation of expert networks if they wish to expand their investigative authority. This point raises the centrality of leadership within more expansive investigative safety boards.

In a short span of less than a decade, the DSB has carved out an international reputation as an innovative and effective investigation agency. Its high standing appears to be closely tied to that of its respected former chairman Pieter van Vollenhoven. The literature on the DSB's formation consistently identifies his central role in creating a political consensus for the establishment of an integrated multi-sector board, and then in building an agile organization. He had been a strong advocate of independent investigations, and went as far as to claim that it helped democracy to function properly (van Vollenhoven 2001). His multi-decade effort to secure the public's right to independent investigations culminated in the founding of the DSB.

Van Vollenhoven epitomizes the policy entrepreneur who successfully fosters bureaucratic autonomy. A member of the Dutch royal family and a lawyer by training, he also had training as a pilot during national service in the Royal Netherlands Air Force. This background led to his participation in several aircraft accident investigations. In 1975, he was appointed as advisor on road safety to the Minister of Transport. Subsequently, as Chairman of the Road Transport Safety Board, he lobbied for the creation of a multi-modal Transportation Safety Board modelled after the NTSB. Although the other transport boards were far more established and reluctant to be merged into a multi-modal board, two serious transport accidents in 1992 generated the momentum to bring about the multi-modal agency.[5] Subsequently, van Vollenhoven also lobbied to extend the

---

[5] The Maritime Court of the Netherlands was set up in 1909, the Civil Aviation Board in 1937, and the Railway Accidents Inquiry Board in 1956.

principle of independent investigations to serious accidents outside the transport sector, eventually carrying the point despite initial pushback from the Ministry of Justice.

Although one might be inclined to attribute van Vollenhoven's success to his privileged position in Dutch society, such an explanation would short-change his consistent cultivation of political support for the principle of independent investigation. Van Vollenhoven worked diligently through the 1980s and 1990s to frame independent investigation as a "right." Domestically, he built a wide political constituency in part by founding a Support Fund to assist victims of traffic accidents and crime. Internationally, he assiduously developed contacts with counterparts in the United States and elsewhere that gave him access to international best practices and solidified his reputation as an expert.[6]

By contrast, the SHK has largely ignored its clear mandate from Parliament to extend its work beyond aviation, preferring to maintain a much narrower focus. While the DSB emerged through the initiative of its founding chairman, the Swedish merger was forced upon reluctant agencies by Parliament. Instead of creating a new agency whose leadership was committed to the multi-sectoral approach, Sweden essentially jammed together several investigation agencies under the authority of an aviation-centric SHK. With the Swedish government turning to a consolidated agency in order to save costs, it is hardly surprising that the SHK has tended to short-change resources and staffing for investigation incidents that fall outside aviation.

CONCLUSION

This review of Commissions of Inquiry and standing accident investigation agencies suggests several lessons for government officials who confront a crisis event. COIs are almost certain to remain an institutional option for governments. Their nature as flexible, independent

---

[6] In 1993, von Vollenhoven worked with the American, Canadian and Swedish transport safety boards to found the International Transportation Safety Association (ITSA), and became its founding chairman. The organization created a forum through which investigation agencies learned from one another's experiences, trained investigators, assisted with complex investigations, and promoted the practice of independent investigation. ITSA has since grown to 14 countries, including agencies from countries like the UK, France, Finland, Japan. In Europe, he worked with German and British agencies to establish the European Transport Safety Council to provide impartial expert advice on transport safety matters to the European Union and member states. For several years in the 1990s, he was also advisor to the European Commissioner for Transport.

investigatory bodies able to draw on experts from industry, academia, labor unions, and NGOs, makes them valuable mechanisms to fashion explanations of how crises occurred and recommendations for how to prevent their reoccurrence. COIs offer distinct advantages over the normal channels of policy formulation, including their relative political independence, their ability to do thorough investigations without the time or resource constraints of other government entities, and the flexibility of their form which allows them to adapt to the needs of each unique crisis situation.

Our survey of major COIs in English-speaking countries additionally suggests several practices that can optimize their effectiveness in investigating major crisis situations.

- **Quick start-up:** Following crisis events, the policy window for implementing the recommendations of a COI is generally very short. Efforts should be made to minimize the start-up time of commissions by quickly appointing commissioners and staff, setting a specific mandate and quickly providing the funding needed to set up a reasonable staff to begin the investigation.
- **Thoroughness and transparency in investigation:** The possession of subpoena power allows COIs to gain access to all relevant facts. The holding of public hearings that allow the public and other decision-makers to feel involved in the investigation and deliberation process increase confidence in the commission's work. This effect is strengthened when the commission makes an effort to include all stakeholders of the crisis in the investigation.
- **Avoidance of political divisions where possible:** Executive-appointed COIs tend to run into fewer partisan divisions that slow down the deliberation process and can hamper the influence of recommendations. However, keeping a political balance among commission members and, most importantly, setting an atmosphere of political independence within the commission, is vital.
- **Provision of genuine autonomy:** Appointing bodies should, when possible, give the COI a flexible mandate as to the form and extent of investigation. Governments should also avoid disbanding the commission before the investigation has been completed.

For all the advantages of COIs, governments across the world have frequently turned to permanent investigation agencies either to address the inadequacies of ad hoc commissions or to meet the recommendations of these commissions. This pattern underscores the inherent structural advantages of a standing investigation agency over commissions, which include:

- **Greater Independence:** As COI members are assembled by the government following a given accident, they can be perceived to be less independent than an investigation agency. Furthermore, it is likely that some COI members will need to be drawn from the regulatory agencies that have responsibilities over safety regulations and compliance. Even if a COI carries out its work with full integrity and objectivity, the perception of conflicting interests may compromise the credibility of its findings. An investigation agency also has the benefit of an established legal framework to safeguard its independence.
- **Efficiency and Legitimacy:** After an accident or broader crisis even, even the most aggressive COI still takes time to hire staff and define its work scope and relationships with other investigating agencies. By the time a commission is established, accident sites may well have been cleared, compelling the commission to rely on secondary investigation reports. The investigation agency's established investigation powers and methodology, and trained investigators also improve public confidence in the findings.
- **Deeper networks:** Safety boards that cultivate relationships with both pivotal industries and regulatory agencies, such as the NTSB, possess clear advantages in quickly compiling a full factual record and in having policy recommendations taken seriously by legislatures and administrative agencies.
- **Capacity to monitor near misses:** Because safety boards remain on the job between accidents or crises, they can develop mechanisms to keep track of data about close calls, and analyze their implications for risk management before the harm caused by an actual accident or crisis. Multi-modal or multi-sectoral boards also can monitor safety trends across many industries.[7]
- **Improved Follow Up:** Commissions of inquiry typically disband after issuing a report, and so cannot monitor the impact of their recommendations on policy-making. Furthermore, investigation boards are able to study series of incidents to undercover more systemic weaknesses.[8]

---

[7] Thus the NTSB has conducted several important studies of fatigue and distraction as problems across modes of transportation.
[8] This overview draws partly on the analysis of Stoop 2004.

Scale and scope are additional important parameters that can determine the success of an investigation agency. We note that the three largest EU economies, the UK, France and Germany, have maintained the single-modal approach. However, for smaller countries like Latvia, the low frequency of major accidents may result in single-modal agencies being too poorly resourced to maintain high quality investigations. Although a railway accident may be completely different from one in aviation, the methods to respond to an accident and how to systematically learn from and create recommendations for change are consistent throughout transport modes and potentially applicable to events in other sectors. Notably, in order to increase the range of investigations with relatively small agencies, both SHK and DSB adopted the NTSB "party system." The agencies' staffs of investigation and accident specialists are teamed with an assembly of experts for each specific investigation. Van Vollenhoven (2001) and Stoop and Roed-Larsen (2009) have asserted that the international and regulatory trend is to move from single modal to multi-modal, and possibly multi-sectoral, investigation agencies since investigations always follow the same procedures, whatever the sector of the accident. Furthermore, they observe that no multi-modal investigation agency has ever split into separate boards.

The historical comparison between the United States NTSB and CSB further suggests that independent investigatory boards will be most successful in sectors where private and public interests are most closely aligned. In order to benefit from an investigative strategy that employs methods similar to the NTSB's "party system," a government agency requires a broad and intensely engaged network of private experts. Comparing the trajectories of the NTSB and CSB suggests that such networks emerge most frequently when private entities consider the aspirations of the public agency as essential to economic growth and development. Thus one might expect a Nuclear Plant Accident Safety Board to encounter greater cooperation from the key plays in the relevant economic sector than a Financial Markets Safety Board or an Offshore Drilling Safety Board. Executives in the nuclear industry perceive more of a "community of fate," since the depth of public anxieties over nuclear power has convinced them that a serious accident anywhere will quickly become the problem of plant managers everywhere (Braithwaite and Drahos 2000; Rees 2009). It is less clear that financial executives or the heads of fossil fuel companies would accept the importance of independent safety investigations in their domains and actively cooperate with their inquiries.

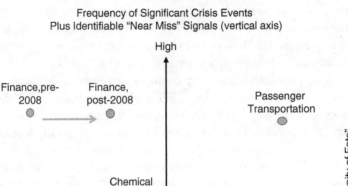

FIGURE 17.9 Mapping economic sectors by frequency of crisis events and degree of perceived community of fate

In Figure 17.9, we offer a schematic representation of selected economic sectors that have recently experienced significant accidents and/or crisis events, roughly sorted with regard two variables – the depth of concern within the sector about safety/systemic stability (represented on the horizontal axis); and the frequency of accidents/crisis events, along with identifiable "near miss" signals (represented on the vertical axis). For regulatory domains that fall within the upper right quadrant, such as passenger and goods transport, there is an especially strong case for establishing permanent safety boards. For those in the lower left quadrant, the lack of frequent events to investigate would suggest a preference for turning to a COI, or perhaps reliance on a safety board like the one in the Netherlands, with cross-sectoral authority. One must remain cognizant, however, that the placement of sectors in Figure 17.9 has a temporal dimension – it holds for a specific moment, and can change. Crisis events might shift the norms and culture within a given sector, making its leaders more concerned about matters of safety/systemic stability, as occurred in air transport during its early decades, and as may have happened at least to some extent within finance after 2008. Over the longer term, it is also possible that effective policy may sufficiently improve safety as reduce the frequency of accidents/crisis events, and even induce a form of policy amnesia. The relative stability of the post-New Deal American financial system, for example, almost certainly inured many leading voices in that industry to the building up of systemic risks since the 1970s.

A final point concerns the centrality of institutional leadership, including the inventiveness of policy entrepreneurs. Despite the presence of both a broad network of private collaborators and a supportive political climate, the effectiveness of NTSB has suffered during moments of internal squabbling and interpersonal conflict (*New York Times* 1974; *Chicago Tribune* 1977). While the NTSB seems to have resolved most of its incapacitating leadership issues, government oversight bodies continue to critique the CSB for mismanagement. A lack of steady leadership not only constrained operations at the Board level, but also affected performance and morale throughout the organization. Experts cite employee satisfaction among the NTSB's greatest strengths. For many at the NTSB, a position at the agency is widely considered the final stop in a long career of transportation safety. Passionate about their work, NTSB staff members often see their jobs as intrinsically valuable and enjoyable, and many choose to stay for years and even decades in the same position. The positive work environment at the NTSB encourages a high level of dedication and motivation among its employees and anchors the stability and effectiveness of the organization. This is not the case at the CSB, where conflict among the board and between the Board and other regulatory agencies, coupled with low staff retention continue to plague the organization (Cheang 2015). Interpersonal drama, constant employee turnover, and mismanagement consistently drain the CSB of important human capital and constrain its ability to adequately fulfill its mission (Fielding et al. 2011; GAO 2008; EPA OIG 2013; Hearings 2014; Hart 2014; Kolly 2014; Sedor and Schulze 2014; Terry and Nantel 2014; Vorderbrueggen 2014). By the same token, effective leadership at the Dutch Safety Board has overcome parochial turf battles and built a powerful coalition that has cemented the DSB's societal legitimacy.

These conclusions provide crucial guideposts for policy-makers considering the establishment of either COIs or independent investigative agencies like the NTSB, CSB, or DSB. When the American Congress created the CSB in 1990, lawmakers were of the opinion that the NTSB's independence was the key to its success. This chapter shows that independence, while a necessary component of effective post-crisis investigation, is hardly sufficient. Instead, at least three additional factors must be present as well. Developing a deeper understanding of what contributed to the NTSB's success is particularly essential in light of recent interest in establishing independent investigatory boards for other sectors such as accounting, finance, and healthcare (Wallace 1994; Fielding et al. 2011; Hart 2012). Without a broad base of expertise fostered by the economic prerogatives of industry, a strategy to incorporate non-governmental constituencies into the data and evidence gathering process, an appropriate investment of time and resources, and quality leadership, an investigatory agency in any sector will be more likely to replicate the weakness of the CSB than the strength of the NTSB.

To make all of this more concrete, we close with a brief consideration of the best way to undertake regulatory "autopsies" of financial shocks. Even before the 2008 Global Financial Crisis, some American social scientists had proposed the establishment of an NTSB-like institution for finance, which would stand ready to investigate the causes of major financial crises, as well as more contained events like hedge fund failures (Getmanksy et al. 2004). After 2008, such calls multiplied among policy experts, though they differed about how to structure such forensic investigation capacity, and even whether it should reside within public agencies (Lo 2010; World Economic Forum 2010). These proposals helped to lay the intellectual groundwork for the creation of the Financial Stability Oversight Council (FSOC) and Office of Financial Research (OFR) in the United States, as well as the European Systemic Risk Board (ESRB) and the international Financial Stability Board (FSB). Born out of a post-crisis impulse to prevent future meltdowns, these new regulatory structures possess deep reserves of technical expertise and extensive access to relevant financial data; they also manifest a strong commitment to systemic stability as a policy goal.

Nonetheless, our overview of safety boards suggests reasons for skepticism about the capacity of these entities to function as effective post-crisis regulatory coroners, since they remain ensconced within regulatory institutions or otherwise incorporate them. Indeed, the voting members of FSOC are all heads of federal agencies or departments with responsibility for financial regulation; the OFR resides within the Department of the Treasury; the ESRB is part of the European Central Bank; and the FSB answers to a governing framework comprised of national financial regulators from leading economies. This approach raises concerns about bureaucratic meddling in investigations, as well about the willingness of investigators to point fingers at the regulatory decision-makers to whom they ultimately report.

At the same time, we would further stress that a fully independent Financial Markets Safety Board, whether framed on a national, continental, or international basis, would also face significant hurdles before it could function to the standard of the NTSB. Any such organization would have to build not only internal technocratic capacity, but also strong linkages to financial firms, so that it would receive appropriate flows of information about "near misses" or outright crises. Pivotal financial firms would have to accept the legitimacy of any safety board for finance and assist with its work. And because the desire to foster innovation and risk-taking represents an additional central goal of financial regulation, our imagined financial safety investigators would confront more complex questions about how to value systemic stability as they craft post-investigation policy recommendations. These are daunting barriers. Overcoming them, without at the same time turning a financial safety board into a lapdog for the financial sector, would surely require abiding political will and the highest caliber institutional leadership. Extending the realm of permanent crisis-investigation bodies, then, has great promise; but it will require not only undoubted technical competence, but also deft coalition-building, creative institutional design, and the sort of policy entrepreneurship that has always characterized the most effective modes of regulatory governance.

References

Andrews, David C. 1993. "Industry Views on Chemical Process Safety," *Process Safety Progress*. 12 (2): 104–105.

Ashforth, Adam. 1990. "Reckoning Schemes of Legitimation: On Commissions of Inquiry as Power/Knowledge Forms." *Journal of Historical Sociology* 3 (1): 1–22. doi:10.1111/j.1467-6443.1990.tb00143.x.

Baxter, Terry. 1995. "Independent Investigation of Transportation Accidents," *Safety Science*. 19: 271–78.

Belke, James C. and Deborah Y. Dietrich. 2005. "The Post-Bhopal and Post-9/11 Transformations in Chemical Emergency Prevention and

Response Policy in the United States." *Journal of Loss Prevention in the Process Industries.* 18: 375–79.

Bennear, Lori S. 2015. "Offshore Oil and Gas Drilling: A Review of Regulatory Regimes in the United States, United Kingdom, and Norway." *Review of Environmental Economics and Policy* 9 (1): 2–22.

Berger, Thomas R. 1999. "Canadian Commissions of Inquiry: An Insider's Perspective." In Allan Manson and David J. Mullan eds., *Commissions of Inquiry: Praise or Reappraise?*, 13–30. Toronto, ON: Irwin Law, Inc.

Bowling, David C. 2005. "The Capture Theory and Federal Investigations of Aviation Accidents," Northern Illinois University (Ph.D. Dissertation).

Bradley, A.W. 2003. "Commissions of Inquiry and Governmental Accountability: Recent British Experience." In Allan Manson and David J. Mullan eds., *Commissions of Inquiry: Praise or Reappraise?*, 31–78. Ontario: Irwin Law, Inc.

Braithwaite, John, and Peter Drahos. 2000. *Global Business Regulation.* New York, Cambridge University Press.

Capps, Lois. 2010. *To Give Subpoena Power to the National Commission on the BP Deepwater Horizon Oil Spill and Offshore Drilling. (2010H.R. 5481).* www.govtrack.us/congress/bills/111/hr5481.

Centa, Robert, and Patrick Macklem. 2001. "Securing Accountability through Commissions of Inquiry: A Role for the Law Commission of Canada." *Osgoode Hall Law Journal* 39(1): 117–60.

Cheang, David. 2015. *Leading Change in the Public Sector: Dr Moure-Eraso's Chairmanship of the US Chemical Safety Board.* Master's Project for Master in Environmental Management. Durham, NC: Duke University. http://bit.ly/1QKuroL.

Chemical Safety and Hazard Investigation Board (CSB). 1998. *Budget Justification Fiscal Year 1999.*

Chicago Tribune. 1977. "Disasters Are Their Business." *Chicago Tribune.* March 15: 1.

Committee on Oversight and Government Reform (COGR). 2014. Whistleblower Reprisal and Management Failures at the U.S. Chemical Safety Board: Staff Report. https://oversight.house.gov/wp-content/uploads/2014/06/CSB-FINAL-REPORT-Redact-version.pdf.

Cooray, L. J. M. 1985. "Royal Commissions Of Inquiry And Statutory Tribunals." In *Human Rights in Australia.* Epping, Australia: ACFR Community Education Project. www.ourcivilisation.com/cooray/rights/chap12.htm#12.0.

Crow, P. 2000. "Chem Board Back on Track," *Oil & Gas Journal.* 98 (1): 29.

Dutch Parliament. 2013. Netherland Food and Consumer Product Safety Authority. www.houseofrepresentatives.nl/dossiers/netherlands-food-and-consumer-product-safety-authority.

Environmental Protection Agency, Office of Inspector General (EPA OIG). 2013. *Semiannual Report to Congress, April 1 2013-September 30, 2013*. EPA-350-R-13-003. www.epa.gov/office-inspector-general/semiannual-report-apr-1-2013-sept-30-2013.

Essers, Loek. 2012. "Report: Dutch Government Was Unprepared for SSL Hack," *IDG News Service*. June 28. www.pcworld.com/article/258505/report_dutch_government_was_unprepared_for_ssl_hack.html.

Farson, Stuart, and Mark Phythian. 2010. *Commissions of Inquiry and National Security: Comparative Approaches: Comparative Approaches*. ABC-CLIO.

Fielding, Eric, Andrew W. Lo, and Jian Helen Yang. 2011. "The National Transportation Safety Board: A Model for Systemic Risk Management," *Journal of Investment Management*. 9(1): 17–49.

Flight International. 2009. "Dutch Find Trio of Shortcomings in Apache Mishap," Feb. 24–March 2.

Getmansky, Mila, Andrew W. Lo, and Shauna X. Mei. 2004. "Sifting through the Wreckage: Lessons from Recent Hedge-Fund Liquidations." *Journal of Investment Management* 2(4): 6–38.

Gilligan, George. 2002. "Royal Commissions of Inquiry." *Australian & New Zealand Journal of Criminology* 35 (3): 289–307. doi:10.1375/acri.35.3.289.

Glassman, Matthew E., and Jacob R. Straus. 2013. "Congressional Commissions: Overview, Structure, and Legislative Considerations." Washington, DC: Congressional Research Service R40076 https://fas.org/sgp/crs/misc/R40076.pdf.

Government Accountability Office (GAO). 2008. Chemical Safety Board: Improvements in Management and Oversight are Needed. GAO-08-864R. www.gao.gov/new.items/d08864r.pdf.

Government Accountability Office (GAO). 2000. Chemical Safety Board: Improved Policies and Additional Oversight are Needed. GAO/RCED-00-192. www.gao.gov/new.items/rc00192.pdf.

Graham, Bob, and William K. Reilly. 2014. "The BP Gulf Oil Spill – Four Years Later." Oil Spill Commission Action. http://oscaction.org/wp-content/uploads/Graham-Reilly-Statement-April-2012.pdf.

Hanlon, Natalie. 1991. "Military Base Closings: A Study of Government by Commission." *University of Colorado Law Review* 62: 331–64.

Hart, Christopher A. 2014. Chair, National Transportation Safety Board. In-person interview.

Hart, Christopher A. 2012. "Transferability of Aviation Mishap Investigation Process to Healthcare?" Presentation to Human Factors in Health Care, Johns Hopkins University, Nov. 12.

Institute for Government Research of the Brookings Institution. 1930. Aeronautics Branch Department of Commerce: Its History, Activities, and Organization. Service Monographs of the United States Government No. 61.

Inwood, Gregory J., and Carolyn M. Johns. 2014. *Commissions of Inquiry and Policy Change: A Comparative Analysis.* University of Toronto Press.

Jacobsson, Eva. 2011. "Accident Investigations: A Comparative Perspective on Societal Safety in Norway and Sweden, 1970–2010." *Scandinavian Journal of History.* 36(2): 206–29.

Johnson, Loch. 1995. "The Aspin-Brown Intelligence Inquiry: Behind the Closed Doors of a Blue Ribbon Commission." *Studies in Intelligence* 48 (3): 1–20. www.cia.gov/library/center-for-the-study-of-intelligence/kent-csi/vol48no3/pdf/v48i3a01p.pdf.

Jupp, Jeff. 2005. "Aviation," in *Accidents and Agenda: Full Sector Reports*, 3–15. London: Royal Academy of Engineering:. www.raeng.org.uk/publications/reports/accidents-and-agenda-full-sector-reports.

Keith, K.J. 2003. "Commissions of Inquiry: Some Thoughts from New Zealand." In Allan Manson and David J. Mullan eds., *Commissions of Inquiry: Praise or Reappraise?*, 153–82. Toronto, ON: Irwin Law, Inc.

Kitts, Kenneth. 2006. *Presidential Commissions & National Security.* Boulder, CO: Lynne Rienner.

Klemin, Alexander, 1927. "Aeronautical Education," *Annals of the American Academy of Political and Social Science* 131: 20–22.

Kolly, Joesph. 2014. NTSB Office of Research and Engineering. In-person interview.

Lo, Andrew W. 2010. "The Financial Industry Needs Its Own Crash Safety Board," *Financial Times*, March 2. www.ft.com/cms/s/0/011cbf2e-2599-11df-9bd3-00144feab49a.html#axzz3P6g0swsg.

Los Angeles Times. 1972. "OK Autonomy, Traffic Safety Board Requests: Federal Panel Asks Congress for Broader Power to Investigate Surface Accidents," *Los Angeles Times*, June 9.

Lieberman, Joseph. Senator (CT). 1989. "A Need for a National Chemical Safety Board." Congressional Record 135. No. 149 (Nov): 14538. http://webarchive.loc.gov/congressional-record/20160314161751/http://thomas.loc.gov/cgi-bin/query/D?r101:16:./temp/~r1010uvHK4.

Lovett, Kenneth. 2014. "Gov. Cuomo Stands by His Decision to Shut down Anti-Corruption Commission." *NY Daily News*, April 25. www.nydailynews.com/news/politics/gov-cuomo-stands-anti-corruption-commission-shutdown-article-1.1768508.

McGarity, Thomas O. 2005. "Resisting Regulation with Blue Ribbon Panels." *Fordham Urban Law Journal* 33: 1157–98.

Moss, Randolf D. 2000. Division of Power and Responsibilities between the Chairperson of the Chemical Safety and Hazard Investigation Board and the Board as a Whole. Opinion of the Office of Legal Council at the US Department of Justice. https://biotech.law.lsu.edu/blaw/olc/chemsafetyboardopinionfinal.htm.

National Aeronautics and Space Administration (NASA). 1987. Implementation of the Recommendations of the Presidential Commission on the Space Shuttle Challenger Accident. http://history.nasa.gov/rogersrep/genindex.htm.

National Transportation Safety Board (NTSB) 1968. *Annual Report to Congress, 1967.*

National Security Preparedness Group. 2011. "The Status of the 9/11 Commission Recommendations." Bipartisan Policy Center. http://abcnews.go.com/images/Blotter/Final%20NSPG%20911%20Report%20Card%20August%202011.pdf.

New York Times. 1922. "Aircraft Industry Forms Corporation," *New York Times.* Jan. 1: 6.

New York Times. 1974. "An Accident Investigator Quits in Dispute of Role of Politics," *New York Times*, Sept. 5: 16.

Orlando Sentinel. 1991. "Bush Sacrificing Environment for a Non-Existent Principle," *Orlando Sentinel.* June 2: G3.

Pollack, Kristina. 1998. "The Swedish Board of Accident Investigation." In Goeters Klaus-Martin, ed. *Aviation Psychology: A Science and A Profession*, 249–55. Burlington, VT: Ashgate.

Post, George B. 1927. "Aspects of Commercial Aviation in the U.S.A.," *Annals of the American Academy of Political and Social Science.* 131: 71–78.

Radio Netherlands News Report. 2012. "Dutch Safety Board: Too Many Medical Mistakes," www.expatica.com/nl/news/Dutch-Safety-Board-too-many-medical-mistakes_324958.html.

Rees, Joseph. 2009. *Hostages of Each Other: The Transformation of Nuclear Safety since Three Mile Island.* Chicago, University of Chicago Press.

Relyea, Harold C. 1996. Government Division, Congressional Research Service. *The National Performance Review.* Dec 6.

Resodihardjo, Sandra L. 2006. "Wielding a Double-Edged Sword: The Use of Inquiries at Times of Crisis." *Journal of Contingencies and Crisis Management* 14 (4): 199–206. doi:10.1111/j.1468-5973.2006.00496.x.

Rowe, Mike, and Laura McAllister. 2006. "The Roles of Commissions of Inquiry in the Policy Process." *Public Policy and Administration* 21 (4): 99–115. doi:10.1177/0952076706021004o8.

Sedor, Joseph and Dana Schulze. 2014. NTSB Office of Aviation Safety. In-person interview.

Spencer, Melissa. "Engineering Financial Safety: A System-Theoretic Case Study from the Financial Crisis." (Master's Thesis in Technology and Policy, Massachusetts Institute of Technology).

Stoop, J. 2004. "Independent Accident Investigation: A Modern Safety Tool," *Journal of Hazardous Materials*, 111(1-3): 39–44.

Stoop, John A. and James P. Kahan. 2005. "Flying Is the Safest Way to Travel: How Aviation Was a Pioneer in Independent Accident Investigation." *European Journal of Transport and Infrastructure Research*. 5(2): 115–28.

Stoop, John A. and Sverre Roed-Larsen. 2009. "Public Safety Investigations—A New Evolutionary Step in Safety Enhancement?" *Reliability Engineering & System Safety, ESREL 2007, the 18th European Safety and Reliability Conference*, 94(9): 1471–79.

Sulitzeanu-Kenan, Raanan. 2010. "Reflection in the Shadow of Blame: When Do Politicians Appoint Commissions of Inquiry?" *British Journal of Political Science* 40 (03): 613–34. doi:10.1017/S0007123410000049.

Swuste, P. 2013. "A 'Normal Accident' with a Tower Crane? An accident analysis conducted by the Dutch Safety Board," *Safety Science*. 57 (August): 276–82.

Tama, Jordan. 2011. "Three Cheers for Blue-Ribbon Panels." *The Wilson Quarterly* 35 (3): 30–34.

Tama, Jordan. 2014. "Crises, Commissions, and Reform: The Impact of Blue-Ribbon Panels." *Political Research Quarterly* 67 (1): 152–64.

Terry, Jane, and Kelly Nantel. 2014. Interview with Jane Terry and Kelly Nantel of the National Transportation Safety Board In-person interview.

The Sun (Baltimore, MD). 1940. Flyers to Carry Fight to Congress, April 30: 13.

US Senate. 1926. 69th Congress S41 *An Act To Encourage and Regulate the Use of Aircraft in Commerce, and for Other Purposes*. Washington, DC: Government Printing Office.

US Senate. 1932. *Safety in Air: Hearings Before the United States Senate Committee on Commerce, Subcommittee on S. Res. 146*, 74th Congress, Second Session, on Feb. 10, 11, 14, 15, 18, 20, 21, Apr. 6, 1936 and 75th Congress, First Session, on Jan.13, Feb. 2, 6, 1937. Washington, DC: US Government Printing Office.

US Senate. 1940. Select Committee on Government Organization. A Resolution Disapproving Reorganization Plan Numbered IV. Hearing May 9. Washington, DC: Government Printing Office: Z10.

US Senate Subcommittee of the Committee on Interstate and Foreign Commerce. Study of Operation of Civil Aeronautics Administration, Hearings. Jan 4, 5, 9, 10, 12, 24, Feb 3, 20, 21, March 5, 17 and May 28, 1956. Washington, DC: Government Printing Office.

van den Akker, J. A., and E. M. R. Lange. 2014. "Towards a European Approach to Food Fraud," *Judicial Explorations* 40(2): 84–96.

van Vollenhoven, Pieter, Chairman, Dutch Safety Board. 2001. "Independent Accident Investigation: Every Citizen's Right, Society's Duty," Speech at the 3rd European Transport Safety Lecture, Jan. 23.

Vietor, Richard H. K. 1990. "Contrived Competition: Airline Regulation and Deregulation," *Business History Review*. 64(1): 61–108.

Vorderbrueggen, John. 2014. National Transportation Safety Board. In-person interview.

Wallace, Wanda A. 1994. "The National Transportation Safety Board as a Prototype for a More Effective Quality Control Inquiry Committee," *Accounting Horizons* 8(1): 90–104.Washington Post. 1941. "Fledgling CAA Struggles Under Weight of Fifth Fatal Crash," *Washington Post*, March 1: 8.

Washington Post. 1997. "Chemical Safety Panel Survives Veto Pen," *Washington Post*, Nov. 4: A15.

Wolanin, Thomas R. 1975. *Presidential Advisory Commissions: Truman to Nixon*. Madison, Wisconsin: University of Wisconsin Press.

World Economic Forum. 2010. *Rethinking Risk Management Practices in Finance: Practices from Other Domains*. New York: World Economic Forum.

Zukowska, J., L. Michalski, and R. Krystek. 2010. "The Value of Independent Investigations Within Integrated Transport Safety Systems," *Scientific Journals*, 20(92).

18

Recalibrating Risk

Crises, Learning, and Regulatory Change

Edward J. Balleisen, Lori S. Bennear, Kimberly D. Krawiec, and Jonathan B. Wiener

It is often observed that crisis events spur new regulation. An extensive literature focuses on the role of disasters, tragedies, scandals, shocks, and other untoward events in stimulating regulatory responses (Baumgartner and Jones 1993; Percival 1998; Kuran and Sunstein 1999; Birkland 2006; Repetto 2006; Wiener and Richman 2010; Wuthnow 2010). We have highlighted numerous examples of arguably crisis-driven regulation in the introductory chapter (Balleisen et al., this volume) and in the several case study chapters in this book. The notion that crises spur regulation has become a "commonplace assertion," and yet one that is "so widely held ... that it remains virtually unexamined in empirical and historical analyses" (Carpenter and Sin 2007, 149). Observing this relationship does not itself explain what causal mechanisms may be driving it (Carpenter and Sin 2007, 154). And the relationship does not always hold (Kahn 2007). We do not claim that all crises spur regulatory change, nor that all regulatory changes arise from crises. Some crisis events do not produce significant regulatory change – perhaps including mass shootings in the United States, and Hurricanes Katrina and Sandy. Some regulatory changes occur without preceding crisis events – such as the Acid Rain Program of the 1990 Clean Air Act.

This volume has sought to enrich the empirical understanding of how the process of crisis stimulus and regulatory response unfolds. Our main question has been not whether, but rather how, regulatory systems change in response to crises. Going beyond the generic assertion that crises spur regulation, we have explored diverse ways in which regulatory change may play out: how different types of regulatory responses may follow from different kinds of crises.

In this volume, we have studied a set of cases in which some regulatory change typically did follow a crisis event, in order to understand how that process led to different types of regulatory changes in different contexts. The case studies in this volume – focusing on oil spills, nuclear power accidents, and financial crashes, with regulatory responses in the United States, Europe, and Japan – illustrate a wide array of crises, institutions, actors, countries, time periods, and policy changes. Although they are not necessarily a representative sample of all crises or all regulatory changes, they do depict considerable variation and help point to several insights. This variation is itself a key point of our study. Instead of seeking to test a single crisis-response relationship, exploring the diversity of crises and of regulatory responses opens a richer array of data for researching empirically how different crises may lead to different regulatory responses, and analyzing normatively which regulatory responses are best designed to respond to each crisis.

In this concluding chapter, we collect key insights and findings from our conceptual and case study chapters. First, we examine the descriptive findings, and we develop several hypotheses to help guide future research on the relationships among the varying attributes of different crises and the varying regulatory changes that may occur in response. Then we examine the normative implications, and we offer recommendations for institutional improvement. We suggest that regulatory systems should seek to avoid "policy regret" after crises by improving "policy resilience," including strategies for both "learning to prepare" for crises and "preparing to learn" from crises.

FINDINGS

A simple model of the crisis-regulation relationship might look like this:

$$\text{Crisis} \rightarrow \text{Regulatory Response}$$

After the adoption of a major regulatory policy, hindsight commentary sometimes speaks in this simple way – attributing the regulation to the crisis, or saying (often lamenting) that governments only act after a crisis. It is as if the regulatory phoenix arose spontaneously out of the fiery ashes of the crisis. But this model has obvious deficiencies. It does not account for the cases in which a crisis occurs but does not yield regulation (Kahn 2007), and the cases in which governments adopt regulation without a motivating crisis. And it does not illuminate the variation across different types of regulatory responses. Thus it cannot explain the actual observed patterns of varying crises and regulatory responses. Further, it speaks as if the crisis and the regulatory response were themselves animate actors; it does not explain the causal mechanisms at work, nor the roles of perceptions, politics, institutions, and other forces that mediate the design and adoption of regulation.

A more sophisticated approach takes account of institutions, interest group politics, public opinion and perceptions, policy entrepreneurs, and other mediating forces and actors in the regulatory system. Thus:

$$\text{Crisis} + \text{System} \rightarrow \text{Regulatory Response}$$

The vast literature on the positive politics of regulation emphasizes the important roles of institutions and actors (for a review, see Wiener and Richman 2010). In this literature, a crisis is often depicted as opening a window of opportunity for advocates or policy entrepreneurs to overcome the ordinary inertia of interest group politics and gridlock, to punctuate the prior equilibrium (Baumgartner and Jones 1993; Kingdon 2003; Birkland 2006; Repetto 2006). In Kingdon's metaphor, policy change may occur at a confluence of the "streams" of problems, politics and solutions. The stronger the resistance of opposing groups (or public opinion), the steeper the "slope" that the advocates of a new policy will have to push up (or have their streams ascend) to achieve their regulatory change (Denzau and Munger 1986). Veto points, where actors in the institutional structure can block a new policy, can be thought of as sharp upturns or bumps in the slope of the policy terrain. A crisis may reconfigure that slope, perhaps lessening its steepness, such as by changing public or elite perceptions of the need for policy change. Research on the psychology of perception finds that recent salient and unusual events tend to be more "available" or overweighted in public perceptions of future risks, compared to statistical estimates or chronic familiar risks (Kahneman et al. 1982; Kuran and Sunstein 1999; Weber, this volume). From infancy, humans respond more strongly to surprises than to familiar conditions (Stahl and Feigenson 2015). Meanwhile, the very characterization of the event as a "crisis" can be contentious; actors compete to frame the narrative of the event as a crisis (or not), and in what way, in order to build support for their preferred policy positions (Stone 1989; Carpenter and Sin 2007, 151, 176–78; Mayer, this volume).

We expand on this approach by proliferating the types of crisis, the types of system, and the types of regulatory response. Hence:

Crisis type x + System type y → Regulatory response type z

This formulation opens questions about which attributes of crisis events and of systems are associated with which attributes of regulatory responses. Going beyond whether crises spur regulation, we can examine an array of different types of regulatory change in response to different crises in different systems and societal contexts. A signal contribution of this volume is thus to expand the study of crises and regulation to recognize the wide array of regulatory changes that may arise. Future research can seek to clarify which types of regulatory responses are associated with which attributes of crises and pre-existing systems.

Our case study chapters illustrate the variety of regulatory responses to different crises. For example, the Exxon Valdez oil spill led to the enactment of the Oil Pollution Act of 1990 imposing prescriptive technology standards (double hulling) as well as a liability regime, whereas the BP Deepwater Horizon oil spill in 2010 led to few new regulatory restrictions, but a reorganization of part of the US Department of the Interior (see chapters 7–9 in this volume). The Three Mile Island nuclear accident in 1979 led to a de facto moratorium on new nuclear power plant licensing in the US, whereas the Fukushima Daichi accident in 2011 led to the shutdown of all of Japan's 50 nuclear power stations and the acceleration of Germany's phase out of nuclear power, but less change in US or French policy (see chapters 10–12 in this volume). In the United States, the Great Depression of the 1930s led to major new legislation requiring disclosure of business risks to investors, and creating the Securities and Exchange Commission (SEC) and the Federal Deposit Insurance Corporation (FDIC), whereas the Great Recession of 2008 led to the enactment of the Dodd-Frank Act, which imposed new capital requirements and orderly liquidation authority, as well as created the Financial Stability Oversight Council (FSOC) and the Consumer Financial Protection Bureau (CFPB) (see chapters 12–16 in this volume). Although several of these examples just mentioned led to the proliferation of regulatory agencies (through the division of existing agencies and/or the creation of new agencies), the September 11th terrorist attacks led to structural change in the opposite direction: the consolidation of numerous federal agencies into the new integrated Department of Homeland Security and the creation of the new Director of National Intelligence (as well as two wars and the enactment of new legislation authorizing national security surveillance measures).

More generally, the types of regulatory responses to a crisis may vary along many dimensions. These responses may be robust or cosmetic. They may be structural (reorganizing government) or instrumental (changing policy tools). They include, for example:

- replacing personnel ("heads will roll")
- tightening the stringency of standards
- changing policy instruments (e.g. prescriptive technology standards, performance standards, taxes, cap and trade, information disclosure)
- reorganizing institutions (e.g. dividing, combining, elevating or demoting agencies)
- creating new agencies
- increasing public or private monitoring efforts
- increasing enforcement efforts
- increasing penalties or liability
- promoting industry self-regulation
- referring the matter to a study commission
- and others.

Major crises in industrialized democracies, then, have spurred widely diverse types of regulatory changes. That in itself is interesting: why so many different types of responses? And why did governments opt for the particular responses selected in each case? Our conceptual and case study chapters suggest several attributes that interact in complex ways to shape regulatory responses to crisis events. These include the attributes of the event and the pre-existing system. Nonetheless, there is seldom a universally applicable relationship between any one of these attributes and a given policy response. Rather, that relationship is typically a complex combination of these attributes, with some attributes receiving greater weight in some crises.

The influential attributes of the crisis event identified in our conceptual and case study chapters include:

- Magnitude and perception. A larger event – causing greater damage – may spur a larger public outcry. But this relationship is not straightforward. The magnitude of an event is perceived relative to the baseline of familiar risks, the availability of the risk in recent memory, and the ability of the public to relate emotionally to the social damages from the crisis (see Weber, this volume; Slovic 2007). Hence as a society becomes safer overall (i.e. as the baseline risk declines), relatively smaller events may be perceived as crises – ironically, a safer society can become more crisis-sensitive. This may help explain simultaneously increasing prosperity and longevity and yet increasing public anxiety about unusual risks. It also appears that experienced events generate more public concern than predicted or described events (Weber, this volume), so the public may respond more strongly to recent episodes that are "available" in the mind than to looming future risks. Moreover, greater numerical magnitude (e.g. a larger total number of deaths) may not always trigger greater public outcry, because people may exhibit "psychic numbing" that reduces their concern for larger numbers of lives lost (Slovic 2007), and because emotional affect may be more important than numerical magnitude in stimulating policy responses. An identified individual victim – a "poster child" – may garner more response than a large number of statistical lives (Small et al. 2007). And global catastrophic crises may be too large for public reaction to spur regulatory response, because of unavailability and psychic numbing, and because the catastrophe itself may destroy relevant institutions and prevent learning (Wiener 2016).

We see evidence of this relationship between magnitude and response in our case studies, particularly responses to the larger offshore oil accidents at Santa Barbara, Exxon Valdez, Piper Alpha and Deepwater Horizon (in terms of quantity of oil spilled, and amplified emotional impact from news media coverage) tending to yield larger policy responses than the smaller accidents in the North Sea. Similarly, the magnitude of the recent financial crises and the Great Depression had more significant impacts on regulation than the smaller financial crises of the 1990s and early 2000s. However, this relationship between magnitude and response is not universally true. In the nuclear accident case studies, the public health consequences of the nuclear power plant accidents at Three Mile Island and even at Fukushima were smaller than those associated with the Chernobyl accident, yet policy responses in most of the countries studied were smaller in response to the Chernobyl accident than to Three Mile Island and Fukushima.

- Proximity and propensity. An event's proximity to the polity may foster greater response. In our oil spill case studies, regulatory responses appeared to recede for spills occurring farther away. Indeed, the European countries had little response to US accidents (even in the case of Deepwater Horizon), while the US had limited response to accidents in the North Sea. Yet for nuclear accidents in our case studies, regulatory responses appeared to correspond more to whether the type of reactor technology and facility design at the accident site was also used in the relevant country considering its policy response. For example, policies in Europe responded more to Fukushima (distant geographically but similar in reactor type and management) than to Chernobyl (close geographically but dissimilar in facility design and management) (Krieger et al., this volume). In Japan, policies responded more to (smaller) domestic nuclear accidents than to (larger) foreign accidents (Kishimoto, this volume), apparently relating propensity to (dis)trust in facility management and flaws in industry's prior risk assessments (Paté-Cornell, this volume). So the attribute of proximity may relate not only to geographic distance, but also to perceptions that the circumstances that led to the crisis pertain to one's own society. This seems to be not so much physical proximity as the propensity for a similar crisis event to happen here to people like us. And in the cases of nuclear accidents, it appears that the lack of proximity or propensity (in terms of similar reactor types) was more important to the response than the magnitude of the event.

- Repetition. A series of multiple similar events may augment public concern. Repeated similar crises may also raise public ire if they seem to show that elites knew about problems (from earlier episodes) but failed to take action to correct them. A series of similar events may also persuade policy experts that they represent a new distribution of risk, not just a bad draw from the prior distribution. On the other hand, multiple distinct events (of different types) may compete for the public's scarce attention (Weber, this volume). This pattern occurred with a series of chemical plant disasters in the 1970s and 1980s; and with barge accidents on American rivers in the 1980s and 1990s (Preston, 2014; Rust 2015). The increasing tempo of regional financial crashes in the 1980s, 1990s, and 2000s has now taken on new analytical light with the Global Financial Crisis of 2008.
- Bad draw vs. new world. A key puzzle facing policy experts (and potentially political elites and the general public) is how to perceive an unusual event. Is it just an outlier, a bad draw from the same prior expected distribution? Or does it convey new information that the state of the world is shifting so that this kind of risk will occur more frequently or with more damages in the future? In her conceptual chapter, Bennear (this volume) highlights this crucial question for further research. We also see it in several of our case studies. Carrigan's Chapter 9 on Deepwater Horizon highlights the relative superficiality of the response to this large magnitude oil spill. Rather than the large-scale new legislation that resulted from the Santa Barbara

and Exxon Valdez spills, the Deepwater Horizon spill resulted mostly in organizational changes. The public opinion on the relative importance of environmental safety and energy exploration quickly settled back to pre-crisis levels. This suggests that the policy elite, and to some extent the public, viewed Deepwater Horizon as a bad draw rather than a revelation of systematically higher risk that required substantial policy responses. By contrast, the depth of the shock from the 2008 financial crisis convinced a significant fraction of economists and policy-makers that reliance on complex derivatives and intensified interconnectedness had combined to create far more substantial risks than policy-makers had previously recognized. After the nuclear accident at Fukushima, policy elites in Japan and in Germany appeared to view the risks as more serious than previously appreciated (despite the low risk of earthquake and tsunami in Germany) and warranting a phase out of nuclear power, whereas policy elites in the US and France appeared to view the risks as more consistent with past estimates and warranting more modest improvements in safeguards (Paté-Cornell, and Krieger et al., this volume).

- Narrative framing. As noted above, actors typically compete to frame the narrative of the event in ways that favor their preferred policies. Thus the occurrence of a crisis is not exogenous – it is a social attribute of the physical events, constructed by actors in the regulatory and political system (Mayer, this volume). For example, public concern may rise if the event is framed as afflicting an individual victim, or as caused by an individual villain, rather than as a mass event, or an inscrutable "act of God." Further, the type of regulatory response may be influenced by how key groups frame the crisis, and the extent to which explanations of the crisis settle on a shared causal story. The policy responses to the Santa Barbara oil spill differed greatly from that of Deepwater Horizon, in part because the Santa Barbara oil spill occurred during, and helped mobilize, the great environmental movement of the late 1960s and early 1970s and was consistent with a broader narrative of environmental degradation due to excessive industrialization. Deepwater Horizon occurred during a period dominated by a narrative of the need for energy independence both for economic and national security reasons, and although it spurred organizational changes at the Interior Department, it did not halt offshore drilling. Three Mile Island was framed in the US as demonstrating the dread risks of nuclear power (about the same time as a popular movie, The China

Syndrome, caught American public attention), and led to distrust in the industry and a de facto moratorium on new licensing of nuclear power plants in the US; by contrast, Fukushima was framed in similar terms of dread and distrust in Japan, but was seen in the US more as a managerial failure to position the backup cooling generators away from the ocean (Kishimoto and Paté-Cornell, this volume). The policy responses to the 2008 financial crisis have taken multiple and often conflicting directions because there have been conflicting causal stories about its origins (Claessens and Kodres, this volume).

Attributes of the system can also mediate the influence of the crisis event on regulatory change. Among these are:

- Institutions, voting rules, veto points. The slope or difficulty of achieving new policy adoption may vary with different institutional structures, such as parliamentary, separation of powers, centralization, federalism, judicial review, and others. Yet it is not always clear how these institutional differences will play out. The nuclear accident cases illustrate the complexity. In Japan, relatively centralized authority over nuclear power may have facilitated the series of institutional reorganizations of the nuclear regulatory agencies in the national government hierarchy, especially after accidents occurring within Japan, although the shutdown after Fukushima persisted in part due to local government opposition to restarting reactors (Kishimoto, this volume). In France, similarly centralized authority and support for nuclear power was not used to reorganize agencies, but rather translated into continuing reliance on nuclear power (perhaps because the accidents mainly occurred outside France); and in Germany, more decentralized politics fostered an accelerated phase out of nuclear power after Fukushima (Krieger et al., this volume). In the US, the Nuclear Regulatory Commission was positioned to be centralized and more independent from politics than cabinet agencies, and the NRC was not restructured after Three Mile Island (perhaps because it had replaced the Atomic Energy Commission just a few years earlier), which may have enabled the launch of an industry self-monitoring body (the Institute for Nuclear Power Operations, INPO) and extensive administrative procedures that, along with economic factors, amounted to a de facto moratorium on new licenses for nuclear power plants (Paté-Cornell, this volume).

- Steeper policy slopes and inertia. The greater the gridlock, polarization, or strength of coalitions with veto threats, the less that crises may spur new regulation. This factor may apply more to legislative than to administrative action: in the US at least, partisan political gridlock may inhibit legislative action but promote executive administrative action instead. If gridlock has increased in the US since the 1970s, that might help explain the greater legislative responses to the earlier Santa Barbara and Exxon Valdez spills, but less legislative response (and more administrative reorganization) after the more recent Deepwater Horizon spill. But growing gridlock might not explain the administrative responses to both TMI and Fukushima, nor the passage by Congress of the Dodd-Frank Act after the 2008 financial crisis. In addition, inertia stemming from formative historical events may constrain the range of policy choices long after new evidence regarding possible risks emerges. For example, as argued by Eichengreen (this volume), the inflation and deficit concerns that had so preoccupied policy debates during the politically formative years of key policy officials, particularly in Germany, may have caused a focus on inflation risk that, in hindsight, appears excessive, while simultaneously failing to account for the possibility of a catastrophic Euro zone crisis.
- Public confidence/trust in institutions and elites. Lower trust may tend toward policy responses that are more prescriptive and less flexible or incentive-based. Thus the Norwegian, and to a lesser extent, UK regulatory regimes for offshore oil are possible only because of the high level of trust between the government, the industry and the unions. As this trust falters, the resilience of the regime is weakened. Trust in engineering elites may partly account for the French policies to support nuclear power (Krieger et al., this volume), while distrust after Fukushima may help explain newfound opposition to engineering elites in Japan (Kishimoto, this volume). In addition, public distrust of political or economic elites may shape policy outcomes. For example, the climate of hostility towards Wall Street and the figure of the small investor wronged by unscrupulous bankers played an important role in shaping American perceptions of the causes of, and appropriate solutions to, the Great Depression. In France, by contrast, revelations of political corruption fueled public outrage of a different sort, influencing the regulatory solutions that eventually emerged (Cassis, this volume).

- Readiness of solutions. If solutions have already been designed in advance, and are ready "off the shelf," they may be deployed more quickly after a crisis (Kingdon 2003). Astute regulatory entrepreneurs may prepare one or more new policies to be ready in anticipation of a future crisis that will open the window of opportunity to move policy proposals through institutional gauntlets and around political obstacles on the policy slope (although these policies may or may not prove to be well-tailored to address the problems that cause the crisis). Remedies that are more palatable and less costly may also garner greater public support for reform (Campbell and Kay 2014). As we see in the Eisner chapter (this volume) the initial decision to rely on a strict liability regime for offshore oil spills had significant impact on subsequent decisions about how to regulate toxic waste. Once the solution was in place in one policy domain it was much more readily adopted elsewhere.

Accordingly, *the same crisis event, if it occurred in different systems or contexts of political, institutional, economic, social, and cultural forces and actors, would likely yield different regulatory responses.* For example, the nuclear accidents at Three Mile Island, Chernobyl, and Fukushima led to different regulatory responses in different countries. And *different crisis events in a single system could yield different regulatory responses.* Multiple types of crisis, and multiple types of system, combine to generate multiple types of regulatory response.

Hypotheses

We can offer some hypotheses that may explain some of the variation in regulatory responses to crises. Our case studies, although diverse, are not a representative sample of all crises, so we cannot be sure that the evidence in our case studies will support unbiased generalizable inferences. But we can draw on our case studies to generate plausible hypotheses about how different types of crisis and different systems can shape the selection among different regulatory responses. These hypotheses could be tested more rigorously in future empirical research employing a broader more representative sampling method. We offer ten interrelated hypotheses below.

1) **Framing the event as an institutional failure (e.g. overlooking important information, conflicts of interest, industry capture)**

may lead toward responses involving structural reorganization (e.g. combining, dividing, elevating, or demoting agencies) and transparency (disclosure). Along these lines, the Japanese government has restructured the nuclear oversight agency several times over the last four decades, each move apparently spurred by an accident in Japan (Kishimoto, this volume). Carrigan (this volume) argues that policy-makers were able to frame Deepwater Horizon as, at least partially, a result of conflicts of interest between the agencies promoting development of offshore resources and those tasked with regulating the safety of those activities, despite the fact that these issues had never been raised in the policy debate prior to Deepwater Horizon. This framing enabled a response that was more organizational than prescriptive.

2) **Framing the event as a failure of industry compliance may lead toward responses involving more stringent standards, penalties or monitoring and enforcement. Relatedly, moral outrage about a crisis event, and the perception of identified villains harming identified victims, may lead toward selection of less-flexible policy instruments.** Market-based incentives, such as taxes and cap and trade systems, even if they reduce social harm more cost-effectively, may be perceived as offering undeserved latitude to those who caused the problem. (By the same logic, market-based incentives may be perceived as most appealing when they do not follow an outrage-sparking crisis – as was the situation for the Acid Rain emissions trading program in the 1990 Clean Air Act Amendments.) In the case of the Exxon Valdez oil spill, the framing of the event as one of gross negligence on the part of a drunken captain and a firm that was aware of his drinking problems, resulted in a far more prescriptive set of regulations in the Oil Pollution Act of 1990 (Eisner, this volume). After the Fukushima accident, outrage at Tepco's perceived mistakes helped spur a more aggressive regulatory response shutting down all 50 nuclear power reactors in Japan and evacuating more than 100,000 people (Kishimoto, this volume) (but after the Three Mile Island accident, despite outcry over human error at the power plant, a key response was the creation of the industry self-monitoring group, INPO (Paté-Cornell, this volume)). Fukushima is especially interesting because the tsunami killed more than 20,000 people, while the ensuing nuclear

power plant accident killed far fewer (some estimates are effectively zero), and yet public outrage at the power plant accident was amplified by the perception of industry errors and hubris, and the repeated graphic images of the reactors catching fire (Kishimoto, this volume). Furthermore, outrage may drive policy responses that neglect countervailing risks (Graham and Wiener 1995) – for example, the evacuation around Fukushima may have cost more lives than the radiation released in the accident (Johnson 2015), and in the short term both Japan and Germany shifted from nuclear to coal which increased their greenhouse gas emissions. In the case of the Great Depression, the framing of the stock market crash as causing harm to identifiable victims – small, individual investors – through the misdeeds of identifiable villains – large banks and Wall Street players – resulted in a level of federal regulation of the financial markets previously unimaginable.

3) **The greater the degree of inertia and gridlock – that is, the steeper "policy slope" to climb – the more likely that regulatory responses will turn out to be comparatively tepid.** Our case studies show that crisis-driven regulatory change not only takes many forms, but also occurs on a spectrum of intensity. Less intense responses are likely, at least in part, to result from institutional inertia, the strength of advocacy coalitions committed to the status quo, cognitive defaults, polarization yielding gridlock, and institutional veto points. In these circumstances, observed regulatory responses may seek to deflect momentary political pressures rather than translate crisis events and insights into new policies. No response, or a cosmetic response, may result. On the other hand, in systems with less steep policy slopes, comparable crises can generate more substantive or dramatic transformations in policy goals and instruments, and more far-reaching institutional restructuring. As already noted, Eichengreen in this volume attributes the inability to anticipate the Euro crisis, and subsequent tepid response, to such factors. Krieger et al. (this volume) observe that German politics were already trending against nuclear power before Fukushima, which then spurred an accelerated phase out.

4) **If policy elites perceive the crisis event as reflecting only a bad draw, the regulatory response may be faster but more deflective. If they perceive the crisis event as portending a new world of higher risk, their response may be slower to develop but more**

fundamental in its rethinking of structures, strategies and delegations of power. Carrigan's analysis (this volume) suggests that policy elites viewed the Deepwater Horizon accident as a bad draw, rather than a reflection of an entirely new risk distribution. As such, decision-makers opted for more modest reforms. Kishimoto (this volume) recounts a series of smaller nuclear accidents in Japan, viewed as bad draws, leading mainly to institutional reorganizations; and then the Fukushima accident, viewed as revealing more serious risks and distrust of the industry, leading to a national shutdown. Paté-Cornell (this volume) points out that there were analytic flaws in the risk assessments used by the Japanese nuclear industry, and management errors such as the location of the backup cooling system at sea level, so that from the US perspective (and the French), the Fukushima accident was more of a bad draw associated with bad forecasting and bad management, warranting a review of safeguards but not a shutdown.

5) **Other things equal, crisis events that cause greater harms, especially as compared to baseline expectations, that occur "closer to home," or that prompt greater fears of potential harms, seem more likely to prompt especially substantive policy adjustments.** This hypothesis must be tempered by the evidence, cited above, that public perceptions exhibit "psychic numbing" to larger numbers of deaths, and show higher concern for an identified individual. Still, our case studies provide some support for these suppositions. During the Great Depression, as Youssef Cassis shows, the industrialized democracy that experienced the deepest economic collapse – the United States – undertook the most thoroughgoing reconstruction of banking and financial regulation. By contrast, Great Britain and France, which encountered less wrenching economic downturns, did far less to change their basic regulatory institutions. Similarly, as our oil spill case studies suggest, major oil incidents in North American waters have prompted considerable regulatory innovation in the US, but little action in Europe, while North Sea oil spills provoked action by European authorities, but not in America. One can see the same sort of pattern in highly publicized accidents at chemical manufacturing plants: the Seveso disaster in 1976 in Italy had far more substantial policy ramifications in Europe than in the US, while the Bhopal disaster that occurred in India in 1984 (at a US-owned

facility) had especially significant impacts on policy in India (and in the US), while only limited significance for European policy-makers. The recent occurrence of failure of coal ash dams in the southeastern United States similarly has had particularly great salience for state regulatory action in the states most directly affected; neighboring states have proved less willing to undertake significant reconsideration of inspection or liability regimes.

6) **On the other hand, one cannot treat the question of what hits "close to home" as purely a matter of geographic distance.** In the aftermath of crisis events, advocacy coalitions, media elites, and policy-makers all make judgments about whether a given crisis truly relates to their worlds, and all struggle to persuade others to adopt their interpretive narrative. **If there is a dominant narrative interpretation of a crisis event within a given society, that narrative will greatly shape policy outcomes.** In some contexts, like the Chernobyl nuclear accident or the Asian financial crises of the 1990s, policy-makers in western Europe and North America discounted the significance of events not just because they occurred far away, but because of other substantive differences – a very different nuclear reactor type from those used outside Russia; distinctive approaches to industrial organization and currency vulnerabilities in East Asia. By contrast, the Fukushima nuclear disaster prompted a dramatic rethinking of nuclear safety regulation in Germany, despite occurring on the other side of the globe, in part because so many Germans view Japan as a technologically sophisticated society with a strong reputation for technocratic risk management (Krieger et al., this volume).

7) **Significant regulatory responses may be more likely when political agents persuasively assign blame to specific perpetrators of (or contributors to) crisis.** This added condition seems to have been a crucial element of the American regulatory response to the Great Depression. The capacity to reconstruct financial regulation was greatly aided by the Pecora Senate Hearings, which created a broad public consensus about the corrosive practices among elite financial institutions and the men who ran them. And yet in the case of the BP Deepwater Horizon disaster, the combination of a rig explosion that took eleven lives, an oil spill of truly enormous proportions, and extensive allegations of slapdash safety practices by BP and Transocean, has arguably had only a limited impact on regulatory change. This crisis did not result in significant

constraints on the expansion of offshore oil drilling, with the exception of a short-term moratorium on new drilling permits, nor the imposition of a far more intrusive publicly enforced safety regime. Instead, decision-makers opted for a more modest regulatory reorganization and the creation of an industry-run safety institute. Perhaps in this case, other factors, such as the framing of the event as one of conflict of interest and the perception by policy elites that the event represented a bad draw, moderated the response. Meanwhile, the nuclear power accidents spurred different responses in different countries; blame on US operators at TMI, Russian operators at Chernobyl or Japanese operators at Fukushima did not necessarily translate into regulatory responses in other countries with different reactor designs and operating cultures.

8) As the conceptual chapter by Tom Birkland and Meg Warnement points out, an additional key issue in assessing the intensity of regulatory responses to crises concerns the relevant time frame for considering crisis-driven policy. On the one hand, immediate reactions to events by the general public often prove ephemeral, as exemplified by surveys of popular attitudes toward nuclear power in several countries in the wake of significant nuclear accidents (discussed in the essays by Krieger et al and Kishimoto). Without significant reinforcement (new developments in the unfolding drama of crisis; effective mobilization by advocacy coalitions), broad public attention to a given disaster may fade rather quickly. Crises generally have more enduring impacts on organizations and experts, including policy-makers at all levels of government, who see a given crisis as falling within their zone of competency and interest. Yet even in these quarters, many experts manifest a tendency to discount the policy salience of an initial event or events, assuming, in the probabilistic language discussed by Bennear's chapter, that it just reflects a bad draw. **When crises do galvanize reforms, the process of developing and implementing post-crisis regulatory policies can take a full decade or even longer.** Assessing the causes of such complex events as deep-sea oil drilling incidents, nuclear accidents, or financial crashes is typically a time-consuming process, taking place over several months even when those in charge of official inquiries feel considerable pressure to move quickly. Forging new legislation can take at least as long, while the business of formulating and

implementing regulatory rules can take years. Indeed, nine years after the onset of the most recent Global Financial Crisis and seven years after the passage of the Dodd-Frank Act in the United States, federal agencies still need to finalize dozens of regulations mandated by the legislation. Thus perhaps it remains too early to say with full confidence that we should view the regulatory response to BP Deepwater Horizon as remarkably modest. That assessment will depend on the longer-term impact of regulatory reorganization in the US Department of the Interior, as well as the actual performance of the new industry safety institute. With events like nuclear accidents or major oil spills, moreover, it may take several decades before experts can ascertain the full set of harms to the environment or human health, since those negative consequences may take that long to manifest themselves. **In such cases, the initial crisis may produce a regular stream of analytical aftershocks, so long as there is credible evaluation of those longer-term impacts. And those later assessments may, of course, feed back into policy deliberations.**

9) Birkland and Warnement make the further crucial observation that one should not think of crisis events as hermetically sealed occurrences. Instead, as they note for American policy about air safety and security, a series of related events can elicit greater policy changes than a single accident or disaster. Each discrete crisis typically prompts new insights about the nature and extent of risks and new ideas about how to respond to them. **Multiple related events thus can have a cumulative impact.** They often produce refinement of policy ideas, burnish the reputation of those experts who have devoted their energies to redressing the risks posed by the crisis, and may lead to heightened public criticism of policy elites who have previously resisted calls to make changes in regulatory policy. To extend the metaphor of the policy slope, with each related crisis event, advocates of regulatory change may sweep away particular obstacles, gain a better vantage point on the path ahead, welcome new coalition partners at their side, and so advance their institutional ascent. A series of similar events may also persuade policy experts that these events represent a new distribution of risk, not just a bad draw from the prior distribution. On the basis of this way of thinking, one might expect that if there were another offshore oil drilling disaster in the Gulf of Mexico, policy-makers would reach back for some of the

stronger regulatory proposals that were not yet adopted after the BP Deepwater Horizon spill. We hasten to caution, however, that the identification of particular crisis events as related occurrences, as indications of a powerful signal rather than ongoing noise, does not occur separately from the give and take of political and policy debate. Interest groups, experts, and decision-makers also tussle over whether a sequence of events actually is closely related. And just as in the case of single crises, they fight over the appropriate story to tell as the best way to characterize those fundamental connections (Mayer, this volume). (Examples of such contests to frame a series of events might include multiple mass shootings across the US, and multiple small earthquakes near wastewater injection from hydraulic fracturing in Oklahoma.) Thus we see a particular need for research into the contested struggles to shape dominant public narratives about crises, as well as how those narratives seek to widen (or constrain) policy alternatives.
10) If solutions have already been designed in advance, they may be deployed more quickly after a crisis (Kingdon 2003). **If regulatory actors (in government, or outside it) have already prepared one or more new policies in anticipation of a future crisis that will open the window of opportunity, that may favor this policy getting adopted.** Still, the new policy may or may not prove to be well-tailored to address the problems that caused the crisis.

RECOMMENDATIONS

A crucial element of the crisis-response process is the way that any society tries to make sense of the causes of the event. This sense-making function can be ad hoc, or driven by policy entrepreneurs, or organized in advance. It can open a window of opportunity for long-overdue reforms. But it can also lead to "policy regret" – whether from failure to prevent the crisis in advance, or from failure to seize the opportunity to effect regulatory change after a crisis, or from hastily designed regulatory responses that turn out to be costly, ineffective, or induce new risks. (For example, after the Fukushima Daichi accident, the shutdown of all nuclear power in Japan and the accelerated phase out in Germany appear to have increased each country's reliance on coal and hence increased its greenhouse gas emissions. And the mass evacuation of the region around Fukushima appears to have yielded few or no health benefits from avoided radiation exposure, while inducing many deaths such as from dislocation of patients from care facilities, see Johnson 2015.)

Instead of policy regret, policy-makers and experts seek what may be termed "policy resilience" – the capacity of policies to prevent crises, and to adapt to those crises that do occur. These objectives entail two types of learning: "learning to prepare" for crises (to prevent them or reduce their damages), and "preparing to learn" from crises (to adjust policies to perform better in the future). We conclude with recommendations for this key element: avoiding policy regret, and seeking policy resilience, by learning to prepare and preparing to learn.

We suggest that regulatory experts (in and out of government) should "learn to prepare" for crisis events in advance. Despite the best efforts of experts and policy elites, some crises will occur – no policy can prevent all risks (as emphasized in the chapters by Bennear; Kousky; and Claessens and Kodres). While regulatory experts must operate under the constraints of scarce time and institutional resources, they nonetheless should – in advance of any specific crisis event – conduct **horizon-scanning exercises envisioning multiple scenarios, play simulation games with unexpected turns and opposing "red teams," and study current data on "near misses" that may be precursors of larger disasters**. They should develop regulatory contingency plans in advance, so that new policies can be ready to propose as solution streams to climb the policy slope toward the window of opportunity (to mix a few metaphors), if the failure of the prior approach is revealed in the crisis.

Similarly, we suggest that regulatory experts should "prepare to learn" from crisis events as they occur. Drawing on our study in the preceding chapter (Balleisen et al., this volume) on "disaster investigation bodies" – such as ad hoc inquiry commissions (like the 9/11 Commission or the BP Deepwater Horizon Inquiry Commission) and standing expert safety boards (like the US National Transportation Safety Board (NTSB), and the Dutch Safety Board) – we suggest that regulatory systems should **create standing (not ad hoc), independent (not regulatory agency) crisis investigation bodies** to investigate crises and their causes, and recommend reforms. These standing bodies have the expertise, experience, working relations, and authority to diagnose the cause of a crisis and recommend successful reforms (whereas one-off ad hoc inquiry commissions must build these capacities anew each time and lack the experiential memory to compare across crises). Their independence from regulatory agencies will likely make their reform recommendations only advisory, but may also enable more candid analyses (whereas agency staff might be hesitant to critique the agency's own past policy). There could be parallel crisis investigation bodies for each policy domain (e.g. transport / NTSB; chemicals / CSB; air quality; water quality; food; climate; finance; violence; terrorism). Or there could be one larger multi-sectoral body for all disaster investigations, like the Dutch Safety Board. The World Bank (2014) has recommended that every country create a "National Risk Board" to assess current and emerging risks (horizon-scanning), advise on priority setting, help resolve conflicts across ministries and risk-risk trade-offs, and prepare for change. (An earlier version of this institutional approach, cited by the World Bank, was advanced in Graham and Wiener 1995, ch. 11.) A related idea would be to create something like an Intergovernmental Panel on Climate Change for a wider set of global risks, as well as something like an Advanced Research Projects Administration (ARPA)-Risk, to study extreme low-probability high-consequence catastrophic risks and their solutions (Wiener 2016). If each country had a "National Risk/Safety Board" (or multiple safety boards for different sectors), and international bodies were created to study global risks, a network of these bodies could be developed to facilitate shared learning across countries and across topical domains.

A key to all of these steps is the objective of building learning into regulatory systems. Rather than just reacting to crises, regulatory experts and agencies should be developing mechanisms to prevent crises from occurring, mitigate their impacts, and learn from crises how to do better over time. We recognize that many regulatory agencies are already attempting to do this, and that they face obstacles in the resources, time, and priority-setting needed to undertake more systematic learning. We suggest that regulatory bodies confronting a crisis should commit themselves to this principle: "don't deflect, reflect." Rather than brushing off the crisis and referring it to a cosmetic or slow study process, regulators should seize the opportunity to learn about the crisis, why it occurred, and how to prevent it (including through expert study commissions).

Going further, efforts to develop "adaptive regulation" would reorient regulation from a one-time decision to an ongoing process of learning and updating (McCray et al. 2010). Adaptive regulations have regular periodic reviews, such as every 5 years. Adaptive regulation would also build in comprehensive monitoring and data analysis, from the outset of each regulatory policy, so that decision-makers would more frequently foresee and prevent crises, and would have the information base to analyze policy impacts and adjustment options thereby reducing policy regret. Adaptive regulation could be more resilient over time, bending and evolving rather than breaking in a disaster. Adaptive regulation does require investment in monitoring and data analysis; it requires criteria on which someone will evaluate performance; and it requires commitments to review these evaluations and adjustment options at periodic intervals. As the Norwegian approach to offshore oil regulation suggests (Engen and Preben, this volume), adaptive regulation may also require a proactive regulator who is constantly on the lookout for problems and mediates solutions as they arise. Nonetheless, we are optimistic that studying crisis events and regulatory responses can help inform better anticipation, prevention, management, and learning.

References

Baumgartner, Frank R. and Bryan D. Jones, 1993. *Agendas and Instability in American Politics*. Chicago, IL: Univ. Chicago Press.

Birkland, Thomas A. 2006. *Lessons of Disaster: Policy Change after Catastrophic Events*. Washington, DC: Georgetown University Press.

Campbell, T. H. and A. C. Kay. 2014. Solution Aversion: On the Relation between Ideology and Motivated Disbelief. *J. Pers. Soc. Psychol.* 107 (5): 809–24.

Carpenter, Daniel and Gisela Sin. 2007. Policy Tragedy and the Emergence of Regulation: The Food, Drug and Cosmetic Act of 1938. *Studies in American Political Development* 21: 149–80.

Denzau, Arthur T. and Michael C. Munger. 1986. Legislators and Interest Groups: How Unorganized Interests Get Represented. *American Political Science Review* 80: 89–106.

Graham, John D. and Jonathan Baert Wiener, eds. 1995. *Risk vs. Risk: Tradeoffs in Protecting Health and the Environment*. Cambridge, MA: Harvard University Press.

Johnson, George. 2015. When Radiation Isn't the Real Risk. *New York Times*, September 22, D3.

Kahn, Matthew E. 2007. Environmental Disasters as Risk Regulation Catalysts? The Role of Bhopal, Chernobyl, Exxon Valdez, Love Canal, and Three Mile Island in Shaping US Environmental Law. *Journal of Risk and Uncertainty* 35: 17–43.

Kahneman, Daniel, Paul Slovic and Amos Tversky. 1982. *Judgment under Uncertainty: Heuristics and Biases*. Cambridge, MA: Cambridge Univ. Press.

Kingdon, John, 2003. *Agendas, Alternatives and Public Policies*, 2nd ed. New York: Longman Classics in Political Science.

Kuran, Timur and Cass R. Sunstein, 1999. Availability Cascades and Risk Regulation. *Stanford Law Review* 51: 683–768.

McCray, Lawrence E., Kenneth A. Oye, and Arthur C. Petersen. 2010. Planned Adaptation in Risk Regulation. *Technological Forecasting and Social Change* 77: 951–59.

Percival, Robert V. 1998. Environmental Legislation and the Problem of Collective Action. *Duke Environmental Law and Policy Forum*, 9: 9–28.

Preston, A. Catherine. 2014. "Reaction and Revision: Regulatory Responses to Post-1870 Chemical Disasters. Senior Thesis, Sanford School of Public Policy, Duke University.

Repetto, Robert, ed. 2006. *Punctuated Equilibrium and the Dynamics of US Environmental Policy*. New Haven, CT: Yale University Press.

Rust, Daniel, 2015. In the Shadow of Tragedy: The Evolution of Safety Coregulation on America's Uninspected Towing Vessels, *Enterprise and Society* 15: 885–920.

Slovic, Paul. 2007. "If I Look at the Mass I Will Never Act": Psychic Numbing and Genocide. *Judgment and Decision Making* 2 (2): 79–95.

Small, Deborah A., George Loewenstein and Paul Slovic. 2007. Sympathy and Callousness: The Impact of Deliberative Thought on Donations to Identifiable and Statistical Victims. *Organizational Behavior and Human Decision Processes*, 102: 143–53.

Stahl, Aimee E. and Lisa Feigenson. 2015. Observing the Unexpected Enhances Infants' Learning and Exploration. *Science*, 348 (April 3): 91–94.

Stone, Deborah A., 1989. Causal Stories and the Formation of Policy Agendas. *Political Science Quarterly* 104: 281–300.

Wiener, Jonathan B. 2016. The Tragedy of the Uncommons: On the Politics of Apocalypse. *Global Policy*. 7(S1): 67–80.

Wiener, Jonathan B., and Barak D. Richman. 2010. Mechanism Choice. In *Public Choice and Public Law*, edited by Daniel Farber and Anne Joseph O'Connell. Northampton, MA: Edward Elgar: 363–98.

World Bank. 2014. World Development Report 2014: Risk and Opportunity – Managing Risk for Development (Washington, DC).

Wuthnow, Robert. 2010. *Be Very Afraid: The Cultural Response to Terror, Pandemics, Environmental Devastation, Nuclear Annihilation, and Other Threats*. Oxford: Oxford University Press.

Index

Alessandrini, Pietro, 413
Alfa Mutual Insurance, 69
Allen, Thad, 219
Allstate Insurance, 69
Alyeska Pipeline Service, 166–167
American Aeronautical Chamber of Commerce, 507
American Bankers' Association (ABA), 31, 371, 375, 377
American International Group (AIG), 64, 73
American Petroleum Institute (API), 171, 218
  Center for Offshore Safety, 219
Amtenbrink, Fabian, 412
Ardant, Henri, 365
Arisawa Committee, 316, 317
Arisawa, Hiromi, 316
Asahi Newspaper Company surveys, 319, 321, 322, 327, 333–334, 340, 341
aviation safety. See also United States government agencies, National Transportation Safety Board (NTSB)
  history of, 505–508

bad draw, 14, 54, 55, 544, 550–552
Badertscher, Brad A., 377
Bang, Paul, 190
Basel Capital Accords, 423, 424, 439, 443, 448, 453–454, 457
Baumgartner, Frank R., 110, 142, 152
Bayesian Rule and Analysis, 46, 53, 88, 129, 247, 248, 262
behavioral economics, 48, 71, 136

affect, 140–141
availability, 136–137, 140
framing, 138–139, 142
illusion of control, 139–140
representativeness, 137–138
Béland, Daniel, 113
Bennear, Lori S., 12–14, 544, 551
Berlusconi, Silvio, 402
Bernanke, Ben, 356
Bhat, Ghauri, 377
Birkland, Thomas, 15–16, 94, 98, 110, 118, 142, 152, 189, 379, 411, 551–552
British Petroleum Deepwater Horizon oil spill, 1, 24, 72, 154, 204
  bad draw, 550
  EU concern and response, 197
  investigatory commissions, 219
  lack of policy change, 175–176, 213, 549
  magnitude of spill, 175, 206
  media coverage, 210, 212
  post-disaster reforms, 215–218, 232, 542
  redundant systems of, 52
British petroleum regulatory regime, 23, See also Norwegian oil industry
  Cost Reductions In a New Era (CRINE), 193
  Cullen Report, 23, 192
  Formal Safety Assessment (FSA), 192–193
Bromwich, Michael, 233
Brown, Gordon, 396
Bruner, Jerome, 132
Buffett, Warren, 65

562

Index

Burks, Jeffrey J., 377
Bush, George H.W., 44
Bush, George W., 135, 209

California Department of Insurance, 70
California Earthquake Authority (CEA), 74
Carrigan, Christopher, 24, 544, 550
Carrothers, Bruce, 31
Carson, Rachel, 153
Carter, Jimmy, 252
Cassis, Youssef, 31, 550
catastrophic risk, 14, 60, 61, 68, 75, 257, 447, 559
Chemical Manufacturers Association, 511, 517
Chernobyl accident, 84
  design flaws and human error, 255
  graphite-moderated RBMK reactor, 246, 249, 255
  public reaction and health effects, 255–256
  response of different countries, 550
Chu, Steven, 219
Cihák, Martin, 463
Claessens, Stijn, 32
climate change, 9, 10, 16, 18, 25, 54, 67, 68, 91, 96, 97, 143, 154, 257, 261, 266, 301, 559
  and risk perception, 90, 96
  Hurricane Katrina and, 67
Clinton, Bill, 44, 207
Commission of Inquiry (COI), 33, 485, 486–487, 528, *See also* standing independent safety boards
  advantages and disadvantages, 496–500, 525
  basic processes, 489–492
  improvements, 525–526
  in the British Commonwealth, 488–489, 491
  in the United States, 488, 489–491
  Ladbroke Grove rail accident, 496
  Piper Alpha oil rig explosion investigation, 496
  recommendations into policy, 492–496
commissions, investigatory, 43
Committee on Payments and Settlement Systems (CPSS), 464
corporate scandals
  fraud bankruptcies, 6
  insurance company collapse, 5
  New York and New Haven Railroad stock, 5
  savings & loan crisis, 6
  South Sea Bubble, 5
crisis events, 1, 17, 33, 528, *See also* Commission of Inquiry (COI); standing independent safety boards
  causality, 172–173
  media and publicity, 6
  public opinion and regulatory response, 541
  regulatory response, 8–11, 12, 18–21, 538–539, 541–557
  regulatory response variations, 548
  terror attacks of 9-11, 6
  volcanic ash crisis, 6
crisis events, system effects
  bad draw, 550
  blame and regulatory response, 550
  close to home, 550
  cumulative effect of multiple events, 552
  institutional structure, 545
  public confidence in institutions, 546
  readiness of solutions, 546, 552
  steeper policy slopes, 545, 550
  time frame of response, 551
crisis events, types of events and response, 554
  "adaptive regulation," 557
  "learn to prepare," 34, 555–557, 558
  "prepare to learn," 4, 34, 558
  magnitude and perception, 544
  narrative framing, 544
  Proximity and propensity, 544
  repetition, 544
  unusual event, 544
Cronkite, Walter, 253
Cullen, William, 496, 500
Cummins, J. David, 76

De Haan, Jakob, 412
deepwater drilling leases, 207–209, *See also* federal regulations, deepwater drilling leases
  expansion of oil production, 209, 214
  government oversight post-disaster, 213–214
  ignorance of disaster possibility, 210–212
  public opinion, 209, 213
  regulation focus, 209
  temporary moratorium on new leases, 215

deepwater drilling leases, committees
  National Academy of Engineering committee, 219
  Outer Continental Shelf Safety Oversight Board, 219
Deepwater Horizon Oil spill. *See* British Petroleum Deepwater Horizon oil spill
Delphi method, 50, 53
Demirgüç-Kunt, Asli, 463
Department of Interior (DOI). *See also* Minerals Management Service (MMS)
  Bureau of Ocean Energy Management (BOEM), 220
  Bureau of Safety and Environmental Enforcement (BSEE), 220
  drilling expansion or moratorium, 213–214, 215
  Minerals Management Service (MMS), 221–222
  Ocean Energy Safety Institute, 219
  Office of Inspector General, 211–212
  Office of Natural Resources Revenue (ONRR), 220
  Offshore Energy and Minerals Management, 223, 224, 225, 227
  role in offshore oil and gas development, 204–205, 206
  role in offshore oil and gas revenue collection, 233
Dillon, Clarence, 357
disaster or catastrophe insurance, 58, 62, 71, 77
  availability heuristic, 71
  catastrophe modeling, 66–68
  extreme events, effect on, 65, 66, 68, 71
  government intervention, 72
  government reinsurance, 61
  insurer reaction to events, amount of exposure, 69–70
  insurer reaction to events, capital management, 70
  insurer reaction to events, conditions of coverage, 69
  insurer reaction to events, premium increases, 68–69
  rates and claims, 60–61
  rating agency requirements, 67
  reassessment and rate increases, 62–64
  risk contagion, 70
  risk transfer, criteria, 59–61
  state regulators and rates, 64
  value-at-risk requirements, 60
Douglas, Mary, 91
Downs, Anthony, 133, 142
Draghi, Mario, 404
Dutch Safety Boards, 485, 501, 518, 521
  Dutch Safety Board (DSB), 521–523
  Transport Safety Board, 521
Dykstra, David A., 378

Easton, Peter D., 377
economic vs. environmental policy, 206–207, 208–209
Eichengreen, Barry, 32, 350, 550
Eisner, Marc Allen, 23, 546
energy management. *See* Department of Interior (DOI); Minerals Management Service (MMS)
Engen, Ole Andreas, 23
Entman, Robert M., 135
environmental and energy policy, 153–154, 232–233, *See also* France; Germany; oil spills
  European energy post-World War II, 273, 274–275, 276–277, 278–281
  Japan, post-Fukushima accident, 338–341
  Norway and US comparison, 186, 188
  nuclear accident effect on energy production, 254, 257, 260–262, 265, 269–270
  policy subsystem, 154
  US fracking and natural gas production, 261
Epstein, Woody, 262
Euro area financial crises, 32, 395, 399–400, *See also* Basel Capital Accords; European Central Bank (ECB)
  "wake-up call hypothesis," 413
  bond rating downgraded, 397, 398, 402, 404
  credit rating agencies role, 412
  efficient-markets hypothesis, 413
  European Financial Stability Facility (EFSF), 398, 404
  European Stability Mechanism, 404
  fear of hyperinflation, 410, 419
  focus on past risk and future expectations, 411
  Greek government and bank crisis, 397–399, 401, 403, 413
  Ireland banking crisis, 399
  Italy and Spain crisis, 401–402, 404, 405

Index

media financial coverage, 415–418
Portugal and Spain bonds downgraded, 398
risks and response, 426–427
stability prior to crisis, 406–407, 409
subprime mortgage crisis in Euro area, 396
wholesale-funding-dependent banks, 402
Euro area financial crises, response, 426–427
  "six pack" of legislative measures, 421
  "two pack" of regulations, 421
  austerity and structural reform policies, 416
  focus on budget deficits, 423
  strengthening the national institutions, 423
European Atomic Agency (Euratom), 277
European Central Bank (ECB), 398–399, 401, 404
  lender of last resort, 402
  Outright Monetary Transactions (OMT), 404
European Coal and Steel Community (ECSC), 276
European Commission
  "cohesion process" failing, 409
  Buti, Deroose, Gaspar and Martins publication, 406–407, 408
  euro, 10th anniversary, 406–407
  German fear of high inflation, 407
  Stability Pact or Stability and Growth Pact, 407
European Union, treaty, statutes and regulations
  Capital Requirements Directive IV, 423
  Capital Requirements Regulation, 423
  European System of Financial Supervision, 423
  Impact Assessment Board (IAB), 9
  Maastricht Treaty, no-bailout clause, 422–423
  Treaty on Stability, Coordination and Governance, 421–422, 423
Executive Orders (US), 98
benefit-cost analysis of regulations, 44
Exelon Nuclear, 253
extreme events. *See also* nuclear power accidents; oil spills
  consumer's risk perception, 71–72
  ex-post analysis, 43

Hurricane Andrew, 63, 64, 69, 73
Hurricane Iniki, 74
Hurricane Katrina, 10, 43, 66, 69, 71, 73, 75
insurance estimation of risk, 65
Love Canal, hazardous waste, 4, 154, 163
mass shootings, 129, 144
Northridge Earthquake, 63, 70, 74
probability of, 53–54
super cats, 66
terrorism, 65–66, 119–120
World Trade Center, September 11, 66, 68, 69, 72, 135
Exxon Valdez oil spill, 151, 165
  effect on petroleum industry, 171
  effect on prevention policies, 172, 542, 549
  industry and company response, 166–167
  investment in safety and environmental performance, 171

Farmer's Insurance, 69
federal regulations, deepwater drilling leases
  Drilling Safety Rule, 217
  Environmental Notice to Lessees (NTL), 217
  Safety and Environmental Management System (SEMS), 218–219
  Safety Notice to Lessees (NTL), 215
  Workplace Safety Rule, 218–219
finance and financial crises, 29–31, 32, *See also* Global Financial Crisis of 2007–08; Great Depression; Lehman Brothers' bankruptcy
  "decommodification," 31, 371–374, 378–379, 382, 384–385, 386, 388–389
  asset valuation, 373–374
  Capital Asset Pricing Model, 92
  excessive risk, 52
  fair value accounting rules, 371, 373, 374–379, 387, 453
  financial crash response, 144
  housing and mortgage market, 373, 382, 383
  mortgage-backed securities (MBS), valuation, 377, 438
  NTSB-like institution for finance, 529–531
  policy-makers' response, 32, 379, 387
Financial Accounting Standards Board (FASB), 371
  FAS No. 157, 374, 375, 376–377, 378

Financial Crisis Inquiry Commission, 43
financial reform, post global crisis. *See also* Basel Capital Accords; Financial Stability Board (FSB)
  "incentives view," 443–444, 457–458, 464–465
  better data, 465–467, 471
  consistency of reforms, 447
  cost-benefit analysis of reforms, 448
  crisis management improvements, 493
  government and NGO cooperation, 470
  independent regulatory governance, 461–464
  liquidity requirements, 457–459
  macroprudential and microprudential policies, 448–451
  migration to reforms, 448
  money market mutual funds (MMMFs), 455
  OTC derivatives markets, 456–457
  procyclicality, 451–454
  repo and securities lending markets, 456
  securitization retention rules, 455
  shadow banking, 453–455
  systemic regulatory reform, 434, 443, 446, 449
  timing of reforms, 448
  too-big-to-fail institutions (TBTF), 433–434
Financial Stability Board (FSB), 439–442, 448, 454–455, 456, 464
Florida Citizens Property Insurance, 74
focusing events, 15–16, 17, 98, 107–108, *See also* extreme events
  advocacy coalitions and media coverage, 112
  agenda setting, 108, 114, 142
  definition, 109–110, 118, 152
  investigations, methods of, 483–485, 486
  long-term perspective, 114–115
  negative attention, 110
  nuclear power regulation after TMI, 112
  policy change opportunities, 109, 113, 120, 152
  proactive problem solving, 121, 181, 189, 192, 201, 218, 379, 557
  regulatory politics and, 120–126, 199, 204–206, 542
Forgas, Joseph P., 131
framing theory, 16
  airline safety example, 115–117, 120–121, 124–125
  effects on response, 549
  financial reforms, 442, 446
  impact of past events, 114
  insight of many disciplines, 130
  media coverage, 6
  policy entrepreneurs role, 8, 110, 120
France
  anti-semitism and small banks, 365
  banking reform, 365, 367
  Bureau of Enquiry and Analysis for Civil Aviation Safety (BEA), 501–502
  centralized state effect on nuclear power, 272, 277–278, 293–296
  Chernobyl, response to, 281–282
  commitment to nuclear power, 269–270, 284
  Depression bank scandals, 358–359
  doctrine Henri Germain, 365
  French Nuclear Safety Authority (ANS), 286
  Fukushima, response to, 286–287, 288
  Great Depression banking crisis, 353
  increased renewable energy, 287
  military and civilian energy needs, 273, 274
  politics and anti-nuclear movement, 280, 281, 284, 287
Frankel, Richard M., 377
Fukushima Daiichi accident, 1, 84, 305, 343
  German environmental response, 246, 260, 542, 550
  Japanese energy and public confidence, 260
  new regulations post-accident, 336–337
  power generation post-accident, 338–341
  pressurized water reactor (PWR), 249, 257
  safety design and management causes, 258–259
  shutdown of Japanese nuclear plants, 542, 549
  site-specific risk analysis, 258
  TEPCO response, 247
  tsunami cause and analysis, 248, 258, 262–266, 332–333
  tsunami defense walls, 338
  US closes older nuclear plants, 262
  worldwide public opinion, 259, 260

Garat, Dominique-Joseph, 358
General Electric, 249, 257

Germany
  Atomic Law, 278, 284–285
  banking reform, 365
  Ethics Committee on energy policy, 288, 289
  Federal Office for Radiological Protection, 283
  Fukushima, response to, 288–291
  nuclear and coal energy, 274, 278, 296
  political resistance to nuclear power, 281, 284–286, 288, 297, 301
  renewable energy, 289
  response to Chernobyl, 283, 293
  response to nuclear accidents, 246, 260, 272
  safety and nuclear power, 271, 278, 280
Glass, Carter, 363, 364
Global Financial Crisis of 2007–08, 31, *See also* finance and financial crises; Financial Accounting Standards Board (FASB)
  asset-backed commercial paper (ABCP), 438
  causes, 435–439
  complexity of financial products, 438
  fear of defaults in major financial institutions, 438
  international financial integration, 438
  regulatory response, 542
  regulatory standing to credit ratings, 385–386
  subprime mortgage crisis, 85, 386, 396, 406, 436, 438
  time frame of response, 551
government intervention. *See also* California Earthquake Authority (CEA)
  Fair Access to Insurance Requirement (FAIR) plans, 73
  federal insurance programs, 75
  federal terrorism insurance programs, 75–76
  French terrorism insurance, 75
  government insurance programs, 73–75, 76
  state insurance regulators, 73
Great Britain, 367
  banking reform, 365
  currency crisis, 352, 359–360
Great Depression, 31, 349–350, 367, 542, *See also* finance and financial crises; Global Financial Crisis of 2007–08

  bank failure crises, 350–353, 361
  currency crisis in Great Britain and Germany, 352
  differing perception across countries, 357
  international monetary system cause, 350
  stock market crash, 350, 549
  strong response in US, 550
Gulf of Mexico oil disaster. *See* British Petroleum Deepwater Horizon oil spill

Hacker, Jacob S., 113
Hall, Peter, 293
Hansen, James, 96
Hardy, Barbara, 130
Hartley, Fred, 155
hazardous waste, 163–164
  Superfund, 164
Hazelwood, Joseph, 166
Hickel, Walter, 155–156, 157–158
Hitachi, BWR, 312
Hollande, François, 287
Hoover, Herbert, 352
Hovden, Jan Fredrik, 194

Institute of Nuclear Power Operations (INPO), 26, 245, 246, 250, 251, 253
insurance. *See also* disaster or catastrophe insurance
  benefits of, 58
International Atomic Energy Agency (IAEA), 335
International Convention on Civil Liability for Oil Pollution Damage, 159, 167
International Fund for Compensation for Oil Pollution Damages, 167
International Organization of Securities Commissions (IOSCO), 455, 464
Iowa Electronic Markets, 467

Jackson, Henry M., 153
Japan, 305–306, 310, *See also* Arisawa Committee; Asahi Newspaper Company surveys
  Chernobyl, response to, 321–323
  domestic accidents, 341, 550
  nuclear power development, 311–313, 314
  nuclear regulatory policy, 313–314, 333, 342, 549
  nuclear safety policy, volcanos, earthquakes and tsunamis, 318, 327–330, 337–338

Japan (cont.)
  probabilistic risk analysis (PRA), 331, 342
  public opinion of nuclear energy, 314, 315, 316, 324
  risk perception after accidents, 305, 306–307, 309, 310, 324–326
  risk perception after foreign accidents, 309
  safety recommendations after TMI, 320, 331–332
Japan, government agencies
  Agency for Natural Resources and Energy (ANRE), 320, 322
  Energy and Environment Council, 338
  Headquarters for Earthquake Research Promotion, 329
  Minister of Economy, Trade, and Industry (METI), 325, 338
  Ministry of International Trade and Industry (MITI), 313, 317–318, 320, 322
  Ministry of the Environment (MOE), 335, 342
  Nuclear and Industrial Safety Agency (NISA), 325–326, 330, 331, 337
  Nuclear Regulation Authority (NRA), 335–336, 342
  Nuclear Safety Commission (NSC), 317, 322–323, 324
Japan, nuclear accidents. *See also* Fukushima Daiichi accident
  JCO nuclear fuel processing facility accident, 323
  Monju fast breeder reactor, 326
  nuclear-powered ship, the *Mutsu*, 315, 316
  Onagawa Nuclear Power Station shutdown, 330
  Tsuruga Nuclear Power Station, 320–321
Johnston, R. Barry, 463
Jones, Bryan D., 110, 142, 152
Jones, Homer, 380

Kahan, James P., 501
Kahn, Albert, 357
Kahneman, Daniel, 88, 136, 137, 138
Kaldor–Hicks principle, 49
Kallaur, Carolita, 207
Kansai Electric Power Company (KEPCO), PWR, 312
Keynes, John Maynard, 367

Kindleberger, Charles, 29
Kingdon, John, 15, 108–110, 172, 419, 541
Kirk, Mark, 377
Kishimoto, Atsuo, 26, 550
Kodres, Laura, 32
Kolly, Joseph, 499
Kousky, Carolyn, 13
Krieger, Kristian, 26, 550

Lamont, Thomas, 357
Laux, Christian, 377
legislation, European
  Capital Requirements Directive IV, 423, 424
  Capital Requirements Regulation, 423
  Single European Act (SEA), 282
legislation, federal
  airline security legislation, 117
  Banking Act of 1935, 361, 379
  Biggert–Waters Flood Insurance Reform Act of 2012, 75, 98
  Clean Air Act Amendments, 153, 161, 511
  Clean Water Restoration Act, 156
  Comprehensive Environmental Response, Compensation, and Liability Act of 1980, 153, 154, 164–165
  Deep Water Royalty Relief Act (DWRRA), 207
  Deepwater Port Act of 1974, 161
  Dodd-Frank Act, 448, 542, 545, 551
  Emergency Economic Stabilization Act of 2008, 376
  Emergency Farm Mortgage Act of 1933, 382
  Emergency Planning and Community Right-to-Know Act (EPCRA), 153, 511
  Energy Policy Act, 208
  Federal Oil and Gas Royalty Simplification and Fairness Act, 208
  Federal Water Pollution Control Act Amendments, 153, 160
  Food and Drug Act of 1938, 112
  Glass-Steagall Act, 361, 364, 367
  Gulf of Mexico Energy Security Act, 208
  Independent Safety Board Act of 1974, 508
  Investment Companies Act of 1940, 361
  National Environmental Policy Act, 153
  Oil Pollution Act (OPA), 167, 169–170, 175, 176
  Oil Pollution Act of 1924, 156

Index

Oil Pollution Act of 1990, 154, 542
Outer Continental Shelf Lands Act (OCSLA), 207–208
Outer Continental Shelf Lands Act Amendments, 162
Resource Conservation and Recovery Act (RCRA), 153, 164
Securities Act of 1933, 361
Securities Exchange Act of 1934, 361
Shipowner's Limitation of Liability Act, 156, 160
Terrorism Risk Insurance Act (TRIA), 76
Trans-Alaska Pipeline Authorization Act, 161
Water Quality Improvement Act (WQIA) of 1970, 159–160
legislation, Japan
  Act on the Regulation of Nuclear Source Material, 313
  Atomic Energy Commission (AEC) Establishment Act, 313
  Atomic Energy Fundamental Act (AEFA), 313, 317
  Electricity Business Act, 313
  Establishment of the Nuclear Regulation Authority, 335
  Nuclear Reactor Regulation Law, 313, 317, 323, 325
  Power Source Development Laws, 311
  Special Measures Concerning Nuclear Emergency Preparedness, 325
legislation, Norway
  Environmental Act, 184
  Norwegian Working Environment Act, 187
legislation, preemption of state law, 160, 162, 165, 167
legislation, state, oil pollution liability regimes, 168
Lehman Brothers' bankruptcy, 2, 29, 395, 396, 426
Leuz, Christian, 377
Lindøe, Preben H., 23, 182
Linowes Commission, 221–222
Lloyd's of London, 171
Löfstedt, Ragnar, 26
Los Angeles Times, 64, 155, 156, 362

MacDonald, Ramsay, 352
Martin, Xiumin, 377
Marx, Sabine, 96

May, Peter, 110
Mayer, Frederick W., 16, 110, 423
McAdams, Dan P., 132
Merkel, Angela, 246, 260, 269, 288, 399
Merrill, Craig B., 377
Milward, Alan, 276
Minerals Management Service (MMS), 204–206
  breakup as symbolic political action, 231–232
  Congressional reorganization plans, 225–227, 228, 231, 232
  disbanded after BP Deepwater oil spill, 220–221, 227
  founding and mission, 221–223, 227
  petroleum production incentives, 207
  reorganization of safety and leasing offshore, 221, 230–231
  replaced by BOEMRE, 217, 220
  Revenue Management Division, 212, 220, 224–225
  royalty and other failures, 224
  Royalty in Kind (RIK) Program ethics failures, 212, 220
Mitchell, Charles, 357–358
Mitchell, George J., 167
Mitsubishi Heavy Industries, PWR, 312
Mitterrand, François, 281
Monnet, Jean, 277
Monte Carlo analysis, 51
Monti, Mario, 402
Morgan, Jack, 357–358
Moss, David A., 162
Moure-Eraso, Rafael, 513
Murkowski, Lisa, 233
Muskie, Edmund, 155

narrative and storytelling, 144
  emotions and, 131
  importance of, 130–131, 132, 141, 547–548
  news stories, 133–135
  shared public narratives, 142
narrative framing, 131, 132–133
National Association of Supervisors of State Banks, 379
National Commission on the BP Deepwater Horizon Oil Spill and Offshore Drilling, 215
National Flood Insurance Program (NFIP), 75

National Partnership for Reinventing Government, 207
Nelson, Barbara J., 142
New York Times, 253
New Zealand Transport Accident Investigation Committee, 485
Nixon, Richard M., 153, 155, 159–160
non-crisis events, 2
"the Nordic Model" of industrial relations, 182
Nordic regulatory regime, 181–182, 184–186, 187, 189, 201
success of regulatory regime, 197
North Sea oil rig accidents, 23, 181, 189–190, 191, 195–196, 200
Norway, Accident Investigation Board (AIBN), 502
Norwegian oil industry
economic policy, 183–184
enforced self-regulation, 186, 187–188, 190–191, 198–200
Health and Safety Executive (HSE), 192, 194–195, 199
increased accidents and risk, 194–195, 196
Ministry of Oil and Energy (MOE), 182
NORSOK, cost-effectiveness, 193
Norwegian Petroleum Directorate (NPD), 182, 183, 190–191, 195
Norwegian Petroleum Safety Agency (PSA), 196
Statoil, 182
trust and power of interested parties, 197–200
nuclear power, 25–26, See also France; Germany; Japan
boiling water reactors (BWR), 257, 264, 312, 314
European policies, 269–271, 272, 291
European public perception post-accidents, 291–293, 297–299, 300
gas-cooled reactor (GCR), 314
phase out or reduction, 84
pressurized water reactor (PWR), 249, 252, 257, 264
nuclear power accidents, 26–28, 245–246, See also Chernobyl accident; Fukushima Daiichi accident; Japan, nuclear accidents; Three Mile Island (TMI) nuclear power plant accident

avoidance through risk analysis, 248
effect on global energy industry, 264
framing and public attention, 28
probabilistic risk analysis (PRA), 247–248, 251–252, 331
probalistic study of tsunami hazards, 262
risk recalibration post events, 246–247, 254, 257, 264–265, 550
site-specific PRA, 262, 265
technical and information improvements post-accidents, 250–251, 264
Nuclear Regulatory Commission (USNRC), 245, 253
PRA increase after Chernobyl, 257
site-specific PRA increase Fukushima, 262

Obama, Barack, 44, 176, 210, 214, 215, 217, 219, 220
Occidental Petroleum, 191
Ognedal, Magne, 192
oil spills, 21, 189, See also British Petroleum Deepwater Horizon oil spill; deepwater drilling leases; Exxon Valdez oil spill; legislation, federal; North Sea oil rig accidents; Piper oil field (UK) fire and explosion; Santa Barbara oil spill; Torrey Canyon oil spill
effect on prevention policies, 172–175
foreign oil spill and action, 159
Helix Well Containment Group, 219
inadequacy of oil spill cleanups, 167
joint venture for containment equipment, 219
Marine Spill Response Corporation (MSRC), 171–172
Marine Well Containment Company, 219
National Contingency Plan, 159, 164, 170
oil spill and pollution control history, 156–158, 159–163
Oil Spill Liability Trust Fund, 169
Polluting Incident Compendium, 173
regulations to prevent future spills, 170–171
Osterhus, Gustav, 379
Oustric, Albert, 358

Papandreou, George, 397, 402
Pareto principle, 49
Partnoy, Frank, 385, 412

Paté-Cornell, M. Elizabeth, 26, 550
Pecora, Ferdinand, 357, 364
Péret, Raoul, 358
Perrow, Charles, 17, 253
Phillips Petroleum, 23, 189, 190
Piper oil field (UK) fire and explosion, 23, 191–193, 496
Polanyi, Karl, 371–372, 388
policy regret, 43, 55
   benefit-cost analysis and, 53, 55
   extreme events, 54
   gaps in benefit-cost analysis, 55
   policy resilience, 554
   static character, 54
political mobilization and collective action, 142–144
Porter, Michael, 183
public policy, development, 152
   effect of disruptions, 152
   Inspector Generals (IGs), 485
   investigations, 485
Pujo, Arsène, 362

Rahall, Nick, 226, 231
Rajoy, Mariano, 404
Rawl, Lawrence, 167
Reagan, Ronald, 44
Reed, John H., 508
reform, post-crisis, 43
regulation
   Office of Information and Regulatory Affairs (OIRA), 98
   specialists and long-term issues, 95–96, 97–98
regulation, chemical industry. *See also* United States, government agencies, Chemical Safety and Hazard Investigation Board (CSB)
   Bhopal, India chemical plant disaster, 4, 511, 517
   Moss Opinion, 513
   West Virginia Union Carbide plant, 511
regulation, food safety, mad cow disease, 6
regulation, pharmaceutical
   ethyl glycol-infused antibiotics, 4
   new disease, 4
   Sulfanilamide, 112
   Thalidomide, 4
regulation, pollution
   Cuyahoga River catching fire, 4
   killer fogs in London, 4
   Love Canal, hazardous waste, 4
   Seveso dioxin accident, 4
   Stratospheric Ozone Hole, 4
regulations, public health, infectious disease epidemics, 4
regulations, safety inspections and rules
   exploding boilers on steamboats, 4
   manufacturing standards in China, 6
   mining disasters, 4
   South Asian clothing factory fires, 6
   Triangle Shirtwaist Factory fire, 4
regulatory crises, definition, 2
Reid, Harry, 228
Reinhart, Carmen, 494
reinsurance, 61–62
   Terrorism Risk Insurance Act (TRIA), 76
   updating of terrorism risk after September 11, 65–66, 68
Renn, Ortwin, 26
Review of Standards and Codes (World Bank), 462
risk and uncertainty, 12–14
   acceptable financial system risk, 444–446
   distinction between, 44
   expected utility framework, 45–46, 47, 51
   expected utility framework, critique, 48
   narrative theory of risk, 130
   present value and discounting, 47–48
   public or expert perception, 24, 53–54
   Regulatory Impact Analysis (RIA), 13
   response to events, 129–130
   subjective probabilities, 46
risk perception, individual, 14–15, *See also* behavioral economics
   causes of changes in perceived risk, 86, 101
   change and predictability, 86–87, 90–91
   control over outcomes, 99–101
   decreased vs. increased risks, 99
   dread risk and unknown risk, 118–120, 139–140
   dual process theory, 88, 94, 101
   experience vs. statistical models, 94
   feelings or emotions and, 88, 89–90
   hormone replacement therapy (HRT), 83
   identified individual victim, 545
   individual action and choice under risk and uncertainty, 83
   levels of and reactions to, 82
   models of risky choice, 92–94
   nuclear power plant accidents, 84, 89
   prospect theory, 92, 95

risk perception, individual (cont.)
  psychic numbing, 545, 553
  psychological processes of, 82
  regulatory responses, 85
  single action bias, 85, 96, 98
  social amplification of risk, 83, 86, 91–92
  uncertainty, 88
risk perception, organizational, garbage can model, 85, 98
risk regulations
  adaptive regulation, 559
  National Risk Board, 559
risk regulations, benefit-cost analysis, 44, *See also* Executive Orders (US)
  errors of, 52
  financial reforms, 448
  political influence, 52
risk-risk tradeoff, 25, 28, 265, 551, 559
  countervailing risk, 552
  side effects, 436
  unintended consequences, 469
risk-risk tradeoffs
  unintended consequences, 21, 472
Rogers, Brooke, 26
Rogoff, Kenneth, 494
Roosevelt, Franklin D., 353
Rosenberg, Beth, 513
Ryggvik, Helge, 195

Sabatier, Paul, 111, 114
safety boards. *See* standing independent safety boards
Salazar, Ken, 213, 215, 217, 219, 220
  order to disband MMS, 227, 228–230
Santa Barbara oil spill, 151, *See also* environmental and energy policy
  concurrent environmental events, 153
  event and media coverage, 155–156
  federal regulation of petroleum industry, 156, 157–158
  impact on other crises, 154
  liability shift to industry, 172
  Union Oil response, 155
Sarkozy, Nicolas, 269, 287
Schmidt, Helmut, 281
Schwartz, Alan, 91
Sedor, Joseph, 499
Seehofer, Horst, 269
Shiroyama, Hideaki, 305
Silent Spring (Carson), 153

Slovic, Paul, 89, 118, 139–140, 545
standing independent safety boards, 500, 528, *See also* Commission of Inquiry (COI)
  advantages over Commissions of Inquiry, 526–528
  leadership importance, 528
  success or failure of NTSB and CSB, 514–517, 528
standing independent safety boards, transportation
  Dutch and Swedish safety boards, 501
  European single-modal safety boards, 501–502
  International Civil Aviation Organization (ICAO), 500
  multi-modal investigatory boards, 501–502, 521, 522
State Farm Insurance, 62, 69–70
Stavisky, Alexandre, 358
Steagall, Henry, 364
Stern, Paul C., 97
Stone, Deborah, 114–115
Stoop, John A., 501
Swedish Safety Boards, 519–521, 523
  State Catastrophe Board, 520
  Swedish Accident Investigation Board (SHK), 520, 521, 524
  Swedish Maritime Accident Investigation Board, 520

TEPCO, 247, 258, 305, 326, 328–329, 330
Three Mile Island (TMI) nuclear power plant accident, 84, 112, 134, 144
  academic response, 253
  effect on European nuclear power, 281
  effect on US nuclear program, 253, 542
  fear of nuclear radiation, 252
  media coverage and surveys, 252–253, 318, 319
  pressurized water reactor (PWR), 249, 252
Thuestad, Olaf, 190
Tokyo Institute of Technology, 262
Torrey Canyon oil spill, 153, 159–160
Toshiba, BWR, 312
tsunami and earthquakes. *See* Fukushima Daiichi accident; Japan
Tversky, Amos, 136, 137, 138

Union of Concerned Scientists, 253
United Kingdom

Air Accident Investigation Branch
   (AAIB), 501
oil industry, 183
United States government agencies, 361, *See also* Department of Interior (DOI); Nuclear Regulatory Commission (USNRC)
   Chemical Safety and Hazard Investigation Board (CSB), 485, 501, 502–504, 511–514
   Civil Aeronautics Administration (CAA), 507
   Civil Aeronautics Board (CAB), 507–508
   Commercial Aviation Bureau, 507
   Comptroller of the Currency, 379
   Consumer Financial Protection Bureau, 462
   Department of Homeland Security, 6
   Federal Deposit Insurance Corporation (FDIC), 379
   Federal Housing Administration (FHA), 383
   Federal Reserve, 5, 350, 351, 352–353
   Federal Reserve, New York, 456
   Inspector Generals (IGs), 485
   National Credit Corporation, 352–353
   National Transportation Safety Board (NTSB), 33, 485, 499, 501, 502–505, 508–510, 517
   Office of Information and Regulatory Affairs (OIRA), 9, 44, 98
   Office of Management and Budget, 44
   Reconstruction Finance Corporation, 353
   Securities and Exchange Commission (SEC), 361, 371
United States, financial reform, 361–362, *See also* finance and financial crises; legislation, federal
   banking reform, 363, 367, 379
   banking reform, depositor losses and FDIC, 364–365
   Home Owners' Loan Corporation (HOLC), 373, 377, 382
   New Deal, 361, 379
   New Deal bond rating reform, 379, 380, 384–386
   Pujo Committee and panic of 1907, 362–363, 367
   Savings-and-Loan crisis of the 1980, 375
   stock market crash, 357–358
   Uniform Agreement on Bank Supervisory Procedures, 379
utility function, collective, 49
   extension of individual choice, 55
   social discount rate of time preference, 49–50

Vollenhoven, Pieter van, 523–524

Wall Street Journal, 166
Warnement, Megan, 15–16, 94, 98, 189, 379, 411, 551–552
Washington Post, 155, 253
Watt, James, 221
Weber, Elke U., 14–15, 96–97, 118, 136, 140
Weber's Law, 90
Wehmer, Edward J., 378
Whitney, Richard, 357
Wiggin, Albert, 357–358
Wildavsky, Aaron, 91
Williams, Bruce, 253
Willis, H. Parker, 361, 363
Women's Health Initiative, 83
Woodin, William, 364
World Association of Nuclear Operators (WANO), 246, 248, 251
World Bank, 462
World Health Organization, 249